12th European Workshop on Modern Developments in Microbeam Analysis

(EMAS 2011)

IOP Conference Series: Materials Science and Engineering
Volume 32

Angers, France
15 - 19 May 2011

ISBN: 978-1-63439-843-5
ISSN: 1757-8981

Printed from e-media with permission by:

Curran Associates, Inc.
57 Morehouse Lane
Red Hook, NY 12571

Some format issues inherent in the e-media version may also appear in this print version.

Printed by Curran Associates, Inc. (2015)

For permission requests, please contact the Institute of Physics
at the address below.

Institute of Physics
Dirac House, Temple Back
Bristol BS1 6BE UK

Phone: 44 1 17 929 7481
Fax: 44 1 17 920 0979

techtracking@iop.org

Additional copies of this publication are available from:

Curran Associates, Inc.
57 Morehouse Lane
Red Hook, NY 12571 USA
Phone: 845-758-0400
Fax: 845-758-2634
Email: curran@proceedings.com
Web: www.proceedings.com

12th European Workshop on Modern Developments in Microbeam Analysis

(EMAS 2011)

IOP Conference Series: Materials Science and Engineering
Volume 32

Angers, France
15 - 19 May 2011

ISBN: 978-1-63439-843-5
ISSN: 1757-8981

IOP Institute of Physics

Table of contents

Volume 32

EMAS 2011: 12th European Workshop on Modern Developments in Microbeam Analysis
15–19 May 2011, Angers, France

Accepted papers received: 17 February 2012
Published online: 7 March 2012

Preface

Papers

IOP Conf. Series: Materials Science and Engineering 32 (2012) 011001 doi:10.1088/1757-899X/32/1/011001

12th European Workshop of the European Microbeam Analysis Society on Modern Developments and Applications in Microbeam Analysis (EMAS 2011), Angers, France, 15-19 May 2011

János L. Lábár[1], Clive T. Walker[2], François Brisset[3], Olivier Dugne[4] and Florence Robaut[5]

[1] Hungarian Academy of Sciences, Research Institute for Technical Physics and Materials Science, Konkoly-Thege M. ut. 29-33, HU-1121 Budapest, Hungary

[2] European Commission, Joint Research Centre, Institute for Transuranium Elements, P.O. Box 2340, DE-76125 Karlsruhe, Germany

[3] CNRS- Université Paris-Sud, ICMMO - Institut de Chimie Moléculaire et des Matériaux d'Orsay, 15 rue Georges Clémenceau, bât. 410, FR-91405 Orsay, France

[4] Commissariat à l'Energie Nucléaire, Centre de Marcoule, P.O. Box 17171, FR-30207 Bagnols-sur-Cèze Cedex, France

[5] Institut Polytechnique de Grenoble, Consortium des Moyens Technologiques Communs, 1260 rue de la Piscine, bât. ENSEEG, FR-38402 Saint Martin d'Hères Cedex, France

Preface

This volume of the IOP Conference Series: Materials Science and Engineering contains papers from the 12th Workshop of the European Microbeam Analysis Society (EMAS) on Modern Developments and Applications in Microbeam Analysis which took place from the 15th to the 19th of May 2011 in the Angers Congress Centre, Angers, France.

The primary aim of this series of workshops is to assess the state-of-the-art and reliability of microbeam analysis techniques. The workshops also provide a forum where students and young scientists starting out on a career in microbeam analysis can meet and discuss with the established experts. The workshops have a very specific format comprising invited plenary lectures by internationally recognized experts, poster presentations by the participants and round table discussions on the key topics led by specialists in the field. This workshop was organized in collaboration with GN-MEBA – Groupement National de Microscopie Electronique à Balayage et de microAnalysis, France. The technical programme included the following topics: the limits of EPMA, new techniques, developments and concepts in microanalysis, microanalysis in the SEM, and new and less common applications of micro- and nanoanalysis.

As at previous workshops there was also a special oral session for young scientists. The best presentation by a young scientist was awarded with an invitation to attend the 2012 Microscopy and Microanalysis meeting at Phoenix, Arizona. The prize went to Pierre Burdet, of the Federal Institute of Technology of Lausanne (EPFL), for his talk entitled "*3D EDS microanalysis by FIB-SEM: enhancement of elemental quantification*".

The continuing relevance of the EMAS workshops and the high regard in which they are held internationally can be seen from the fact that 74 posters from 18 countries were on display at the meeting and that the participants came from as far away as Japan, Canada and the USA. A selection of participants with posters was invited to give a short oral presentation of their work in three dedicated sessions. The prize for the best poster was an invitation to participate in the 22nd Australian Conference on Microscopy and Microanalysis (ACMM 22) at Perth, Western Australia. The prize was awarded to G. Samardzija of the Jozef Stefan Institute, Ljubljana, for the poster entitled: "*EPMA-WDS quantitative compositional analysis of barium titanate ceramics doped with cerium*".

This proceedings volume contains the full texts of 5 of the invited plenary lectures and of 23 papers on related topics originating from the posters presented at the workshop. All the papers have been subjected to peer review by a least two referees.

January 2012

Acknowledgements

On behalf of the European Microbeam Analysis Society I would like to thank all the invited speakers, session chairs and members of the discussion panels for making the meeting such a great success. Special thanks go to François Brisset and Luc Van't dack who directed the organisation of the workshop giving freely of their time and talents. As was the case for previous workshops, the EMAS board *in corpore* was responsible for the scientific programme. The technical exhibition, which occupied 130 sq.m of floor space, was outstanding. It was very encouraging to see new instruments on display, including a FEG electron microprobe as a first worldwide presentation. Moreover, almost all the companies that exhibited provided financial support, either by sponsoring an event or by advertising.

Below, in alphabetical order, is a list of exhibiting companies and sponsors of the workshop.

- Ametek GmbH, Edax Business Unit
- Bruker Nano GmbH
- Cameca
- Carl Zeiss NTS
- Commissariat à l'Energie Atomique
- European Institute for Transuranium Elements (Germany)
- Elexience
- FEI Company
- Fondis Electronic SA
- Gatan (France)
- GN-MEBA
- Jeol (Europe) SAS
- L'Oréal, Direction Générale Recherche et Innovation
- NanoMEGAS sprl
- Oxford Instruments SAS
- Probe Software, Inc.
- SAMx
- Target-Messtechnik
- Thermo Fisher Scientific

Clive T. Walker
EMAS President

Peer review statement

All papers published in this volume of *IOP Conference Series: Materials Science and Engineering* have been peer reviewed through processes administered by the proceedings Editors. Reviews were conducted by expert referees to the professional and scientific standards expected of a proceedings journal published by IOP Publishing.

Chemical characterisation of scale formation of high manganese steels (Fe-Mn23-C0.6) on the sub-micrometre scale: a challenge for EPMA

E Augustyn[1,4], B Hallstedt[2], B Wietbrock[3], J Mayer[1], A Schwedt[1] and S Richter[1]
[1] RWTH Aachen University, GFE - Central Facility for Electron Microscopy, Ahornstrasse 55, DE-52074 Aachen, Germany
[2] RWTH Aachen University, Institute of Materials Chemistry, Aachen, Germany
[3] RWTH Aachen University, Institute of Metal Forming, Aachen, Germany

E-mail: augustyn@gfe.rwth-aachen.de

Abstract. Fe-23Mn-0.6C steel oxidation in the initial state of annealing in synthetic air and an Ar / 4 % H_2 / 7 % H_2O atmosphere (600, 800 and 1000 °C for 20 min) was investigated. The chemical structure of the oxide scale and near-surface region of the oxidized sheet was characterized by electron probe microanalysis (EPMA). The results were correlated with the crystallographic structure of the sheet measured by electron backscatter diffraction (EBSD). Special preparations techniques i.e., focussed ion beam (FIB) milling and cross-section polishing by Ar-ions were applied to satisfy the requirements of quantitative X-ray and EBSD analyses. Moreover, carbon concentration profiles were measured on samples prepared by ion beam techniques and the conventional way with lubricants and polishing materials. Finally, a model of the near-surface region was derived by combining the results of quantitative analysis and X-ray mappings and was compared with thermodynamic calculations.

1. Introduction

In recent years, high manganese steels (HMS) with 15 - 30 wt% Mn are strongly developed because of their attractive combination of high formability and strength. For this reason they represent very promising materials for the automotive industry. The content of C, Al and Si directly affects the mechanical properties of HMS and results in the appearance of specific deformation mechanisms such as transformation induced plasticity (TRIP) or twinning induced plasticity (TWIP) [1-4]. Due to the technological processes, such as hot and cold rolling, high manganese steels can be held at high temperatures for long durations (e.g., hours). This can result in substantial formation of oxide scales, which influence the surface quality of the sheet and have to be removed by cost and time consuming procedures. Therefore, studies of oxidation in the initial state of annealing are essential.

Electron probe microanalysis (EPMA) has been used in general as a tool to study bulk materials with a lateral resolution of about 1 µm. The development of special procedures however allows the layered structures or subsurface regions to be characterized [5, 6]. In order to examine the sub-multilayered structure of the near-surface region a combination of EPMA and FIB specimen preparation can be applied.

[4] To whom any correspondence should be addressed.

Moreover, the depth profile analysis along the FIB bevel combined with Monte Carlo simulation allows quantitative information on the sub-micrometre layers to be obtained [7-11].

In the presented paper, the oxidation of a new TWIP steel containing of 23 wt% Mn and 0.6 wt% C was examined. The Fe-Mn-C alloys have been designed within the framework of the collaborative research centre (SFB761) 'Stahl - ab initio', funded by the German Research Foundation (DFG). The main goal of this work was to characterize the structure of the oxide scale and to determine the elemental distribution in the near-surface region. Moreover, carbon concentration profiles were measured on samples prepared by ion beam techniques and the conventional way with lubricants and polishing materials. Finally, a model of the near-surface region was derived by combining the results of quantitative analysis and X-ray mappings and was compared with thermodynamic calculations.

2. Experimental

The Fe-23Mn-0.6C steel samples (cold-rolled and re-crystallized sheets) were oxidized in a furnace for 20 minutes under air and an Ar / 4 % H_2 / 7 % H_2O atmosphere between 600 and 1000 °C.

The chemical composition of the near-surface region as well as the depletion (i.e., decarburized) zone were studied using Camebax SX 50 microprobe with a tungsten cathode, equipped with 4 wavelength-dispersive X-ray spectrometers and secondary and backscattered electron detectors. The measurements were performed at an accelerating voltage of 10 kV for specimens after annealing at 600 °C, and at 15 kV for sheets after oxidation at 800 and 1000 °C. For the X-ray microanalysis the following spectral lines were used: C Kα (PC2 crystal), O Kα (PC1 crystal), Fe Kα1 (LIF crystal), and Mn Kα1 (LIF crystal). Iron carbide (Fe_3C), iron oxide (Fe_2O_3), and pure Fe and Mn were used as standards. To avoid the contamination effects during carbon measurements an oxygen jet was applied to clean the surface.

For EPMA a metallographic cross-section was prepared. To obtain an overall flat surface and a gap-free region the samples were plated by nickel before mounting in a conductive resin. Afterwards the samples were grinded on 400 - 2500 grade SiC abrasive paper, polished (3 - 0.25 μm diamond paste) and cleaned in an ultrasonic bath. In a last step the samples were cleaned in a plasma cleaner. Alternatively, the FEI Strata 205 focused ion beam (a gallium-ion beam with an energy of 30 kV) was used for the preparation of specimens after oxidation at 600 °C. Good results could be obtained with the ion beam almost parallel to the oxidized surface (~ 3°). The obtained shallow bevel into the sheet revealed the layered structure. In addition, the sheet oxidized at 1000 °C was ground on 600 grade paper and polished with Ar-ions (6 kV) using a Jeol SM-09010 cross-section polisher.

The EBSD measurements were made on Jeol JSM-7000F scanning electron microscope with a field emission gun (FEG) and an Edax-TSL Pegasus combined EDS/EBSD system for chemical and crystallographic analysis.

The morphology of the oxidized surface was studied by scanning electron microscopy (SEM) using a Zeiss DSM 982 Gemini with a field emission gun. To avoid charging a thin conductive layer of Pt/Pd was sputtered on the surface.

3. Results and discussion

3.1. Oxidation at 600 °C in an Ar / 4 % H_2 / 7 % H_2O atmosphere – Combined FIB specimen preparation and EPMA

Figure 1 shows a BSE micrograph of the plain cross-section of the sheet after oxidation for 20 minutes at 600 °C in an Ar / 4 % H_2 / 7 % H_2O atmosphere. The thickness of the formed oxide scale is estimated to be below 0.5 μm. Moreover, the X-ray maps of O Kα, Fe Kα, Mn Kα and C Kα reveal the presence of an inner α-Fe layer of thickness also < 0.5 μm. In case of plain sample cross-section, however, the EPMA analysis of the oxide scale as well as the α-Fe layer, especially the quantitative analysis, is limited by the lateral resolution of the measurement. In order to demonstrate the depth distribution of the emitted characteristic X-rays Monte Carlo simulation was performed. In case of a

(Fe,Mn)O type oxide (ρ = 5.75 g/cm^3) at a beam energy of 10 kV the depth of emission (after absorption correction) for O Kα line has the maximum at about 0.55 μm (figure 2).

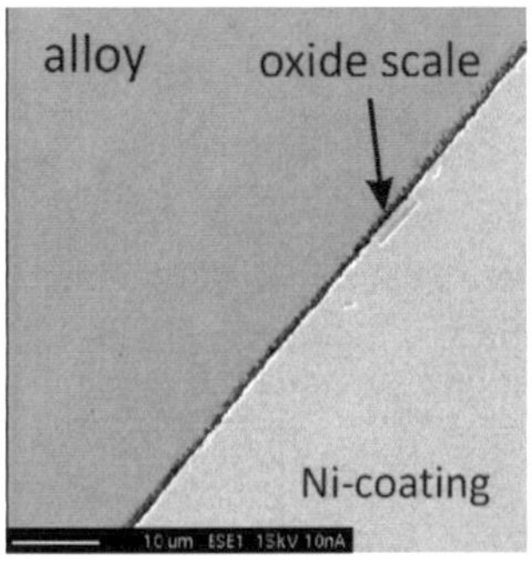

Figure 1. BSE micrograph (plain cross-section).

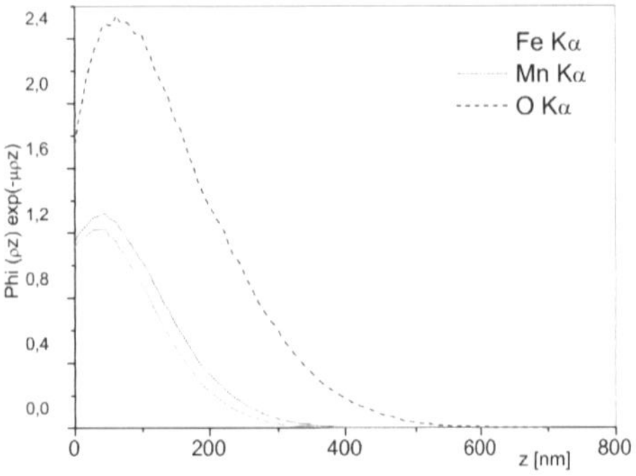

Figure 2. The depth of emission of Fe Kα, Mn Kα and O Kα characteristic X-rays in (Fe,Mn)O.

Therefore, in order to study the layered structure the FIB specimen preparation method was combined with EPMA. Figure 3 shows a diagram of the specimen after FIB preparation. Due to the FIB technique features (slope control) the thicknesses of sub-micrometre layered structure is enlarged. The X-ray maps for O Kα, Fe Kα, Mn Kα and C Kα acquired across an area of the milled bevel are shown in figure 4. The presence of the oxide scale and α-Fe layer can clearly be distinguished which is not the case when X-ray maps are produced on plain cross-sections. The α-Fe layer can also be characterized by the quantitative analysis of micron-sized areas and the average composition was calculated to be: 95.6 wt% Fe, 4.31 wt% Mn and 0.09 wt% C. For the α-Fe layer both X-ray mapping and quantitative analysis reveal a depletion in carbon. The application of plasma cleaning and oxygen jet allows minimizing the contamination effect. However, the measured content of carbon in ferrite phase is more than the theoretical value (0.008 wt% at room temperature). The higher content of carbon can be caused by an incomplete transformation of ferrite phase which has a core of retained austenite (figure 5).

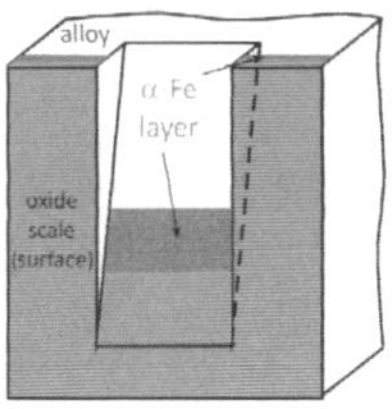

Figure 3. The specimen after FIB preparation.

Figure 4. BSE micrograph and X-ray maps for O Kα, Fe Kα, C Kα and Mn Kα (10 kV, 20 nA).

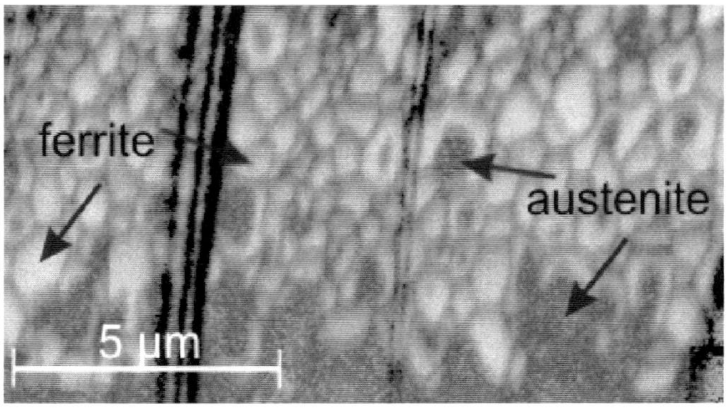

Figure 5. EBSD result (image quality) - ferrite layer (FIB specimen preparation).

In order to get the reliable quantitative results for carbon, the contamination produced by both sample preparation techniques, FIB and metallography was measured. It was found that the contamination level on the surface of a FIB specimen is lower than that on the surface of a cross-section produced by conventional metallography (figure 6).

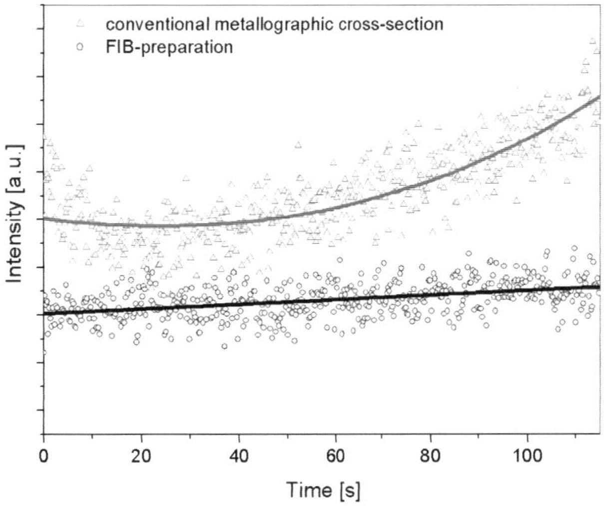

Figure 6. Measurement of C Kα intensity versus time (conventional cross-section and FIB preparation). The curves are introduced to emphasise the trend.

Moreover, a qualitative line scan along a bevel was also performed. However, the depth profile analysis needs a special interpretation. To provide a reliable composition, especially for the oxide scale, the EPMA results have to be evaluated using Monte Carlo simulation software [7]. Figures 7 and 8 show the simulated and experimental k-ratios measured along the FIB bevel. The composition of the oxide scale, α-Fe layer and alloy can be reconstructed by the comparison of measured k-ratios with those calculated by Monte Carlo simulation (table 1). The application of the technique mentioned above allows the oxide scale to be identified as mangano-wustite - (Fe,Mn)O. Additionally, table 2 shows calculated theoretical concentration of oxygen, manganese and iron in the various phases.

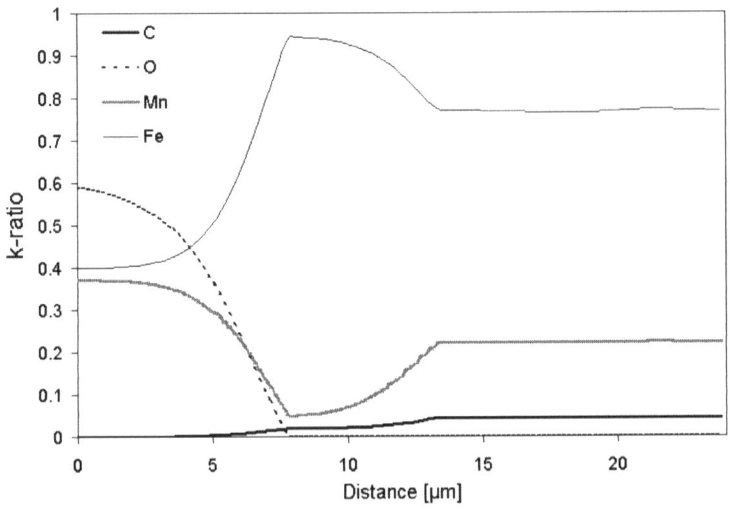

Figure 7. K-ratios calculated by Monte Carlo simulation.

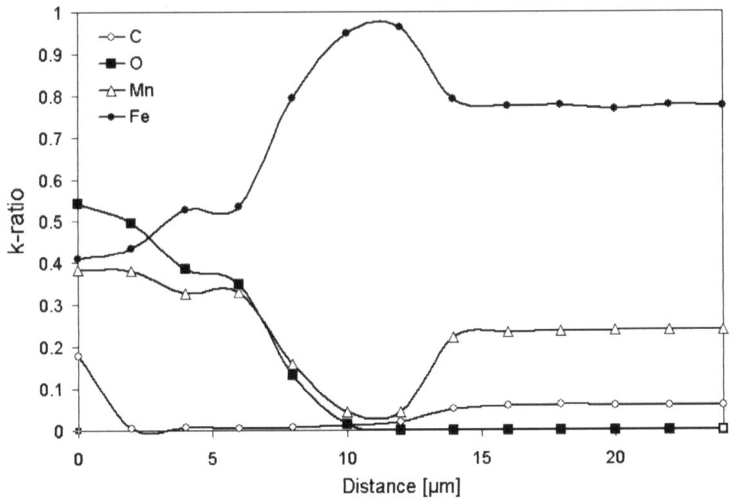

Figure 8. Experimental k-ratios measured along the FIB bevel.

Table 1. The composition reconstructed by comparison of measured k-ratios with those predicted by Monte Carlo simulation.

	wt%			at%		
	Fe	Mn	O	Fe	Mn	O
oxide scale	42.31	39.91	17.78	29.20	27.99	42.81
	Fe	Mn	C	Fe	Mn	C
α-Fe layer	95.29	4.57	0.13	94.79	4.63	0.58
steel	76.99	22.71	0.29	75.91	22.76	1.34

Table 2. Calculated theoretical concentration of oxygen, manganese and iron in the various phases.

Oxide (y = 0.5)	wt%			at%		
	Fe	Mn	O	Fe	Mn	O
$(Fe_{1-y}Mn_y)O$	39.11	38.48	22.41	25.00	25.00	50.00
$(Fe_{1-y}Mn_y)_3O_4$	36.39	35.80	27.80	21.43	21.43	57.14
$(Fe_{1-y}Mn_y)_2O_3$	35.17	34.60	30.23	20.00	20.00	60.00

3.2. Oxidation at 1000 °C in an Ar/4%H$_2$/7%H$_2$O atmosphere – Cross-section polished by argon ions

The oxide scale structure formed in an Ar / 4 % H_2 / 7 % H_2O atmosphere at 1000 °C is complex since both outer (external) and inner (internal) oxidation is observed (figures 9 and 10). In this case, besides the conventional cross-section, another preparation technique was also applied: the sample cross-section was additionally polished by argon ions. As seen from figure 11 the latter results in a lower contamination level than conventional sample preparation. The thickness of the outer scale is about 5 μm. Both the BSE micrograph and the qualitative line scans reveal the chemical homogeneity of the external layer. According to the quantitative analysis the average composition of that oxide scale is: 48.5 wt% Mn, 28 wt% Fe and 23.5 wt% O, which corresponds to mangano-wustite - (Fe,Mn)O. However, the oxide layer contains more manganese compared with the oxide scale after oxidation at 600 °C. According to the BSE micrograph and qualitative line scans the thickness of the inner region is between 5 and 10 μm. The region of inner oxidation consists of a nearly pure iron film as well as zones of Mn-rich oxide. Moreover, the composition of the brighter region on the BSE micrograph corresponds to the ferrite while the darker corresponds to the (Mn,Fe)O oxide.

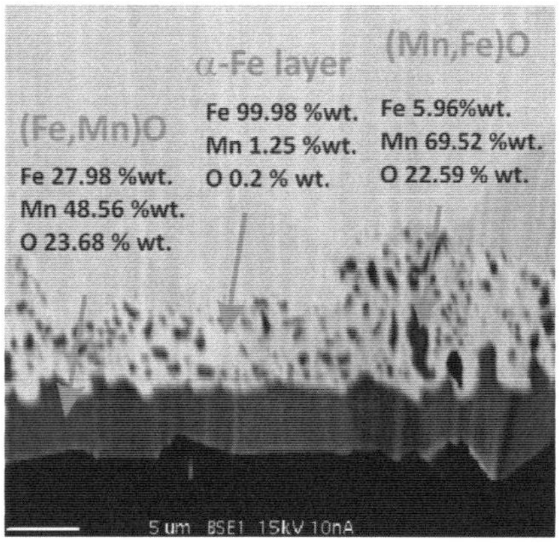

Figure 9. BSE micrograph with the results of quantitative analysis.

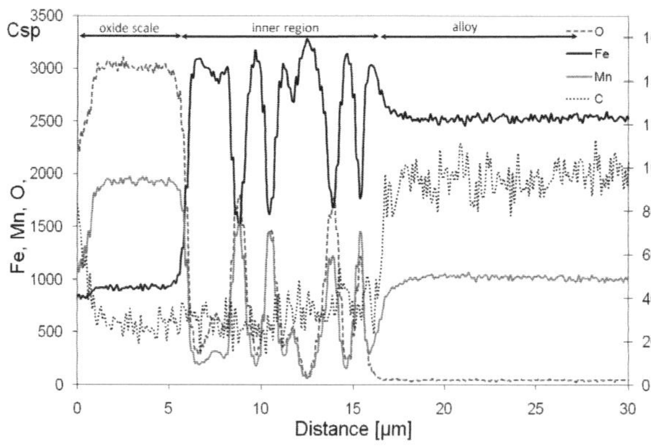

Figure 10. Line scans for the net intensity of Fe Kα, Mn Kα, O Kα and C Kα across the near-surface region.

To obtain further information about the morphology of the oxidized surface SEM was performed (figure 12). Incidentally, the shape and size of mangano-wustite grains were verified by EBSD. The EBSD examination was performed for the sheet after argon ions preparation. Figure 13b shows the phase distribution in the near-surface region. EBSD measurements reveal the outer continuous layer and some zones of the internal region to be wustite (cubic, Fm3m, a = 4.312 Å) while the rest of the

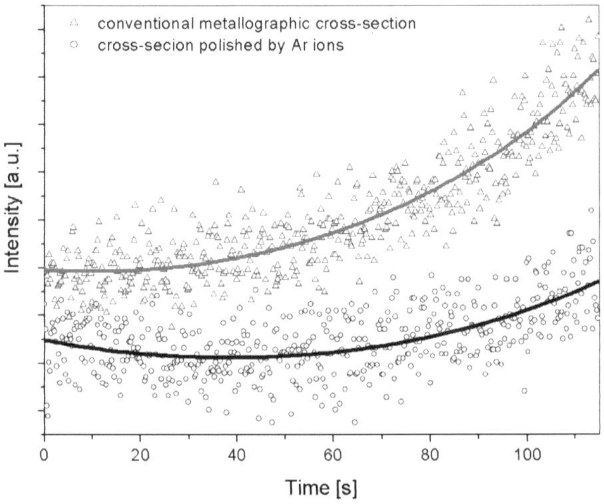

Figure 11. Measurement of C Kα intensity versus time (conventional cross-section and Ar-ions preparation). The curves are introduced to emphasise the trend.

internal band is ferrite (cubic, Im3m, a = 2.87 Å). Additionally, between the austenitic (cubic, Fm3m, a = 3.65 Å) substrate and the ferritic band a thin layer was found at the interface which is indexed as bcc by EBSD, but shows a poor image quality and small grains. This points too, a thin layer of bct martensite which was found at the austenite/ferrite interface and is too thin to be detected by EPMA. It is known [12] that both the mangano-wustite and MnO have also the same NaCl crystal structure as wustite. Thus, the EBSD results correlate well with the BSE micrograph and qualitative analysis (EPMA).

Figure 12. SEM micrograph of the surface after oxidation.

3.3. Oxidation at 800 °C in synthetic air

According to the X-ray mapping, the oxide scale that is formed in air atmosphere can be characterized as multilayered (figure 14). The structure of the oxide scale is more complex than that on the sheet after oxidation in an Ar / 4 % H_2 / 7 % H_2O atmosphere at 1000 °C. It can be observed that the oxide scale consists mainly of two regions: (I) - an external oxide layer with two layers having different oxidation states, (II) - an inner region with manganese-rich zones as well as nearly pure iron. The combination of X-ray images and results of quantitative analysis clearly shows two different oxides in region (I): Fe-rich $(Fe,Mn)_3O_4$ as a continuous layer and Mn-rich (Mn,Fe)O (figure 14 and table 3). Moreover, in this region, the external sub-layer exhibits a higher manganese content than the sub-layer beneath. In the case of region (II) zones of Mn-rich (Mn,Fe)O (occasionally Fe-rich (Fe,Mn)O) appear to be embedded within the continuous ferrite layer. In contrast with the oxide scale formed in

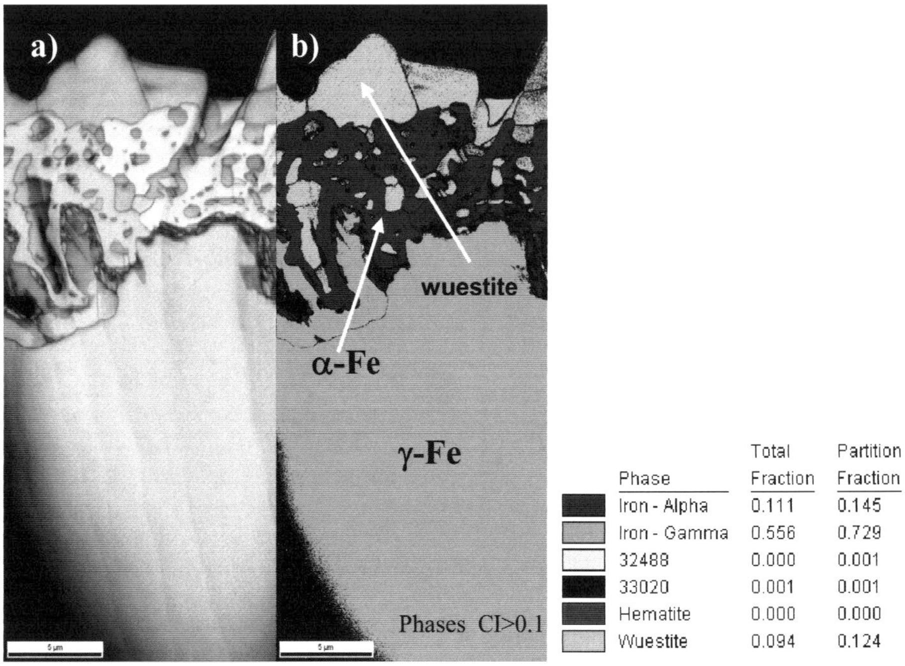

Figure 13. EBSD results; a) image quality; b) phase distribution image.

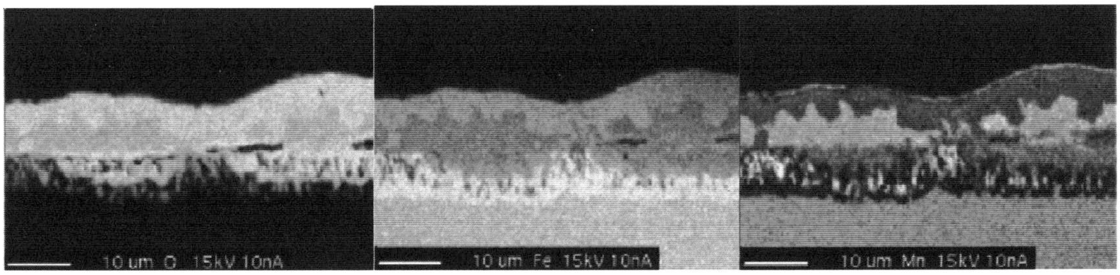

Figure 14. X-ray maps for O Kα, Fe Kα and Mn Kα.

Table 3. Results of quantitative analysis*.

Phase		Fe [wt.%]	Mn [wt.%]	O [wt.%]	Oxide type
1	outside scale	58.41	12.46	27.03	$(Fe, Mn)_3O_4$
2		36.16	40.64	21.59	(Mn,Fe)O
3	inner region	52.25	18.21	27.97	$(Fe, Mn)_3O_4$
4		68.24	7.00	23.89	(Fe,Mn)O
5		8.75	67.76	18.73	(Mn,Fe)O
6		97.65	3.12	0.44	α-Fe layer

* the location of the phases is shown in figure 14.

an Ar / 4 % H_2 / 7 % H_2O atmosphere, for the near-surface region after oxidation in air, the voids exist between regions (I) and (II) (figure 15). Additionally, a zone with a higher content of oxygen than the oxide layer lying above it is present between regions (I) and (II). The presence of Fe-rich $(Fe,Mn)_3O_4$

and voids in the inner region is connected with partially spalling during the oxidation process. The peeling of the oxide scale generally causes that the oxidation process occurs in two steps. In the first step the $(Fe,Mn)_3O_4$ layer is formed and probably because of stresses caused by volume mismatches, (i.e., the oxide occupies more volume than the metal it replaces) part of oxide scale is spalled. In the second step, the process of oxidation is continued from that layer. This explains the existence of an oxide with a higher oxidation state between regions (I) and (II).

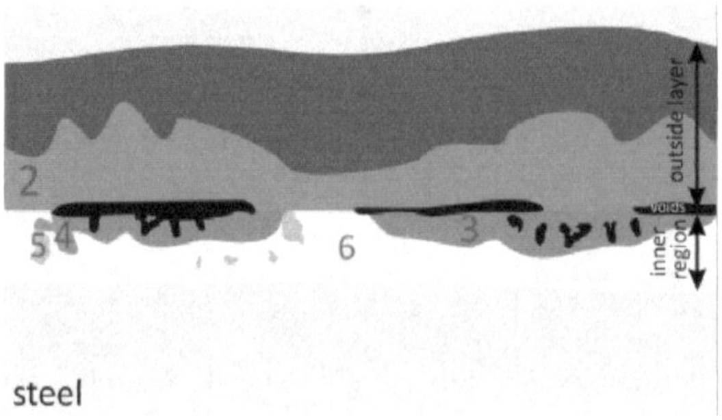

Figure 15. Model of the near-surface region.

Thermodynamic calculations were performed in the Fe-Mn-O system in order to simulate the oxidation process in a simplified fashion. Phase fractions and compositions of the phases were calculated at a fixed metal ratio as a function of the independence oxygen partial pressure as suggested by Kjellqvist [13]. This simulates a case where oxygen diffuses through the oxide scale and into the metal, but the metal atoms are considered immobile in the direction perpendicular to the surface. Although this is not quite true in most cases, it still gives a reasonable view of the oxidation process as long as one of the metals is not oxidized selectively. The thermodynamic description of the Fe-Mn-O system from Kjellqvist and Selleby [14] was used for the calculations. The calculated phase fractions for Fe-23%Mn at 800 °C are shown in figure 16a and the compositions (Mn content) of the phases are shown in figure 16b. Since no kinetic information is used the widths of the phase regions of figure 16a do not reflect the widths found in the actual oxidation process. The diffusion rates (and therefore the growth rates) decrease in the order (Fe,Mn)O, $(Fe,Mn)_3O_4$, $(Fe,Mn)_2O_3$. In particular $(Fe,Mn)_2O_3$ grows very slowly and would be expected to form only a very thin layer on top of the $(Fe,Mn)_3O_4$ layer. Austenite (fcc, γ) with a low Mn content transforms to ferrite (bcc, α) or martensite during cooling.

There is general agreement with the experimental observations but there are also some differences. The sequence Fe-rich metal (ferrite) together with Mn-rich (Fe,Mn)O followed by (Fe,Mn)O poorer in Mn, and Mn-poor $(Fe,Mn)_3O_4$ is found both experimentally and in the calculation. $(Fe,Mn)_2O_3$ is not found experimentally. It is to be noted, however, that when the layer is very thin identification by EPMA is difficult.

The measured compositions (table 3) are different from the calculated compositions, suggesting that the metals also diffuse perpendicular to the surface. In particular, Fe-rich (Fe,Mn)O found in the inner region is not expected upon the calculations.

4. Summary

The oxidized surface of a HMS-steel was successfully characterized by EPMA. For the first time it could be shown, that ion beam preparation techniques deliver more reliable carbon concentration profiles than conventional preparation techniques using lubricants and polishing materials. Another challenge was to improve the lateral resolution of EPMA, which was realized by the application of a special FIB-technique. The measurement across the milled bevel revealed clearly the presence of

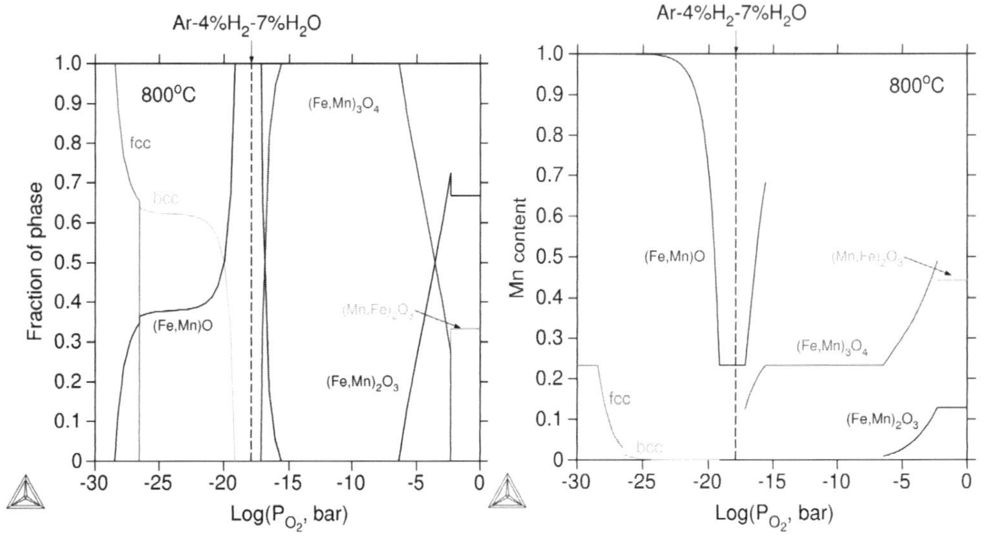

Figure 16. a) Calculated phase fractions as a function of oxygen partial pressure the Fe-23%Mn at 800 °C. The dashed line shows the oxygen partial pressure of Ar with 4 % H_2 and 7 % H_2O. $(Fe,Mn)_2O_3$ is the Fe_2O_3-based (hematite) solid solution and $(Mn,Fe)_2O_3$ is the Mn_2O_3-based (bixbyite) solid solution.
b) The corresponding compositions of the phases expressed as in terms of Mn content. Note that in single phase regions the composition of the phase is the nominal composition.

depletion in carbon. No conventional cross-section preparation could deliver such reliable EPMA results in terms of carbon values and high spatial resolution. Additionally, for the specimen after annealing at 600 °C in an Ar / 4 % H_2 / 7 % H_2O atmosphere the combination of EPMA, FIB-technique and Monte Carlo simulation results in the identification of the oxide phase as mangano-wustite - (Fe,Mn)O. At a higher temperature (1000 °C) the same phase was formed in the external scale as well as in the internal oxidation region, but with a different Mn/Fe ratio. Furthermore, due to the negligible damage produced by ion beam polishing we succeeded in performing EBSD analysis, which has verified the EPMA results.

Finally, the EPMA results for the specimen annealed at 800 °C in air reveal that the oxide scale had a more complex structure than that formed in an Ar / 4 % H_2 / 7 % H_2O atmosphere. The composition for the external oxide scale: $(Fe,Mn)_3O_4$ and (Mn,Fe)O, and for the internal region: ferrite layer and (Mn,Fe)O zones are consistent with thermodynamic calculations.

Acknowledgement
The authors gratefully acknowledge the financial support of the Deutsche Forschungsgemeinschaft (DFG) within the Collaborative Research Center (SFB) 761 "Stahl - ab initio".

References
[1] Wietbrock B, Xiong W, Bambach M and Hirt G 2011 *Steel Research Int.* **82** no. 1
[2] Perez P, Perez F J, Gomez C and Adeva P 2002 *Corros. Sci.* **44** 113-27
[3] Xiong W, Wietbrock B, Saeed-Akbari A, Bambach M and Hirt G 2011 *Steel Research Int.* **82** no. 2
[4] Gong Y F, Kim H, Kim S and De Cooman B C 2010 *Mater. Sci. Forum* **654-656** 258-61
[5] Phung T M, Jensen J M, Johnson D C, Donovan J J and McBurnett B G 2008 *X-ray Spectrom.* **37** 608-14
[6] Willich P and Bethke R 1995 *Anal. Chem.* **353** 389-92

[7] Karduck P and von Richthofen A 1995 *Microsc. Microanal. Microstruct.* **6** 421-32
[8] Von Richthofen A, Matsuo M, Karduck P and Ammann N 1994 *Microchim. Acta* **114/115** 511-23
[9] Ammann N and Karduck P 1994 *Surf. Interf. Anal.* **22** 54-9
[10] Richter S, Kyrsta S, Hajas D, Schneider J and Mayer J 2005 *Microsc. Microanal.* **11** 2
[11] Lesch N, Aretz A, Pidun M, Richter A and Karduck P 2000 *Microchim. Acta* **132** 377-82
[12] Perrot P 2008 *11D4: Iron Systems, Part 4.* Berlin-Heidelberg: Springer Verlag). doi: 10.1007/978-3-540-78644-3_18
[13] Kjellqvist L 2009 *Ph.D.-thesis.* (Stockholm: KTH)
[14] Kjellqvist L and Selleby J 2010 *Phase Equilib. Diffus* **31** 113-34

Microstructure and mechanical properties of nanocrystalline Ni-Mo protective coatings

A Bigos[1], E Beltowska-Lehman and P Indyka
Polish Academy of Sciences, Institute of Metallurgy and Materials Science,
25 ul. Reymonta, PL-30059 Krakow, Poland

E-mail: agnieszkabigos21@gmail.com

Abstract. Nickel-molybdenum alloys are of interest due to their functional properties, such as high hardness, corrosion resistance, as well as low wear and friction coefficients. Thus, they provide an excellent alternative to hard chromium coatings obtained from toxic electrolytes. Characterized coatings were electrodeposited on the steel substrates, from an aqueous citrate complex solution, containing nickel and molybdenum salts, in a system with a rotating disk electrode (RDE). The effect of cathodic current density on microstructure, chemical and phase composition, and crystallite size of the coatings was determined by the scanning and transmission electron microscopy, energy-dispersive X-ray spectrometry, and X-ray diffraction. The influence of microstructure as well as chemical composition on functional properties of coatings were also determined.

1. Introduction

Hard chromium coatings, characterised by enhanced functional properties, are widely used in many industrial applications. However, according to the European Union legal restriction (Directive: 2000/53/EC), they should be eliminated from the manufacturing process due to toxicity of electrolytes (containing carcinogenic hexavalent chromium) used for their preparation [1]. Ni-based alloys with refractory metals are proposed as a promising substitute for chromium coatings, characterized by good corrosion properties in many aggressive environments and catalytic activities for hydrogen evolution [2, 3]. Moreover, the addition of molybdenum to the nickel deposit should significantly improve their hardness, wear, and friction resistance.

Among the methods for obtaining Ni-based coatings, electroplating is a relatively simple and low-cost technique, which enables uniform covering of substrates with defined shapes. In addition, it eliminates the problem of conventional thermal processes associated with large differences in the melting points of metals (e.g., Ni-1455 °C, Mo-2620 °C) and their limited mutual solubility [4].

The mechanism of electrodeposition process is still not clearly understood, despite a few hypotheses presented in the literature [5, 6]. It is known that pure Mo cannot be electrodeposited in a metallic state from an aqueous solution of its salts. The process is hindered by the formation of a molybdenum oxide layer of lower valences on the cathode surface. Only co-deposition of the alloy with iron-groups metals (Fe, Co, Ni) is possible. According to Brenner that phenomenon is called induced co-deposition [7].

[1] To whom any correspondence should be addressed.

Microstructure and chemical composition are the main factors influencing the functional properties of electrodeposited coatings. Thus, for optimising electroplating parameters to produce adherent and uniform electrodeposits with appropriate micro-mechanical properties, the impact of chemical composition on microstructure changes should be discussed in detail. Hence, the aim of the present study was to determine the correlation of the microstructural characteristics, chemical (influence of molybdenum content) and phase compositions with coating mechanical and tribological properties such as microhardness, Young modulus, wear resistance, and roughness. To achieve this goal, the scanning electron microscopy (SEM), transmission electron microscopy (TEM), as well as X-ray diffraction (XRD) techniques for the analysis of Ni-Mo deposits were applied.

2. Experimental details

Micrometre-thick Ni-Mo alloys were electrochemically deposited from the citrate bath solution containing $NiSO_4$, Na_2MoO_4 of pH 8.5 (adjusted by an addition of ammonia). No detergents, wetting agent, or brightener were used. The cathode was a low carbon steel disc of 0.028 dm^2, rotating at 640 rpm, whereas a 0.05 dm^2 Pt spiral was used as an anode. Electrolysis was carried out in a 0.75 dm^3 cell in a model system with a rotating disk electrode (figure 1). Current density was controlled by a PAR 273A potentiostat/galvanostat with the correction of an ohmic drop by the current interrupt method. Ni-Mo alloys were deposited at room temperature under galvanostatic conditions in the range of current density 0.5 - 5 A/dm^2.

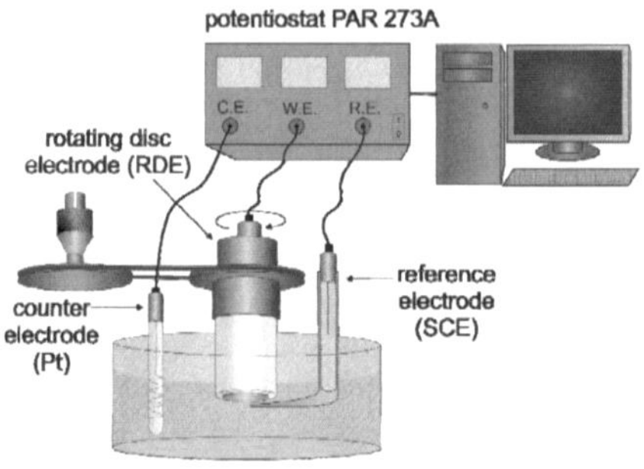

Figure 1. Experimental system with a rotating disk electrode (RDE).

The microstructure of the deposits was characterized by electron microscopy techniques: scanning FEI ESEM XL30 (accelerating voltage conditions of 20 kV), and transmission FEI Tecnai G^2 (200 kV), equipped with an Edax energy-dispersive X-ray spectrometer. The thin foils for TEM investigation were prepared by focussed ion beam (FEI Quanta 2D Dual Beam with Omniprobe lift-out system). Concentrations of elements were determined by energy-dispersive X-ray spectroscopy (EDS) in the SEM equipped with an Edax Gemini 4000 X-ray spectrometer. ZAF correction was used to improve the accuracy of X-ray microanalysis. Phase composition and crystallite size measurements were carried out by X-ray diffraction, using a Co K_α line (diffractometer Philips PW 1710 with X'Pert system). The surface roughness measurements were conducted by TOPO 01P profilometer from the sample area of 1.2 mm x 1.2 mm. The coating hardness was examined using the instrumented indentation technique according to the ISO 14577-1 standard (Micro-Combi-Tester produced by CSEM Instrument). The measurements were performed with a Vickers diamond indenter (for each coating at least 7 indentations) under the load of 50 mN applied at 100 mN/s. The duration time at maximum load was 5 seconds. The wear properties of electrodeposits were evaluated using a ball-on-disk apparatus with alumina balls of 1 mm in diameter as the counter body. The investigations were conducted at established test parameters: wear track of 5 mm diameter, rotation speed of 60 rpm, normal load of 1 N and 20000 as the total number of cycles.

3. Results and discussion

The process of electrochemical deposition is determined mainly by cathodic current density. That crucial operating parameter controls the chemical composition and microstructure as well as the rate of the electrodeposition process and current efficiency.

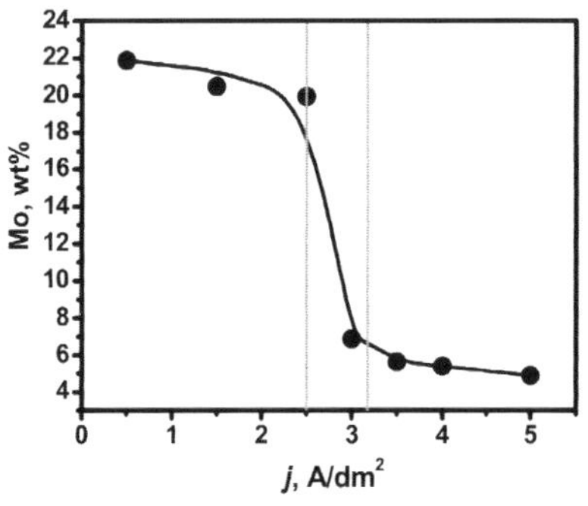

Figure 2. Mo content (wt%) in function of current density.

Figure 2 shows the dependence of the chemical composition of the coatings (determined by the EDS technique) on an applied current density. As can be seen, an increase of current density is strictly correlated with decrease in Mo content (from about 22 to 5 wt%) (table 1).

Table 1. Chemical composition of Ni-Mo alloys electrodeposited at different current densities determined by EDS (analyses were performed at 20 kV).

wt%	j, A/dm^2							
	0.5	**1.5**	**2.5**	**3**	**3.5**	**4**	**4.5**	**5**
Mo	22.2±0.4	21.0±0.4	19.0±0.8	6.9±0.3	5.6±0.2	5.4±0.2	5.1±0.2	4.9±0.5
Ni	77.8±1.6	79.0±1.6	81.0±1.6	93.1±1.9	94.4±1.9	94.6±1.9	94.9±1.9	95.1±1.9

The SEM cross-section examination revealed that all deposits (regardless of the current density at which they were deposited) were crack free, homogenous, and well adherent to the steel substrate as presented in figure 3.

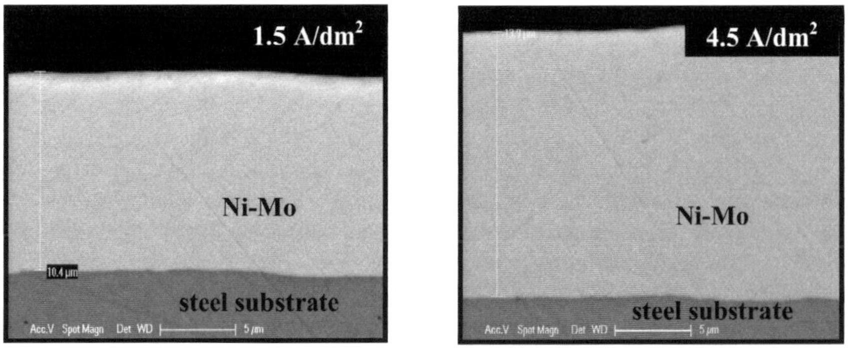

Figure 3. SEM images of cross-section of metallic Ni-Mo coatings electrodeposited at 1.5 A/dm^2 and 4.5 A/dm^2, at an RDE speed of 640 rpm.

However, scanning electron microscopy observations of the Ni-Mo surface show marked influence of current density on the microstructure of coatings (figure 4).

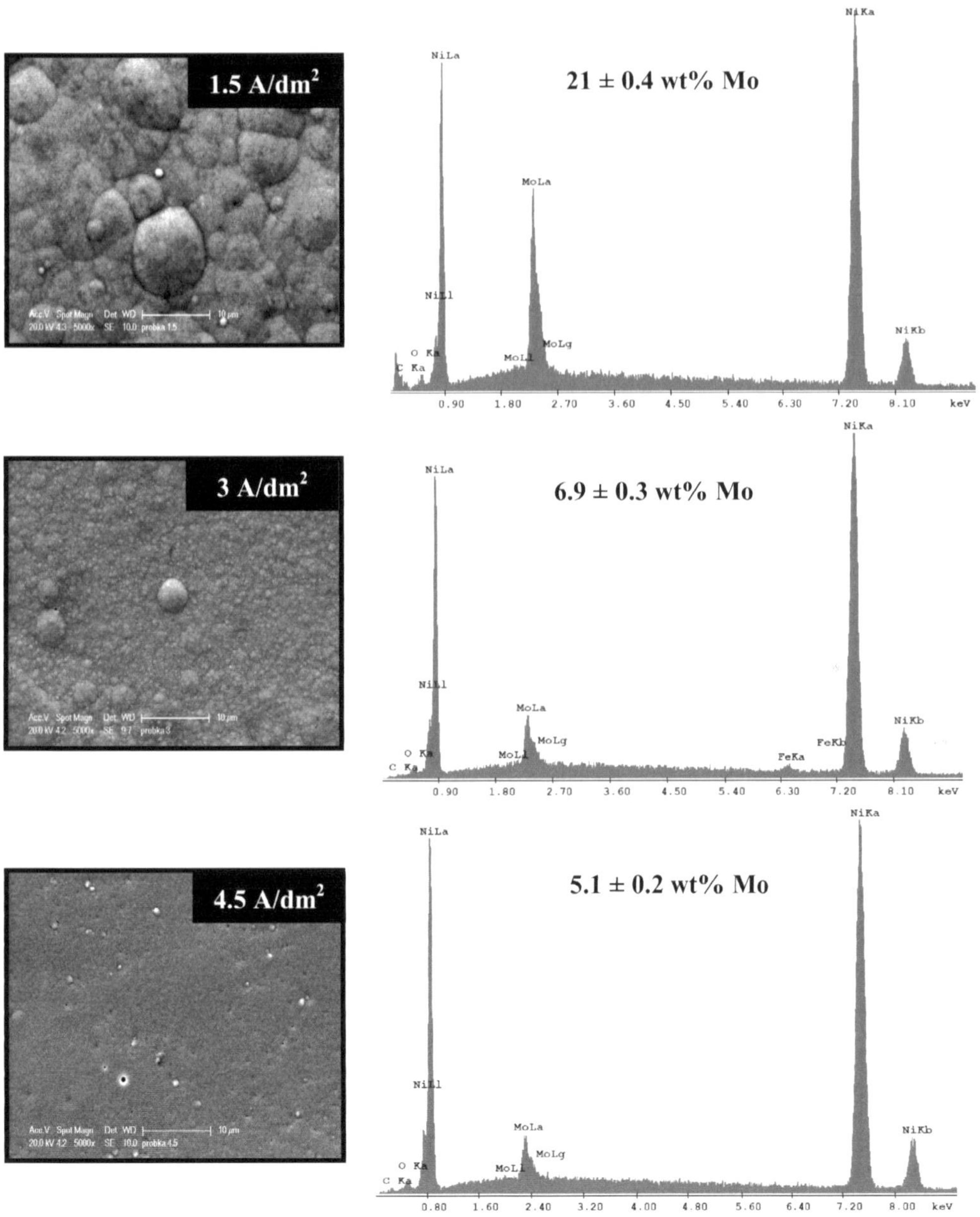

Figure 4. SEM (SE) images of surface morphology and EDS spectra of Ni-Mo coatings electrodeposited at 1.5 A/dm^2, 3 A/dm^2, and 4.5 A/dm^2, at an RDE speed of 640 rpm.

As seen, the molybdenum content in deposited alloys strongly affects the surface topography. In the first range of low current densities (0.5 - 2.5 A/dm^2) the Mo content is relatively high (about 20 - 22 wt%) and coatings are characterized by nodular morphology. When the density range increases to 2.5 - 3.5 A/dm^2 the size of the nodules are considerably reduced, which correlates with the decrease in Mo content from about 19 to 6 wt%. At a current density above 3.5 A/dm^2, coatings have smooth and compact morphology with a relatively low content of Mo (about 5 wt%).

The EDS spectra of exemplified deposits obtained at three different current densities (figure 4) confirm that the complete reduction of Mo(VI) complex ions occurs at the steel cathode. Indeed, the oxygen content in such coatings is at the background level. The obtained Ni-Mo coatings do not contain oxides.

Moreover, deposit morphology change is correlated with the decrease of the roughness coefficient R_a from a value of 0.27 μm to 0.15 μm up to 3 A/dm^2 (figure 5).

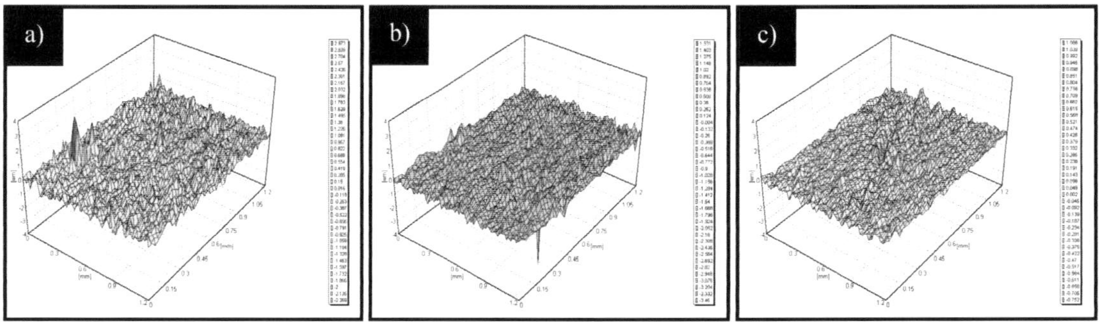

Figure 5. 3D surface profile of Ni-Mo coatings deposited at different current densities a) 1.5 A/dm^2, b) 3 A/dm^2 and c) 5 A/dm^2.

TEM observations of the Ni-Mo cross-section prepared by FIB technique showed that they are characterized by an equiaxial nanocrystalline microstructure with a grain size below 20 nm (for coatings deposited at 5 A/dm^2) (figure 6). The selected area electron diffraction pattern phase analysis indicated that the coatings revealed an α-Ni(Mo) solid solution [8, 9].

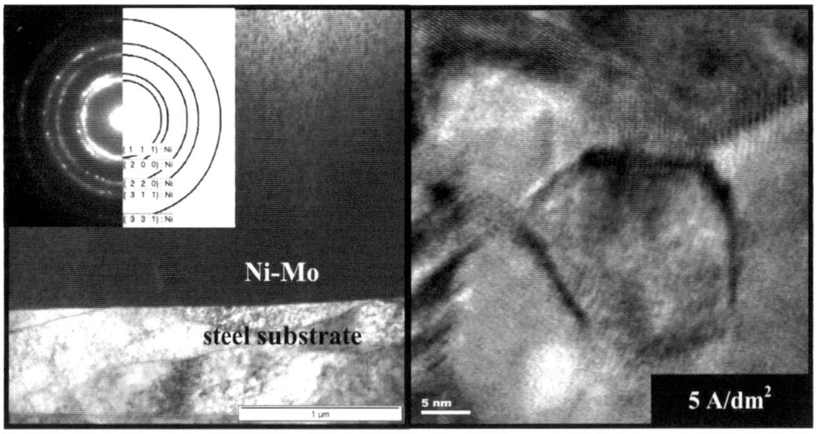

Figure 6. TEM images (bright field with the electron diffraction pattern (SAED) left, high resolution right) of an example of Ni-Mo coatings obtained at a current density of 5 A/dm^2.

XRD shows that with increasing molybdenum content in the electrodeposits, the lines of the patterns increasingly broadened, their intensities decreased, and the (111) preferred orientation became less pronounced. The grain size (the size of the coherent domains) was calculated from the line broadening, using Scherrer's equation. They decrease slightly (from 10 to 4 nm) when the molybdenum content increases from about 5 to 22 wt% (Figure 7).

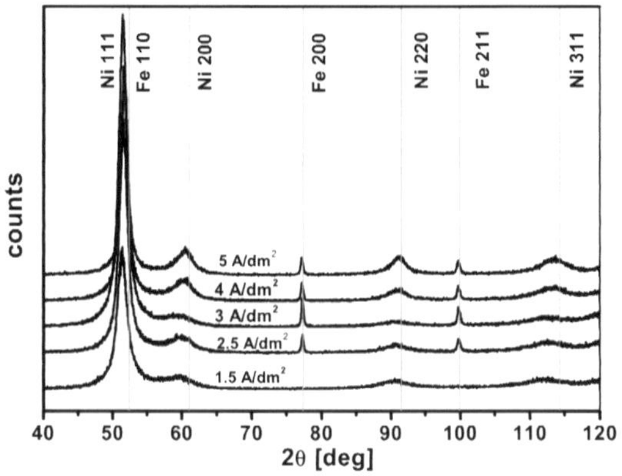

Figure 7. XRD pattern of Ni-Mo coatings deposited at different current densities (1.5 - 5 A/dm^2).

Microstructure and chemical composition of electrodeposited Ni-Mo alloys are closely related to their mechanical properties [10].

Microhardness and elastic modulus values, analyzed by microindentation test, rise (from 6.6 to 7.8 GPa and 184 to 258 MPa, respectively) with an increase of current density up to 3 A/dm^2. Further rise of this operating parameter leads to stabilisation of their values till 4 A/dm^2 (7.6 GPa and 182 GPa) and slightly drop above 5 A/dm^2. The deformation mechanism and adhesion to the steel substrate were investigated by scratch test. For coating deposited under low current density, any form of cracks and delimitations were observed up to a maximal load 30 N. For stiffer coatings prepared with 3.5 A/dm^2, the first cohesive cracks appeared when the load exceeded 18 N, which correlates with an increased value of the elastic modulus. At current densities above 4 A/dm^2, again a plastic character of deformation was observed.

Results of tribological tests indicated that a significant increase of wear resistance was found for electrodeposits prepared under current densities higher than 3 A/dm^2. Moreover, the friction coefficient is the lowest (0.23) for the Ni-Mo alloys obtained with high current densities.

4. Conclusion

- Electrodeposition of alloys from the designed citrate bath solution containing considerable excess of Ni(II) relative to Mo(VI) content results in obtaining a good quality Ni-Mo coatings: crack free, compact, and well adherent to the steel substrate.
- Cathodic current density is an operating parameter which controls the efficiency of the electrodeposition process as well as the chemical composition and microstructure of coatings.
- The nanocrystalline Ni-Mo coatings of smooth surface microstructure and relatively low molybdenum content, obtained at the highest examined current densities, are characterized by the best micromechanical and tribological properties. Thus, the Ni-Mo alloys obtained under these conditions are promising materials from the viewpoint of their potential application as an alternative to hard chromium coatings.

Acknowledgments
The results presented in this paper have been obtained within the project "KomCerMet" (contract no POIG.01.03.01-14-013/08-00 with the Polish Ministry of Science and Higher Education) in the framework of the Operational Programme Innovative Economy 2007-2013.

References
[1] Brooman E W 2000 *Met. Finish.* **6** 38-43
[2] Ghahin G E 1998 *Plat. Surf. Finish.* **8** 8-14
[3] Hille R 2002 *Trends Biochem. Sci.* **7** 360-7
[4] Massalski T B 1990 *Binary alloy phase diagrams* (The United States of America: ASM International)
[5] Chassaing E, Yu Quang K and Wiart R 1989 *J. Appl. Electrochem.* **19** 839-44
[6] Podlaha E J and Landolt D 1996 *J. Electrochem. Soc.* **143** 893-9
[7] Brenner A 1963 *Electrodeposition of alloys* (New York, London: Academic Press)
[8] Labar J L 2008 *Microsc. Microanal.* **14** 287-95
[9] Labar J L 2009 *Microsc. Microanal.* **15** 20-9
[10] Kot M, Bełtowska-Lehman E, Bigos A, Indyka P, Morgiel J and Rakowski W 2010 *Mater. Eng.* **3** 373-6

Characterisation of high temperature refractory ceramics for nuclear applications

P D W Bottomley[1,3], Th Wiss[1], A Janssen[1], B Cremer[1], H Thiele[1], D Manara[1], M Scheindlin[2], M Murray-Farthing[1], P Lajarge[1], M Menna[1], D Bouexière[1] and V V Rondinella[1]

[1] European Commission, JRC, Institute for Transuranium Elements (ITU), Hermann-von-Helmholtz Platz 1, P.O. Box 2340, DE-76125 Karlsruhe, Germany

[2] now at Russian Academy of Sciences, Joint Institute of High Temperatures IVTAN, Moscow, Russia

E-mail: paul.bottomley@ec.europa.eu

Abstract. The ternary oxide ceramic system UO_2-ZrO_2-FeO is a refractory system that is of great relevance to the nuclear industry as it represents one of the main systems resulting from the interaction of the Zircaloy cladding, the UO_2 fuel and the structural elements of a nuclear reactor. It is particularly the high temperature properties that require investigation; that is, when substantial overheating of the nuclear core occurs and interactions can lead to its degradation, melting and result in a severe nuclear accident. There has been much work on the UO_2-ZrO_2 system and also on the ternary system with FeO but there is still a need to examine 2 further aspects; firstly the effect of sub-oxidized systems, the UO_2-Zr and FeO-Zr systems, and secondly the effect of Fe/Zr or Fe/U ratios on the melting point of the U-Zr-Fe oxide system. Samples of UO_2-Zr and UO_2-ZrO_2-FeO were fabricated at ITU and then characterized by optical microscopy (OM) and X-ray diffraction to determine the ceramic's structure and verify the composition. Thereafter the samples are to be melted by laser flash heating and their liquidus and solidus temperatures determined by pyrometry. This programme is currently ongoing. The frozen samples, after testing, were then sectioned, polished and the molten zone micro-analytically examined by OM & SEM-EDS in order to determine its structure and composition and to compare with the existing phase diagrams. Examples of results from these systems will be given. Finally, a reacted Zr-FeO thermite mixture was examined, which had been used to generate high temperatures during tests of reactor melt-concrete interactions. The aim was to assess the reaction and estimate the heat generation from this novel technique. These results allow verification or improvement of the phase diagram and are of primary importance as input to models used to predict materials interactions in a severe nuclear accident.

1. Introduction

The ternary oxide ceramic system UO_2-ZrO_2-FeO is a refractory system that is of great relevance to the nuclear safety as it represents one of the main systems resulting from the interaction of the Zircaloy cladding, the UO_2 fuel and the structural elements of a reactor when substantial overheating of the nuclear core occurs and interactions can lead to its degradation and melting. There has been

[3] To whom any correspondence should be addressed.

much work on the UO_2-ZrO_2 system and some on the ternary system with FeO but there is still a need to examine further: the effect of Fe/Zr or Fe/U ratios on the melting point and the effect of sub-oxidized systems (e.g., UO_2-Zr).

This paper reports the first steps of a work programme in this field at ITU. This work also represents a contribution to the SARNET 2 (Severe Accident Network of Excellence 2) Work Programme supported by the European Commission and forms part of the work package on Molten Corium Concrete Interaction (MCCI) of severe accident in nuclear reactors, which looks at not only in-vessel melts but also melts interacting with the concrete basement of the reactor building.

2. Systematic study of the Fe-Zr-U-O system

2.1. Introduction to sample preparation

Samples of UO_2-Zr and UO_2-ZrO_2-FeO were fabricated and then characterized by optical microscopy (OM) and X-ray diffraction to determine the material's structure and verify the composition. Thereafter liquidus and solidus temperatures are determined by pyrometry while melting the samples by laser flash heating. The frozen samples will then be sectioned and the molten zone micro-analytically examined using scanning electron microscopy with energy-dispersive X-ray spectrometry (SEM-EDS) in order to determine its structure and composition. The objective of this study was to both improve the existing phase diagrams as well as provide data for modelling.

2.1.1. U-Zr-Fe-O sample preparation. The following samples were fabricated by the fuels fabrication unit at ITU [1]. These were prepared from UO_2, ZrO_2, FeO and Fe powders as specified in table 1 below.

Table 1. Compounds used in the fabrication of the prototypical Fe-U-Zr-O system.

Compound/ Granulometry	Manufacturer	Purity [%]	Batch
UO_2 (40 μm mean particle size)	Cogema	NA	TU2 E1697 - 2986
ZrO_2 (< 40 μm mean particle size)	Alfa Aesar	> 99.7 %	4314936
FeO (2 mm mean particle size)	Aldrich	> 99.9 %	S40086-6, S40430-339
Fe (< 10 μm mean particle size)	Aldrich	> 99.9 %	129U60

The compositions of the system that were made are given in table 2 (with actual figures in brackets). The uranium dioxide powder was made from depleted uranium with approximately 0.3 % ^{235}U content. The UO_2 was determined by thermogravimetry to have an O/M ratio of 2.024 before sintering (this has been determined to reduce to 2.00 after sintering at 1300 °C under Ar / 6 % H_2 gas for 6 h). There was a systematic variation of the FeO/(UO_2+ZrO_2) from 8:1 to 1:4.5 with the UO_2/ZrO_2 ratio held at 1:1. This was to examine both rich and poor mixtures of iron oxide. Approximately 1 wt% Fe powder was also added to the mixture in order to ensure that there was no oxidation of FeO to higher oxides in the mixtures during the preparation [2]. The compositions are also shown on the ternary oxide diagram as given in figure 1. This shows that the compositions cross a zone of three phases: FeO and ZrO_2+UO_2 solid solutions.

Table 2. Nominal molar compositions of UO_2, ZrO_2 and FeO prepared for investigation (figures in brackets give actual values).

BATCH	UO_2 (mol%)	ZrO_2 (mol%)	FeO (mol%)
1	10 (10.000)	10 (9.998)	80 (80.002)
2	20 (19.968)	20 (20.061)	60 (59.971)
3	30 (29.986)	30 (30.019)	40 (39.995)
4	40 (40.022)	40 (39.973)	20 (20.005)
5	45 (44.989)	45 (44.954)	10 (10.057)

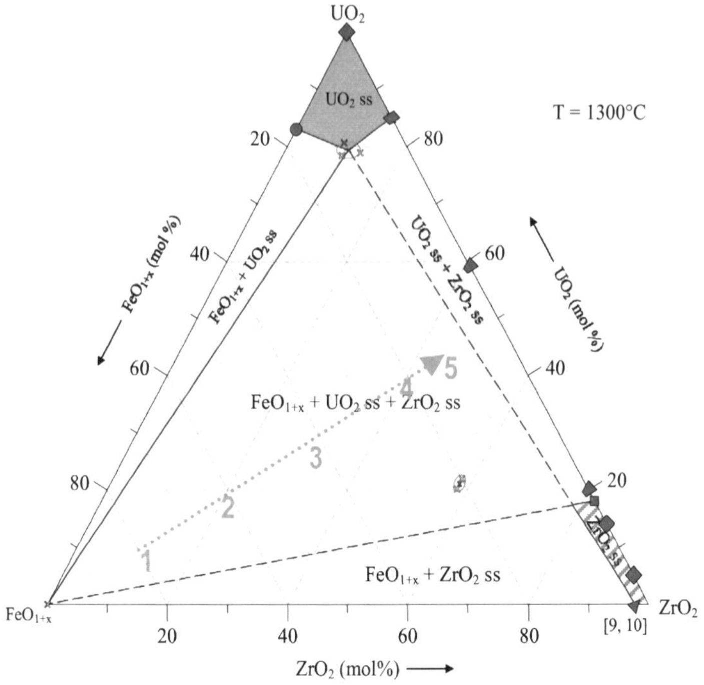

Figure 1. FeO-UO_2-ZrO_2 ternary oxide system showing the positions of the batch compositions 1 to 5 across the single mixed oxide phase [3]

The powder starting materials were mixed along with ca. 0.3 % zinc stearate as a lubricant. The batches were mixed together for one hour by rotation in a blender and then pressed into pellets of 4 mm high by 9.75 mm diameter at circa 600 MPa. The pellets were then heated for 6 hours in a molybdenum crucible at 1200 °C under an argon atmosphere. The height and weight of the pellets after sintering was on average 3.2 mm and 1.4 g.

Figures 2a and 2b shows the sintered pellets for batches 1 ($10UO_2$-$10ZrO_2$-80FeO, in mol%) and 5 ($45UO_2$-$45ZrO_2$-10FeO, in mol%) respectively. They have a similar dull grey appearance and some pores and cracks were visible on the surface after the sintering process.

Figure 2. Sintered discs from a) batch 1 (FeO-rich mixture) and b) batch 5 (FeO-poor mixture).

2.1.2. U-Zr-O sample preparation. The UO_2-Zr samples were manufactured from depleted UO_2 (ITU stock -Barleon) and Zr metallic spheres (99.2 % purity Zr +Hf from LS Technik GmbH) with a particle size of < 250 μm. The UO_2 powder was previously checked by thermogravimetry to have an O/M ratio of 2.024. The Zr metal spheres and the UO_2 powder were weighed and mixed with 0.5 % zinc stearate before mixing for 15 hours with a roller. They were then pressed at 350 MPa in the form of pellets and then sintered under Ar atmosphere for 6 hours at 1560 °C [4]. The compositions of interest were UO_2 mixed with - 5 mol% Zr, - 10 mol% Zr and -30 mol% Zr and were the completion of an earlier examination of the UO_2-metallic Zr system.

2.2. Characterisation of the samples

After preparation the samples were then characterized by visual inspection and by x-ray diffraction. Thereafter a sample was taken for sectioning, mounting and polishing for examination by optical microscopy to check the uniformity and density of the samples.

2.2.1. Optical microscopy of U-Zr-Fe-O system. The mounted samples of the pellets were examined by high and low power microscopy. Figures 3a to e shows the low power micrographs of the samples.

The larger agglomerations of the FeO can be seen in the FeO-rich mixtures since they have an initially larger grain size and retain their size because of their greater fusibility (FeO fuses at circa 1500 °C while the ZrO_2-UO_2 system does not fuse below 2500 °C). The more detailed microstructure can be seen in the figures 4a and 4b for batch 1 (10UO_2-10ZrO_2-80FeO, in mol%) composition, where the FeO-rich zones and the mixed UO_2-ZrO_2 zones can be seen. There are also metallic precipitates of iron from the Fe addition to the mixture (figure 4b); these are mostly located within the FeO agglomerates or at the boundary areas between the FeO and UO_2-ZrO_2 rich areas. These analyses were confirmed in the later SEM-EDS analyses of the unmelted areas in the samples after testing (see Section 2.2.5.3).

The same microstructural features are seen in the 40UO_2-40ZrO_2-20FeO (in mol%) composition in figure 5 and in the 45UO_2-45ZrO_2-10FeO (in mol%) composition in figure 6. The main differences are observed in the macrostructure with the proportion of large FeO agglomerates increasing with the FeO in the initial composition.

2.2.2. Optical microscopy of the U-Zr-O system. The microstructure of the UO_2-5Zr (in mol%) composition is given in figure 7. The 2-phase matrix is composed of a U-rich oxidic phase (UO_2 with some Zr in solid solution) and a Zr-rich metallic phase with an α-Zr(O) structure and containing some U in solution. The sample had a theoretical density (TD) of the order of 82 - 85 % after sintering. In the Zr-rich metallic phase some secondary precipitates are seen that are presumably a U-rich metallic

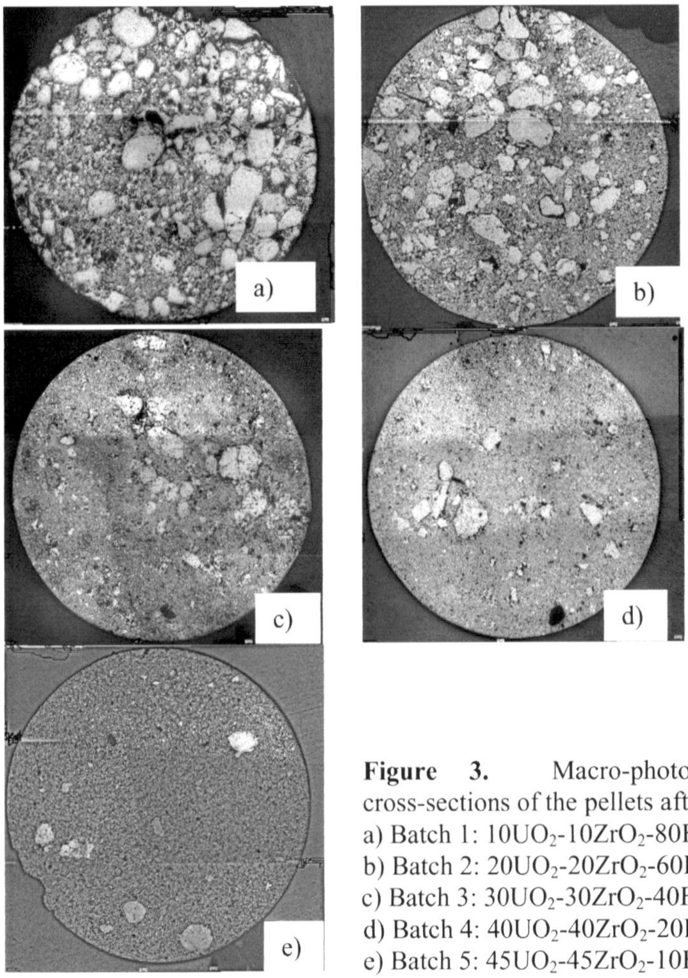

Figure 3. Macro-photographs showing the polished cross-sections of the pellets after sintering:
a) Batch 1: $10UO_2$-$10ZrO_2$-80FeO (mol%).
b) Batch 2: $20UO_2$-$20ZrO_2$-60FeO (mol%).
c) Batch 3: $30UO_2$-$30ZrO_2$-40FeO (mol%).
d) Batch 4: $40UO_2$-$40ZrO_2$-20FeO (mol%).
e) Batch 5: $45UO_2$-$45ZrO_2$-10FeO (mol%).

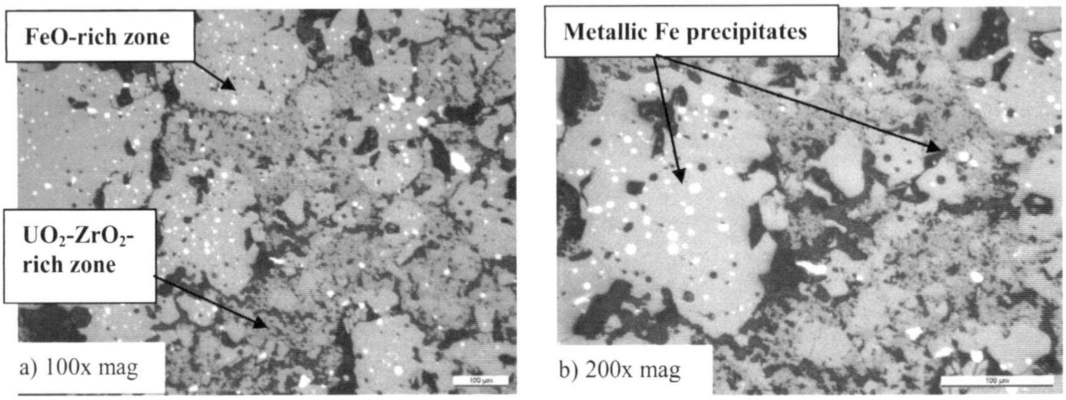

Figure 4. a) and b) Micrographs for the batch 1 composition of $10UO_2$-$10ZrO_2$-80FeO, in mol%.

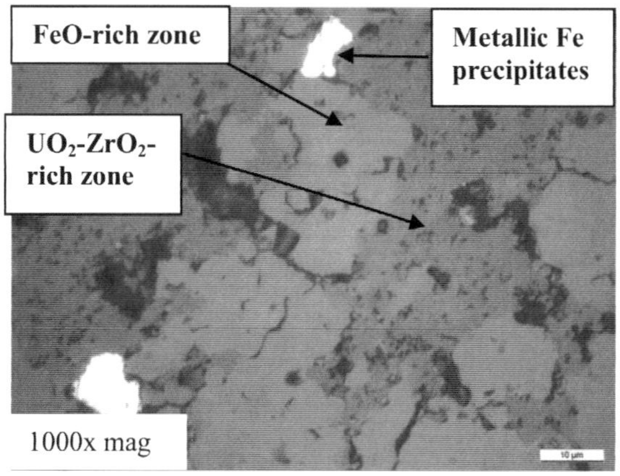

Figure 5. High power micrograph of $40UO_2$-$40ZrO_2$-20FeO (in mol%).

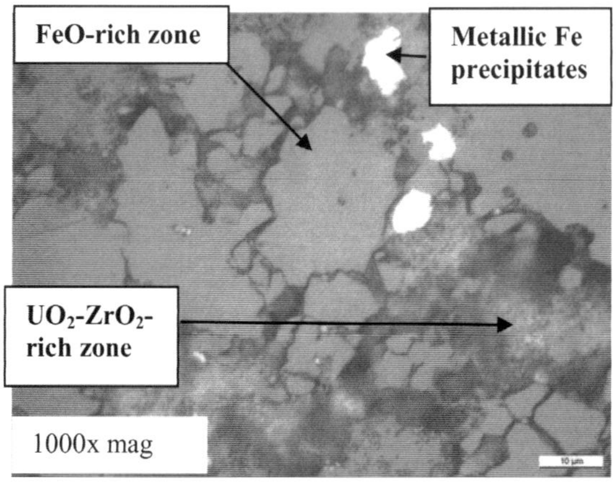

Figure 6. High power micrograph of $45UO_2$-$45ZrO_2$-10FeO (in mol%).

Figure 7. a) low power; and b) higher power optical micrographs of the UO_2-5Zr (in mol%) sample.

phase that formed on cooling. A similar sample was sent for melting point determination by laser flash technique followed by examination of the sample after testing to verify the overall composition of the melt and its phase structure.

2.2.3. X-ray diffraction. Samples of the UO_2-ZrO_2-FeO pellets were also analyzed by X-ray diffraction. Although metallic Fe precipitates were visible under the optical microscope these were at the level of 1wt% and so at the lower limit of detection for X-ray diffraction. These results confirmed the presence of original components, UO_2, ZrO_2 and FeO, with no higher oxides present and the presence of Fe precipitates also confirmed that no higher Fe oxides would be formed.

2.2.4. Laser flash melting point determination. This uses a continuous high power YAG laser to rapidly heat a planar surface of the sample to melting point within 1 s and monitor the cooling curve to observe phase changes. The sample is monitored by a high-speed, 2-wavelength pyrometer (operating at wavelengths (λ) of 643 nm and 488 nm with 1 measurement every 10 μs) and a multi-wavelength spectrometer (measuring between 550 and 910 nm wavelengths with 1 measurement every 1 ms) in order to follow the heating and cooling curve. The 2-wavelength pyrometer is able to give the brightness temperature and the multi-wavelength pyrometer is able to determine the emissivity of the sample at the chosen wavelength (λ = 643 nm) and then the true sample temperature can be recalculated from the temperature trace of the 2-wavelength pyrometer.

The cooling arrests indicate the phase changes, typically melting points or liquidus/solidus temperatures. In addition there is a green light (λ = 488nm) diode laser aimed at the sample surface and the reflection of this signal is monitored via a telescopic system. This is to register any variation in the surface of the sample and enables any movement on the surface due to liquefaction to be followed. The system is schematically shown in figure 8. More details of the technique are given in refs [5, 6].

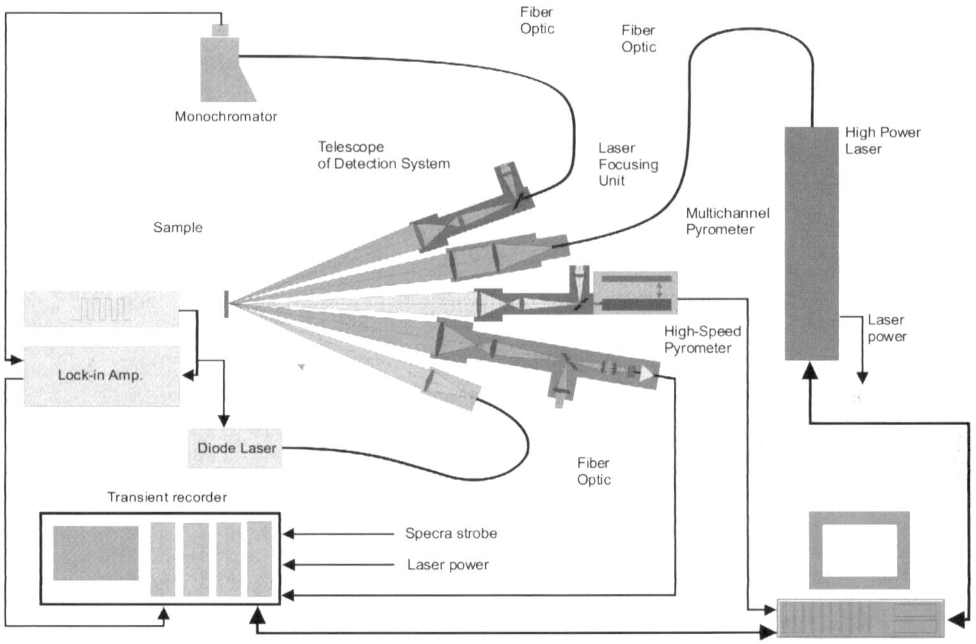

Figure 8. Diagram of the laser flash facility at ITU for determination of liquidus and solidus temperatures of high temperature refractory systems.

The melting point determination for the UO_2-5Zr (in mol%) is given in figure 9. This indicates the original brightness temperature measurements from the 2-wavelength pyrometer and also the true temperatures after correction for the emissivity determined by the multi-wavelength pyrometer. This shows a transition at 2900 K (± 30 K) (2627 ± 30 °C), which is attributed to the liquidus temperature for this composition, with a second possible transition at 284 5K (± 30 K), attributed to the solidus temperature. The detection telescope of the green diode also (blue line) shows the vibrations on the surface of the sample due to liquefaction on the surface. These oscillations correspond to the molten surface movements, and their onset and cessation match well with the melting and solidification of the sample.

The U-Zr-Fe-O system is currently under investigation and only the first results are available. In this paper we will concentrate on Fe-richest sample being investigated: $45UO_2$-$45ZrO_2$-10FeO (in mol%). The liquidus temperature for this sample is 2360 K (± 30 K) (2087 °C) while a very low solidus temperature of 1560 K (± 30 K) (1287 °C) was recorded. This wide melting range would be anticipated given the melting of pure FeO is ~1648 K [7].

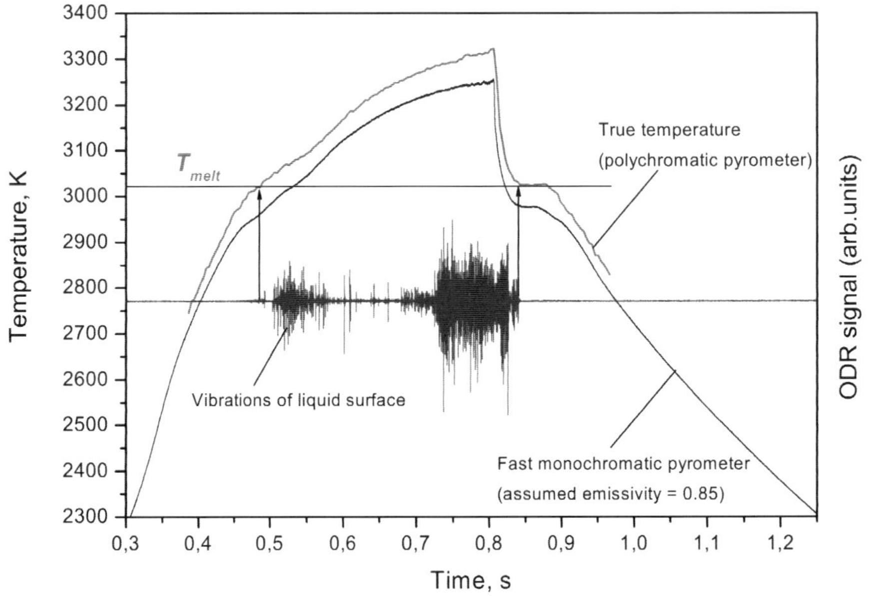

Figure 9. Laser flash heating thermogram for the UO_2-5Zr (in mol%) sample.

2.5. Post test sample examination

2.5.1. Introduction. The sample was sectioned and embedded in resin and polished; and was then examined by optical microscopy and scanning electron microscopy with energy-dispersive X-ray analysis (Vega Tescan 5130 LS SEM combined with an Oxford Instruments 7830 EDS system using a 10 mm^2 Si(Li)-detector). The accelerating voltage was usually 30 keV and the current usually 2 nA, with an acquisition time of typically 200 s. The objective was to examine the melted and non-melted areas of the sample and to check the amount of melt and verify its composition & condition.

2.5.2. Examination of the U-Zr-Fe-O system. The sample shows a fine 2-phase eutectic structure in the melted zone with precipitates of metallic iron mainly at the edge of the molten zone (see figure 10).

Non-melted zones as seen on the left hand side of figure 10b energy-dispersive X-ray analyses were also carried out here and indicated zones of mixed U,Zr oxides $(U,Zr)O_2$ with some FeO present, oxidic zones rich in FeO containing Zr and U and finally metallic precipitates of pure Fe. Oxygen could only be measured in the FeO-rich precipitates. The absence of an O-signal was noted in the metallic precipitates (which was also evident from their bright lustrous appearance under the optical microscope while the other poorly reflective phases were ceramic (i.e., oxidic) in nature).

However with the difficulty of oxygen measurements with the Zr or U-rich precipitates and for consistency, oxygen was not included and only the metallic composition was measured.

The main eutectic structure of the molten zone is seen in figure 11a has a uniform composition; at high magnification it shows a very fine eutectic structure with phases less than a 1 μm thick. This confirms the optical microscopy which also showed a fine melt structure. Values are given in the figure 11c below correspond to the initial components – UO_2-rich particles, ZrO_2-rich particles as well as larger FeO particles and fine precipitates from the additional Fe from the sintering of the samples. The_FeO grains are relatively pure, containing only 1 at% U or Zr, whereas the UO_2-rich and ZrO_2-rich particles contain much more Fe. This reflects the phase diagram at 1300 °C (figure 2)

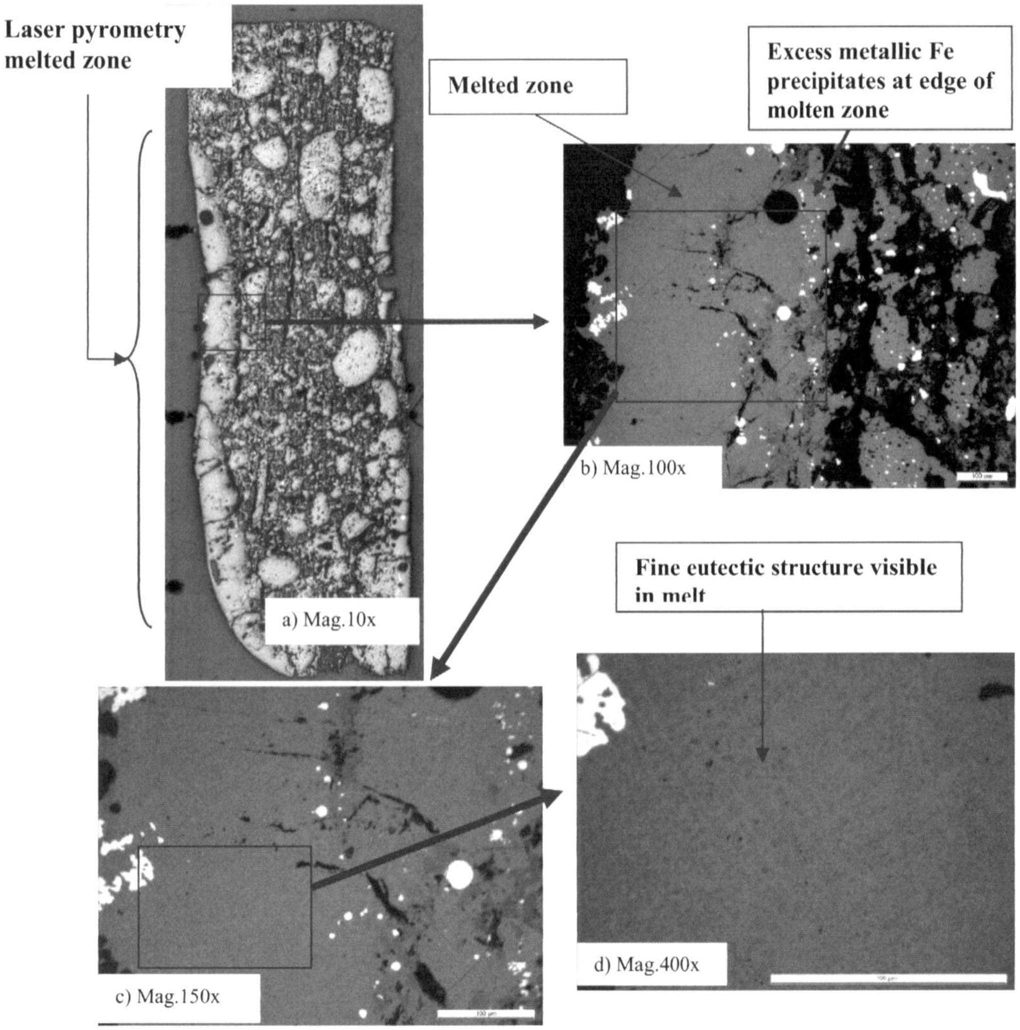

Figure 10. a-d. OM micrographs of the $10UO_2$-ZrO_2-80FeO (in mol%) showing the melted zone of the left-hand lower surface and the non-melted material below it. Metallic Fe precipitates are visible at the edge of the zone as well as a very fine eutectic structure in the main melt zone.

where the solubility of FeO in UO_2 is ~17 mol% and for FeO in ZrO_2 is ~5 mol%. The 2 analyses of the surface suggest the overall composition of the melt in shifted towards the Fe-rich end. Another 3rd zone was also taken with slightly more and coarser, dense precipitates present and the area analysis of a typical zone (see marked area in figure 12) indicated higher U and Zr contents (~7 at%) than in the fine eutectic melt (~5 at%).

2.5.3. Examination of the U-Zr-O system. SEM micrographs and positions of the point analyses of the melt for the UO_2-5Zr (in at%) samples are shown in figures 13. Figure 13a shows a low magnification micrograph in which the melt zone (~1 - 2 mm wide) where the laser beam melted the material to about 100 μm depth. Beneath that, the material is effectively unmelted. Figure 13b is a backscattered electron (BSE) image of the unmelted zone that shows brighter UO_2-rich grains of typically 20 - 30 μm diameter with smaller, secondary precipitates in the grains (1 - 2 μm) and at the grain

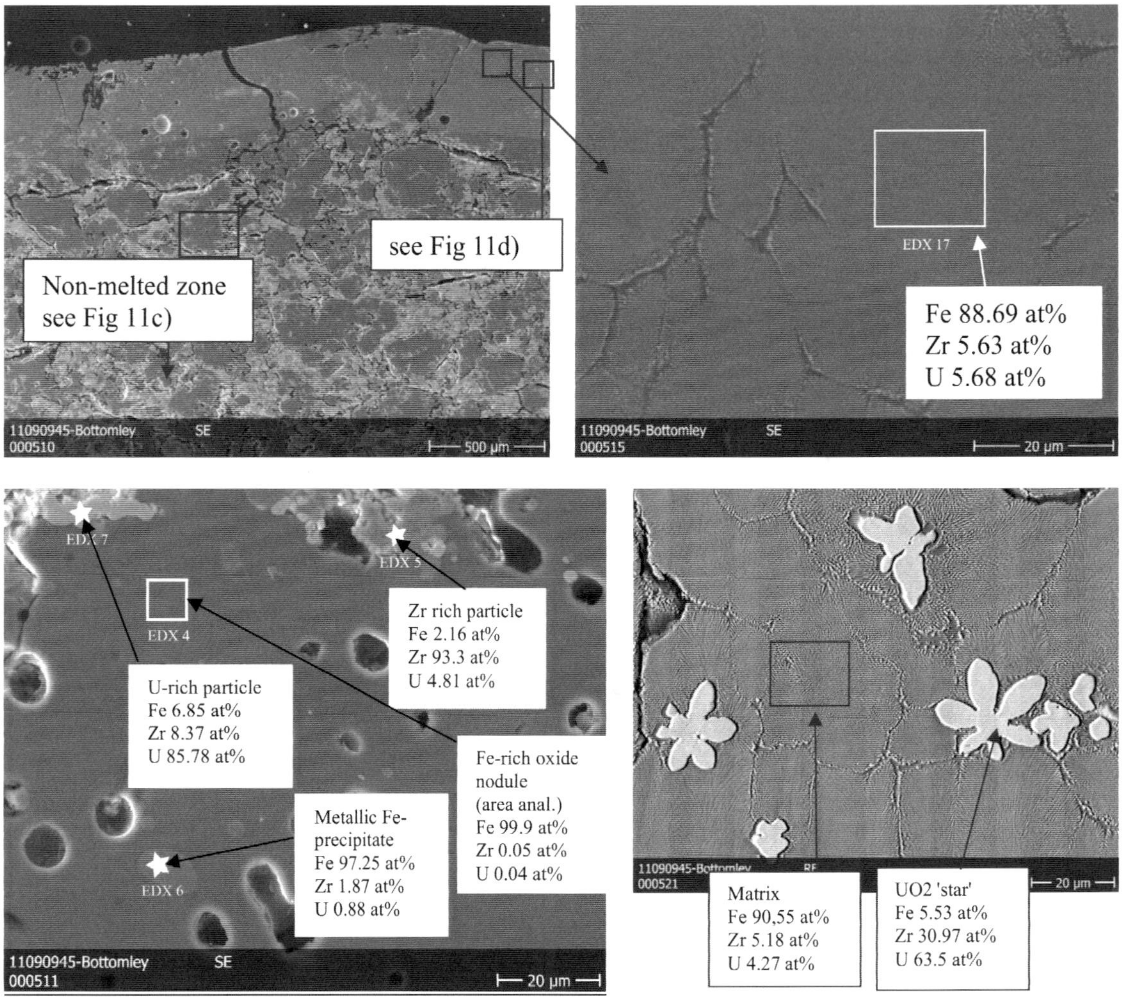

Figure 11. a) SEM micrograph of the 500 µm-deep molten zone with a uniform structure; b) closer view of the molten phase showing the very fine eutectic structure (Fe Kα-, Zr Lα- and U Mα-lines were used for the analysis); c) Higher magnification (700x mag.) of the unmelted zone showing the main components of the mixture; d) (700x mag., back-scatter mode) second area of the molten zone also with a very fine eutectic structure and stars of U-rich particles.

boundary (~10 µm). These are darker (i.e., less dense) Zr-rich phase. An area EDS analysis of this zone gave a U 90.3 at% and Zr 9.7 at% composition and indicated a Zr-richer composition than expected in the sintered pellets. Only metal elements were measured as the O-signal was difficult to measure reliably. Figures 13c and d show the melted zone at higher magnification; in a dense melt with uniform grains and secondary precipitates at the grain boundaries are seen. The precipitates are U-richer that formed on cooling and solidification.

Table 3 shows an average of 4 point analyses (see figure 13d) in the middle of the melt of 94.7 % U and 5.3 % Zr for the metal elements. This indicates that the overall composition of the melted zone is similar to the original composition.

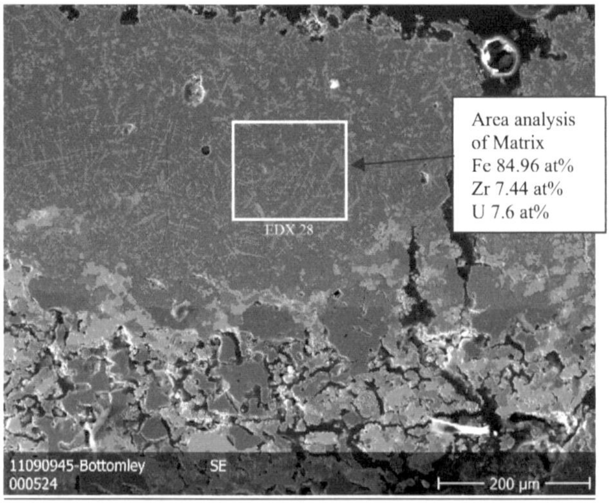

Figure 12. Low power SEM micrograph (mag. 10x) showing a melt with a coarser, dense precipitate in the melt.

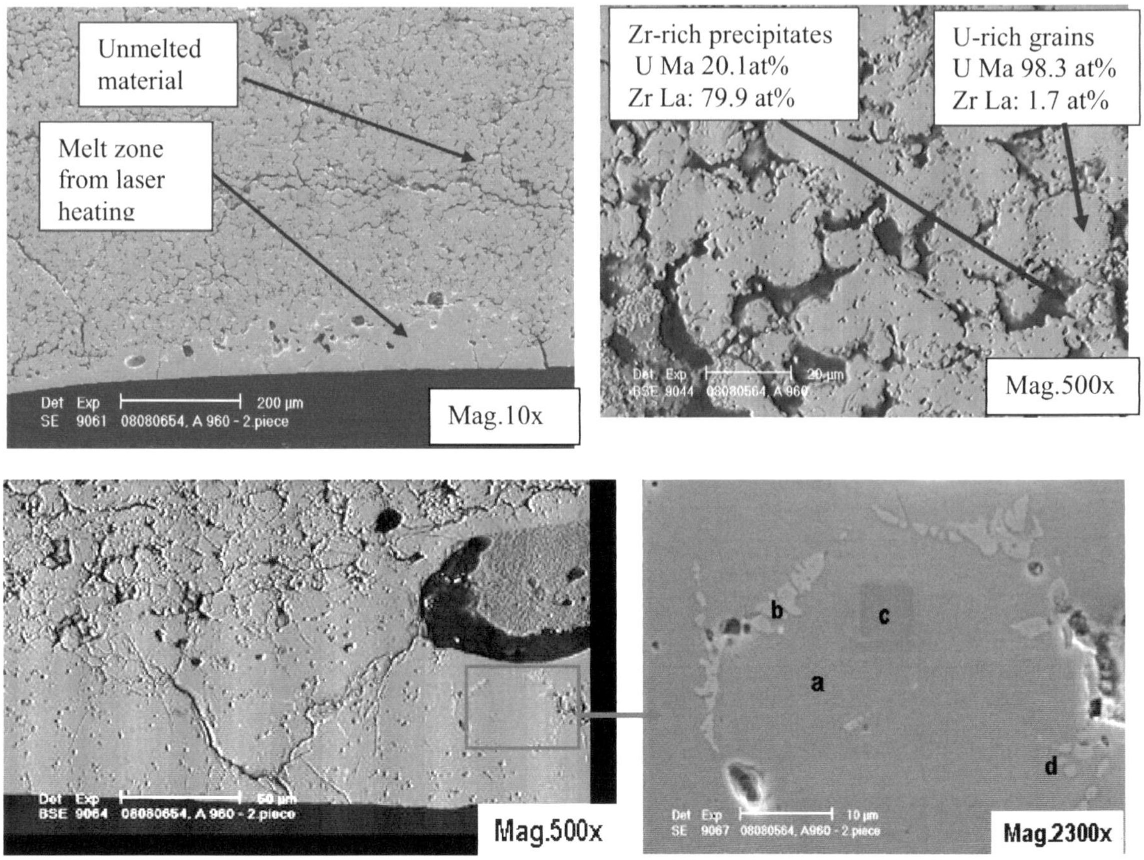

Figure 13. a) Low power SEM image of the UO_2-5Zr (in at%) after laser flash testing; the area of melting is approximately 1 mm wide and 100 µm deep. Below this, the material retains its original structure (see figure 10b); b) SEM backscattered image of non melted fuel of the UO_2-5Zr sample. The larger, lighter-coloured, UO_2-rich grains have small, darker Zr-rich precipitates; c) Backscattered image of cross-section of unaffected zone and melted zone (corium) of the UO_2-5Zr sample; d) High power micrograph of dense melted corium of the UO_2-5Zr sample.

Table 3. Point analyses of the molten phase of the UO_2-5Zr (in at%) sample cross-section after laser flash melting.

Peak	Point a	Point b	Point c	Point d	mean	Std. deviation.
Zr_L	7,7	3,2	4,8	5,6	**5,3**	**± 1,9%**
U_M	92,3	96,8	95,2	94,7	**94,7**	**± 1,9%**

Finally, a cross-sectional area with unmelted and melted zones were subjected to mapping of the U_M- and Zr_L-lines. This is given in figure 14 for the UO_2-5Zr (in at%) sample after laser flash pyrometry testing.

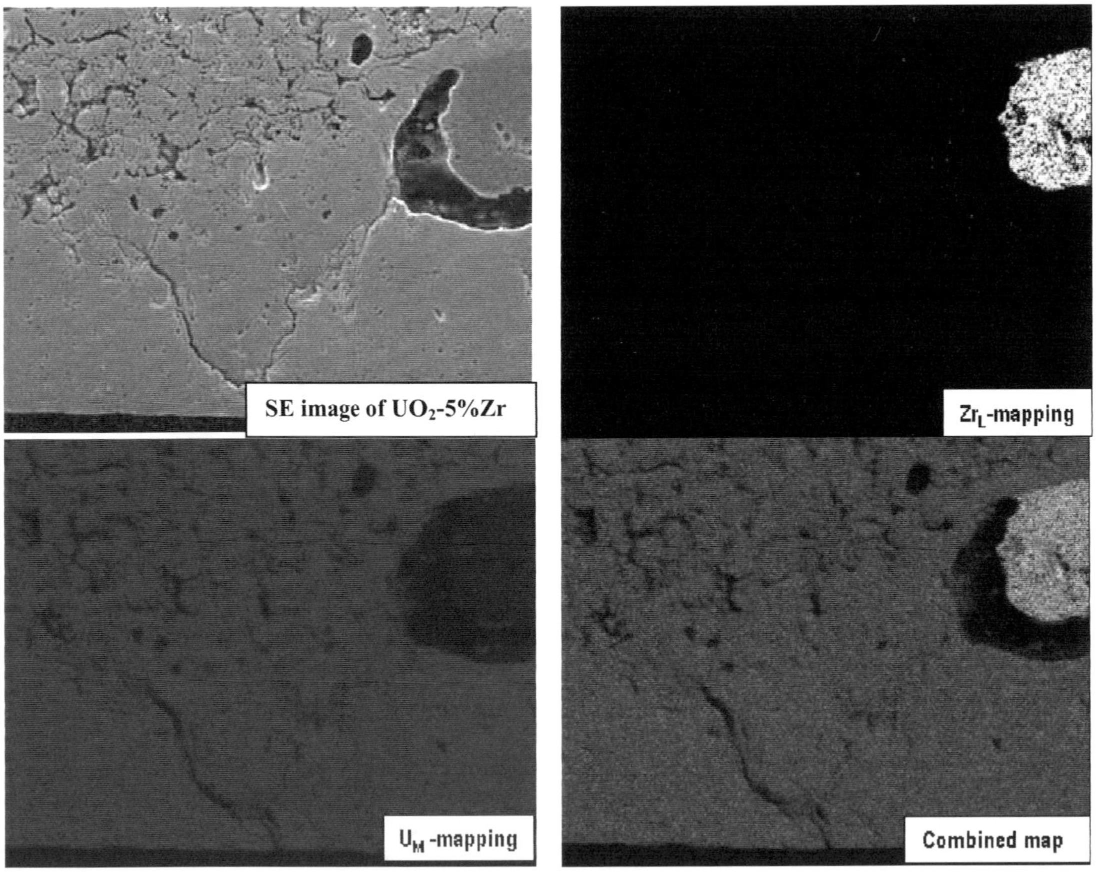

Figure 14. Secondary electron image of the UO_2-5Zr (in at%) sample in an area with unaffected and melted zones along with individual Zr_L and U_M peak mappings and a combined U and Zr mapping of the sample.

The secondary electron image shows clearly the more porous unaffected and the dense melted areas. The individual mappings show a similar level of signal for the U in both areas. The Zr is heterogeneously distributed in the unaffected area and a Zr-rich particle is seen in the edge of the molten zone where it has not completely melted.

In the fully melted area a uniform intensity of the each of the 2 signals is noted indicating a homogeneous distribution of the Zr and U in the corium and not greatly altered from that of the unaffected zone.

2.6. Melting point determinations

2.6.1. U-Zr-Fe-O system. In the first examinations, the $10UO_2$-$10ZrO_2$-80FeO (in mol%) sample appears to show a melt composition that is shifted towards FeO richer contents. However it clearly shows a wide melting range which it is noted that UO_2-rich and ZrO_2-rich phases are dispersed between the Fe-rich phases in a heterogeneous structure prior to melting. After laser flash melting the molten zone appears to be a 2-phase, FeO-rich and $(U,Zr)O_2$-rich, mixture with a fine eutectic structure. EDS analysis shows a slight shifting in the melt of the 10 % UO_2 - 10 % ZrO_2 - 80 % FeO sample towards FeO enrichment of the order of 3 - 5 mol%. The results also show a wide melting range as expected with oxides of differing refractory nature. The wide melting range of the FeO-rich mixtures indicates the possibility of the materials relocating for a much wider range of temperatures (and hence length of time) after initial melting, so that the damage in the event of a severe nuclear accident is consequently greater. However, the remaining samples need to be analyzed before the phase diagram can be confirmed.

2.6.2. U-Zr-O system. The melting point temperatures were assigned to liquidus temperatures as no other transformations were observed in the thermograms of the UO_2-5Zr and UO_2-10Zr (in at%) samples. They are displayed in green on the pseudo-binary UO_2-Zr diagram below in figure 15 [8]. This Kleykamp & Pejsa diagram is based on phase analyses measurements of samples annealed at lower temperatures. Considerable work has been carried out in this field [9-12], but the higher temperature boundaries remain uncertain. The points in red and blue represent earlier experimental campaigns carried out at ITU.

Overall there is a good agreement on the tendency of the liquidus line. The latest (green) laser flash values for low Zr values give the same values to those predicted from the tie lines drawn for this diagram (within experimental error), confirming earlier results indicating that the liquidus line would descend to meet a solidus temperature of approx. 2400 K at about UO_2-40Zr (in at%).

3. Example of a complex sample in a large scale test (SAROV Institute)

The SAROV Institute in Nizhny Novgorod, Russia has also performed large scale corium concrete interaction (MCCI) tests using 60 kg of material as part of an international project in which ITU also participated [13]. This test examined the effects of a molten corium of fused reactor materials on concrete of the basement of the nuclear reactor (the so-called ex-vessel severe accident scenario). This technique, which was developed by the SAROV Institute, used a FeO-Zr thermite mix in a concrete vessel. After ignition, the FeO-Zr powder generates enough heat to form a molten mix of ZrO and Fe from the self-oxidation of the starting materials. This melt then reacts with, and degrades, the concrete vessel. In this example ITU had a small sample of the material that was taken from the concrete vessel surface and analyzed this as a contribution to the project. In figure 16 below, the specimen in cross-section can be seen to have very mixed metallic and ceramic phases (figure 16a). SEM-EDS analyses both of the surface as well as the cross-section show that the metallic phases were principally Fe, while there was a great variety of ceramic phases. These ceramics were principally Fe oxide but also mixed oxides with concrete materials and with either Zr or Fe. In figure 16b a dark interfacial region of mixed Si, Ca and Fe oxides is noted between a metallic Fe nodule (upper half) and an outer pure FeO layer (lower half). The pure FeO layer also includes a region with of mixed Fe, Zr oxide with traces of Ca and Si from the concrete. A nodule of FeO is also seen within the metallic Fe nodule, showing that locally conditions may not have been equilibrium.

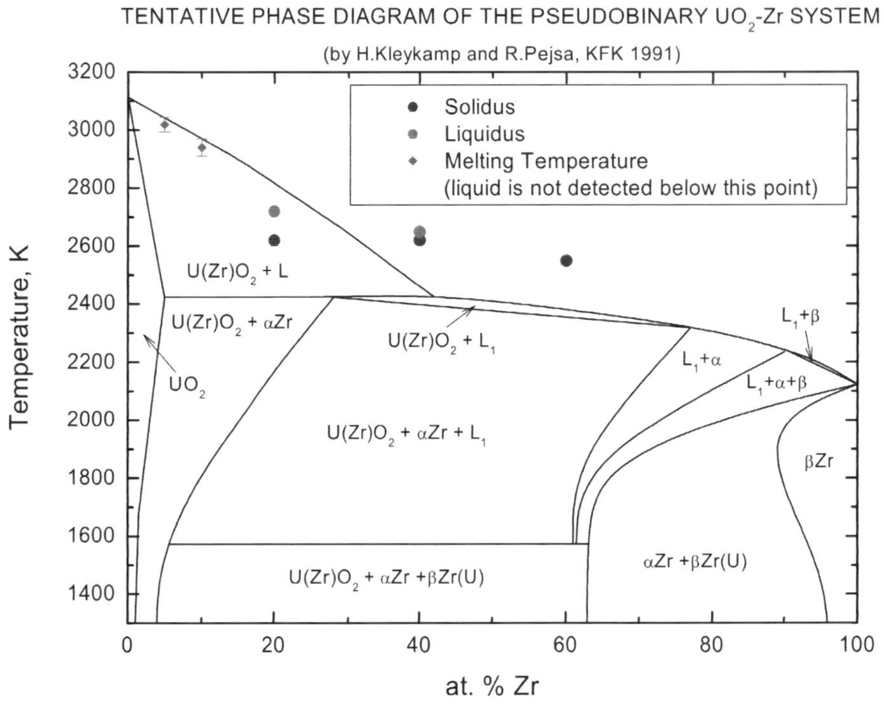

Figure 15. Pseudo-binary diagram of the UO_2-Zr system [8] showing the melting point determined for UO_2-5Zr and UO_2-10Zr (in at%) in this work.

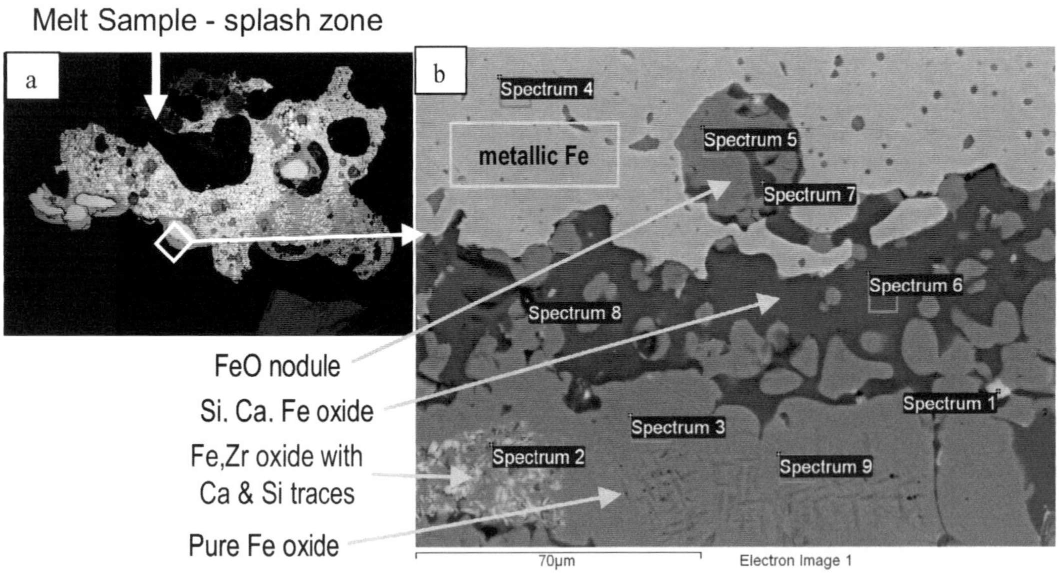

Figure 16. Sarov Institute large scale test for a molten core concrete interaction (MCCI) sample of FeO-Zr thermite ignited in concrete.

The overall analysis showed that the thermite reaction had generally gone to completion and that the reactions simulated the typical reactions with concrete expected under severe accident conditions. The test successfully eroded a uniform thickness of the concrete [13]. The test showed that this technique could be used to simulate the interaction between reactor material melt and the concrete materials of the building (ie molten corium concrete interactions or MCCI). It is hoped that this technique can be used to perform even larger scale test and provide important date for modelling of the reactor scale case.

4. Conclusions

1) Good quality data is needed on high temperature refractory oxide systems for validation of codes and improvement of the phase diagram in severe nuclear accident research.
2) A series of UO_2-ZrO_2-FeO ceramics have been fabricated and are undergoing characterisation. The objective is to understand fuel rod-reactor vessel interactions. FeO phase remains as larger agglomerates after sintering compared to UO_2 and ZrO_2. The presence of metallic iron precipitates in the sample (< 1 % by XRD) indicates that the system remains close to stoichiometric UO_2, FeO (and ZrO_2) with no higher oxides forming. This is important to know as 1) this represents the conditions expected (at least initially) in a severe accident, and 2) to model the correct system.
3) Laser flash pyrometry has been successfully used to measure high melting points for a sub-oxidized UO_2-Zr system (for 5 and 10 at% Zr) important for nuclear safety and has confirmed the melting range for these compositions and has helped to determine the interaction mechanisms. The compositions of the molten zones have been checked and certain shift in composition compared to the original pellet composition noted.
4) In addition to the smaller, systematic studies ITU has also examined specimens from large integral tests. Experience with analyses of simpler systems such as U-Zr-Fe-O is clearly an aid to evaluating the more complex systems encountered in practice.
5) The MCCI test carried out by the SAROV Institute shows the typical structures and mixed phases formed in a Fe-Zr-O melt after its interaction with concrete simulating the late stages of severe nuclear accident.

References

[1] Menna M, Bottomley D, Bouexière D, Eloirdi R, Gardeur S, Lajarge P, Murray-Farthing M and Pagliosa G 2011 *Fabrication and characterisation of UO_2-ZrO_2-FeO/Fe_2O_3 pellets for SARNET2.* (Karlsruhe: European commission, JRC, Institute for Tranuranium Elements) ITU Technical Report, JRC-ITU-TPW-2010/20

[2] Almjashev V I et al. 2009 Eutectic crystallization in the $FeO_{1.5}$–UO_{2+x}–ZrO_2 system. *J. Nucl. Mater.* **389** 52

[3] Almjashev V I, Barrachin M, Bechta S V, Bottomley D, Defoort F, Fischer M, Gusarov V V, Hellmann S, Khabensky V B, Krushinov E V, Lokuph D B, Mezentseva L P, Miassoedov A, Petrov Yu B and Vitol S A 2010 Phase equilibria in the FeO-UO_2-ZrO_2 system in the FeO-enriched domain. *J. Nucl. Mater.* **400** 119

[4] Boshoven C and Somers J 2006) *Fabrication of Zr/UO_2 pellets for the phase diagram Measurements.* (Karlsruhe: European commission, JRC, Institute for Tranuranium Elements) ITU Technical report JRC-ITU-TN-2006/40

[5] Manara D, Ronchi C and Sheindlin M 2002 *Int. J. Thermophys.* **23** 1147

[6] Manara D, Ronchi C and Sheindlin M 2002 *Proc. 14th European Conference on Thermophysics (London, 1-4 September 2002)*

[7] Darken L S and Gurry R W 1946 *J. Am. Chem. Soc.* **68** 798-816

[8] Kleykamp H and Pesja R 1992 *Chemical and X-ray diffraction analysis on selected samples from the TMI-2 reactor core, Appendix H.* in: OECD-NEA Report OECD/CSNI/R(91)9 Vols. 1 & 2 (April 92)

[9] Chevalier P Y et al. 1998 *J. Nucl. Mater.* **257** 213-55

[10] Skokan A 1984 *High temperature relations in the U-Zr-O system, 5th International Meeting on Thermal Nuclear Safety (Kalrsruhe, September 9-13, 1984)*
[11] Ackermann R J et al. 1978 *J. Am. Ceram. Soc.* **61** 275
[12] Hofmann P and Politis C 1979 *J. Nucl. Mater.* **87** 375
[13] ISTC Project #3831 Project Development Grant MCCI Intermediate size test. (Moscow: Sarov Institute). Final report, January 2010

IOP Conf. Series: Materials Science and Engineering **32** (2012) 012004 doi:10.1088/1757-899X/32/1/012004

Using a controlled pressure field emission gun scanning electron microscope to acquire EBSD data on non-conductive samples

F Brisset[1]
CNRS - Université Paris-Sud, ICMMO - Institut de Chimie Moléculaire et des Matériaux d'Orsay, 15, rue Georges Clémenceau, bât. 410, FR-91405 Orsay Cedex, France

E-mail: francois.brisset@u-psud.fr

Abstract. The present work is dealing with electron backscattered diffraction. It is well known that EBSD requires most of the time and up to now, both an important high voltage and a large beam current. However, in such conditions, acquiring EBSD maps on non-conductive samples require to either coat the sample with a conductive thin layer or to use a controlled pressure SEM and then introduce some gas in the sample chamber. The idea here is to present some data comparisons between both solutions. Patterns, orientation images and simulations were carried out to try to access meaningful quantitative results.

1. Introduction

Electron backscattered diffraction (EBSD) is used for many years now to access, mostly on metals but not only of course, top surface grain orientation or texture measurements by example; these materials being either deformed, re-crystallized, etc., or not. However, this technique can be used on any sample as soon as its top surface (first 10^{ths} of nanometres (figure 1) [1-2] has its atomic arrangements well enough crystallized in the diffracted electron zone. This includes of course metals but also semi-conductive or non-conductive samples like typically ceramics or glasses. Indeed, as seen in figure 1 simulations, BSE electrons are at 70° tilt coming from the top surface of samples, even at a high voltage (25 kV) and a quite low mean Z material (Al_2O_3 compound in figure 1).

Acquiring EBSD data require usually a high voltage (typically between 20 and 30 kV) and a large probe current (typically from 1 to 20 nA). High voltage is used to generate enough diffracted backscattered electrons, accordingly to the Bragg law, to be visualized by phosphorescent screens but also to help primary electrons to penetrate enough inside the sample "surface" and then get rid of the deformed part of the sample surface, when needed. High current allows to have more diffracted BSE as well but the probe should keep a size not too important as regard to the microstructure to be analyzed; it is one reason why Schottky FEG-SEMs are generally a very nice tool for EBSD.
These working conditions are however, incompatible, with the acquisition of EBSD maps on non-conductive or badly-conductive samples or eventually cross-sectioned conductive samples included in resins as charging effects are going to be observed, generating deformation of maps, etc.

[1] To whom any correspondence should be addressed.

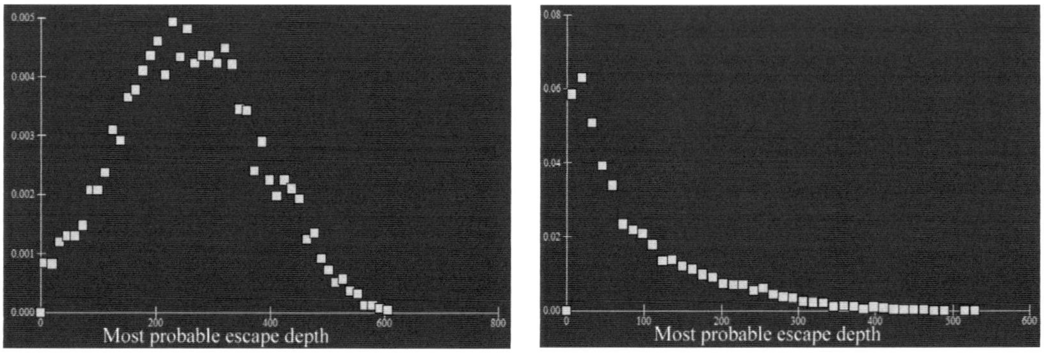

a) (b)

Figure 1. Monte-Carlo simulations showing the Most Probable Escape Depth (MPED) (X-axis in nm, Y-axis probability) of backscattered electrons. (a) The MPED is close to 300 nm for a 0° tilt, and (b) to 20 - 30 nm at 70 tilt, both for Al_20_3 at 25 kV. Simulations done using MC2 [1].

[3-4]. As an example, figure 2 shows a typical charging effect on a non-conductive sample at high voltage and current and at 70° tilt (SE image). These effects are going to become particularly important when acquisition of EBSD maps require a lot of electrons. This is the case when recording at very high speed (more than 300 or 500 patterns per second, by example) as it is now possible with the present generation cameras, but it is obviously the case, as well, at lower acquisition speed.

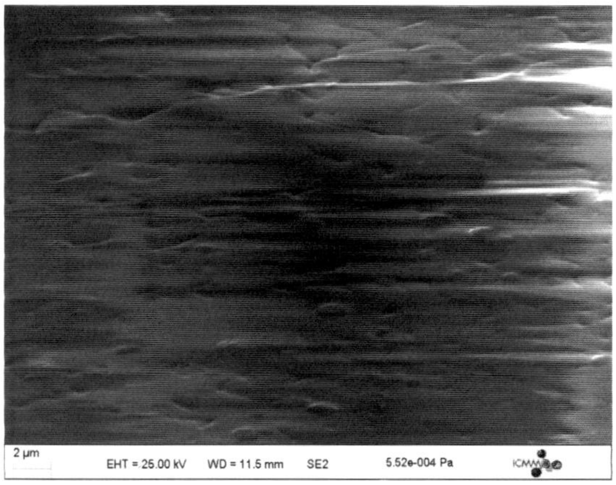

Figure 2. Charging effect on Al_20_3 sample observed in high vacuum (70° tilt, 25 kV and 4 nA).

2. Coating effect

Avoiding charging effects is possible and solutions are then either to deposit a conductive coating on the sample surface or to introduce some gas pressure in the SEM chamber. The first solution is generally used because it does not require a special SEM and because most of the laboratories have a carbon or metal coating equipment. It is either done by evaporation or sputtering in the case of C or Au/Pd. However, in EBSD, it is recommended to use carbon as this layer will less degrade the quality of the Kikuchi patterns or index quality factor IQ (figure 3), but it does as shown by Nowell (figure 4) on a re-crystallized Ni acting like a control sample [5]. Figure 3 points out the fact that using a low-Z coating degrades less the Kikuchi patterns even if, as seen here, the carbon coating (figure 3c) is thicker than the Au/Pd deposit (figure 3a-b). Figure 4, however shows that the coating effect is not without effect on the quality index as this one decrease significantly as the thickness of the coating increase. As a lower quality index could generate some difficulties for the indexation of the patterns, users have then to be careful and coat with the less possible thickness that avoids the charging effect and this will become essential when sample crystallization is not perfect. The later solution requires a controlled pressure SEM and could have the advantage of a better control of the process.

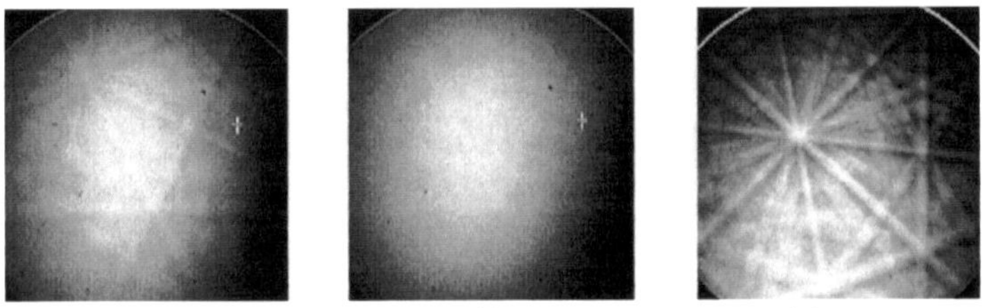

(a) (b) (c)

Figure 3. EBSD patterns (without background correction) with different coatings thicknesses and materials on Zr O_2, (a) 1 Å of Au/Pd, (b) 7 Å of Au/Pd, (c) 20 Å of C.

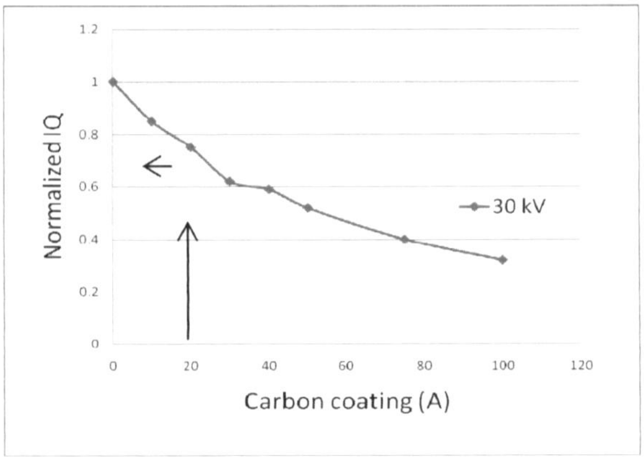

Figure 4. Decrease of index quality factor as regard to the carbon coating thickness at 30 kV on a Ni re-crystallized sample [5]. IQ is normalized at 1 for no carbon coating. At 20 Å of C, IQ = 75 %.

3. Gas effect

As for the C deposit, introducing gas in the chamber will discharge excess of electrons coming from the probe. Indeed, positive ions generated by the beam interaction with the gas will allow the acquisition of correct electron and EBSD maps without damageable effects. As a matter of fact, some high energy electrons will react with the gas to create positive ions which will help compensating on the sample surface the negative accumulation of charge (figure 5).

However, gas decreases as well the quality of patterns with the increase of its pressure even if the decrease of the pattern quality is not very important (figure 6). To corroborate this, one can check figure 7 that shows at 30 kV the EBSD maps which can be acquired on a re-crystallized Ni sample (similar as the one shown on the figure 4). It points out that just a few patterns are badly indexed and there is no significant difference between high vacuum and 60 Pa of N gas in the chamber.

This is confirmed by simulations of electrons scattered using electron flight simulator (EFS) [6]. They show (figure 8a) that only a low percentage of electrons are scattered, especially at these high voltages. Figure 8b reproduces the normalized IQ for the EBSD patterns as regard to the gas pressure obtained in the same condition as for the figure 4. It points out the decrease of the IQ. However, to

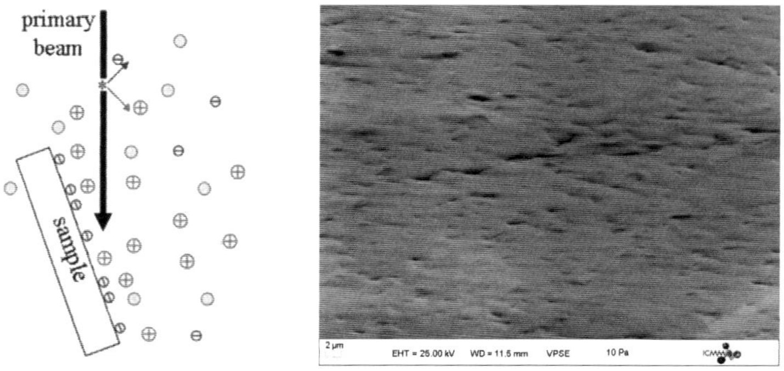

(a) (b)

Figure 5. a) Electron interactions with gas which will avoid charging effect; b) no charging effect observed on Al_2O_3 with a low gas pressure (10 Pa, 70° tilt, 25 kV and 4 nA).

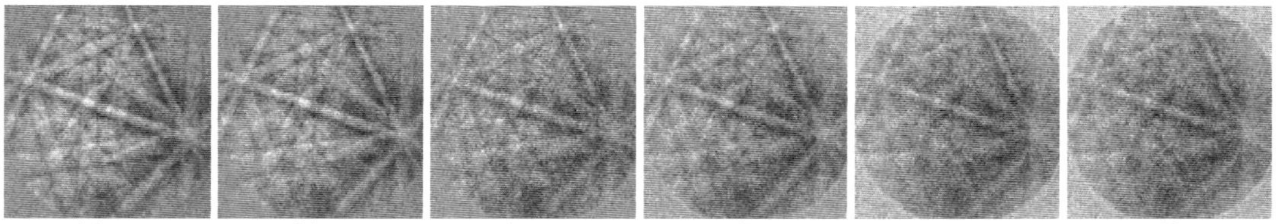

(a) (b) (c) (d) (e) (f)

Figure 6. EBSD patterns versus the gas pressure, from left to right (a-f) 10^{-4}, 10, 20, 30, 40, 60 Pa.

(a) (b) (c) (d) (e)

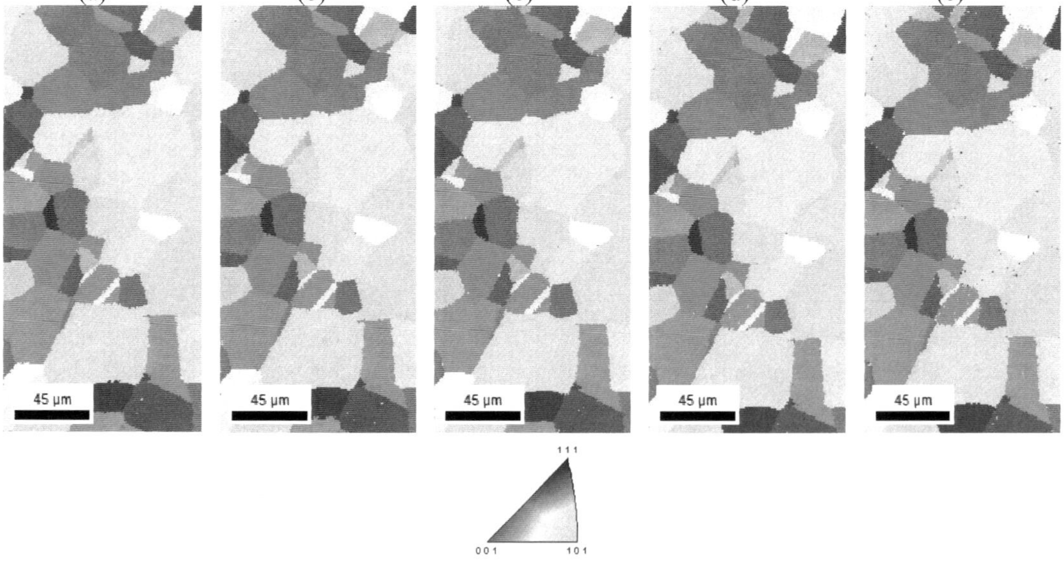

(f)

Figure 7. EBSD maps as regard to the gas pressure, from left to right (a-e) 10^{-4}, 10, 30, 40, 60 Pa and respectively 99.95, 99.92, 99.85, 99.80 and 99.09 % of good indexed patterns, and (f) orientation map colour code.

get a 25 % decrease of the IQ (like with 20 Å of carbon coating - figure 4) one needs to use a nitrogen gas pressure of 35 Pa. This means that if EBSD maps on non-conductive samples are acquired at less than 35 Pa, it should be possible to observe an IQ reduction less than 25 % and then possibly get better results than using a 20 Å carbon conductive layer. Note here that 20 Å of C is a usual thickness of Carbon deposition to avoid charging effect on badly or non-conductive samples for EBSD.

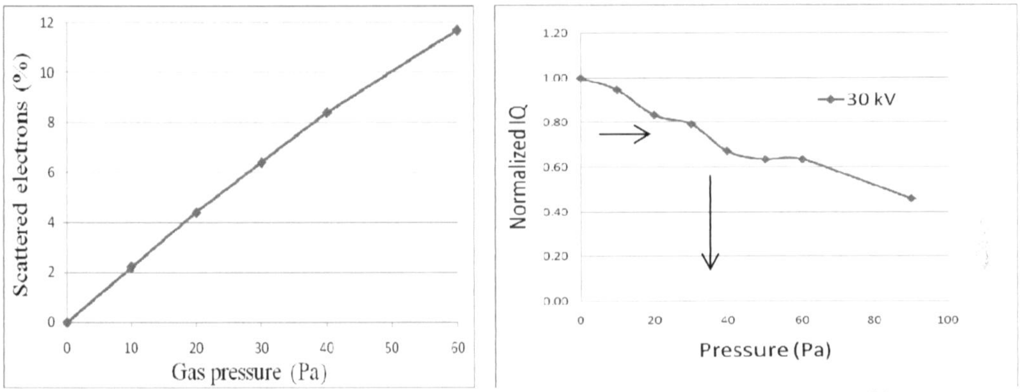

(a) (b)

Figure 8. a) Simulated electrons scattered versus gas pressure obtained with EFS [6] at WD = 25 mm, that WD (working distance) correspond at the all over distance for the high energy electrons (distance from the polar piece to the sample then to the EBSD camera screen); and b) normalized IQ as a function of the gas pressure. Arrows show that at IQ = 75 % nitrogen pressure is equal at 35 Pa.

4. EBSD maps under nitrogen gas

In order to verify if it was possible to acquire EBSD maps at these low N gas pressure, various materials have been tested including ceramics and re-crystallized glass. These examples are presented on figures 9 to 10 and data are acquired at either 15 or 20 Pa of nitrogen gas. Figure 9a is an example of an orientation map for a partly re-crystallized glass whereas figure 9b is a zircon sample; both have been acquired under a nitrogen gas pressure of 20 Pa, at 25 kV high voltage and 4 nA of beam current.

At last, figure 10, an alumina sample has been tested at 15 Pa, it is clear that it works as nicely as with the two previous samples. None of these images present artefacts which could be created by charging effects and they seem to prove that it is a very interesting choice to use gas when a controlled pressure SEM is available instead of a carbon conductive coating on non-conductive samples. On these samples some studies are done as regard to the grain size and the grain size evolution after heat treatments. On the last one some texture components were searched out as well.

5. Conclusion

It is concluded that at a pressure of 15/20 Pa of nitrogen gas charging effects do not appear and EBSD acquisitions are possible without reducing significantly maps quality as a very low percentage of electrons are scattered. This allows avoiding uncertain conductive coating deposition as the pressure value is easily controlled on the SEM sample chamber.

Then to acquire EBSD maps, a low gas pressure seems to be a very good alternative for no or badly-conductive samples.

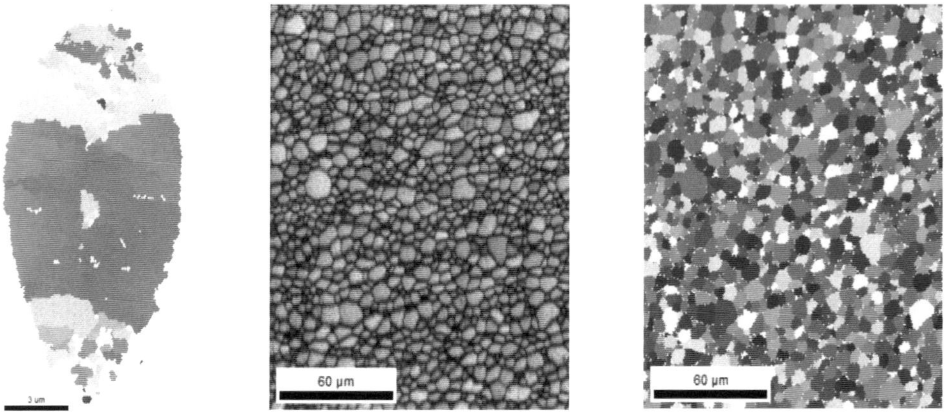

(a) (b1) (b2)

Figures 9. Examples of EBSD maps on non-conductive samples with a low nitrogen gas pressure inside the SEM chamber (25 kV, and 4 nA). a) Re-crystallized glass (SiO_2) at 20 Pa; b1-2) ZrO_2 image quality and grain image maps at 20 Pa.

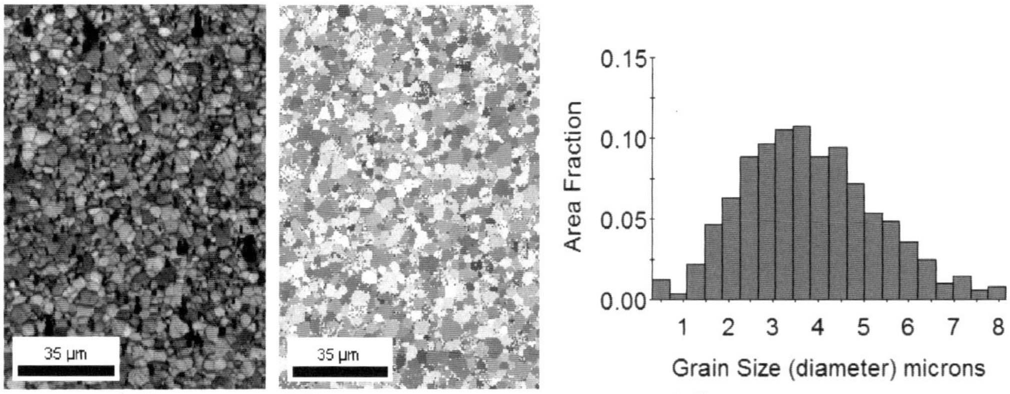

(c1) (c2) (c3)

Figure 10. Examples of EBSD maps on non-conductive samples with a low nitrogen gas pressure inside the SEM chamber (25 kV, and 4 nA), (c1-3) Al_2O_3 respectively image quality map, orientation map and grain size distribution at 15 Pa.

References

[1] Pouchou J-L, JLP-soft, MC2 software, Monte-Carlo simulation - Statistical scattering of electron trajectories, jlpouchou@yahoo.fr
[2] Brisset F and Nowell M 2010 Sample preparation techniques for EBSD. in: 7th Workshop SEMAT/UM - Caracterização Avançada de Materiais - EBSD (University of Minho, Guimaraes, Portugal, 2-5 March)
[3] Kunze K, Buzzy St, Löffler J and Burg J P 2008 *EMC 2008* **1** 575-576, doi: 10.1007/978-3-450-85156-1_288
[4] El-Dasher B S and Torres S G 2009 *Electron backscattered diffraction in materials science.* doi: 10.1007.978-0-387-88136-2-25
[5] Nowell M *Sample preparation.* OIM Academy, (EDAX-TSL, Utah)
[6] Electron Flight Simulator software, V 3.1-E, http://www.small-world.net/efs.htm

IOP Conf. Series: Materials Science and Engineering **32** (2012) 012005 doi:10.1088/1757-899X/32/1/012005

Prototypic corium analysis: a round robin for SEM and EDS characterisation

L Brissonneau[1,2], C Journeau[2], P Piluso[2], M Kiselova[3], S Bakardjieva[4], T Wiss[5], P W D Bottomley[5] and H Thiele[5]

[1] C.E.A., DEN, STPA, LIPC, Cadarache, FR-13108 St. Paul lez Durance, France
[2] C.E.A., DEN, STRI, LMA, Cadarache, FR-13108 St. Paul lez Durance, France
[3] UJV, Integrity and Technical Engineering Division, Husinec-Řež 130, CZ-25068 Řež, Czech Republic
[4] Czech Academy of Sciences, UACh, Husinec-Řež 1001, CZ-25068 Řež, Czech Republic
[5] European Commission, JRC, Institute for Transuranium Elements (ITU), Hermann-von-Helmholtz Platz 1, P.O. Box 2340, DE-76125 Karlsruhe, Germany

E-mail : laurent.brissonneau@cea.fr

Abstract: In case of a nuclear reactor severe accident, the core could melt, forming a high temperature mixture called corium. A uranium oxide-containing sample from a corium-concrete interaction test has been analyzed by scanning electron microscopy (SEM) and energy-dispersive X-ray spectrometry (EDS) at three different laboratories. In all the measured windows, the composition lay on or close to the line connecting the initial corium melt composition to the concrete average composition. The round robin exercise confirmed that SEM and EDS analysis can be used confidently to perform fast and good quality local corium analyses and provide local quantitative compositions in the metallic elements of corium with an uncertainty of 10 % of the measured value. Differences up to 25 % of the measured value were found for the oxygen content, although two of the laboratories provided very close results.

1. Introduction

In case of a nuclear reactor severe accident, the core could melt, forming a high temperature mixture of fuel (UO_2), partially or totally oxidized cladding (Zr, O) and structural materials (mainly steel) that could melt through the steel pressure vessel and eventually react with the concrete of the basement. The study of this mixture, called corium, is required to improve our knowledge of severe accident phenomena. Therefore, experiments using prototypic corium (mixtures having the same chemical composition than those expected in severe accident conditions, but a different isotopic composition, using e.g., natural or depleted uranium) have been conducted in several research laboratories [1-5].

Materials analysis can be of a great help as it can provide information on the different reactions that occur during the accidental sequences [4, 6-10] and it can help the modelling of the interaction between corium and concrete, by improving the existing thermodynamic databases [11-16] of the major species interacting in corium (U, Zr, Fe, Ca, Si, O…).

[2] To whom any correspondence should be addressed.

In current pressurized water reactors, the concrete constitutes the ultimate barrier and it is thus of great interest to characterize the molten core concrete interaction (MCCI). Several MCCI tests were performed in the PLINIUS experimental platform at CEA Cadarache in order to assess the rate of penetration of corium in concretes of different compositions [5]. In order to understand the differences in the observed ablation profiles, it is important to take into account the complex phenomena that occur when a high temperature multi-element melt meets and reacts with a complex composite material such as concrete.

Scanning electron microscopy (SEM) equipped with energy-dispersive X-ray spectrometry (EDS) analysis of polished metallographic samples of the corium/concrete "pool" is the preferred method, as it provides information, quite rapidly and easily, on the composition, phases and microstructure of the mixture.

But corium is a complex material containing heavy (U) and light elements (Ca, Si as well as O), so that the chemical analysis is complex. As few laboratories are able to work on prototypic corium and reliable results are needed to validate the interaction models and thermodynamic databases, it was decided to perform a round robin between three different European laboratories on the SEM-EDS analysis of a VULCANO MCCI [5] test sample. Thus the purpose of the test is not to compare intrinsic performances of the different apparatuses but rather to assess the reliability of the experimental chemical compositions and to give confidence to the users of the data by quantifying their typical scattering.

In a first section, we will shortly present the corium-concrete interaction experiment, the corium sampling procedure and the SEM/EDS apparatuses. Then the analysis results will be presented, followed by a comparison on the relative element analysis and then a short discussion on the corium concrete mixing deduced from the results in a final section.

2. Experimental

The round robin was performed on a sample of the VB-U6 Vulcano test [5]. In this test, a mixture of 62 % UO_2 - 34 % ZrO_2 - 2 % Fe_2O_3 - 2 % CaO was melted in an arc furnace and poured into a limestone-rich concrete crucible with the following average concrete composition (wt%): 42 % CaO, 25 % SiO_2, 25 % CO_2, 1 % Al_2O_3, 1 % Fe_2O_3, 4 % H_2O, 3% others, and held at 2000 °C with induction heating during several hours.

Different samples were extracted, at the interface between the corium and the concrete at the bottom (axial interface) and at the side (radial interface) of the pool, as well as in the core of the corium melt (see figure 1). The selected sample for the round robin VBU6_8 was extracted at the axial interface.

2.1. Sample preparation

The original sample (figure 1) was cut in two parts, cold mounted and polished down to 1 μm polish in a glove box. The sample was coated with a slight carbon layer for the first SEM examination, in Cadarache. For practical reasons (SEM specimen holder size limits), the mounted sample was then cut into four parts at ITU, Karlsruhe and a selected area examined before being sent for further examination at UJV, Řež.

2.2. SEM/EDS

The three different EDS and SEM system used for the round robin test in the CEA Cadarache, the ITU Karlsruhe and the UACH Řež are detailed in table 1. The acceleration voltage varies from 20 kV to 30 kV, working distances from 10 to 25 mm, operating pressures were low or average (in UACH). Due to smaller working distance (15 mm) analysis areas of ITU are smaller 0.35 mm^2 compared to analysis in Cadarache (1 to 4 mm^2) and in UACH (0.35 to 2 mm). In CEA analyses, probe current could vary from 3 to 5 nA from one analysis to the other as a function of the sample electrical conductivity.

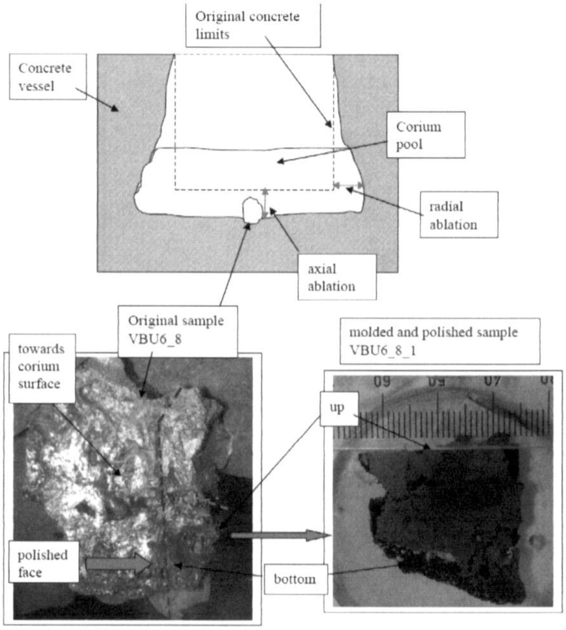

Figure 1. Scheme of sample extraction from corium concrete interaction experiment, cut and molding of the analyzed sample.

Table 1. Characteristics of the three apparatus participating in the round-robin.

	CEA Cadarache	ITU Karlsruhe	UACH Rez
SEM	Cambridge S360	Vega Tescan 5130 LS	Philips XL CP30
Filament	W	W	W
Operating voltage	20 kV	30 kV	30 kV
Working distance	25 mm	15 mm	25 mm
Probe current	2-3 nA	0.5-2 nA	5 nA
Operating pressure	$6\text{-}8.10^{-4}$ Pa	$< 9.10^{-3}$ Pa	3.10^{-4} Pa
Detector	SE, BSD	SE, BSD	Robinson
EDS	Si-Li , Oxford EDS 7060	Si-Li, Oxford EDS-7830	Si-Li, EDAX
Detector area	10 mm^2	10 mm^2	10 mm^2
Software	ISIS 300	INCA energy 250	Genesis EDAX v3.6
Correction method	ZAF	ZAF	ZAF
calibration	Co standard	Cu-Al standard	semi-quantitative
Acquisition time	≈ 200 s	≈ 200 s	200-250 s

All operators used their software internal calibration (using only Co or Cu-Al standards for optimisation) and did not try to optimize the analytical conditions, as the aim of the test was to assess if fast and (how?) rather reliable analysis of corium samples could be performed using EDS, whatever could be the analyst (or the lab).

EDS analyses were performed at different locations in the all zones by all laboratories. For the analysis performed in UACH, first analyses were performed at different locations than those performed in CEA; in a second stage, the location of the CEA-analyzed zones were indicated to UACH (without the results) and these analyses were repeated. The location of the analyses performed in ITU was free with respect to those performed in CEA, nevertheless they could be compared with analyses from comparable, adjacent locations. It also provided a check that the results do not depend on the exact analysis point. The procedure for the three analyses can thus be considered as a blind test. Strictly speaking in a round robin test, exactly the same zones should be analysed by all the participants. Because of the difference in the configurations, it was not always possible (lower areas in ITU). But in a sense, not only the apparatus but also the operators were tested in this round robin. Particular choices of an operator to analyse or not a zone can potentially affect the result, if slight heterogeneity exists. Even if the question of heterogeneity was questioned in CEA work by much more analyses done than by other partners, it was of interest to test "new eye" operators.

Oxygen was not deduced from stoichiometry but measured directly. For CEA analyses, the sums of element weight compositions vary from 95 % to 105 %. It was considered as rather good quality results, considering that even when analyses were performed with less stable filaments, leading to variations of the sums between 85 % and 115 %, the final normalized results were very close to results with more stable filaments. The presented results are normalized to 100 % for sake of clarity.

3. Results

The corium sample VBU6_8 can be divided in three different kinds of zones: the corium-rich zone; the corium-poor zone and the "concrete" zone (see figure 2).

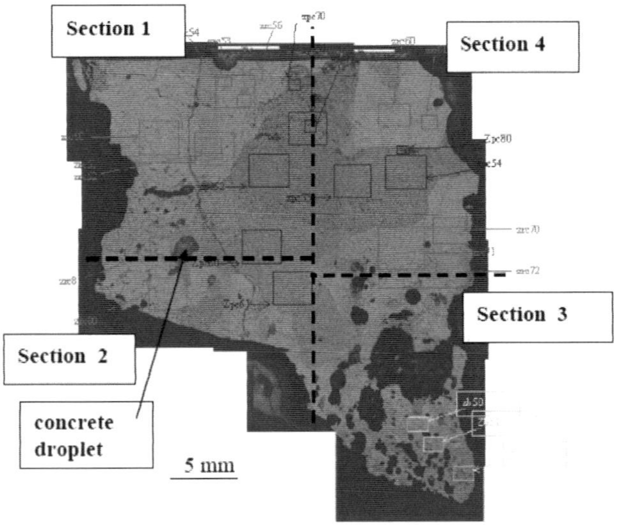

Figure 2. Composite macrograph of the sample VBU6_8 from the bottom of the crucible (axial interface); the section was cut into 4 sections. Section 1 was selected for the round robin. Different locations of analysis are noted in red for corium rich zones, blue for corium poor zones and green for near concrete zones.

The corium-rich zones appear to be brighter on the back-scattered electron (BSE) micrographs as they are mainly composed of heavy elements such as U and Zr (more than 65 %). On the contrary, the corium-poor, concrete-rich zones (between 45 % and 65 % UO_2-ZrO_2) are darker, as they are composed of lighter elements as Fe, Si and Ca. On this sample, the corium-rich zones are 5 to 10 mm large and surrounded a corium-poor zone. Bubbles or light element-rich droplets (between 10 % to 40 % UO_2-ZrO_2) were found in some places in the corium-rich zone. At CEA Cadarache all the plotted windows seen on figure 2 were analyzed. The sample was then cut into 4 sections which are marked in figure 2. The 3-way comparison (CEA-UACh-ITU) was carried out in the corium-rich and corium-poor zones in section 1 in the top left-hand corner of the sample. At the right bottom of the sample is a zone which delimits the corium and the concrete. Bubbles and light element rich droplets can be found in some place in the corium poor and rich zone. Bubbles are very numerous in the concrete rich zone.

The white dendrites or nodules in BSE micrographs (figure 3) are uranium-zirconium-calcium oxide in a "concrete-rich" matrix. In the corium-rich zones, the density of uranium-rich nodules is higher than in the corium-poor zones. Small secondary precipitates of the corium-rich phases are seen in the concrete-rich phase matrix.

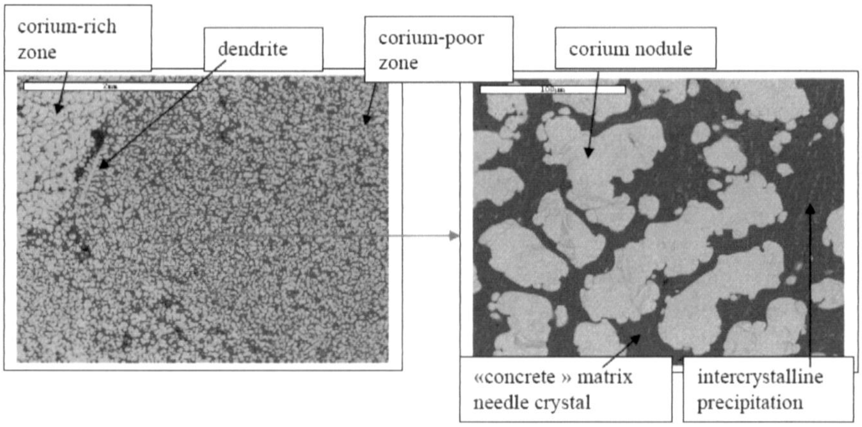

Figure 3. (left) Micrograph showing a white corium-rich zone at the upper left-hand side and a darker concrete-rich zone elsewhere. (right): High magnification micrograph of the concrete-rich zone showing the dense corium nodules in the low density concrete-rich matrix.

3.1. EDS results according to different laboratories

3.1.1. Zone analysis. In the jointly-analysed section, corium-rich and corium-poor zones were present.

In Cadarache, the dimensions of the analyzed zones were between 1 and 4 mm², while they were 0.35 mm² in ITU analysis, and between 0.35 and 2 mm² in UACH analysis.

The different analyses of bright, corium-rich zones are reported in table 2 and for the dark, corium-poor zones in table 3. These analyses have been repeated for at least four areas. Trace elements as Na and K, mainly present in the matrix, have been detected and reported by CEA and UACH, but only sporadically by ITU. ITU had decided to exclude these elements from the results but they always amounted for less than 1 % and will have little effect on the overall results. Traces of Hf have also been found in ITU analysis, which is not reported in table 2. ITU decided to include oxygen in the results of only few analyses; they are not reported in table 2 or in table 3 but are discussed later.

The results are found to be in good agreement between the different labs. The variations between analyses are much greater for Ca, Si and Fe than for U or Zr. But considering two corium-rich zone areas, it can be seen that there is more scatter between areas (Zrc52 and Zrc55 for example) than between laboratories (i.e., between two laboratory analyses of the same areas ; for that reason, two different areas in the analyses have been included as being relevant to the interpretation of the comparison). This pattern was observed with all the other results.

In Cadarache, exactly the same zones or phases were also analysed at different times (15 days) in order to evaluate the uncertainty on one apparatus in these conditions. Very good reproducibility of the results was obtained, less than 0.1 wt% point, except for oxygen (about 1 wt% point).

Table 2. Analyses of a corium-rich zone (all the results are normalized to 100 %, both including and excluding O). The CEA and UACH zones have the same location while ITU's location is adjacent to the others. The Zrc55 zone is bigger than the Zrc52 one.

	Zrc 52 CEA		ITU	Zrc 52 UACH		Zrc 55 CEA		ITU	Zrc 55 UACH	
Element	wt%	wt% / O	wt% / O	wt%	wt% / O	wt%	wt% / O	wt% / O	wt%	wt% / O
O	23.6			17.2		24.2			18.4	
Mg	0.1	0.1	0.1			0.1	0.1	0.2	0.2	0.2
Al	0.2	0.2	0.2	0.5	0.5	0.2	0.3	0.4	0.7	0.9
Si	2.2	2.9	2.8	1.7	2.0	3.0	3.9	5.7	3.9	4.8
K				0.8	0.9				0.6	0.8
Ca	4.5	5.8	5.7	3.2	3.9	5.6	7.3	9.3	6.0	7.3
Fe	2.2	2.9	3.1	2.0	2.5	2.5	3.2	5.0	3.2	3.9
Zr	21.6	28.3	29.1	23.7	28.6	21.3	28.1	26.3	22.8	27.9
U	45.7	59.8	58.7	51.0	61.6	43.3	57.1	53.2	44.1	54.1

Table 3. Analyses of a corium-poor zone (all the results are normalized to 100 %, both including and excluding O). The CEA and UACH zones have the same location while ITU location is adjacent.

	Zpc 52 CEA		ITU	Zpc 52 UACH		Zpc 52 CEA		ITU	Zpc 52 UACH	
Element	wt%	wt% / O	wt% / O	wt%	wt% / O	at%	at% / O	at% / O	at%	at% / O
O	30.16			21.7		65.76			56.6	
Na	0.07	0.1	0.2			0.1	0.3	0.6		
Mg	0.2	0.3	0.3			0.28	0.8	0.8		
Al	0.69	1.0	1.0	1.1	1.4	0.89	2.6	2.5	1.7	3.8
Si	8.13	11.6	11.9	8.2	10.4	10.1	29.5	29.3	12.2	28.1
K			0.1	0.5	0.7			0.2	0.6	1.3
Ca	10.5	15.0	16.2	9.9	12.6	9.14	26.7	28.0	10.3	23.8
Fe	6.31	9.0	8.4	6.2	7.9	3.94	11.5	10.4	4.6	10.7
Zr	14.16	20.3	21.5	16.9	21.6	5.42	15.8	16.3	7.8	17.9
U	29.79	42.6	39.6	35.6	45.4	4.37	12.8	11.5	6.3	14.4

Firstly, it can be concluded that corium samples are so heterogeneous that the quality of the analysis depends less on the system used and more on having sufficient analyses on large enough areas to obtain accurate average compositions. For the concrete zone of section 3 (see figure 2), the spatial heterogeneities (due to the heterogeneous nature of concrete) were too high to allow a satisfactory direct comparison between laboratories results.

3.1.2. Phase analyses. Two main phases are present in this sample, nodules of oxides rich in U and Zr, and a matrix, an oxide rich in concrete elements, Fe, Si and Ca. The compositions of the phases may vary from one point to another and do not appear to be fixed, especially between corium-rich zones and corium-poor zones. The results are summarized in Table 4. Other phases can be found in the concrete matrix as $(Fe_{0.90}Si_{0.05}Al_{0.05})O_{1.5}$ or $(Si_{0.45}Ca_{0.15}Al_{0.30}Fe_{0.05})O_{1.8}$, in the form of faceted nodules. Good agreement is found between the three laboratories, except for the oxygen content which is systematically lower in the UACH analyses than in CEA and ITU analyses. It will be discussed in section 4.1.2, by comparing the relative amounts of the elements that the agreement between these three analyses is sufficient enough for experiment interpretation (as long as accurate oxygen content is not required).

Table 4. Typical phase analyses in corium-rich zones and corium-poor zones.

Zone	Institute	Nodules	Matrix
Corium rich	CEA	$(U_{0,4}Zr_{0,5}Ca_{0,1})O_{2,2}$	$(Ca_{0,4}Fe_{0,1}Si_{0,5})O_{1,7}$
	ITU	$(U_{0.4}Zr_{0.5}Ca_{0.1})O_{2.6}$ $(U_{0.35}Zr_{0.45}Ca_{0.15}Fe_{0.05})O_x$	$(Ca_{0.3}Fe_{0.2}Si_{0.4})O_x$
	UACH	$(U_{0.4}Zr_{0.5}Ca_{0.1})O_{1.7}$	$(Ca_{0.3}Fe_{0.15}Si_{0.5}Zr_{0.05})O_{1.2}$
Corium poor	CEA	$(U_{0.40}Zr_{0.45}Ca_{0.15})\ O_{2+x}$	$(Ca_{0,4}Fe_{0,1}Si_{0,5})O_{1,7}$
	ITU	$(U_{0.48}Zr_{0.36}Ca_{0.15})O_{2.5}$	$(Ca_{0.35}Fe_{0.1}Si_{0.5})O_{1.8}$
	UACH	$(U_{0.4}Zr_{0.5}Ca_{0.1})O_{1.7}$	$(Ca_{0.3}Fe_{0.15}Si_{0.5}Zr_{0.05})O_{1.2}$

4. Discussion

4.1. Comparison of analyses

4.1.1. General view of the window compositions. The results of more than 50 window-analyses performed at CEA on three different samples have been plotted on a ternary composition diagram (figure 4), with $(U,Zr)O_2$, SiO_2 and CaO compositions as the three apices (it should be noted that such an analysis relies only on the ratios of the metallic elements and hence is not dependent on the uncertainty of the oxygen content). It is found that all the analyses can be plotted not far from a (full red) line passing through the original compositions of the melted corium and of the concrete.

It could be concluded that there is no significant segregation of Si and a slight Ca segregation from the mortar phases to the corium phases. The main difference in the compositions in the corium-rich and corium-poor zones would result from different dissolution or proportions of the mortar phases mixed with the corium phases (or vice versa).

However, it can be noted that the CaO/SiO_2 ratio is different from other analyses in most of the analyses concerning zones with high concrete fraction (UO_2-ZrO_2 content lower than 20 %) that correspond to the darker 'droplets of concrete' as shown in figure 2. These droplets have been found to be depleted in calcium compared to the average concrete composition: $CaO/SiO_2 \approx 0.55$, instead of 0.63 as indicated by the solid line in figure 4 for corium rich and corium poor zones. Now, by passing a line from corium rich zone composition to "concrete drop" composition (dash line in figure 4), it can be seen that this new line fits better the corium-poor zone analyses than the solid line. In fact, the corium poor zones which compositions are above this line (i.e., silica enriched) correspond also to concrete droplets in concrete rich zones, when those under the dash line correspond rather to real corium poor zone as shown in the left micrography in figure 3. Hence, corium-poor zone (45 – 65 % $(U,Zr)O_2$) could be partly formed by dissolution of "concrete droplets" in the corium rich zones.

It suggests that two different dissolution mechanisms of the concrete by the corium may be active: direct dissolution of concrete by corium (solid line) and the formation of silica-rich droplets, which are transported by Archimedean forces in the pool, and eventually dissolved (leading to corium poor zone slightly enriched in silica, dash line). As the viscosity of the silica droplets must be high, they are only slowly enriched in corium or other concrete components.

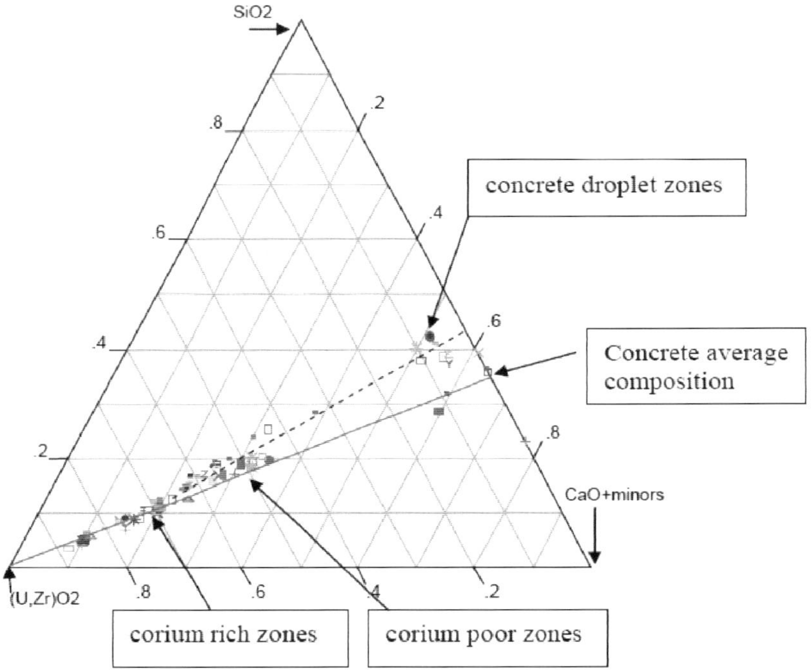

Figure 4. Pseudo-ternary plot of the area compositions (in mass) measured by CEA from VB-U6 different samples. Each point represents a window analysis. The red line passes from the original corium composition to the original concrete composition. The dash black line passes from corium rich zone composition to "concrete droplet" composition.

4.1.2. Comparison between elements. Except for small U, Zr content (corium penetrating into the concrete) all data points fit on a constant U/Zr ratio line. ITU results slightly underestimate U compared to CEA and UACH.

To provide a deeper understanding of the composition in the samples, the contents of the different elements have been plotted versus the uranium or silicon contents. The majority of the plotted results comes from CEA analyses on several samples, and is compared with the results from UACH and ITU from the Zone 1 sample (see figure 5).

There is clearly a good linear correlation in nearly all cases; only for iron versus uranium (or silicon – not shown in figure 5) is this not apparent. However, Fe oxides originate from the corium and, to a lower extent, from the concrete while in the concrete it is not homogeneously distributed. This may explain the loss of correlation for Fe at the low U concentrations.

The results from ITU and UACH fit well with the results of the CEA, for the zirconium versus uranium correlation, even if results from ITU slightly underestimate uranium concentration with respect to CEA and UACH results. Concerning iron vs. uranium, ITU and UACH are within the experimental range of CEA results. For the aluminium versus silicon plot, UACH underestimates aluminium content with respect to CEA and ITU. Nevertheless, for all the metallic elements, the spread between analyses has been found to be of about 10 % of the measured values.

Concerning oxygen (figure 6), it was found that U/O ratio followed a clear linear trend for all analyses. The measurements by CEA and ITU (both performed with Oxford EDS probes but of different series: Oxford EDS-7000 systems of CEA and ITU compared to EDAX (USA) system for UACH-Rez) were extremely well aligned. On the other hand, there is a notable discrepancy between UACH and CEA or ITU results. This is not surprising as uranium and oxygen have large differences

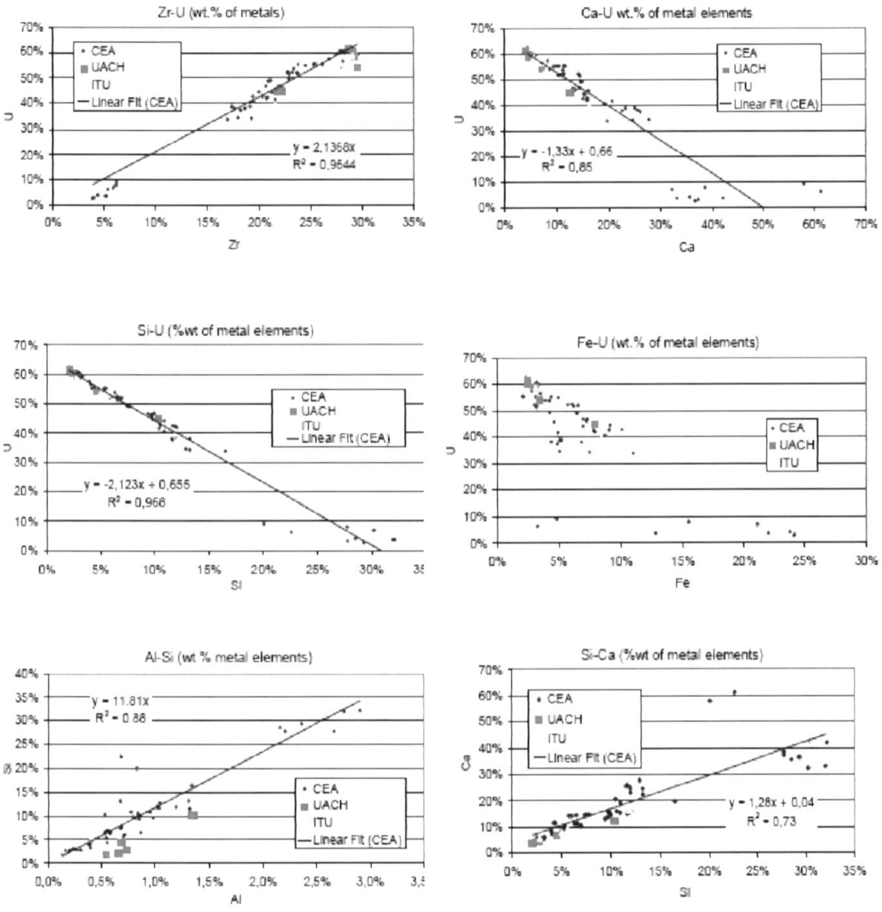

Figure 5. Composition correlations of elements against the main elements U and Si for all analyses from CEA, UACH and ITU.

in their atomic number and that it is a well-known fact that oxygen analysis by EDS is very difficult when heavy elements are present and there are very high absorption factors for oxygen [17, 18]. It must be stressed that there seem to be a systematic error (similar to a bias) between these two sets of analyses rather than a random error. It is probably associated with differences in the measurements and analysis carried out by the individual EDS systems or detectors. The analysis software is different for all three laboratories while the detector crystal and its size are identical so that these are unlikely to be the cause.

4.2. Insights from the round robin

The round robin confirmed that SEM and EDS analysis can be used confidently to perform fast and good quality corium characterisation. Due to the relatively good homogeneity of the zones in the corium pool, the choice of an individual window size or position within a zone only affects slightly the results of the measurement. The uncertainty on the composition of the metallic elements is of the order of 10 % of the measured values or lower.

Concerning the measurements of the lightest and most difficult element to measure: oxygen, a difference of the order of 25 % of the measured values has been observed between UACH and CEA or ITU. It must be noted that, contrary to what could be previously thought, it is not a random error but rather a bias that is more likely to be associated to the details of the EDS technique. At this stage, it is

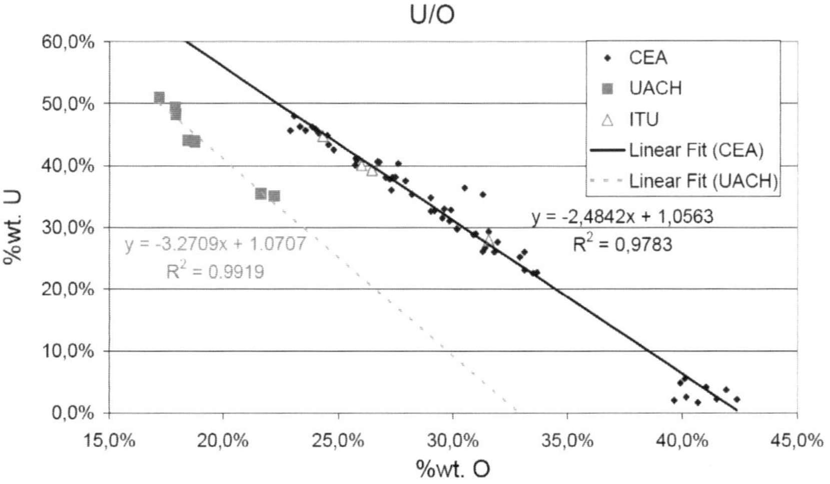

Figure 6. Composition correlations of uranium vs. oxygen for all analyses from CEA, UACH and ITU.

not possible to establish which are the correct values; further analyses, using a complementary method such as wavelength-dispersive X-ray spectrometry analysis (WDS), would be necessary. A better assessment of the oxygen content would require the analysis of reference samples of ZrO_2 and UO_2 samples, based on previous WDS measurements. More likely, accurate oxygen content could not be obtained by EDS but should require dedicated WDS analysis. Such an analysis would be also obviously interesting to compare the results obtained for the other elements. Because of the various contents in minor elements in some phases, as Ca in$(U,Zr)O_2$ phase, XRD would certainly be of no help for the determination of the oxygen content.

5. Conclusion

A corium sample from a prototypic corium-concrete interaction experiment has been successively analyzed in three European laboratories.

In all the measured windows, the composition lay on, or very close to, the line connecting the initial corium melt composition to the concrete average composition, showing the absence of chemical segregation during the interaction process. Nevertheless, distinct zones with different concentrations in corium have been found. In terms of microstructures, two major phases have been found, corium-rich nodules within a concrete-rich matrix. Their relative proportions differ within the zones.

The round robin exercise confirmed that SEM and EDS analysis can be used confidently to perform fast and good quality corium analyses and provide quantitative local compositions in the metallic elements of corium with an uncertainty of 10 % of the measured value. Concerning the oxygen content, CEA and ITU results were well aligned while UACH values were 25 % lower. In the absence of further analyses, we can only conclude that the uncertainty is clearly not random but is rather similar to a bias, and so the EDS analyses can provide reliable information on the relative amounts of oxygen between two phases or areas.

As all the analyses are currently made without calibration with respect to a "corium" standard, future work will involve the synthesis and characterisation of $(U,Zr)O_2$ standards. In the meantime, a comparison of analyses on standard samples, e.g., pure stoichiometric ZrO_2, would have been helpful in understanding such differences. A similar round robin including WDS analysis would be very interesting to estimate the measurement errors induced by EDS, in particular in the case of oxygen.

Acknowledgements

This round-robin exercise was started within the French-Czech Barrande project no. 10783UH and has been pursued within the French Foreign Ministry ECO-NET project number 18555VE.

It is part of the SARNET (Severe Accident Research NETwork of excellence) joint programme of activities, funded by the European Commission 6th Framework Programme (Contract FI6O-CT-2004-509065).

The work and efforts of the VULCANO experimental teams and of all the analytical laboratory staffs are gratefully acknowledged.

References

[1] Hofmann P, Hagen S, Schanz G, Skokan A 1989 Reactor core materials interactions at very high temperature. *Nucl. Technol.* **87** 146-86

[2] Bechta S B, Khabensky V B, Vitol S A, Krushinov E V, Granovsky V S, Lopukh D B, Gusarov V V, Martinov A P, Matinov V V, Fieg G, Tromm W, Bottomley D and Tuomisto H 2006 Corrosion of vessel steel during its interaction with molten corium, Part 1: Experimental. *Nucl. Engng. Des.* **236** 1810-29

[3] Magallon D 2006 Characteristics of corium debris bed generated by large-scale fuel coolant interaction experiments. *Nucl. Engng. Des.* **23** 1998-2009

[4] Journeau C, Sudreau F, Magne S and Cognet G 2001 Physico-chemical analyses and solidification path reconstruction of multi-component oxidic melt spreads. *Mater. Sci. Engng.* **299A** 249-66

[5] Journeau C, Piluso P, Haquet J F, Boccacio E, Saldo V, Bonnet J M, Malaval S, Carenini L and Brissonneau L 2009 Two-dimensional interaction of corium with concretes: the VULCANO VB test series. *Ann. Nucl. Energ.* **36** 1597-613

[6] Skokan A and Hollek H 1987 The significance of chemical reactions between reactor materials under core melting conditions. *Nucl. Engng. Des.* **103** 107-13

[7] Bottomley P D W, Brémier S, Papaioannou D and Walker C T 2002 EPMA and X-ray diffraction of the degraded fuel bundle from the Phebus FPT1 test. *Microchim Acta* **139** 27-38

[8] Piluso P, Trillon G and Magallon D 2005 Material effect and steam explosion at high temperature (T>2300 K). *Int. J. Thermophys.* **26** 1095-114

[9] Min B T, Song J H, Park Y S and Kim J G 2006 A physical and chemical analysis of fast quenched particles of UO_2 and ZrO_2 mixture. *J. Nucl. Mater.* **358** 243-54

[10] Piluso P, Trillon G and Journeau C 2005 The UO2-ZrO2 system at high temperatures (T> 2000 K): importance of the metastable phases under severe accident conditions. *J. Nucl. Mater.* **344** 259-64

[11] Chevalier P Y, Fischer E and Cheynet B 2004 Progress in the thermodynamic modelling of the U-O-Zr ternary system. *Calphad* **28** 15-40

[12] NUCLEA Database (Thermodata 2003-2008), (©Thermodata 2003-2008), <http://hal.archives-ouvertes.fr/hal-00165418>

[13] Barrachin M, Chevalier P Y, Cheynet B and Fischer E 2008 New modelling of the U-O-Zr phase diagram in the hyper-stoichiometric region and consequences for the fuel rod liquefaction in oxidising conditions. *J. Nucl. Mater.* **375** 397-409

[14] Guéneau C, Baichi M, Labroche D, Chatillon C and Sundman B 2002 Thermodynamic assessment of the uranium-oxygen system. *J. Nucl. Mater.* **304** 161-75

[15] Bechta S V, Krushinov E V, Almjashev V I, Vitol S A, Mezentseva L P, Petrov Yu B, Lopukh D B, Khabensky V B, Barrachin M, Hellmann S, Froment K, Fischer M, Tromm W, Bottomley D, Defoort F and Gusarov V V 2007 Phase diagram of the UO_2-Fe_{1+x} system. *J. Nucl. Mater.* **362** 46-52

[16] Petrov Yu, Udalov Yu, Subrt J, Bakardjieva S, Sazavsky P, Kiselova M, Selucky P, Bezdicka P, Journeau C and Piluso P 2011 Phase equilibriums in melts of the system uranium oxide-iron oxide in air. *Glass Phys. Chem.* **37** 212-29
[17] Gorzkowski E P, Watanabe M, Scotch A M, Chan H M and Harmer M P 2004 Direct measurement of oxygen in lead-based ceramics using the ζ-factor method in an analytical electron microscope. *J. Materials Sci.* **39** 6735-41
[18] Cliff G and Lorimer G W 1975 The quantitative analysis of thin specimens. *J. Microscopy* **103** 203-7

IOP Conf. Series: Materials Science and Engineering **32** (2012) 012006 doi:10.1088/1757-899X/32/1/012006

Application of EBSD method for the investigation of microstructure and crystallographic orientation in $RE_2Zr_2O_7$ TBC

B Chmiela[1], M Sozańska and G Moskal
Silesian University of Technology, Department of Materials Science and Metallurgy, ul. Krasinskiego 8b, PL-40019, Katowice, Poland

E-mail: bartosz.chmiela@polsl.pl

Abstract. Modern aero engine turbine blades made of nickel-based superalloys are covered by thermal barrier coatings (TBC) for thermal and oxidation protection. A new generation of TBCs consist of a bond coat (thin layer of MCrAlY, where M may be Ni, Co, Fe) followed by a ceramic top coat of $RE_2Zr_2O_7$ (RE - rare earth element). In this paper we present the possibility of the electron backscatter diffraction (EBSD) method for characterisation of the microstructure and crystallographic orientation of a new TBC consisting of a $Gd_2Zr_2O_7$ top coat and a NiFeCrAlY bond coat after long thermal exposure (1100 °C, 500 h). During thermal exposure, a thermally grown oxide (TGO) layer forms at the bond coat/top coat interface. The TGO is mainly composed of Al_2O_3. But, there is a possible reaction between $Gd_2Zr_2O_7$ and Al_2O_3, leading to Gd-Al-O phases. Phase composition plays an important role in controlling the stress evolution, TGO deformation and crack propagation. Application of SEM-EDS-EBSD techniques allows direct characterisation of the chemical composition, phase composition and crystallographic orientation in the ceramic top coat and TGO layers. This paper presents the possibilities of using the EBSD method for phase identification ($Gd_2Zr_2O_7$, spinel $Ni(Al,Cr)_2O_4$, $GdAlO_3$ and other phases) and orientation analysis of grains in the TGO layer.

1. Introduction

Aero engine turbine blades are made of nickel-base superalloys that have high creep resistance and fatigue strength. For superalloys, excellent mechanical properties are determined by the proper microstructure, which is achieved with the correct chemical composition and phase composition. But, the chemical composition of modern superalloys makes these alloys prone to topologically close-packed phase formation and hot corrosion under service conditions. Therefore, various protective coatings are applied. Ceramic thermal barrier coatings (TBCs) deposited using the air plasma spraying (APS) technique are commonly used to protect many parts of aero engines (e.g., turbine blades, vanes, combustion chambers) from high temperatures and hot corrosion [1, 2]. Application of the APS technique provides strong adhesion of the TBC to the substrate and a porous microstructure with lower thermal conductivity and improved strain accommodation. Classic TBCs consist of a bond coat (thin layer of MCrAlY alloy, where M may be Ni, Co, Fe) followed by a ceramic top coat (yttria stabilized zirconia - YSZ: ZrO_2 with 7 wt% Y_2O_3). During the deposition

[1] To whom any correspondence should be addressed.

process and service conditions, a thermally grown oxide (TGO) layer forms at the top coat/bond coat interface. The TGO is a result of oxidation of the upper part of the bond coat and other chemical reactions. The TGO layer mainly consists of α-Al_2O_3 (alumina) and spinels $Ni(Al,Cr)_2O_4$ [1-4]. Due to the presence of yttrium in the bond coat, the TGO may contain some phases composed of this element [2].

Several investigations into ceramic materials with better properties than YSZ led to the application of lanthanide zirconates $RE_2Zr_2O_7$ (RE = La, Nd, Sm, Gd) with the pyrochlore structure [5-8]. $RE_2Zr_2O_7$ compounds are characterized by thermal conductivities that are lower than that of YSZ [8] and very good hot corrosion resistance and thermal shock resistance. Previous studies on experimental TBCs containing lanthanide zirconates revealed the possibility of detrimental chemical reactions at the top coat/TGO interface that simulate actual thermal conditions during service [9-11]. Those reactions lead to the formation of new phases and are associated with volume changes [9], which contribute to the premature failure of the TBC. In particular, the reaction between gadolinium zirconate and alumina leads to the formation of the $GdAlO_3$-phase that has the perovskite structure [9-11]. Therefore, in order to understand the degradation mechanisms of TBCs, investigating the phase composition and distribution should become standard.

This paper proposes the possibility of applying scanning electron microscope, energy-dispersive X-ray spectroscopy and electron backscatter diffraction (SEM-EDS-EBSD) techniques to evaluate the microstructure, chemical composition, phase composition and crystallographic orientation [12] in experimental TBC, which is fabricated by air plasma spraying a $Gd_2Zr_2O_7$ top coat on the NiFeCrAlY bond coat.

2. Materials and experimental procedure

An experimental TBC with a ceramic $Gd_2Zr_2O_7$ top coat was used in this study. A metallic bond coat layer composed of NiFeCrAlY alloy (140 μm thick) was deposited onto the substrate of a CMSX-4® superalloy by the vacuum plasma spraying technique. The top coat was deposited by the APS technique using a $Gd_2Zr_2O_7$ powder. The thickness of the top coat was about 180 μm. One of the specimens was examined in the as-deposited state, and the others were subjected to thermal exposure at a temperature of 1100 °C and for various times: 2 h, 10 h, 48 h, 175 h and 500 h. After thermal exposure all specimens were air cooled.

Examinations of the microstructure, chemical composition and phase composition were performed on the cross-sections of the specimens. To obtain the cross-sections, specimens were cut carefully using a precision cutter to avoid the risk of mechanical surface damage. Then, the specimens were placed in a quick-setting resin, ground and polished. Each of these specimens was first ground on waterproof abrasive papers at successively decreasing grit sizes. Next, specimens were polished in a diamond suspension (9 μm, 3 μm, 1 μm and 0.25 μm grit sizes) and Al_2O_3 suspension (0.05 μm grit size). For EBSD analysis the specimen was additionally vibratory polished (diamond suspension: 0.05 μm grit size) for 5 hours to obtain as low surface roughness as possible [13].

The microstructure, chemical composition, phase composition and crystallographic orientation were characterized using an SEM (Hitachi S-3400N) equipped with an EDS (Thermo NORAN (System Six)) and an EBSD detector (INCA HKL Nordlys II (Channel 5)). Detailed chemical composition measurements of the TBC cross-sections were performed using the EDS technique. Variable energy of the primary electron beam was used, depending on the results obtained for primary electron energy equal to 15 keV, to guarantee the excitation of spectral lines of all analysed elements. If a given element was not identified in a chosen area, the energy of the primary electron beam was decreased to minimize the excitation volume in the material and maintain an overvoltage equal to about 2. For the cross-sectional TBC specimen treated with a thermal exposure for 500 h, phase identification through EBSD and the phase composition map were based on results from a literature survey and the EDS investigation. The specimen subject to a 500 h exposure was chosen because it had the thickest TGO layer and greatest probability of forming all potential phases. Phase identification was based on comparing the experimental Kikuchi pattern with the theoretical pattern.

Orientation analysis in micro areas of the TBC also was carried out. The energy of the primary electron beam was 20 keV, which is the optimal value for most materials [13]. Due to the small sizes of the analysed areas, the beam scanning mode was used. Additionally, to avoid the risk of overlooking small precipitates of various phases, a beam step size of 0.05 μm was utilized.

Due to the use of a non-conductive material, all SEM-EDS-EBSD investigations were performed under low vacuum conditions in the SEM chamber. The gas in the chamber was air and the working distance was 10 mm. After many trials the best pressure value was determined to be 50 Pa. This pressure effectively prevents the charge effect of the specimen surface and allows the collection of good quality of Kikuchi patterns. SEM observations were performed in two modes: backscatter electron mode, revealing differences in chemical composition (BSE COMP), and environmental secondary electron mode (ESED), revealing the morphology of the material.

3. Results and discussion

3.1. SEM investigation

Figure 1 presents SEM micrographs of the cross-sections of the TBC containing gadolinium zirconate. The TGO layer appeared during the deposition process (figure 1a), and its thickness increased with time of thermal exposure. After 500 h of thermal exposure, the thickness of the TGO was about 6 μm (figure 1f). Systematic degradation of the top coat was clearly visible – many cracks appeared in the top coat and at the top coat/TGO interface.

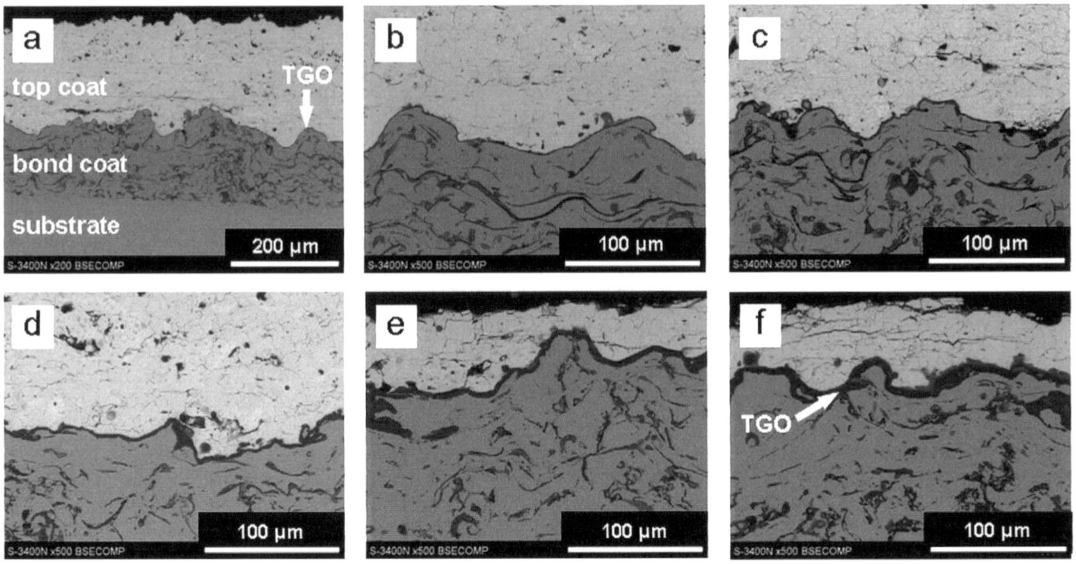

Figure 1. SEM micrographs of cross-sections of the TBC: a) as-deposited; b) after 2 h of thermal exposure: c) after 10 h of thermal exposure; d) after 48 h of thermal exposure; e) after 175 h of thermal exposure; and f) after 500 h of thermal exposure.

The characteristic porosity just above the TGO layer was visible (figure 2). The most probable reason for this phenomenon is a chemical reaction between gadolinium zirconate and alumina due to the negative volume change.

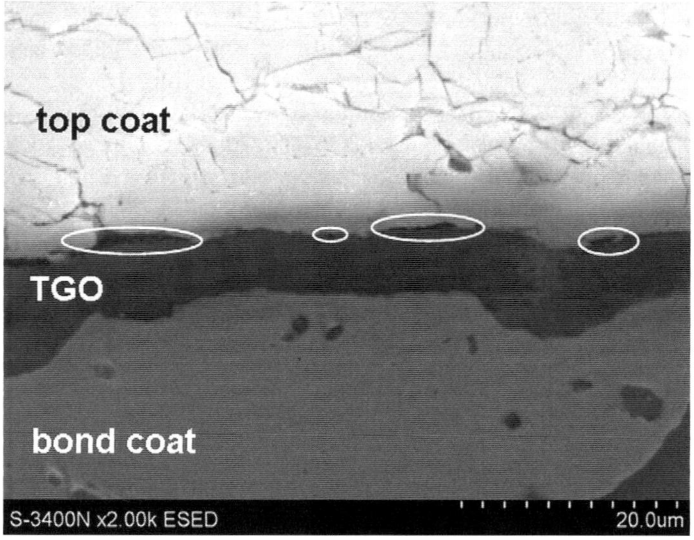

Figure 2. Pores at the top coat/TGO interface.

Cracks running in various directions (parallel and perpendicular to the top coat surface) join together, leading to spallation and delamination of the top coat after 500 h of thermal exposure (figure 3).

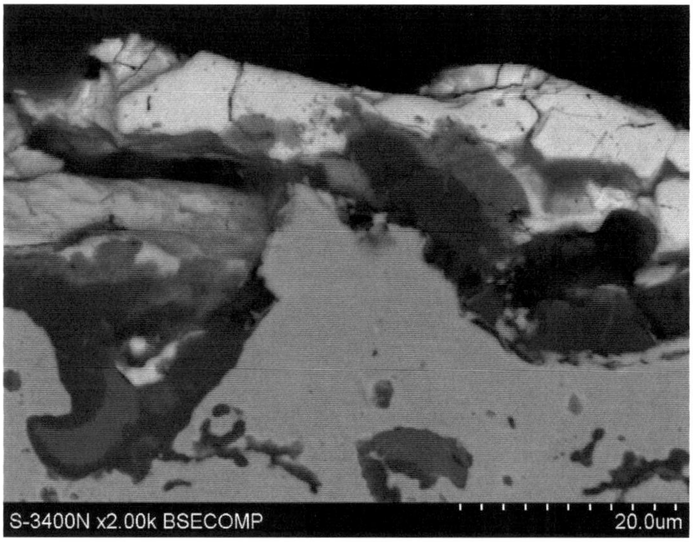

Figure 3. Spallation of the top coat after thermal exposure (1100 °C, 500 h).

Back scattered electron images clearly reveal the differences in chemical composition of various areas in the TGO layer and EDS analysis confirms these findings.

3.2. EDS investigation

EDS analyses were carried out on the cross-sections of the as-deposited specimen and thermally exposed (500 h) specimen. To estimate the skirt effect in low vacuum, several analyses (spot mode) in the TGO layer after 500 h of thermal exposure were carried out (figure 4). Based on the literature data and the knowledge about the TBC deposition and degradation processes, it is obvious that the TGO layer consists mainly of $(Al,Cr)_2O_3$ and $Ni(Al,Cr)_2O_4$ compounds. EDS results confirm this findings – there are only Al, O and Cr signals in the middle part of the TGO (figure 4c). Analyses near the top coat/TGO interface and near the TGO/bond coat interface reveal the presence of elements from the top coat (Gd, Zr), spinel (Ni) and bond coat (Ni, Fe, Y, Mo). Because the analysis in point 2 (figure 4c) reveals only aluminium, oxygen and chromium, it is safe to say that scattering of the electron beam (skirt effect) in quite small. Therefore, EDS analysis in given low vacuum conditions seems to be reliable.

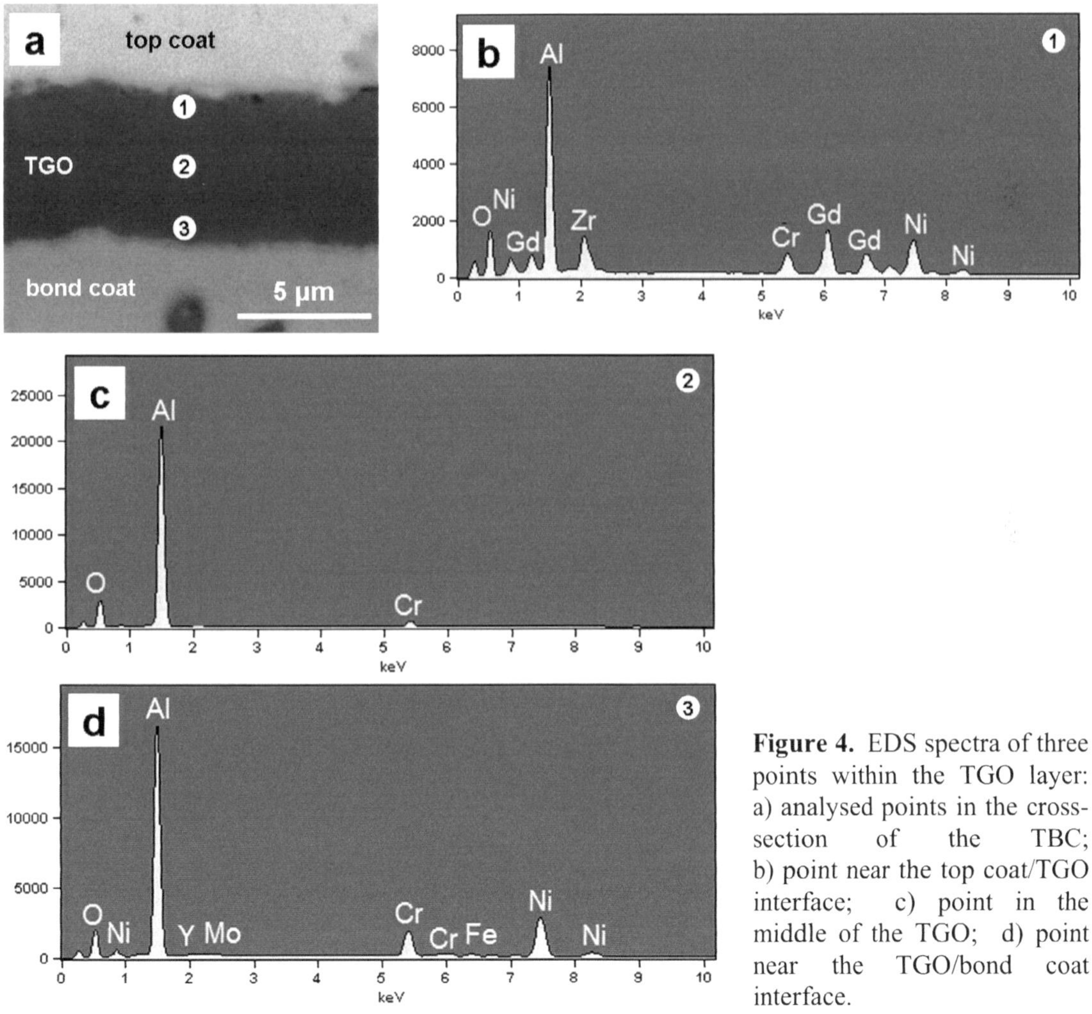

Figure 4. EDS spectra of three points within the TGO layer: a) analysed points in the cross-section of the TBC; b) point near the top coat/TGO interface; c) point in the middle of the TGO; d) point near the TGO/bond coat interface.

Analysis results of the top coat and the bond coat of the as-deposited coatings are shown in table 1. Quantitative results were the mean of five analyses in the spot mode. Because a standardless analysis was performed, light elements (carbon and oxygen) were not taken into account in the quantitative analysis. The presence of carbon probably resulted from the preparation (polishing using the diamond suspension, which could fill small pores). The ratio of Gd to Zr concentrations corresponded to stoichiometric $Gd_2Zr_2O_7$, and it may be considered as binary system of ZrO_2-Gd_2O_3 containing 33 at% Gd_2O_3 [5-7, 9].

Table 1. Chemical composition of the as-deposited TBC – metallic elements (at%).

	Al	Cr	Fe	Ni	Y	Zr	Mo	Gd
top coat	-	-	-	-	-	49.7	-	50.3
bond coat	11	22.6	2.2	62.1	0.7	-	1.4	-

EDS analysis after 500 h of thermal exposure revealed characteristic changes in the chemical composition of the top coat and TGO that are related to diffusion processes in a ceramic material and the resultant chemical reactions. The composition of the upper part of the top coat was almost identical to that before thermal exposure. But, in the lower part of the top coat (near the TGO), there was a characteristic depletion of gadolinium and zirconium and an enrichment of aluminium (table 2).

Table 2. Chemical composition of the top coat layer – metallic elements (at%).

	Al	Zr	Gd
upper part	-	49.1	50.9
lower part	19.6	39.2	41.2

Those changes suggest the progression of diffusion and reactions between $Gd_2Zr_2O_7$ and Al_2O_3. At the thermal exposure temperature (1100 °C), reaction (1) is probably occurring and leading to the formation of the $GdAlO_3$-phase – gadolinium aluminate [9]:

$$Gd_2Zr_2O_7 + Al_2O_3 \rightarrow 2GdAlO_3 + 2ZrO_2 \tag{1}$$

Proof of this reaction can be found from EDS and XRD analyses [10]. Figure 5 presents EDS spectra performed on the top coat and TGO layers. In some areas of the top coat/TGO interface, a phase with a qualitative composition connected to the $GdAlO_3$-phase (figure 5c) was identified. One should note that a phase of this composition occurred locally and did not form a continuous layer. This observation implies that at the applied temperature, the intensity of reaction (1) is not too high, because even after 500 h of thermal exposure, the amount of this phase is small. Moreover, there was a negative volume change from reaction (1) (about 9 %), which led to the formation of additional porosity [9]. One might surmise that the rate of $GdAlO_3$ formation increases with time, while just below the top coat, the decreased thickness of the top coat due to spallation and the increased number of cracks increase the thermal diffusivity and coating temperature. Furthermore, EDS analyses of the TGO layer made identification of the other phase easier. Qualitative analysis of selected areas and results from the literature suggested the presence of the following phases: $(Al,Cr)_2O_3$ (figure 5d), $Ni(Al,Cr)_2O_4$ (figure 5e) and $Y_3Al_5O_{12}$ (figure 5f). The $(Al,Cr)_2O_3$-phase is sapphire, which is alumina with some Al^{3+} cations substituted with Cr^{3+} cations. The $Ni(Al,Cr)_2O_4$-phase is a spinel, and $Y_3Al_5O_{12}$-phase is yttria-alumina garnet (YAG), forming due to the yttrium content in the bond coat layer [14].

3.3 EBSD investigation

Based on the literature and SEM-EDS results for the microstructure and phase composition, the several phases were chosen for EBSD analysis (table 3). In the area surrounding the top coat/TGO interface, a phase composition map was constructed using a phase component in the software (figures 6a and 6b). This map presents the phase distribution in the cross-section of the TBC. Raw maps were submitted to standard cleaning procedure, according to instruction for users (Oxford

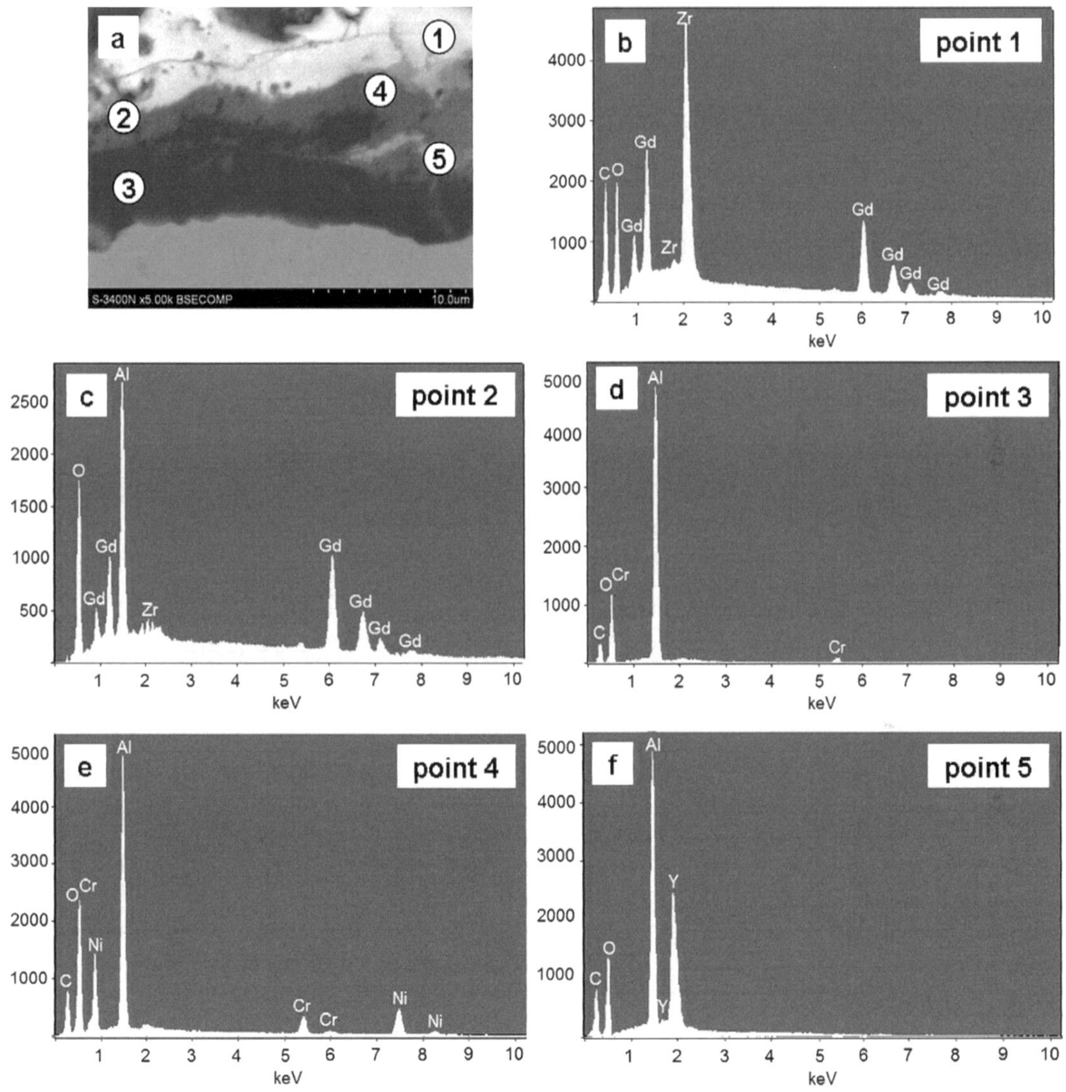

Figure 5. EDS spectra of various phases of the TBC after thermal exposure (1100 °C, 500 h): a) analyzed areas in the cross-section of the TBC; b) gadolinium zirconate; c) gadolinium aluminate; d) sapphire; e) spinel; and f) yttria-alumina garnet.

Instruments HKL Technology, 2006). The cleaning procedure consisted of several stages: removing wild spikes (isolated points that have been incorrectly indexed) and then removing zero solutions (low level extrapolation). Finally, the orientation averaging filter (Kuwahara filter) was applied with a 3x3 size.

Just below the top coat (aqua colour), small precipitates of the $GdAlO_3$ phase (red colour) were visible. The amount of this phase was small, and there was no continuous layer. These results indicate that during given conditions of thermal exposure, the reaction between $Gd_2Zr_2O_7$ and Al_2O_3 occurred and produced a noticeable yield only in selected areas. This reaction may be a result of the increased thermal diffusivity of the top coat that is caused by spallation and delamination. Under the

Table 3. Phases investigated by EBSD.

Phase name	Chemical formula	Crystallographic system	Space group	Unit cell parameters (nm)		
				a_0	b_0	c_0
Gadolinium zirconate	$Gd_2Zr_2O_7$	cubic	$Fm\bar{3}m$	0.526	0.526	0.526
Gadolinium aluminate (GAP)	$GdAlO_3$	orthorhombic	$Pbnm$	0.530	0.744	0.525
Alumina	Al_2O_3	trigonal	$R\bar{3}c$	0.475	0.475	1.298
Spinel	$NiAl_2O_4$	cubic	$Fd3m$	0.805	0.805	0.805
Yttria alumina garnet (YAG)	$Y_3Al_5O_{12}$	cubic	$Ia\bar{3}d$	1.201	1.201	1.201

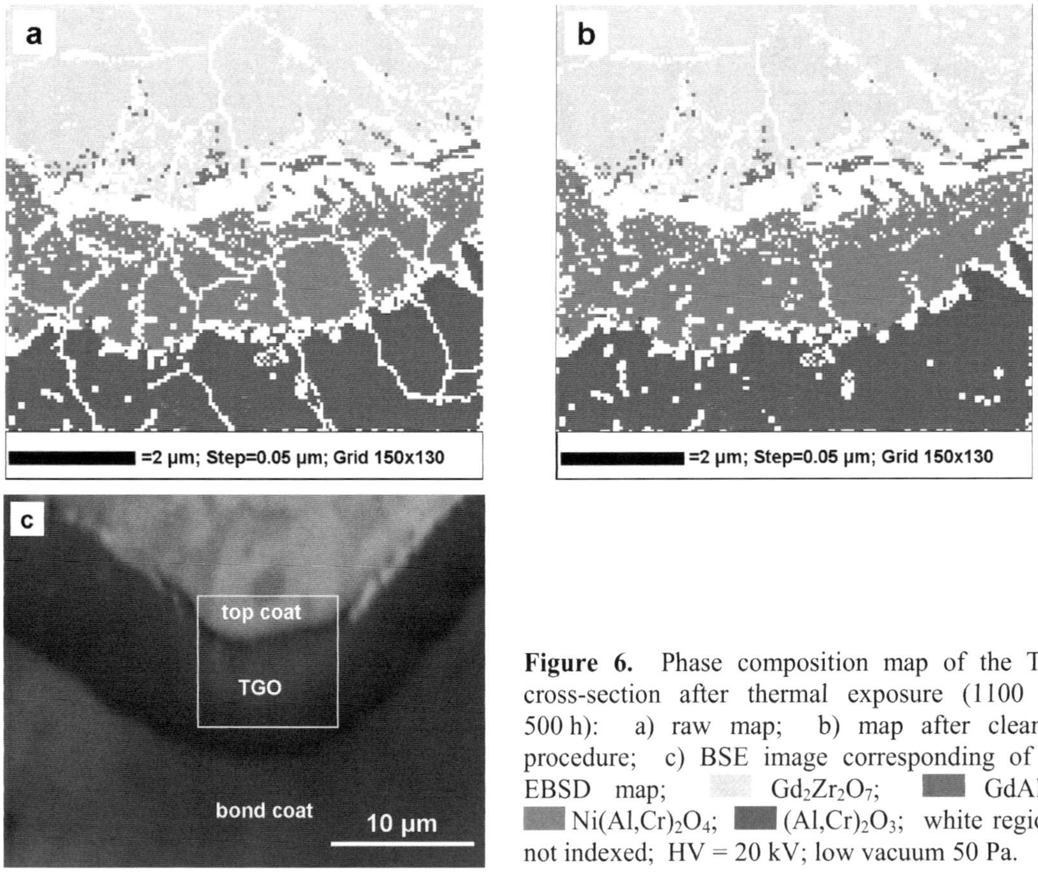

Figure 6. Phase composition map of the TBC cross-section after thermal exposure (1100 °C, 500 h): a) raw map; b) map after cleaning procedure; c) BSE image corresponding of the EBSD map; $Gd_2Zr_2O_7$; $GdAlO_3$; $Ni(Al,Cr)_2O_4$; $(Al,Cr)_2O_3$; white regions: not indexed; HV = 20 kV; low vacuum 50 Pa.

$GdAlO_3$-phase precipitations, mainly spinel phase (fuchsia colour) was identified. The last zone within the TGO layer consisted of sapphire (teal colour) and small amounts of spinel. Identification of these phases confirms the predictions made on the basis of the literature and EDS analyses. White regions on the phase map denote non-indexed areas. These areas were not indexed probably due to overlapping patterns, insufficient quality of the specimen surface (too high roughness) and stresses or strains in the crystal lattice [13].

The phase distribution results in the cross-section aids in formulating the supposed mechanism of TGO formation in the TBC that contains gadolinium zirconate. A very thin TGO layer was present in the as-deposited coating, as a result of applying the top coat by air plasma spray. During thermal exposure in an air atmosphere in the furnace, $Gd_2Zr_2O_7$ serves as an ionic conductor, and due to the presence of micro-cracks, oxygen diffuses to the top coat/TGO interface. In addition, the oxygen diffusion causes oxidation of aluminium and chromium in the bond coat and growth of $(Al,Cr)_2O_3$ in the upper part of the bond coat. Then, $Gd_2Zr_2O_7$ reacts in selected areas with $(Al,Cr)_2O_3$, leading to the $GdAlO_3$-phase, while the reaction between nickel, oxygen and sapphire leads to the spinel phase. Additionally, oxidation of yttrium and reactions between yttrium oxide and alumina lead to small YAG precipitates [1, 14].

In selected areas of the cross-section of the top coat that are free of cracks, orientation analysis was performed. Raw orientation map in all Euler colouring scheme is shown in figure 7. The same cleaning procedure as for phase map was applied.

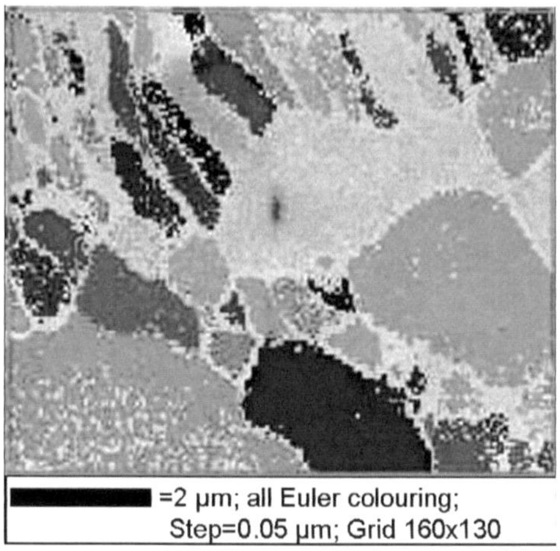

Figure 7. Raw orientation map of the cross-section of the top coat, after thermal exposure (1100 °C, 500 h), with all Euler colouring scheme; HV = 20 kV; low vacuum 50 Pa.

Figure 8 presents two orientation maps with inverse pole figure (IPF) colouring scheme, which allows the crystallographic orientations to be interpreted in terms of the specimen coordinate system (in this case in terms of the specimen Z axis – perpendicular to their surface). The maps revealed that the top coat layer was characterized by a fine-grained microstructure.

APS top coat do not display optimal mechanical properties in various directions. For this reason, in contrast to the electron beam physical vapour deposited (EB-PVD) top coat, the APS top coat layer is characterized by low deformability (especially in the direction parallel to the substrate surface), which supports crack formation, thus decreasing the lifetime of the TBC [15, 16]. But, the APS technique is more readily available and cheaper than the EB-PVD technique.

The orientation analysis revealed high angle boundaries (HABs) in the top coat; the misorientation between neighbouring grains exceeded even 30° and 40° (figure 9).

4. Conclusions

Thermal barrier coatings (TBCs) are characterized by low thermal conductivity and low thermal diffusivity. Application of new compounds – lanthanide zirconates – lead to coatings with a thermal conductivity lower than that for YSZ coatings. Unfortunately, TBCs with the $Gd_2Zr_2O_7$ compound display many drawbacks. Low service life of the coating containing $Gd_2Zr_2O_7$ is apparent. After long thermal exposure (1100 °C, 500 h), the top coat in selected areas is almost completely destroyed. Large amounts of cracking result from the mismatch of thermal expansion coefficients between the top

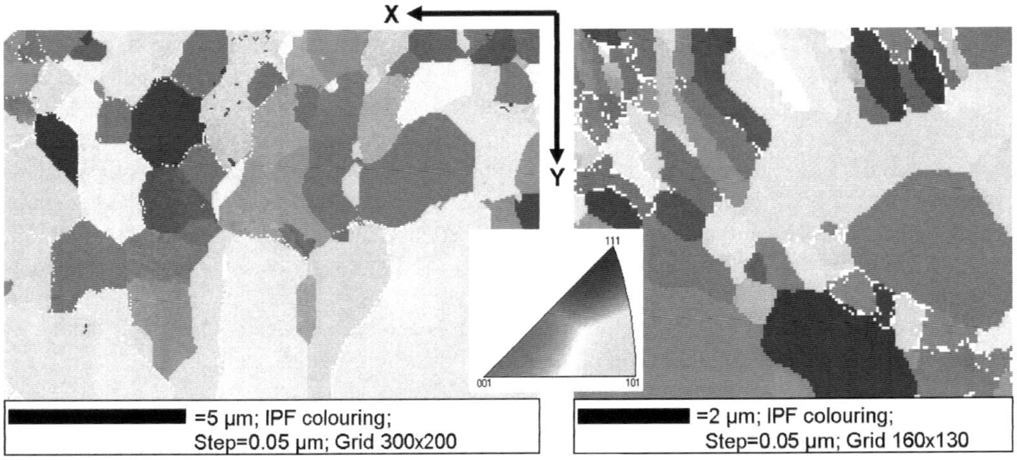

Figure 8. Orientation maps of the cross-section of the top coat, after thermal exposure (1100 °C, 500 h), with the IPF colouring scheme; HV = 20 kV; low vacuum 50 Pa.

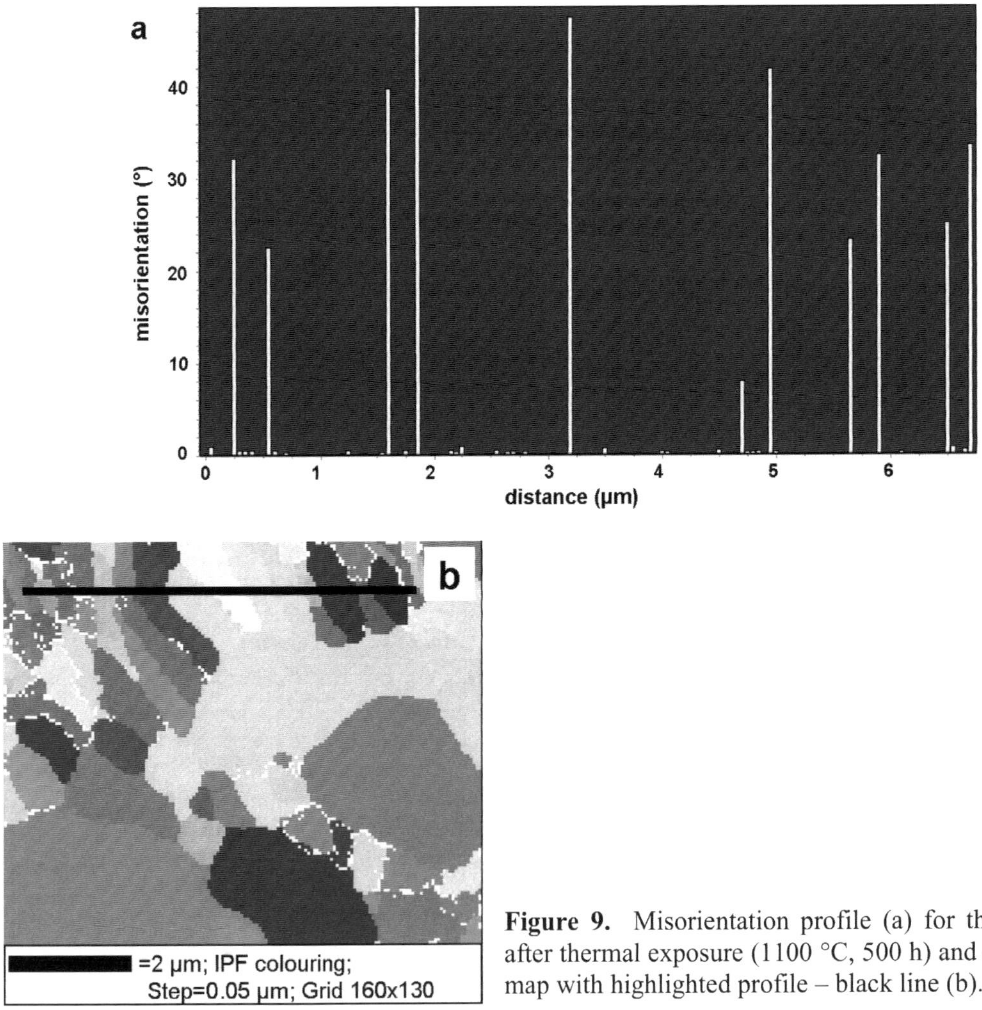

Figure 9. Misorientation profile (a) for the top coat after thermal exposure (1100 °C, 500 h) and orientation map with highlighted profile – black line (b).

coat and the bond coat and too low deformability of the top coat. Additionally, reaction between gadolinium zirconate and sapphire (accelerated by spallation of the top coat) causes increased porosity at the top coat/TGO interface. Therefore, TBCs with gadolinium zirconate may be used at a temperature lower than 1100 °C; otherwise, it will be necessary to use an additional ceramic layer (YSZ) between the top coat and the bond coat, consequently decreasing the diffusion rate and preventing detrimental reactions [15, 16].

Applying the SEM-EDS-EBSD techniques provides a complete description of the microstructure, chemical composition, phase composition and orientation of the TBC. A non-conductive TBC requires the use of the low vacuum mode in the SEM; then, it is not necessary to deposit an additional conductive coat, which usually makes EDS and EBSD analysis more difficult. By performing EDS (when wavelength-dispersive X-ray spectrometry is not available), qualitative analysis of the chemical composition and correct quantitative analysis of heavy elements are possible. In addition to XRD analyses and literature data, EBSD investigations can be performed. The EBSD technique can be used to obtain a phase composition map, which is impossible with X-ray diffraction, of the TBC cross-section. Many difficulties related to specimen preparation are revealed in some non-indexed areas, but the fraction of correctly indexed points are sufficient to perform the analysis. The EBSD technique helps to evaluate applying the crystallographic orientation, which influences the mechanical properties of the ceramic top coat. Correct application of the EBSD technique requires the knowledge of as much information about the material as possible (especially about the chemical composition and most likely about the phase transformations in the material).

Acknowledgements
Financial support of Structural Funds in the Operational Programme-Innovative Economy (IE OP) financed from the European Regional Development Fund-Project no. POIG.0101.02-00-015/09 is gratefully acknowledged.

References
[1] Tamarin Y 2002 *Protective coatings for turbine blades.* (Ohio: Materials Park)
[2] Bose S 2007 *High temperature coatings.* (Manchester, Connecticut, USA: Elsevier Science & Technology Books)
[3] Lee C H, Kim H K, Choi H S and Ahn H S 2000 *Surf. Coat. Techn.* **124** 1
[4] Czech N, Fietzek H, Juez-Lorenzo M, Koralik V and Stamm W 1999 *Surf. Coat. Techn.*. **113** 157
[5] Wang Ch 2006 *Experimental and computational phase studies of the ZrO_2-based systems for thermal barrier coatings.* (Stuttgart: Max-Planck-Institut für Metallforschung)
[6] Leckie R M R 2006 *Fundamental issues regarding the implementation of gadolinium zirconate in thermal barrier systems.* (Santa Barbara: University of California)
[7] Wilde P J and Catlow C R A 1998 *Solid State Ion.* **112** 173
[8] Suresh G, Seenivasan G, Krishnaiah M V and Murti P S 1997 *J. Nucl. Mater.* **249** 259
[9] Leckie R M, Krämer S, Rühle M and Levi C G 2005 *Acta Mater.* **53** 3281
[10] Wu J, Padture N P and Gell M 2004 *Scr. Mater.* **50** 1315
[11] Lakiza S, Fabrichnaya O, Wang Ch, Zinkevich M and Aldinger F 2006 *J. Europ. Cer. Soc.* **26** 233
[12] Fujiyama K, Nakaseko H, Kato Y and Kimachi H 2010 *J. Sol. Mech. Mater. Eng.* **4** 178
[13] Faryna M 2003 *Analysis of orientation reliationships in ceramic based composites.* (Krakow: Institute of Metallurgy and Materials Science, Polish Academy of Sciences)
[14] Haynes J A, Rigney E D, Ferber M K and Porter W D 1996 *Surf. Coat. Techn.* **86-87** 102
[15] Moskal G 2009 *Arch. Mater. Sci. Eng.* **38** 19
[16] Vassen R, Kassner H, Stuke A, Mack D E, Jarligo M O and Stöver D 2010 *Mater. Sci. Forum* **631-632** 73

IOP Conf. Series: Materials Science and Engineering **32** (2012) 012007 doi:10.1088/1757-899X/32/1/012007

SEM-EDS analysis of ancient gold leaf glass mosaic tesserae. A contribution to the dating of the materials

A Conventi[1,3], E Neri[2] and M Verità[1]
[1] Università IUAV di Venezia, Laboratorio Analisi Materiali Antichi, Palazzo Badoer, San Polo 2468/B, IT-30125 Venice-Murano, Italy
[2] Università Cattolica del Sacro Cuore, Milan, Italy

E-mail: alberto.conventi@iuav.it

Abstract. Metal leaf (gold, silver or their alloys) glass tesserae began to be used in wall mosaics in the first century AD (the first examples are in Rome) and their use has been uninterrupted up to day.
The metal leaf could be obtained from circulating coins, jewellery or refining. According to various techniques that have changed over the centuries, the leaf was hot fixed between two glass layers.
From an archaeological point of view, it is interesting to know when and where these tesserae were made, if they were new made or if they were reused tesserae recovered from earlier dismantled mosaics.
The determination of the glass composition of the tesserae is not of great help in this connection, for the same kind of glass was used over long periods. Available information is still scanter for glasses produced between the 1st to 8th centuries when the batch of raw materials (a natural soda called natron and a silica-lime sand) was melted in large tank furnaces and chunks of raw glass were transported all over the Mediterranean to be remelted and shaped into manufacts in small pot furnaces.
The SEM-EDS analysis is proposed in this study as a useful tool to investigate the composition of both the glass and the gold alloy in leaf tesserae from mosaics of the 1st - 9th centuries. The comparison of the composition of the gold leaf of the tesserae with that of circulating gold coins (for which an important analytical data base is available), adds further information to the glass analysis, allowing us to improve the dating of the tesserae and increase the knowledge that may result from scientific analyses. The results demonstrate that good quantitative analyses of the metal leaf can be performed and that metal leaves made of pure gold or gold-silver alloys were used.

1. Introduction

A mosaic consists of cube-shaped small pieces of different materials and colours, known as tesserae, which are held in place by mortar and generally separated by visible interstices. Glass is an ideal material for tesserae because it provides a rich polychrome decoration, is easy to cut from poured glass cakes and is a durable material. The earliest examples of wall and vault decoration with glass tesserae date to the 1st century AD. The technique of colouring glass tesserae with a metal leaf dates back to the same period; the first examples are in Rome [1].

---

[3] To whom any correspondence should be addressed.

The metal leaf tesserae are real composite materials made of a leaf of gold or silver hot fixed between two glass layers (figure 2). The glass embedded in the mortar (the support) is 5 to 10 mm thick, while a thin sheet of blown glass (the cartellina, usually 0.3 to 0.7 mm thick) protects the fragile metal foil and adds to its brilliance [2]. These tesserae are made also today by laying the beaten square metal leaf (less than 1 μm thick) on a small piece of blown glass (less than 1 mm thick). Molten glass is then poured on the leaf; the whole piece is heated in the furnace and then pressed to form a cake and to ensure good adhesion of the three layers. The tesserae are then cut from these cakes.

From an archaeological point of view, it is interesting to know when and where these tesserae were made and the technology adopted to prepare them. In this paper, scanning electron microscopy equipped with energy-dispersive X-ray spectrometry (SEM-EDS) analysis is applied to determine the quantitative composition of gold leaves of tesserae from mosaics of the 1st to 9th centuries. The aim was to verify whether the metal composition corresponds to the composition of contemporary circulating gold coins, jewels or was prepared from refined gold. In the first case, due to the fact that the composition of the gold coins changed during the considered period, the analyses of the metal will add to the scant information obtained by the glass analysis, allowing us to improve the dating of the tesserae and to identify their provenance.

2. Experimental

2.1. Gold leaf tesserae

More than thirty gold leaf tesserae from mosaics in Italy (Rome, Ravenna and Milan) dated to between the 1st and 9th centuries A.D. were analyzed. The tesserae were sampled during restoration of in situ mosaics or during archaeological excavation. In Table 1 the provenance of the analyzed tesserae is reported.

Table 1. Provenance of the analyzed tesserae. (*): archaeological sites.

mosaic	century	n. of samples	reference
Rome*	1st	2	[1]
Aquileia*	1st-4th	2	[15]
Ostia, Rome*	4th	4	[16]
Rome, St Pudenziana	4th	1	[17]
Rome, St Sabina	5th	4	
Ravenna, St Apollinare Nuovo	5th	4	
Milan, St Lorenzo	5th	2	
Milan, St Giovanni alle Fonti	5th-6th	6	
Ravenna, St Vitale	6th	6	
Rome, Sts Cosma e Damiano	7th	2	[18]
Rome, St Stefano Rotondo	7th	4	[19]
Rome, St Cecilia	9th	3	[20]

When intact tesserae were available, fragments including the support, gold leaf and *cartellina* were cut with a rotating diamond wheel and embedded in cross-section in acrylic resin in a teflon mould.

The discs were ground and polished down to 3 μm grain size diamond pastes. When only the support or the *cartellina* was available (weathered tesserae), samples were stacked to a stub, so that the surface originally in contact with the gold leaf was exposed to the electron beam. In secondary or backscattered electron mode, also small residues of the gold leaf can be easily detected (figure 1).

Figure 1. Small residues of the gold leaf on a *cartellina* (the large bright part of the image).

2.2. Methods

After observation with an optical microscope in reflected light, samples were carbon coated to be analyzed by scanning electron microscopy (Philips XL 30). The quantitative X-ray microanalysis of the glass of the *cartellina* and of the support was done by wavelength-dispersive X-ray microanalysis using a Cameca SX-50. Various reference glasses of certified composition were analyzed under the same experimental conditions to verify the accuracy of the method.

The microprobe was equipped with three wavelength-dispersive X-ray spectrometers (PET, LiF and TAP crystals). Twenty elements were quantified: X-ray Kα-lines were analyzed except for Pb and Bi (Mα-lines), Sb, As and Sn (Lα-lines). Operating conditions were: accelerating potential 15 kV, beam current 20 nA (major and minor components) or 100 nA (trace elements) respectively. A 40x50 μm scanning electron beam and limited counting time (10 s for major and minor elements, 20 to 30 s for traces) were employed to ensure that no significant alkali drift (ion migration) occurred during the irradiation. The net X-ray intensities (peak minus background) were quantified by means of a PAP correction programme supplied by Cameca. X-ray microanalysis applied to the glass analysis is discussed in [3].

Quantitative analysis of the metal leaf was performed by energy-dispersive X-ray microanalysis (Edax). In a SEM image at 4000 magnifications of the tesserae cross-sections, the electron beam was scanned across an area 5 μm long and as wide as the gold leaf (less than 1 μm). The aim was to investigate during the analysis a metal volume as much representative as possible of the gold foil as a whole. The analysis was repeated in three areas for each sample and the average value was considered. Special care was taken to carry out the analysis on areas free of cavities (bubbles) and deposits formed at the glass-gold boundary during the hot forming of the glass cake. A standardless ZAF correction of the raw data was performed. To verify the accuracy of the method, Au-Ag and Au-Cu reference alloys were analyzed under the same experimental conditions as for metal leaves. The calculated concentration values agreed within 5 % with the concentrations of the certified samples.

During analysis SEM operates at 20 kV, with beam intensity adjusted to give an input rate of approximately 1500 count per second (cps) and counting live-time of 200 s. These analytical conditions allowed us to calculate a lower limit of detection (determined as three standard deviation of the background on a ternary reference alloy: Au 96 %; Ag 2.5 %; Cu 2.5 %) of 0.10 wt% for silver (Ag, Lα) and 0.20 wt% for copper (Cu, Kα). The lead (Pb less than 300 ppm) and platinum (Pt less than 500 ppm) contents found in Roman gold coins [4] are under the lower limit of detection of the analytical method and were not investigated.

3. Results and discussion

3.1. Glass analysis

The analyses demonstrate that the composition of the *cartellina* (thickness varies between 0.5 and 0.7 mm) is practically the same as that of the support for most of the tesserae. The optical quality, however, is different: the *cartellina* glass is homogeneous while the support glass is heterogeneous and contains many bubbles (figure 2).

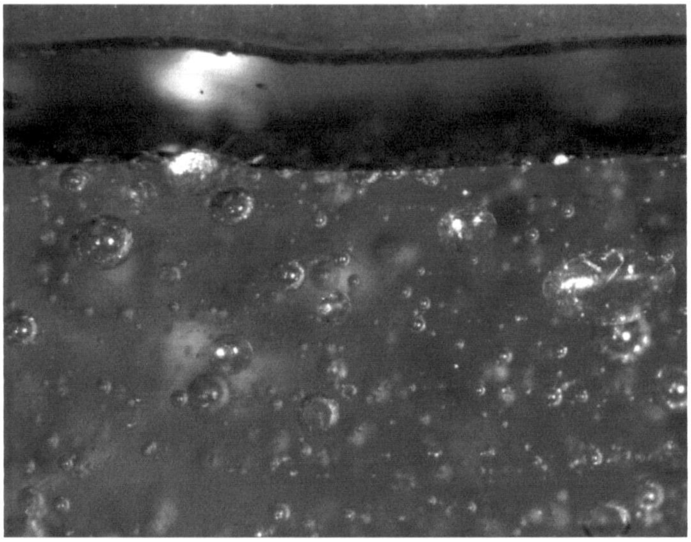

Figure 2. Optical micrograph of the polished section of the *cartellina* and support glass (gold leaf tessera from the mosaic of St Stefano Rotondo, Rome). Short side of the micrograph: 2.5 mm.

A thickness in the range 0.3 to 0.7 mm was measured for most of the *cartellina*. Only rarely was the *cartellina* deliberately coloured slightly purple or well decolorized on a light green support.

The *cartellina* glass composition differs from that of the support only in few tesserae. For instance, in a tessera from Aquileia (AQ659g) the two layers have the same glass composition (iron included: Fe_2O_3 0.50 %) except for decolourants. The *cartellina* is better decolourized with manganese and antimony (MnO 0.28 %; Sb_2O_3 0.32 %), as compared to the support, which is decolourized only with manganese (MnO 1.10 %).

As expected for this period, the main glass components are the oxides of sodium, calcium and silicon (soda-lime-silica glass), while the oxides of potassium and magnesium are each below about 1.5 % and phosphorous below 0.2 %. This composition is in agreement with the dominant glass type from the Roman time until the 8th - 9th centuries AD (later, glass was made by melting silica and plant ash), which was made by melting a batch of a sodium carbonate mineral from Egypt, called natron, mixed together with a natural silica-lime sand in which quartz and calcium carbonate were present in suitable ratios to make glass [5]. In this period the melting of the batch of raw materials was carried out in a limited number of places located near the sources of the raw materials (the Levant,

Egypt), in tank furnaces where several tons of glass could be melted. Once the melting was completed, the furnace was left to cool and then demolished. Blocks of transparent glass were recovered (raw glass), traded throughout the Mediterranean region and Western Europe and distributed to workshops where they were remelted and made into objects. The analytical investigations confirm this model; in fact only a few glass groups can be distinguished by their composition in the period between the 1st and 9th centuries AD [5]. The division of the production process among a few primary melting centres and many secondary fabrication workshops restricts the contribution that chemical analyses can give to the understanding of the place of production and the circulation of the finished glass tesserae.

The final appearance and shade of metal-leaf tesserae depend on various parameters such as the metal used (gold or silver), the colour of the support and of the *cartellina*. Because the metal leaves are so thin, they often fail to provide a uniform layer and discontinuities occur, through which the glass support may be seen. Thus the colour of the support can influence the chromatic effect of the tesserae. Though it has been poorly studied, the use of different types of gold tesserae in the same mosaic seems to have been a widespread practice in Late Antiquity and Byzantine ages. Preliminary investigations in Milan [6] and Ravenna [7] contexts confirm the use of two kinds of gold leaf tesserae in the same mosaic: one made with a colourless glass and the second with a yellow-green glass. This confirms a deliberate choice according to an aesthetic principle intended to give the gold leaf a different luminosity: a silvery effect on a colourless support and a warmer one on a yellow-green glass. The analyses of the Milan gold leaf tesserae reveal that yellow green glass was obtained by using sand with high amounts of contaminants (high iron, titanium and alumina concentrations) and only partially decolourizing the glass with manganese. These compositional characteristics are compatible with a natron glass type termed HIMT (high iron, manganese and titanium), which was widespread in Europe and in the Mediterranean area between the 5th and 7th centuries [5]. Instead, the perfectly colourless glass of other tesserae was obtained from purer sand (the iron content is about one half as compared to the yellow green gold tesserae) and antimony alone or in association to manganese was used for the decolouration. This type of glass was hold to be in use between the 2nd and the 4th centuries AD [8], but recent studies demonstrated that antimony was still in use at least until the 9th century [9].

3.2. Gold leaf analysis

A text of the 1st century AD by the physician Dioscorides Pedanius Anazarbei (*De materia medica libri sex*) reports that gold leaves were obtained by beating gold between two copper sheets to avoid their rupture. Pliny (*Naturalis Historia, XXXIII, XI*) writes that from one ounce (27.264 g) 750 leaves four inches side length (7.4 cm) could be obtained for marble, wood and copper gilding. The *Manoscritto di Lucca* (Greek tradition, late 8th century) shows that this technique was still in use in the late 8th century. This manuscript specifies for the first time (chapter 15) that the gold leaf used in mosaics is held between two layers of glass, a thin layer and a thicker one [10].

The thickness of the gold leaf was measured by SEM in SE mode in the polished cross-sections. Due to the irregular beating of the leaf in a single tessera, important differences in the thickness of the metal leaf were found. Thickness varies between 0 μm (areas where the gold leaf is torn) and 0.9 μm; most of the measurements are in the range 0.4 ± 0.2 μm. This thickness is compatible with the 0.325 μm thickness calculated for medieval leaves obtained by beating a florin (3.53 g) to obtain (following medieval treatises) 100 leaves of about 7 cm width and an area of about 53 cm^2 [10]. This leads to the conclusion that in the period between the 1st and 9th centuries, by beating 19.3 g of gold (one cubic centimetre), 2.5 to 3.0 m^2 of gold leaf for mosaic tesserae were obtained.

Ancient texts do not mention the source of the gold. We have to wait for the *Il libro dell'arte* by Cennino Cennini (chapter 139) written in the late 14th century to learn that gold coins (*ducato* from Venice, in circulation in the second half of the 13th century) were beaten to make gold leaves. 13th century (second half) documents report the use of gold florine and ducate to make gold foils [11]. The hypothesis cannot be excluded that circulating gold coins were beaten to obtain gold leaves also during the late Roman period. Since gold coins were common currency in those days, they represented an easily accessible source of gold guaranteed by the mint. Pure gold was preferred instead of gold alloys mainly because it is malleable and ductile, therefore ideal for making the thin sheets required for the gold tesserae, but also because it is brilliant and resistant to weathering.

It is well known that high purity was generally a characteristic of circulating gold coins, while the goldsmith's art was practised be remelting gold items or gold coins. In this context, it is reasonable to assume that gold beaters preferred to obtain gold leaves directly from coins rather than remelting jewellery items and purify gold alloys. The composition of circulating gold coins varies in time and sometimes also according to the mint of coinage. For instance, in the 8th and 9th centuries the average gold content of circulating coins was about 96.4 % [12].

Of the 33 gold tesserae analyzed in this work, in 9 cases the metal leaf was made of pure gold, in 8 samples gold varied in a restricted range between 99 % and 99.5 % and silver made up for the difference 0.5 to 1 %. In the remaining 16 tesserae the gold amount was less than 99 %, with a lowest concentration of 91 %, and silver completing the alloy. Small amounts of copper (Cu 0.5 to 1 %) were detected only in two tesserae from Aquileia.

These results seem to exclude the use of jewellery items as a source of the metal of the gold leaf mosaic tesserae. In jewellery, gold alloys containing copper and silver were mainly used: the gold content was between 90 and 96 % in the late Roman period [13] and below 91 % during the Byzantine period [4]. Instead, these results indicate marked similarities with gold coins circulating in the considered period. Analyses of gold coins circulating between the 1st and 9th centuries are published and both the period and the mint of coinage are well established for ancient gold coins.

Coins are almost invariably amenable to close dating because they are usually inscribed with the name of a known ruler. As illustrated in figure 3, which is based on the measurements performed by Callu et al. with several analytical techniques [14], in the period 63-253 AD the silver content never exceeded 1 %. Between 254 and 261 (Valeriano) the silver content increased to 7.8 %. This level decreased to 2.9 % during 263-283 (Claudio) and later under 1 %. Since 317 until 345 the silver content is between 2 - 3 %, and between 346 until 368 it increased up to 5.1 %, then it decreased again to less than 1 %.

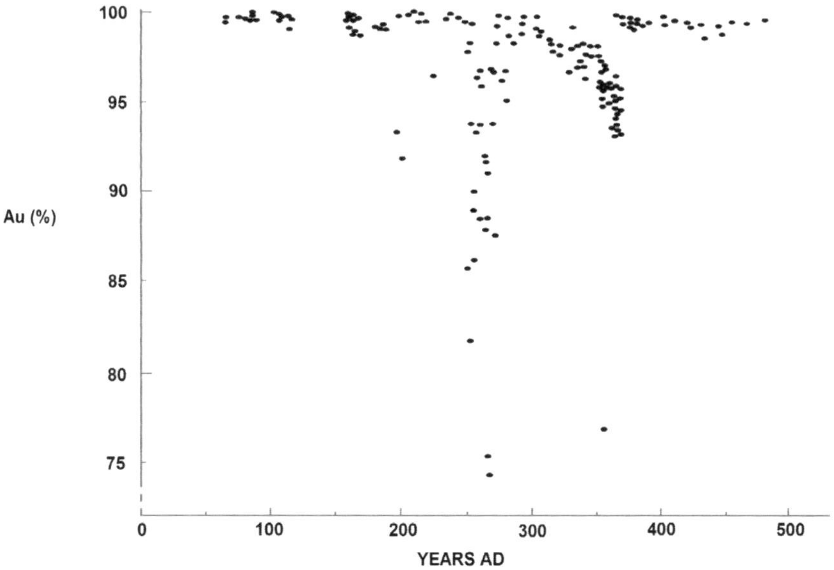

Figure 3. Composition of gold coins circulating in the early centuries from Callu et al. [14].

A contribution arises from the analysis of the gold leaf to distinguish reused tesserae recovered by dismantling ancient mosaics. Reuse has always been a common practice, owing to the high cost of mosaic tesserae. It is difficult to state from analytical data whether a tessera was expressly made for a given mosaic or is a reused one, because the same natron glass has been used over very long periods of time. One can observe that tesserae with a gold leaf of different composition were used for certain mosaics. As an example, the chemical analyses of four tesserae from the 5th century mosaic of St. Sabina in Rome are reported in table 2.

Table 2. Glass (wt% of oxides) and gold leaf (wt% of the elements) composition of tesserae from the 5th century mosaic of St. Sabina, Rome.

	SiO_2	Al_2O_3	Na_2O	K_2O	CaO	MgO	SO_3	P_2O_5	Cl	TiO_2	Fe_2O_3	MnO	Sb2O3	Au	Ag
SSA-E(sc) colourless	68.8	2.05	19.0	0.44	5.8	1.00	0.22	0.12	1.15	0.08	0.57	0.53	0.20	99.6	0.4
SSA-D(sc) colourless	68.7	2.11	19.6	0.64	5.9	0.54	0.28	0.16	1.08	0.05	0.39	0.13	0.42	92.5	7.5
SSA-G(sc) amber yellow	67.4	1.84	19.9	0.43	6.7	0.74	0.29	0.14	1.15	0.08	0.53	0.61	0.20	99.3	0.7
SSA-F(sc) green	69.1	2.89	15.8	0.61	7.9	0.89	0.16	0.21	0.89	0.06	0.52	0.93		99.6	0.4

Important differences in glass composition are found only for the SSA-F tessera, showing a larger content of sodium ad calcium and no antimony. The gold leaf analysis clearly shows that the SSA-D tessera, similar to the SSA-E both for glass decolouration technique (manganese and antimony) and composition, has a different origin as demonstrated by the different silver content (Ag 7.5 % for SSA-D, only 0.4 - 0.7 % for SSA-E and the other tesserae).

4. Conclusions

An experimental method for the analysis of the gold leaves of glass mosaic tesserae by energy-dispersive X-ray microanalysis was set up. Lower limits of detection of 0.1 % and 0.2 % were calculated for silver and copper, respectively.

Analyses carried out on more than thirty mosaic tesserae of the period 1st to 9th centuries demonstrate that the leaves were made of pure gold or gold-silver alloys in a restricted range of composition, with a silver content less than 9 %. These data strongly suggest that circulating gold coins were beaten to make the gold foils.

In conclusion, these results seem to indicate that a comparison between the composition of the gold leaves and that of the coins circulating in different times and areas will be possible in order to obtain additional information on the dating and provenance of this kind of mosaic tesserae. In particular, comparison of compositional data can help with the identification of reused tesserae, that is, tesserae recovered from earlier mosaics or of different provenance. Further research on this subject is in progress by the authors.

References

[1] Bartoli M, Cousi V and Felici F 2008 Il mosaico parietale del ninfeo di Lucullo sotto il palazzo nuovo della biblioteca hertziana a Roma. Nuove acquisizioni e progetto di restauro. *Proc. XIV Colloquio AISCOM (Spoleto, 2008)* pp. 509-519

[2] Verità M 2000 Technology and deterioration of vitreous mosaic tesserae. *Rev. Conservation* **1** 65-76

[3] Verità M, Basso R, Wypyski M T and Koestler R 1994 X-ray microanalysis of ancient glassy materials: a comparative study of wavelength dispersive and energy dispersive techniques. *Archaeometry* **36** 241-51

[4] Oddy W A and Niece S L 1986 Byzantine gold coins and jewellery. A study of gold content. *Gold Bullettin* **19** 19-27

[5] Freestone I C 2005 The Provenance of ancient glass through compositional analysis. *Mater. Res. Soc. Symp. Proc.* **852** 008.1.1-14

[6] Conventi A, Neri E and Verità M 2010 Glass mosaic tesserae from the 5th-6th century baptistery of San Giovanni alle Fonti, Milan, Italy. Analytical investigations. *Proc. Conference: the New Light on Old Glass: Byzantine Glass and Mosaics. (London, 27-29 May 2010)* ed. L James; in press

[7] Carbonara E, Muscolino C and Tedeschi C 1999 La luce del mosaico: le tessere d'oro di Ravenna. Tecniche di fabbricazione e utilizzo. *Proc. VI Colloquio AISCOM (Venezia)* pp. 709-16

[8] Jackson C 2005 Making colourless glass in the Roman Period. *Archaeometry* **47** 763-80

[9] Verità M 2010 Indagini analitiche degli smalti dell'altare di Sant'Ambrogio a Milano: transizione verso una nuova tecnologia vetraria nel IX secolo. *Proc. National A.I.Ar. Conference 2009: Riflessioni e trasparenze. Diagnosi e conservazione di manufatti vetrosi (Ravenna 2009)* ed. M Vandini (Bologna: Patron) pp. 13-26

[10] Merzenich C 1996 Dorature e policromie delle parti architettoniche nelle tavole d'altare toscane fra Trecento e Quattrocento. *Kermes*, Dossier IX **26** 51-71

[11] Travaini L 2005 Coins, gold-beaters and painters. How gold was used in wall paintings: some exemples from the Scrovegni chapel. in: *Giotto in the Scrovegni Chapel: materials used in the painting technique. Studies and research by Istituto centrale per il Restauro.* ed. G Basile. *Bollettino d'Arte*, special volume, 145-52

[12] Oddy W A 1988 The debasement of the provincial Bizantine gold coniage from the seventh to ninth centuries. in: Studies in early Byzantine gold coinage. eds. W Hahn and W Metcalf. *ANS Numismatic Studies* **17** 135-42

[13] Yeroulanou A 1999 *Diatrita:gold pierced-work jewellwry from the 3th to the 7th century.* (Athens, Benaki Museum)

[14] Callu J P, Brenot C, Barrandon J N and Poirier J 1985 Aureus obryziacus. in: *Or monnayé.* C E Babelou. pp. 81-111

[15] Verità M 2006 Tessere vitree a foglia d'oro nei mosaici di Aquileia. *Quaderni Friulani di Archeologia* **XVI** 7-12

[16] Verità M, Arena M S, Carruba A M and Santopadre P 2009 Materiali vitrei nell'opus sectile di Porta Marina (Ostia antica). *Bollettino ICR* 78-94

[17] Verità M 2003 Analisi delle tessere musive. in: *Il mosaico di Santa Pudenziana a Roma, il restauro, a cura di V. Tiberia.* (Perugia: Ediart) pp. 178-199.

[18] Verità M, Profilo B and Vallotto M 2002 I mosaici della Basilica dei Santi Cosma e Damiano a Roma: studio analitico delle tessere vitree. *Riv. Stazione Sperimentale del Vetro* **32** no. 5 13-24

[19] Verità M and Santopadre P 1993 Le analisi delle tessere musive vitree: esempi d' applicazione nelle indagini preliminari al restauro del mosaico absidale della Cappella dei Santi Primo e Feliciano in S. Stefano Rotondo a Roma. *Arte Medioevale, II serie* **VII**, no. 1, 197-228

[20] Verità M and Santopadre P 2009 Mosaico absidale della basilica di Santa Cecilia a Roma. Studio delle tessere vitree. in: *Restauri a Santa Cecilia, a cura di Daila Radeglia.* (Florence: Edifir) pp. 263-72

IOP Conf. Series: Materials Science and Engineering **32** (2012) 012008 doi:10.1088/1757-899X/32/1/012008

Smart pile-up consideration for evaluation of high count rate EDS spectra

F Eggert[1,3], T Elam[2], R Anderhalt[1] and J Nicolosi[1]
[1] Edax Inc., Ametek Materials Analysis Division, 91 McKee Drive, US-07430 Mahwah, New Jersey, U.S.A.
[2] now at: University of Washington, Applied Physics Laboratory, Seattle, US-98105-6698 Washington, U.S.A.

E-mail: frank.eggert@ametek.de

Abstract. This work describes a new pile-up consideration for the very high count rate spectra which are possible to acquire with silicon drift detector (SDD) technology. Pile-up effects are the major and still remaining challenge with the use of SDD for EDS in scanning electron microscopes (SEM) with ultra thin windows for soft X-ray detection. The ability to increase the count rates up to a factor of 100 compared with conventional Si(Li) detectors, comes with the problem that the pile-up recognition (pile-up rejection) in pulse processors is not able to improve by the same order of magnitude, just only with a factor of about 3. Therefore, it is common that spectra will show significant pile-up effects if count rates of more than 10000 counts per second (10 kcps) are used. These false counts affect both automatic qualitative analysis and quantitative evaluation of the spectra. The new idea is to use additional inputs for pile-up calculation to shift the applicability towards very high count rates of up to 200 kcps and more, which can be easily acquired with the SDD. The additional input is the 'known' (estimated) background distribution, calculated iteratively during all automated qualitative or quantitative evaluations. This additional knowledge gives the opportunity for self adjustment of the pile-up calculation parameters and avoids over-corrections which challenge the evaluation as well as the pile-up artefacts themselves. With the proposed method the pile-up correction is no longer a 'correction' but an integral part of all spectra evaluation steps. Examples for the application are given with evaluation of very high count rate spectra.

1. Introduction

The use of silicon drift detectors (SDD) in electron microscopes comes with the ability to acquire spectra with very high count rates. But these high count rate spectra are the major challenge with this quite new detector technology. Energy-dispersive X-ray spectrometers (EDS) are always equipped with a pulse processor to perform the measurement of an individual detected X-ray photon energy. Both types, analogue and digital pulse processors, use a spectrometric channel with optimized parameters to achieve best energy resolution. There is always a need for a processing time (shaping time) to measure the photon energy. It is required that during this time period no other detected X-ray should influence the measuring process to avoid measurement mistakes. Therefore, a fast channel is used which is optimized more for time resolution to recognize any double or multiple events which may occur during a single photon processing time. These multiple event measurements will be

[3] To whom any correspondence should be addressed.

excluded from collection and will not be included in the spectrum. This is pile-up rejection. The limit with pile-up rejection is reached if the X-rays arrive in a shorter period than even the fast channel is able to resolve. These events in the spectrum may be called sum-peaks (two or more X-rays are detected at the same time). But there is also another limit: good time resolution comes with poor energy resolution. Therefore, the energy threshold for the fast channel is set much higher than for spectrometric processing to avoid triggering with noise events which would finally lead to rejection of all event measurements. The consequence of the higher threshold is blindness for very soft X-rays in the fast channel and, therefore, the pile-up rejection is not effective for those soft energy counts, also if the coincidence is together with high energy X-rays. Modern digital pulse processors are equipped with several fast channels which are optimized for different energy regions to keep the number of incorrect measured pulses as low as possible.

The use of very short shaping times with SDD also brings benefits in pulse processor pile-up rejection. But this improvement is only about a factor of three. This means that, if clean spectra were obtained with 3 kcps using a conventional Si(Li) detector, SDD should not be used at count rates any higher than about 10 kcps. Nevertheless the practice is to use them with really huge count rates, up to 100 kcps and more. But these spectra, which are acquired beyond these limits, have many artefacts and are often called 'dirty spectra' from a point of view of normal spectral evaluation.

Figure 1 demonstrates the readily visible influence of pile-up process in all spectral parts. Qualitative and quantitative evaluation procedures are always faced with pile-up artefacts and often produce bad and even wrong results. Sum peaks are detected as characteristic element lines and cause 'false-positives' in element identification. The Bremsstrahlung background is affected by smeared pile-up components. All quantitative results are influenced by count losses of major lines and pile-up overlays with small lines of trace specimen components. Therefore, qualitative and quantitative results depend on count rate, if no correction is applied.

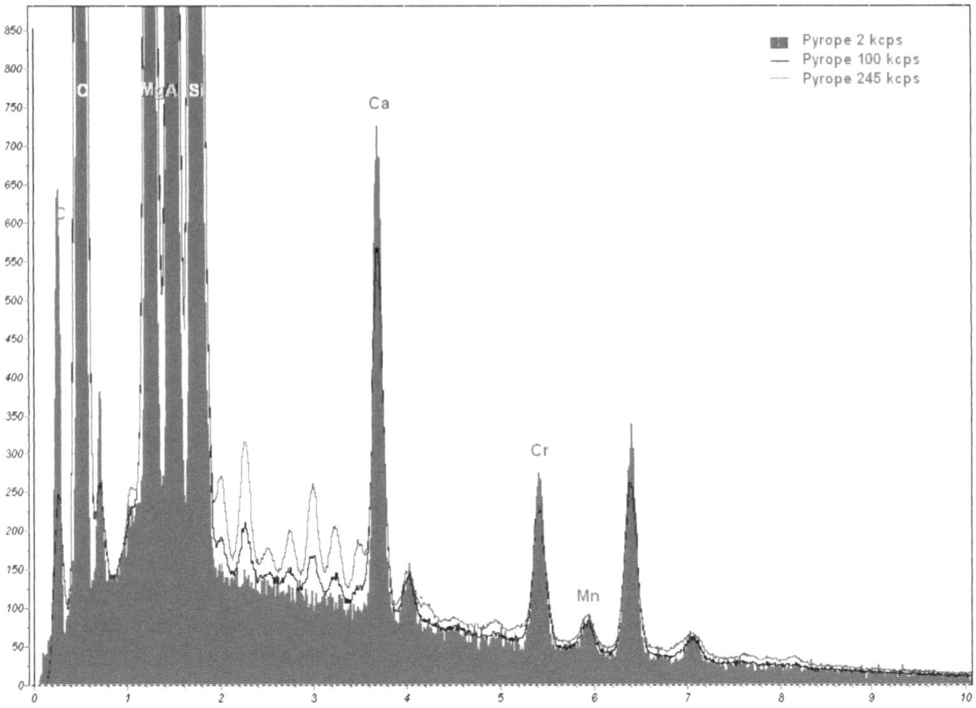

Figure 1. Measured spectra of same Pyrope specimen with 2, 100 and 245 kcps in comparison.

The pile-up correction topic is the subject of wide discussion [1-3]. The best results will be achieved with consistent optimizing of the pulse processor with regard to the post processing of remaining pile-up events [2].

2. Known solution

A straightforward correction can be applied immediately after each spectrum acquisition [1, 2]. The total count rate in combination with all known pulse processor parameters are used to calculate all remaining spectral pile-up artefacts with probability based models. This pile-up distribution will be subtracted from the raw spectrum. However, there are limits in using this approach, because all parameters have uncertainties, and because of the limits due to statistical fluctuations. Sometimes artificial over-corrections are visible in a spectrum, which also cause problems in spectra evaluation (see figure 2). An intimate knowledge of the pulse processor operation is required with all of its important parameters. But the high absolute stochastic fluctuations of sum peaks will be directly coupled into the remaining background area which has lower noise level. Therefore, this kind of pile-up rejection usually works up to about 40 … 50 kcps without remaining artefacts.

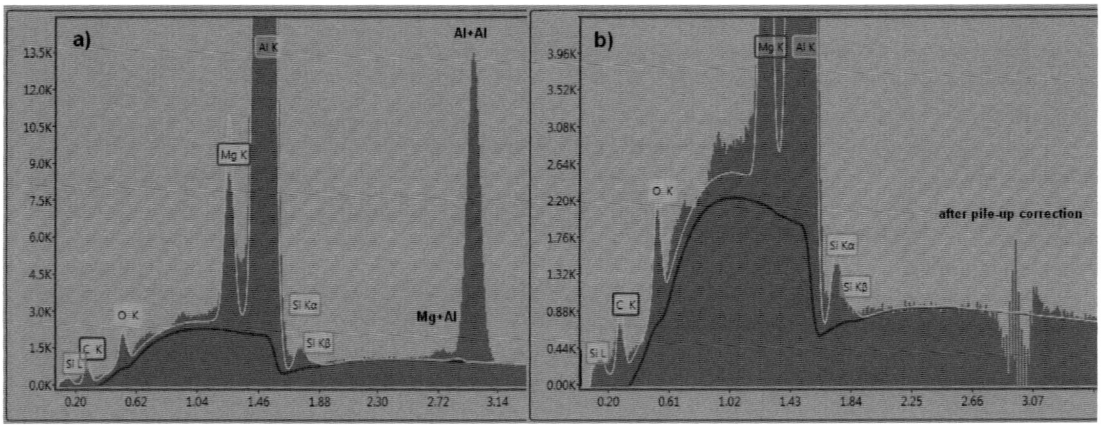

Figure 2. a) Spectrum of Aluminium dominated specimen acquired with 108 kcps. b) Same spectrum but after pile-up correction with statistical fluctuations in corrected area. The residual artefacts are due to a large Al-sum peak subtraction (more than 10 times higher than the background level).

The basic theory is that the probability of an un-rejected pile-up pulse P^{pu}_{C12} (event in channel $C1+C2$) is calculated in the following way:

$$P^{pu}_{C12} = P_{C1} \times P_{C2} \times CR \times PPR \tag{1}$$

where P_{C1} is the probability of an event in energy channel $C1$ considering all X-ray counts being measured, P_{C2} is the same for events in channel $C2$, ***CR*** is count rate, and ***PPR*** is the average pulse pair resolving time (determined by the fast channel). Therefore, ***CR*** ***x*** ***PPR*** describes the probability that another measured X-ray quantum is detected before the fast channel is open to recognize additional events. During the ***PPR*** even the fast channel is not able to resolve events. These counts are very close to each other with regard to the spectrometric channel processing time and, therefore, produce sum channel (energy) events. If the parents of a sum event are from two characteristic lines, then this is visible as a sum-peak. But Bremsstrahlung background X-rays can also produce sum events, either with each other or with characteristic line counts. This leads to a continuum and small structured part of the pile-up distribution. The calculation of un-rejected pile-up counts $\boldsymbol{N}^{pu}_{C12}$, which are measured in channel $\boldsymbol{C}1+\boldsymbol{C}2$, is then the multiplication with the total spectrum counts $\boldsymbol{TC}$:

$$N_{C12}^{pu} = P_{C1} \times P_{C2} \times CR \times PPR \times TC \tag{2}$$

In practical terms the equation can be transformed using the acquired spectrum for the probability distribution ($\boldsymbol{N}_{C1}$ is the channel content in channel $\boldsymbol{C}1$):

$$N_{C12}^{pu} = N_{C1} / TC \times N_{C2} / TC \times CR \times PPR \times TC$$

$$N_{C12}^{pu} = N_{C1} \times N_{C2} \times CR \times PPR / TC \tag{3}$$

The last part is a scalar value equal for all channels. $\boldsymbol{N}_{C1}$ and $\boldsymbol{N}_{C2}$ are actually measured counts in channel $\boldsymbol{C}1$ and $\boldsymbol{C}2$ of the spectrum. Therefore, the calculation of $\boldsymbol{N}^{pu}_{Cik}$ for each channel (energy) combination is a double loop over all spectrum channels i and k. The calculated counts are added to the channel they come from and subtracted from the wrong positions. Following this approach, triple events can also be predicted.

But there must be a second part of the model which describes the probability that a soft X-ray is detected during the processing time of another count. The fast channel is blind if the energy of the photon is below the recognition threshold. Now the pulse pair resolving time of the spectrometric channel is the relevant resolving time, which is usually much higher than for the fast channel. The remaining unresolved events in the fast channel are very close in time, compared to the spectrometric processing time. Therefore, it is valid to assume exactly the sum of energies with single unresolved events. Energy variations (small losses) in the sum due to a small time shift can be neglected but cause small shifts and altered shapes of sum peaks, making the pile-up processes more complex. There is a special case for soft X-rays, because all pile-up events during the processing time are not recognized. The question, how often they lead to pile-up, must be solved. The different time shifts cannot be ignored, because all shifts up to the main channel processing time will occur. Resulting energies of these pile-up events are somewhere between the higher energy parent and the sum energy. This is classically named 'continuous pile-up' which should not be confused with the basic sum pile-up with continuous radiation events.

There are limits to the pile-up correction practicality because of the stochastic nature of all X-ray emission processes, the simplifications made in the probability based model, and also the knowledge of all important parameters. For example the pulse pair resolving time can depend on the energies and the energy recognition threshold for pile-up rejection is not a sharp value, because several fast channels are used. This explains the difficulties with the correction as demonstrated in figure 2.

3. The benefits of additional inputs for pile-up correction

The new idea is to get beyond the limits of the applicability of pile-up effect calculations by using additional inputs to avoid over-corrections, oscillations and other difficulties. A '*Smart Pile-up Consideration*' is introduced for the first time, which fulfils the task. The main innovation is to avoid an explicit correction after spectra acquisition, but to consider the pile-up effect internally in all qualitative and quantitative iteration loops. That makes it possible to adjust the pile-up content calculation iteratively and with additional consideration of calculated background. It is possible to adjust the pile-up content estimation (the pile-up distribution) in the spectrum automatically with this additional knowledge. The major parameters of the model are the mean pulse-pair resolving time (PPR) and a mean energy threshold for event recognitions in the fast channel pile-up rejection logic. The pile-up probability can be calculated with these values following (eq. 3). The basic assumption is that the probability for a pile-up event does not depend on the energy of the involved X-rays (except the pile-up rejection blindness for soft energies). Therefore, the entire pile-up probability (the considered pile-up distribution) is self adjustable due to inspection and evaluation of spectra regions with major pile-up peaks. Over-corrections are caught this way by considering the calculated background. Therefore, the background calculation and pile-up distribution estimation are interdependent. But the pile-up estimation improves successively with the iterations of the calculated background. The introduced method is an integrated part of iterative spectra evaluation and makes it possible to evaluate 'dirty spectra' with count rates of 150 kcps and even higher, without significant changes in analytical results. The correction will never actually be done in practice which means that the original spectrum will be unmodified but the artefacts will not be misinterpreted.

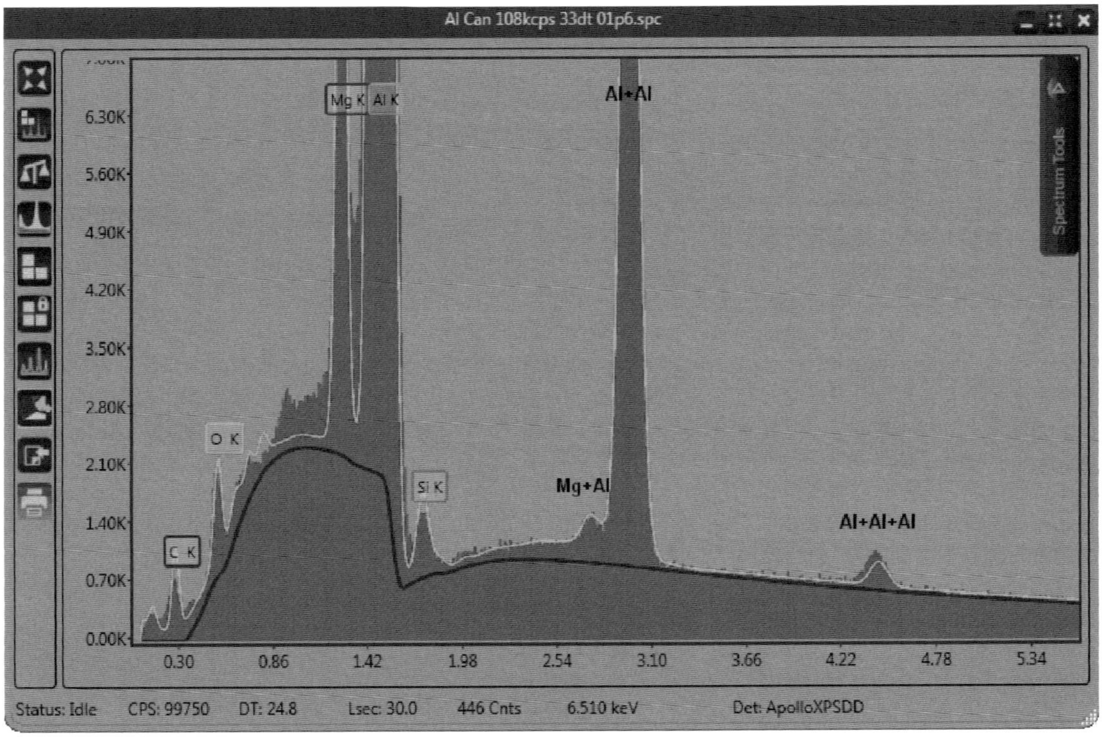

Figure 3. The same spectrum as in figure 2 (108 kcps count rate) without application of pile-up correction but evaluated with *Smart Pile-up Consideration*. The reconstructed spectrum includes also the calculated pile-up distribution and gives an excellent fit with sum peaks and a reasonable fit with the triple Al pile-up. Also the pile-up contributions of Bremsstrahlung radiation are visible in comparison to the calculated background (blue line).

The pile-up distribution is always recalculated with each new background calculation. There is a choice to display the reconstructed spectrum without the pile-up distribution as in figure 4a. Then the reconstruction is as close as possible to the 'clean' spectrum without pile-up. Figure 4b demonstrates the reconstruction with pile-up distribution included. Now the reconstruction is as close as possible to the real acquired spectrum for better comparison.

Because it is never a physical spectrum correction artificial structure issues never occur in a residual spectrum which would make the correction process questionable. *Smart Pile-up Consideration* tries to consider the pile-up spectrum with each evaluation loop. Any consideration is better than doing nothing to avoid false positives with automated element identification and poor

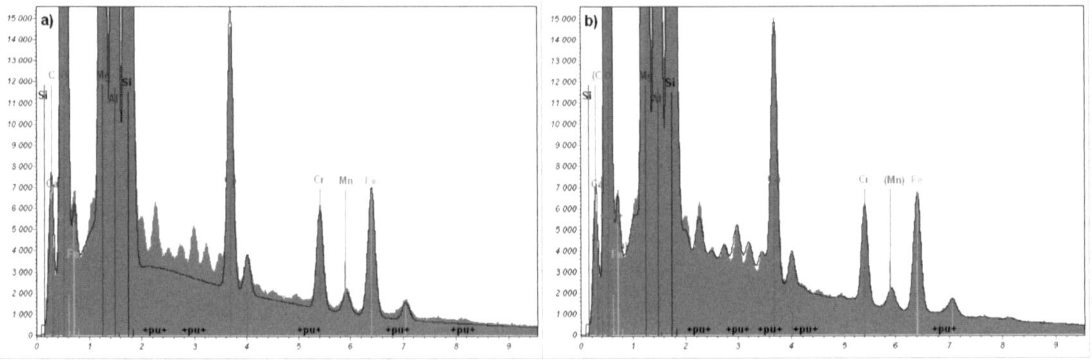

Figure 4. Pyrope spectra evaluation with included *Smart Pile Up Consideration* method, without pile-up in blue line reconstruction (a) and with added the internally used pile-up distribution (b). The pyrope spectrum was acquired with 143 kcps count rate.

quantitative results. Nevertheless, the additional input of the calculated background estimation makes possible not only the self-adjustment of the scalar parameter product value in eq. (3), but is also able to avoid channel-wise subtractions of pile-up counts (stochastic influenced) that are significantly below the background level.

Figure 5 shows the calculated pile-up distribution in comparison to the acquired spectrum (a) and also the internally corrected spectrum (b). The flat regions are influenced by bad overcorrection due to the known Bremsstrahlung background level, which cannot undercut a statistically reasonable value. However, it is much better to evaluate the blue line spectrum in figure 5b despite the artificial parts in the spectrum, than to deal with the original acquired spectrum, which is overloaded by many pile-up events (the difference between the red and blue spectra).

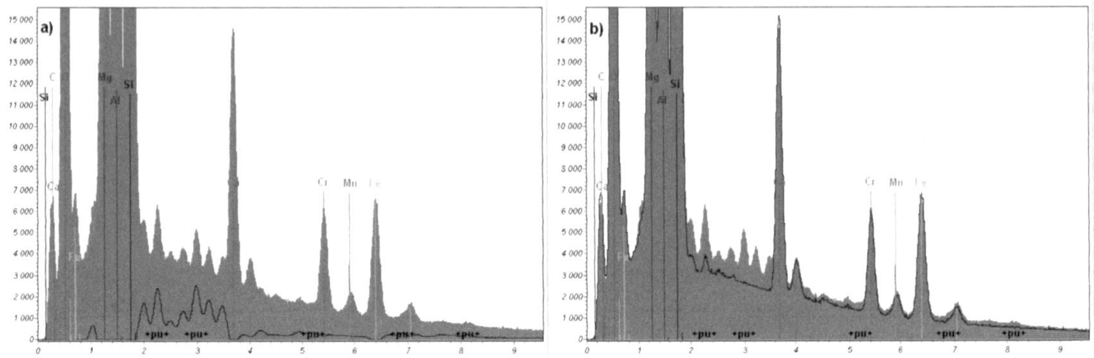

Figure 5. The same spectrum as in figure 4 in comparison to the calculated pile-up distribution (a) and the real corrected spectrum (b), which is used in a hidden manner for all evaluations. Both compared spectra are recalculated with each new background fit and are usually not visible during qualitative and quantitative spectra evaluations.

The additional input of background knowledge into the pile-up correction makes it possible to evaluate spectra with very high count rates, beyond the practical limits of usual pile-up correction models. The background and other quantitative calculations are already an iteratively processed and integrated part in modern computing environments even for qualitative EDS spectra evaluations, like with EXpertID [4]. Therefore the additional consideration of the pile-up spectral distortions with evaluation of the background fits very well the entire spectra evaluation approach as an iteratively processed and integral part.

4. Application example for smart pile-up considerations for qualitative analysis

The goal of qualitative spectral evaluation is to determine which elements are present in the sample area. The EXpertID method [4] uses multiple spectrum reconstructions and quantitative evaluations to solve this task. Therefore, the results will be dependent on the count rate if pile-up contributions are not taken into account. This is because the spectral artefacts may be identified as characteristic element lines.

Figure 6a shows a mica spectrum without pile-up content (count rate: 4 kcps). Pile-up counts are large and easily visible with a count rate of 209 kcps (figure 6b). A qualitative evaluation with EXpertID would result with the additional elements Tl, Lu, Pr, Ce, Sb, Ag, P, Sr and Kr, all false positives. But the evaluation with *Smart Pile-up Consideration* has similar results (figure 6b) to the evaluation of the 4 kcps spectrum (figure 6a) with regard to element list and spectrum reconstruction. A range between very small count rates and 200 kcps is applicable with only one setting of the parameters for pile-up consideration, due to self adjustment. There is no significant influence to automatic qualitative evaluation results even though the count rates are extremely different.

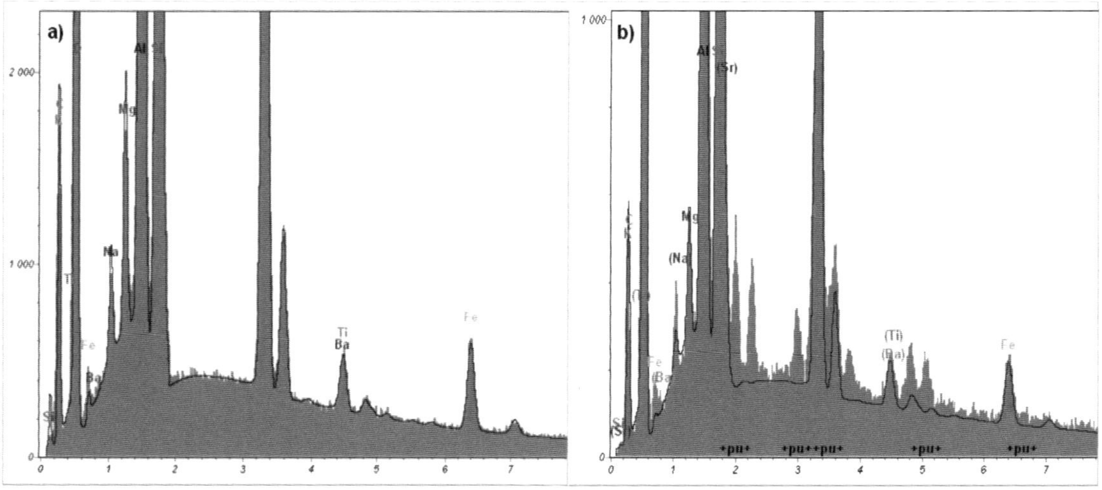

Figure 6. Results of automated qualitative analysis (EXpertID) with two spectra of a mica specimen using different count rates, 4 kcps (a) and 209 kcps (b). The qualitative evaluation result and reconstruction (blue line) of the spectra are similar, despite the big difference due to pile-up content. The pile-up distribution is not included into 209 kcps spectrum reconstruction (for this see figure 7, where it is displayed included). The spectra regions with big pile-up contents are marked with +pu+.

5. Application example for smart pile-up considerations for quantitative analysis

The mica spectrum acquired at very high count rates demonstrates that the pile-up effects will influence the quantitative results if these effects are not considered. For illustration the elements S and Ca were included additionally into evaluation, despite the fact that they are not present in the specimen (and not found by EXpertID). Not only do the pile-up peaks influence the results for elements which are overlapped, but the background is also increased by about 10 % (2.5 and 4 keV) and even 20 % (8 keV). This influences significantly the weight fractions calculation using P/B methods. The count rate losses for Al and Si are much more visible with non-normalized results. There are spectral regions which have a net loss of counts, and other regions, where the pile-up effects result in important additional components which are superimposed on the true spectrum (see figure 7).

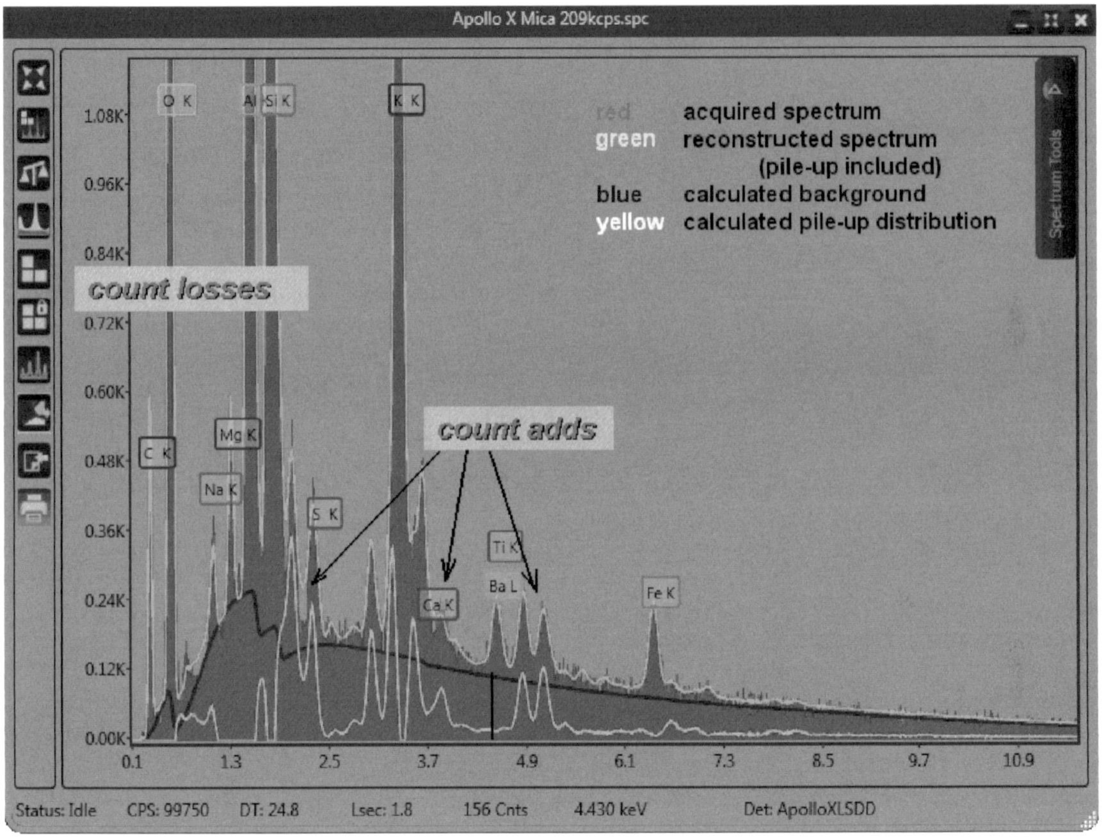

Figure 7. Mica spectrum with 209 kcps count rate. In addition to the original spectrum are displayed the reconstructed spectrum (cyan), the background (blue) and the considered pile-up distribution (yellow) for comparison. Spectral regions are marked which have a loss of counts to other areas where the pile-up causes additional counts.

When *Smart Pile-up Consideration* is applied to the 209 kcps spectrum the quantitative results are not significantly modified by these pile-up effects. Even the null-test elements S and Ca are quantified near or below the MDL despite the huge pile-up overlaps (see figure 8). Table 1 demonstrates selected results for 4 kcps in comparison to 209 kcps spectrum without and with applying the proposed method. The result column is highlighted, where pile-up was not considered. The elements Al and Si are underestimated because of the count losses. S and Ca are detected above the detection limit. Ti and Ba do overlap and the deconvolution is much influenced by the pile-up counts. Therefore, the Ba content is about 4 times higher than it should be and Ti seems to be not in specimen. Both are big analytical mistakes.

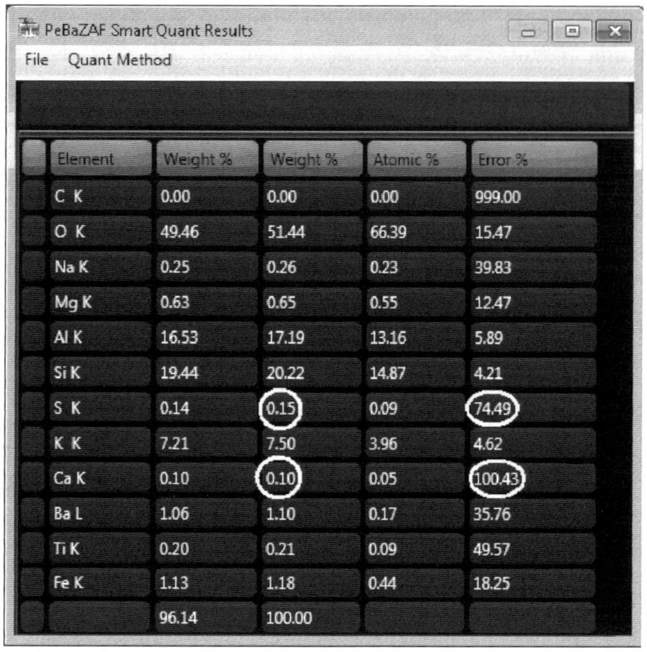

Element	Weight %	Weight %	Atomic %	Error %
C K	0.00	0.00	0.00	999.00
O K	49.46	51.44	66.39	15.47
Na K	0.25	0.26	0.23	39.83
Mg K	0.63	0.65	0.55	12.47
Al K	16.53	17.19	13.16	5.89
Si K	19.44	20.22	14.87	4.21
S K	0.14	0.15	0.09	74.49
K K	7.21	7.50	3.96	4.62
Ca K	0.10	0.10	0.05	100.43
Ba L	1.06	1.10	0.17	35.76
Ti K	0.20	0.21	0.09	49.57
Fe K	1.13	1.18	0.44	18.25
	96.14	100.00		

Figure 8. Standardless quantitative results with mica 209 kcps spectrum with *Smart Pile-Up Consideration* applied. The elements S and Ca are not in specimen and are calculated near the detection limit, despite big pile ups would lead to other results (see figure 7 and table 1). All results compare close to the evaluation of the 4 kcps spectrum without pile-up.

Table 1. Selected quantitative results for Mica 4 kcps spectrum without pile-up contents, Mica 209 kcps spectrum without and with pile-up consideration calculation during quantitative evaluation. The weight fractions are calculated with a standardless P/B-model [5].

element line	wt%	*rel. error %*	wt%	*rel. error %*	wt%	*rel. error %*
	4 kcps		**209 kcps**		**209 kcps**	
			no pile-up consideration		**with pile-up consideration**	
Al K	**17.43**	*5.5*	16.05	*5.7*	**17.19**	*5.8*
Si K	**21.09**	*3.7*	19.2	*4*	**20.22**	*4.2*
S K	**0.02**	*100*	0.69	*11*	**0.15**	*75*
Ca K	**0.06**	*72*	0.42	*20*	**0.1**	*100*
Ti K	**0.32**	*18*	0.07	*88*	**0.21**	*50*
Ba L	**0.64**	*25*	2.57	*14*	**1.1**	*36*

All of the quantitative mistakes with 209 kcps spectrum evaluation compared to the 4 kcps spectrum are significantly reduced using the pile-up consideration (last two columns in table 1). S and Ca are below or near detection limits again (note the errors). Al and Si are near the target values, when the reported uncertainties are considered. Also Ba and Ti results are significantly improved. Note the statistical influences of the pile-up counts and systematic errors due to corrections affect the result uncertainties. This is calculated considering the pile-up counts which were subtracted during internal spectrum correction. The error is about three times higher for Ti. Indeed it is still the better choice to avoid pile-up effects if possible. But often the spectra are already acquired in a spectral mapping measurement with no opportunity to repeat at the same specimen point. Now these spectra are possible to evaluate, almost like there was no pile-up problem in spectra acquired with SDD detectors at very high count rates.

6. Conclusions

One of the most interesting benefits of SDDs with high count rate capability was already published even with the very first publication about application of SDD technology for electron microscopes [6]: "Originally designed as position sensitive detectors for particle tracking, silicon drift detectors are nowadays used for high count rate X-ray spectroscopy, operating close to room temperature". An element map was presented which had been collected at 250 kcps. But those detectors were still equipped with a beryllium window and not yet applicable for quantitative spectral analysis. One of the reasons was the unsolved pile-up problem with these very high count rates. The proposed method with a smart pile-up consideration may be the first attempt to solve the pile-up problem entirely for extreme count rate application in scanning electron microscopes (SEM) and even includes the low energy X-ray pile-up rejection problem of pulse processors with usual ultra thin detector windows. The remaining problem with SDD technology application in SEM has been solved. The requirements are a reliably working pulse processor which does not produce additional artefacts and a specimen homogeneous in the scanned specimen surface area. The only theoretical limit is random counting statistics. Using the introduced method, the spectrum evaluation gives analytically reliable qualitative and quantitative results almost independently of the count rate of the acquired spectrum.

References

[1] Statham P J 2006 *Microchimica Acta* **155** 289
[2] Elam T, Anderhalt R, Sandborg A, Nicolosi J and Redfern D 2008 *Microsc. Microanal.* **14** (Suppl. 2) 1260
[3] Mott R B 2008 *Microsc. Microanal.* **14** (Suppl. 2) 1180
[4] Eggert F 2010 *IOP Conf. Ser.: Mater. Sci. Eng.* **7** 012007
[5] Eggert F 2006 *Microchim Acta* **155** 129
[6] Strüder L, Meidinger N, Stötter D, Kemmer J, Lechner P, Leutenegger P, Soltau H, Eggert F, Rohde M and Schülein T 1998 *Microsc. Microanal.* **4** 622

IOP Conf. Series: Materials Science and Engineering 32 (2012) 012009 doi:10.1088/1757-899X/32/1/012009

Contribution of EPMA to airborne pollen analysis

F Guimarães[1,4], L Duque[2], H Ribeiro[2], R Sousa[2] and I Abreu[2,3]
[1] National Laboratory of Energy and Geology, Rue da Amieira, P.O. Box 1089, PT-4466-956 S. Mamede Infesta, Portugal
[2] University of Porto, Geology Centre, Environment and Society Group, Porto, Portugal
[3] University of Porto, Faculty of Sciences, Biology Department, Porto, Portugal

E-mail: fernanda.guimaraes@lneg.pt

Abstract. As a component of aerosol, pollen is found in suspension with other mineralogical and anthropogenic particles that can adhere to the pollen wall. The aim of this study was to determine possible alterations that pollen grains suffer under different meteorological conditions in the coastal city of Porto, Portugal. For this study, 2 airborne pollen types were taken into account: Poaceae and *Alnus* spp sampled in July 2010 and February 2011, respectively. Quantitative analysis and X-ray map analysis were performed with a field emission electron probe microanalyzer (EPMA) equipped with 5 WDS spectrometers. Samples were mounted over adhesive carbon tape and carbon coated. Control pollen showed mainly the presence of C, O, N, P, K, Mg, S, Cl and Ca in different quantities, depending on the considered species. Airborne pollen consistently revealed a higher content of Mg and Cl than control pollen, while Si, Ca, K and P contents varied differently according to the samples. Our results show that during their "flight", pollen grains acquire an external coating, become heavier and change significantly their composition.

1. Introduction

Aerosol is a combination of liquid droplets and small solid particles, suspended in a gas or a mixture of gases. Particles from aerosol, also called particulate matter, have different sources, sizes and compositions. Sea-salt, mineral dust, particles derived from carbonaceous combustion, and biological particles (e.g., pollen grains) are some examples of particles that make up atmospheric aerosol [1].

Types, sources, chemical composition and size range are the variables commonly studied by works that focus particulate matter [1, 2]. Among these studies some draw attention to the study of biological airborne material namely pollen.

The pollen grain is the male gametophyte of seed plants, being part of the sexual reproduction cycle. They are biological particles that can be dispersed by wind along several kilometres and consequently can be found in suspension in the atmosphere with natural and anthropogenic pollutants. Pollen grains are a seasonal air pollutant and their presence and amount in the atmosphere depends on a wide range of factors including meteorological (temperature, rain, humidity, wind, etc.), biological (physiological state of plants, plants distribution, etc.), and topographical issues.

Several techniques, such as SEM-EDS [3, 4] and Raman spectroscopy [5, 6], have been employed to characterize pollen wall composition as well as adhered particles. Also, EPMA was used by

[4] To whom any correspondence should be addressed.

Cerceau-Larrival et al [7, 8] to characterize pollen and study the influence of pollution on its composition. Presently, no WDS quantitative microanalysis study has been performed on pollen directly sampled from the atmosphere.

A combined Monte Carlo simulation and EPMA analysis method was successfully applied to the analysis of aerosol particles, in particular to the analysis of light elements [9]. However, the low density of pollen grains makes their analyses more difficult.

The aim of this study was to determine possible alterations that pollen grains suffer under different meteorological conditions in the coastal city of Porto, Portugal, using EPMA.

2. Methods

This study was conducted in Porto which is the second largest city of Portugal. This city is limited on the west by the Atlantic Ocean and on the south by the Douro River. The annual average temperature is around 15 °C, annual air humidity is between 75 % and 80 %, and the total annual mean precipitation varies between 1000 mm and 1200 mm. Prevailing winds are from the W and NW in summer and from the E and SE in winter [10, 11].

Airborne pollen sampling was performed using a 7-day volumetric spore trap (Burkard Co. Ltd.) set on the roof of the Faculty of Sciences in Porto. This sampler is composed by a built-in vacuum pump that allows air sampling at 10 litres per minute (simulates human inhalation) and has a 2x14 mm intake orifice through which the sampled particles are impacted on a double-sided carbon tape supported on a clockwork-driven drum (rotating once every 7 days). A vane tail keeps the intake orifice facing the wind. After exposure the tape was cut into daily segments and mounted on slides.

Control pollen samples were collected in public gardens or sidewalks. In the laboratory, anthers were separated from extraneous materials, dried at 27 °C, gently crushed and the pollen thus released was passed through different grades of sieves and preserved at -20 °C.

For this study, two airborne pollen types were taken into account: Poaceae and *Alnus* spp, sampled in July 2010 and February 2011, respectively. Different atmospheric conditions were observed in these two days (table 1). The 6th of July was much hotter and less humid than the 3rd of February. Wind speed was higher on the 6th of July and wind direction was also different in the two days.

Table 1. Atmospheric conditions on the sampling dates.

	Temperature (°C)	Relative humidity (%)	Wind speed (m/s)	Wind direction
	Average		Maximum	
6 Jul 2010	28.83	47.52	7.03	W - N - E
3 Feb 2011	10.28	73.34	5.83	N - E

Qualitative, quantitative and intensity X-ray map analyses were performed with a Jeol JXA-8500F field emission electron probe microanalyzer (EPMA) equipped with 5 wavelength-dispersive X-ray spectrometers (WDS) and an Oxford INCA X-Act energy-dispersive X-ray spectrometer (EDS). To maintain similar conditions of analysis, control pollen was also mounted over adhesive carbon tape and carbon coated. Carbon coating thickness was kept constant in all samples. Since the density of biological material is around 1 [12] and analyzed pollen size ranges between 20 and 30 µm, EDS qualitative analyses were performed at 10 kV, 10 nA, during 15 s; WDS quantitative analyses were performed at 6 kV, 15 nA; X-ray maps were obtained on beam mode and performed at 6 kV, 25 nA and 25 ms dwell time. The accelerating voltages used on WDS analyses ensure that the volume of excitation caused by the beam is smaller than pollen size. In this work, we tried to avoid the irradiation of external particles on WDS analyses, by not analysing pollen grains with visible particles on the top of their surface, in order to focus on the pollen itself. The matrix correction method used was XPP. Chosen standards for element analyses were: N-BN; Cl-vanadinite; Na-albite; S-pyrite; O and Ca-calcite; P-apatite; Mg-MgO; Si-jadeite; K-orthoclase. Five spectrometers were employed for the analyses. The crystals used for measurements were: N and O- LDE1; Si- TAP; Cl and P-PETH; Ca, S and K- PETJ; Na and Mg- TAPH. Before analysis, a peak search routine was performed to determine the O peak-position, because standard chemical composition is different from pollen chemical composition.

3. Results

We started by acquiring EDS spectra for both control and airborne pollen, in order to ascertain which major and minor elements were present. Acquired spectra revealed variability in intensity peaks, which was small for control and large for airborne samples. Figure 1 compares one example of airborne Poaceae pollen spectrum to one of control Poaceae pollen. Figure 2 compares one example of control *Alnus* pollen with airborne *Alnus* pollen. Control pollen showed mainly the presence of C, O, N, P, K, Mg, S, Cl and Ca in different quantities, depending on the considered species. There are differences in composition between control and airborne pollen, being the most evident an increase in Mg- and Cl-contents, in both pollen types, and in Na and Al in airborne *Alnus* pollen.

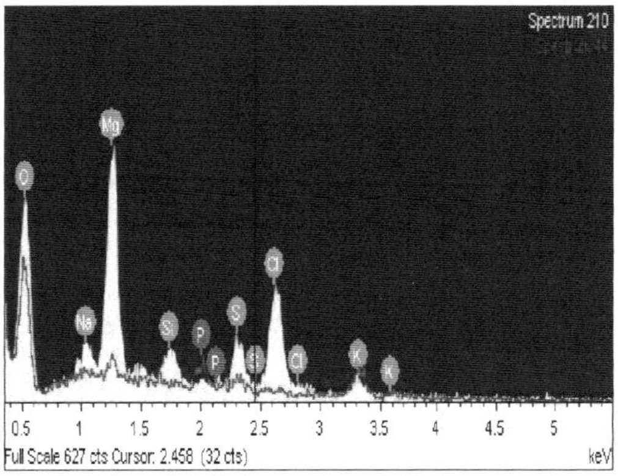

Figure 1. EDS spectra of airborne (yellow) and control (red) Poaceae pollen. Beam conditions: 10 kV, 10 nA.

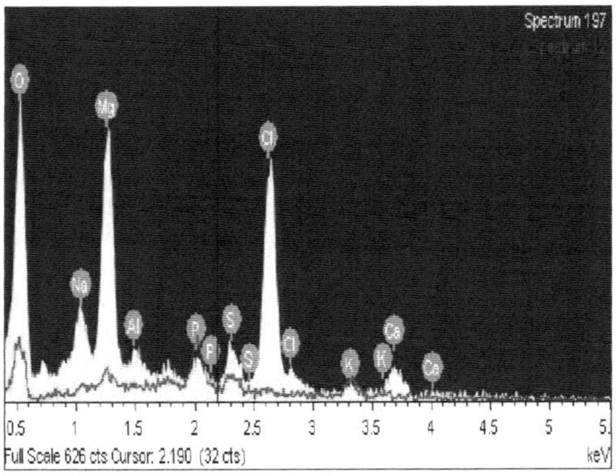

Figure 2. EDS spectra of airborne (yellow) and control (red) *Alnus* pollen. Beam conditions: 10 kV, 10 nA.

It was also possible to perform subsequent WDS quantitative analyses to a set of pollen grains, which allowed us to confirm these alterations (figures 3 and 4). Airborne pollen consistently revealed a higher content of Mg and Cl than control pollen. These increases were statistically significant ($p < 0.05$) aside from the Cl increase in Poaceae pollen. K and N contents varied differently according to the pollen type.

There was a small amount of Si in control *Alnus* pollen whereas in control Poaceae pollen Si was not detected. A significant increase in Si-content was observed in airborne Poaceae pollen collected on the 6th of July ($p < 0.05$), while no increase was observed in airborne *Alnus* pollen sampled on the 3rd of February. On the other hand, in Poaceae pollen, the Ca- and P-contents remained similar between control and airborne samples, whereas in airborne *Alnus* pollen a significant increment in Ca-content and a significant decrease in P-content were observed ($p < 0.05$).

Finally, it was possible to perform X-ray maps for Mg, Cl, Ca and Si, elements chosen because they had shown the greatest differences in the overall of the quantitative analyses. These maps allowed comparing the changes in the elemental distribution between control and airborne pollen.

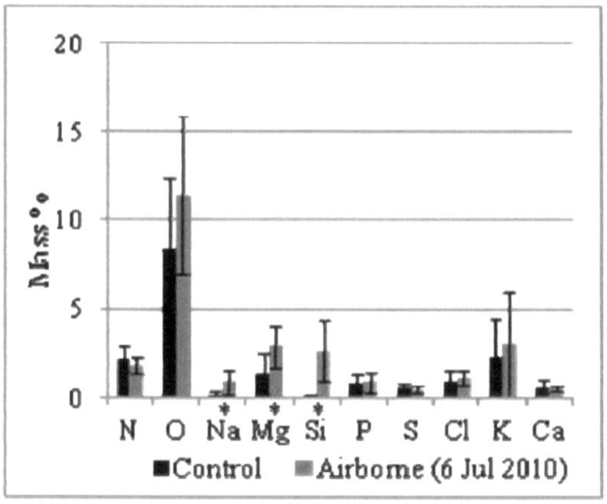

Figure 3. Mass % of analyzed elements of airborne and control Poaceae pollen (mean ± SD, n = 13). Beam conditions: 6 kV, 15 nA. Asterisks above the elements correspond to significant differences between control and airborne pollen (Mann-Whitney U-test, $p < 0.05$).

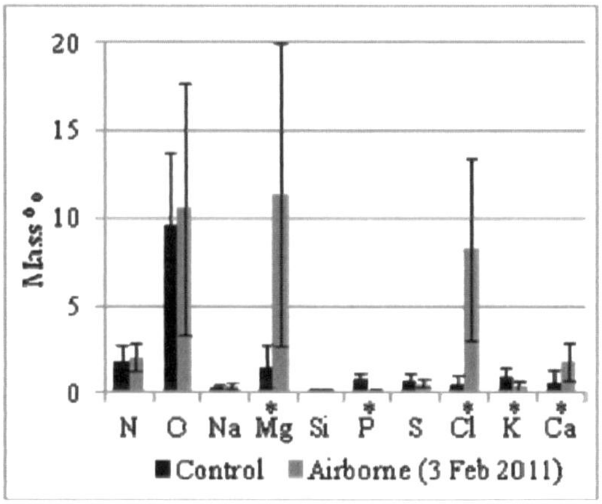

Figure 4. Mass % of analyzed elements of airborne and control *Alnus* pollen (mean ± SD, n = 15). Beam conditions: 6 kV, 15 nA. Asterisks above the elements correspond to significant differences between control and airborne pollen (Mann-Whitney U-test, $p < 0.05$).

Maps are presented for just one pair (control/airborne) of each pollen type in figures 5-7. One may observe the different compositions of the pollen wall and also some attached and surrounding particles. An increase in the Mg- and Cl-content was observed on the pollen wall of both Poaceae and *Alnus*. Particles around the pollen grains also have a great amount of these elements in their composition. Si is not a component of Poaceae pollen but can be seen on the airborne pollen wall and on the particles nearby. On the other hand Ca is a component of Poaceae pollen wall and in this case it seems masked by other elements. The opposite situation occurs for *Alnus* pollen in what Si and Ca are concerned: there is an increase in Ca-content while intrinsic Si appears to be concealed.

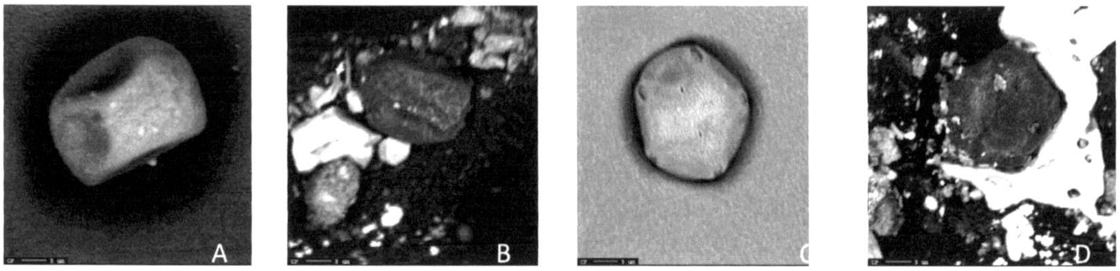

Figure 5. BSE images obtained during mapping; A) control Poaceae; B) airborne Poaceae; C) control *Alnus*; D) airborne *Alnus*.

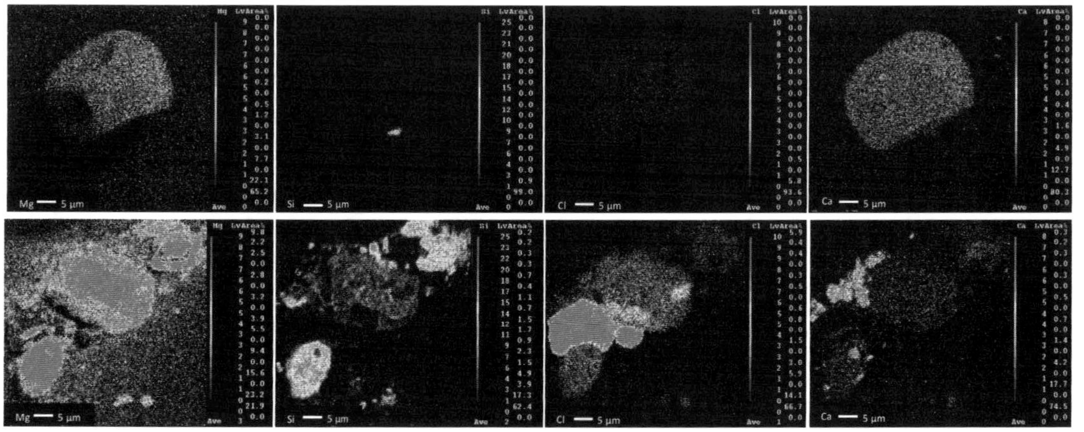

Figure 6. Intensity X-ray maps of control (above) and airborne (below) Poaceae pollen. Beam conditions: 6 kV, 25 nA.

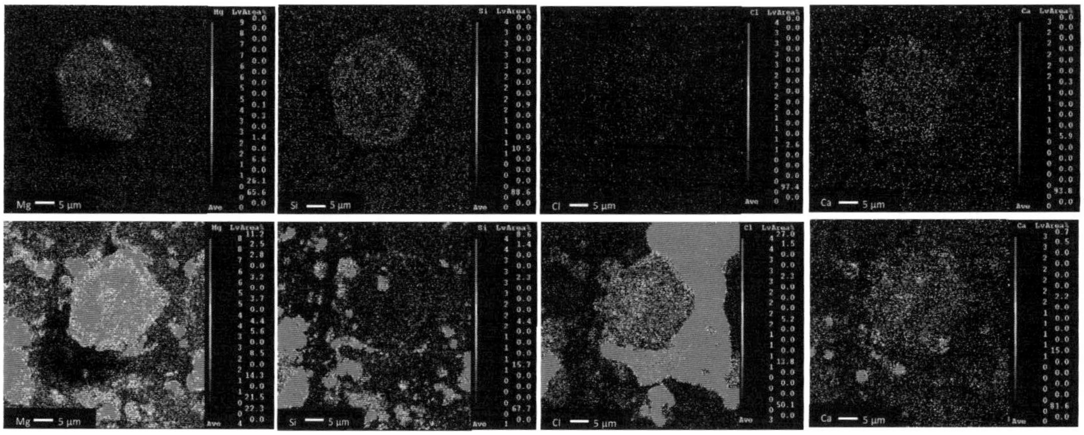

Figure 7. Intensity X-ray maps of control (above) and airborne (below) *Alnus* spp. pollen. Beam conditions: 6 kV, 25 nA.

4. Discussion

EPMA allowed us to analyze and to compare pollen grains in order to understand what happens during their travel through the air of Porto. EPMA is a technique used for the study of biological materials with several described difficulties [13, 14]. The use of a low voltage beam, on WDS analyses, allows a good spatial resolution. The systematic errors resulting from the rounded shape geometry of pollen were minimized, because control and airborne pollen were analyzed under the same conditions. Since there are many difficulties in this type of analyses, our results were interpreted in terms of relative compositions of control and airborne pollen rather than as absolute values.

Among the major constituents of the pollen wall are C, H, O, N, P, K, Mg, S, Cl and Ca. N, P and K (together with C, H and O) are the main constituents of organic matter. Also, a major presence of other elements, such as Na, Al and Si, was observed on airborne pollen. Airborne pollen grains seem to acquire an external layer that in some cases covers the surface completely and, therefore, in our analyses some elements present on the pollen can be virtually reduced or occulted. Cerceau-Larrival et al [8] call it an "inversion" (when elements change order of dominance) or a "decrease in the number of elements" (when certain elements that appeared on the analyses of freshly shed pollen do not appear in polluted pollen). In this work, pollen may also have acquired an external layer, hiding some elements and changing their order of dominance.

In our study, the external layer is composed mainly of Mg, Cl and Si or Ca depending on the pollen species considered. This airborne pollen coating seems to be related to sea spray (mainly in winter), due to Porto's location, since Mg^{2+}, Cl^{-} Na^{+} and SO_4^{2-} represent about 95 % of sea water salinity. Worobiec et al. [15] also described dust particles covered by a layer of sea-salt. Niemi et al. [16] reported that in samples collected between the English Channel and Antarctica, the amount of Mg-sulphate found in aerosols of the marine boundary layer is the greatest near the coast of Portugal. This can explain the large amount of Mg found on our samples. This coating can also be related to re-suspension or long-distance transport of crustal components (Si, Ca and others) [1].

The different alterations of composition observed on the two pollen types can be associated to differences in meteorological conditions (table 1), since samples were gathered in different dates and, in this case, different seasons (summer and winter). Pollution may also have an effect on these alterations. The way the substances adhere to or react with the surface of the pollen may also be related to its wall ornamentation, but in this case this does not seem a suitable explanation to the differences observed, because Poaceae and *Alnus* pollen have similar wall textures (granular exine) [17]. A follow-up of this study is currently being carried out to confirm our hypotheses.

5. Conclusion

EPMA is a useful method for the identification of pollen grain major elements as well as the adsorbed material to airborne pollen.

Our results showed that during its "flight", pollen acquires an external coating, becoming heavier, and its wall composition is significantly changed. Mg and Cl are the main components of this coating. Weather conditions are important to the extent of these modifications.

Acknowledgments

This work was supported by FEDER funds through the Operational Program Competitiveness Factors - COMPETE and National Funds through FCT - Foundation for Science and Technology in the project Ref. PTDC/AAC-AMB/102796/2008 and POCI 2010. Third author benefits from a scholarship (SFRH/BDP/43604/2008) financed by QREN-POPH and FCT.

References

[1] Giere R and Querol X 2010 *Elements* **6** 215-22
[2] Grobety B, Giere R, Dietze V and Stille P 2010 *Elements* **6** 229-34
[3] Majd A, Chehregani A, Moin M, Gholami M, Kohno S, Nabe T and Shariatzade M A 2004 *Aerobiologia* **20** 111-8
[4] Shahali Y et al. 2009 *J. Physics: Conf. Ser.* **151** 012027
[5] Laucks M L, Roll G, Schweiger G and Davis E J 2000 *J. Aerosol Sci.* **31** 307-19
[6] Schulte F, Lingott J, Panne U and Kneipp J 2008 *Anal. Chem.* **80** 9551-6
[7] Cerceaularrival M T, Cauneaupigot A, Bocquel C, Garbonnierjarreau M C, Derouet L and Verhille A M 1994 *J. Trace Micropr. Techn.* **12** 37-50
[8] Cerceaularrival M T, Nilsson S, Cauneaupigot A, Berggren B, Derouet L, Verhille A M and Carbonnierjarreau M C 1991 *Grana* **30** 532-45
[9] Szaloki I, Osan J, Ro C U and Van Grieken R 2000 *Spectrochim. Acta B Atom. Spectr.* **55** 1017-30
[10] Monteiro A 1997 *O clima urbano do Porto: contribuição para a definição das estratégias de planeamento e ordenamento do território.* (Lisbon: Fundação Calouste Gulbenkian)
[11] Santos F D, Forbes K and Moita R 2002 *Climate change in Portugal: scenarios, impacts and adaptation measures.* (Lisbon: Gradiva)
[12] van Hout R and Katz J 2004 *J. Aerosol Sci.* **35** 1369-84

[13] Heinrich K F J 1981 *Electron beam X-ray microanalysis.* (New York: Van Nostrand Reinhold Company)
[14] Reed S J B 1993 *Electron microprobe analysis.* (Cambridge, U.K.: Cambridge University Press)
[15] Worobiec A, Szaloki I, Osan J, Maenhaut W, Stefaniak E A and Van Grieken R 2007 *Atmos. Environ.* **41** 9217-30
[16] Niemi J V, Tervahattu H, Virkkula A, Hillamo R, Teinila K, Koponen I K and Kulmala M 2005 *Atmos. Res.* **75** 301-21
[17] Wodehouse R P 1935 *Pollen grains.* (New York and London: McGraw-Hill Book Company)

Microstructural characterisation of electrodeposited coatings of metal matrix composite with alumina nanoparticles

P Indyka[1], **E Beltowska-Lehman and A Bigos**
Polish Academy of Sciences, Institute of Metallurgy and Materials Science,
25 ul Reymonta, PL-30059 Krakow, Poland

E-mail: nmindyka@imim-pan.krakow.pl

Abstract. In the present work a nanocrystalline Ni-W metallic matrix was used to fabricate Ni-W/Al_2O_3 composite coatings. The MMC (metal matrix composite) coatings with inert α-Al_2O_3 particles (30 - 90 nm) were electrodeposited from aqueous electrolytes under direct current (DC) and controlled hydrodynamic conditions in a system with a rotating disk electrode (RDE). The chemical composition and microstructure of electrodeposited composites mainly control their functional properties; however, the particles must be uniformly dispersed to exhibit the dispersion-hardening effect. In order to increase the alumina particles incorporation as well as to promote the uniform distribution of the ceramic phase in a matrix, outer ultrasonic field was applied during electrodeposition. The influence of embedded alumina nanoparticles on structural characteristics (morphology, phase composition, residual stresses) of the resulting Ni-W/Al_2O_3 coatings was investigated in order to obtain a nanocomposite with high hardness and relatively low residual stresses. Surface and cross-section morphology and the chemical composition of deposits was examined in the scanning electron microscope, the EDS technique was used. Microstructure and phase composition were determined by transmission electron microscopy and X-ray diffraction. Based on microstructural and micromechanical properties of the coatings, the optimum conditions for obtaining crack-free homogeneous Ni-W/Al_2O_3 composite coatings have been determined.

1. Introduction

MMC (metal matrix composite) materials which are composed of inert particles embedded into a metal matrix are widely used in the industry sectors where hard and wear-resistant coatings play important roles [1]. The MMC structure allows the interaction in phase boundary between the ductile matrix and embedded particles to appear, resulting in their specific characteristics which differ from the component properties [2]. The presence of ceramic particles could prevent migration of grain boundaries and dislocation motion resulting in enhanced thermal stability and hardness of composites [3]. The coatings demonstrate metallic features (e.g., electric and thermal conductivity, plasticity) and modifier properties (e.g., hardness, wear resistance). One of the MMC preparation methods is electrochemical deposition [4]. The co-deposition process proceeds in a single technological step in which the ceramic particles are encaptured and built-up by the growing matrix to form a composite. The co-deposition mechanism for the dispersion of inert particles into metallic coatings is still not entirely understood despite numerous theories presented in the literature (e.g., electrophoresis, mechanical entrapment, adsorption, and convective-diffusion [5]), which results mainly from many

[1] To whom any correspondence should be addressed.

factors influencing the electrodeposition process and their interaction. The most commonly used composite coatings consist of a nickel matrix containing dispersed Al_2O_3 particles [6-8]. Ni-based nanostructured alloys containing refractory metal (e.g., tungsten) addition are known to exhibit superior mechanical and chemical properties compared to nickel coatings [9, 10]. Hence, in the present work nanocrystalline metallic Ni-W matrixes were used to fabricate Ni-W/Al_2O_3 composite coatings. However, Ni-W alloys are difficult to obtain by conventional thermal methods due to the large difference in melting point of components (Ni -1455 °C, W - 3410 °C) and a limited mutual solubility [11]. Compared to the metallurgical processes the formation of the galvanic coatings proceeds under disproportionately lower temperatures. In the present study, the effects of processing parameters (current density and ultrasound frequencies) on the Ni-W/Al_2O_3 composite microstructure and micromechanical properties were examined. With the aim of the optimal design of the Ni-W/Al_2O_3 composite, the relationship between the operating parameters and functional properties has been determined. Microstructure and chemical composition are the main factors influencing the functional properties of electrodeposited coatings. Thus, for optimising electroplating parameters to produce adherent and uniform composites with appropriate micro-mechanical properties, the impact of chemical composition on microstructure changes should be discussed in detail. Hence, the aim of the present study was to determine the correlation of the microstructural characteristic, chemical (influence of tungsten, alumina content), and phase composition with coating mechanical and tribological properties such as microhardness and Young modulus. To achieve this goal, electron microscopy (SEM, TEM) as well as X-ray diffractometry techniques were applied.

2. Experimental details

Ni-W/Al_2O_3 nanocomposite coatings were electrochemically deposited from aqueous electrolyte solutions containing sodium tungstate and nickel sulphate complexed by citrate ions, into which Al_2O_3 nanopowder in the amount of 40 gdm^{-3} was added. The electrolysis was carried out at a temperature of 60 °C under a galvanostatic regime (4 - 8 Adm^{-2}) in a system with a rotating disc electrode (RDE). The electrodeposition process was also carried out in an ultrasonic field of variable frequency (35 kHz and 130 kHz). The ferritic steel cathode of 0.028 dm^2 in an area rotating at 340 rpm was supplied by a potentiostat/galvanostat PAR 273A. The MMC and alloy coatings morphology and microstructure were characterized by scanning electron microscopy (SEM) technique using an ESEM FEI XL-30 microscope (accelerating voltage conditions of 20 kV). The chemical composition of deposits was determined by energy-dispersive X-ray spectrometry (EDS) analysis in the SEM equipped with an Edax GEMINI 4000 spectrometer. The ZAF correction method was used to improve the accuracy of the X-ray microanalysis. TEM studies of the coatings were performed using a FEI TECNAI G2 (accelerating voltage conditions of 200 kV). TEM images of MMC and metallic matrix coatings with the corresponding selected area diffraction patterns analysis were studied. The thin foils for transmission electron microscopy (TEM) investigation were prepared by the focussed ion beam technique (FEI Quanta 2D Dual Beam with Omniprobe lift-out system). Phase composition, crystallite size, and macro-residual stress analysis were carried out by the X-ray diffraction method using a Bruker D8 Discover with polycapillary optics and a position sensitive detector. Diffraction spectra were analyzed using the Rietveld method (MAUD software [12]). Microhardness and Young's modulus were measured using the instrumented indentation method on MCT-CSM Instruments apparatus. Vickers geometry diamond, and 50 mN maximum load with 100 $mNmin^{-1}$ loading and unloading rates were applied. Duration time at maximum load was 5 seconds. Elastic modulus was calculated from the unloading curve according to the Oliver and Pharr method [13].

3. Results and discussion

Ni-W/Al_2O_3 composite coatings were deposited from a plating bath in which Al_2O_3 powder was dispersed. This is the thermodynamically unstable, heterogeneous two-phase system, in which stability depends mainly on the type, size, shape, and structure of the ceramic powder and on plating bath variables. Thus, first the parameters of dispersive phase were characterized. The commercial

IOP Conf. Series: Materials Science and Engineering **32** (2012) 012010 doi:10.1088/1757-899X/32/1/012010

Al_2O_3 powder produced by Taimicron has been used as an original material. Figure 1 presents XRD patterns and a TEM image of the Al_2O_3 nanoparticles as well as a sector graph of their grain size. As seen, powder is characterised by isometric shaped crystallites. The XRD spectra showed a trigonal α-Al_2O_3 phase only. An average grain powder diameter was estimated to about 80 nm on the basis of the isotropic size-strain model of the Rietveld method analysis. Figure 2 presents the XRD patterns of Ni-W/Al_2O_3 coatings containing these particles, electrodeposited under different current densities in the range of 4 - 8 Adm^{-2}. Analysis of diffraction patterns reveals that the alloy matrix is composed of a nanocrystalline face-centred-cubic single phase (solid solution of tungsten in nickel). As seen in figure 2 the presence of alumina in the deposits is detected by the X-ray diffraction. The grain dimension (the size of the coherent domains) of the coatings matrix was calculated using Rietveld algorithms. The average matrix grain size is estimated on the level of about 10 nm. It increases slightly when the tungsten content decreases i.e., with increasing current density (table 1). As seen, the effect of Al_2O_3 on the crystallite size of the Ni-W matrix is practically negligible, however, a slight diminishing effect is observed.

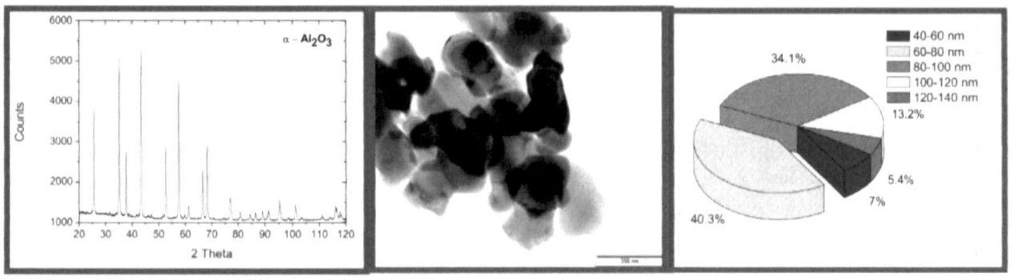

Figure 1. XRD patterns (Co Kα), TEM image of the Al_2O_3 powder and a sector graph of grain size.

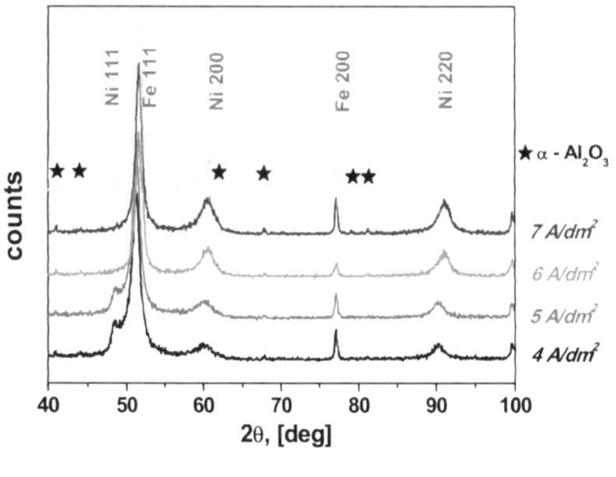

Figure 2. XRD patterns (Co Kα) of Ni-W/Al_2O_3 composites deposited at different current densities.

Compared to the metallic matrix, all examined composite coatings were characterized by relatively high tensile macro-residual stresses (determined by the sin2ψ XRD method) on the level of about 1 GPa (even up to about 1.4 GPa), which decrease with an increase of the current density, i.e., when lowering the tungsten content in the deposit (table 1). Stresses in metallic Ni-W coatings are much lower (50 - 450 MPa). Moreover, the change from tensile to compressive residual stresses for deposits obtained at current densities lower than 7 Adm^{-2} has been observed.

Table 1. Characterisation of metallic Ni-W and composite Ni-W/Al_2O_3 coatings (10 μm) electrodeposited at different current density (j): б – macro-residual stresses; k – crystallite size.
* The alumina concentration was calculated in accordance with compound stoichiometry on the basis of Al Kα line intensity measured using the EDS technique.

	Ni-W			**Ni-W/Al_2O_3**			
***j*, A/dm^2**	**W, wt%**	***б*, MPa**	***k*, nm**	**W, wt%**	**Al_2O_3, wt%***	***б*, MPa**	***k*, nm**
4	44.0±0.9	447±67	8	35.6±0.7	6.4±0.3	-------	6
5	42.4±0.8	217±76	10	34.4±0.7	5.4±0.2	1360±102	7
7	40.5±0.8	114±33	10	28.5±0.6	5.6±0.2	1460±80	6
8	37.6±0.7	-209±58	13	23.3±0.5	7.0±0.3	781±186	5

The resulting composite coatings are characterized by a compact microstructure, shiny-grey appearance, and good adhesion to the steel substrate. The Al_2O_3 distribution in the matrix was approximately uniform. However, a tendency to agglomeration has been observed. The use of an additional ultrasound field of a relatively low frequency (35 kHz) during electrolysis resulted in a homogeneous distribution of ceramic particles in the matrix phase. Figures 3 and 4 exemplify the SEM microstructure of the surface and a cross-section of Ni-W/Al_2O_3 composite samples. TEM studies confirmed the XRD results concerning the phase composition and the average metallic Ni-W matrix grain size (figure 5). The electron diffraction pattern phase analysis [14, 15] indicated that the composite matrix consists of α-Ni(W) solid solution. For Ni-W/Al_2O_3 material, they demonstrated a lack of columnar growth of the composite coating, the presence of defects in the form of micro-twins, and a very good dispersion of the Al_2O_3 ceramic phase throughout the whole volume of composite (figure 6). Functional properties of composites are mainly controlled by their composition and microstructure. They depend on the size, shape, and structure of the co-deposited ceramic particles but also on their distribution in the metallic matrix. Indeed, a Ni-W/Al_2O_3 coating obtained under the additional outer ultrasonic field of 35 kHz frequency is characterised by more favourable micromechanical properties compared to coatings deposited without ultrasounds. For example, composite coatings (10 μm thickness) deposited at 7 Adm^{-2} under ultrasound fields are characterized by higher microhardness (HIT) and plasticity index (HIT/EIT, EIT - elastic modulus) i.e., 8.5 GPa and 0.04 respectively, compared to 6.5 GPa and 0.03 for coatings without ultrasound treatments. They exhibit also substantially higher wear resistance (2 to 4 times).

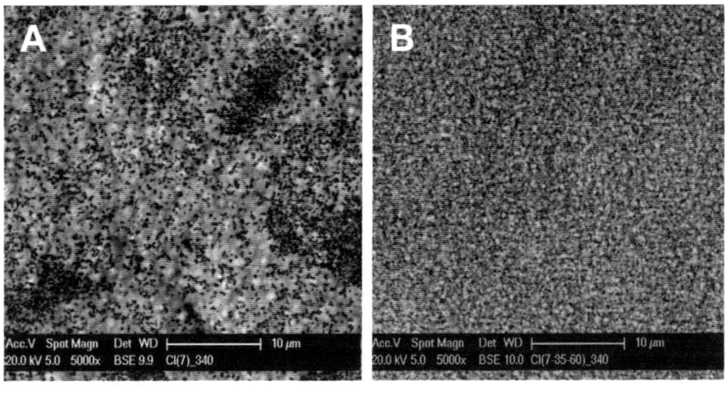

Figure 3. SEM (BSE) images of surface morphology of Ni-W/Al_2O_3 coatings electrodeposited at 7 A/dm^2 and RDE speed of 340 rpm: A) without ultrasonic field; B) with ultrasounds of 35 kHz.

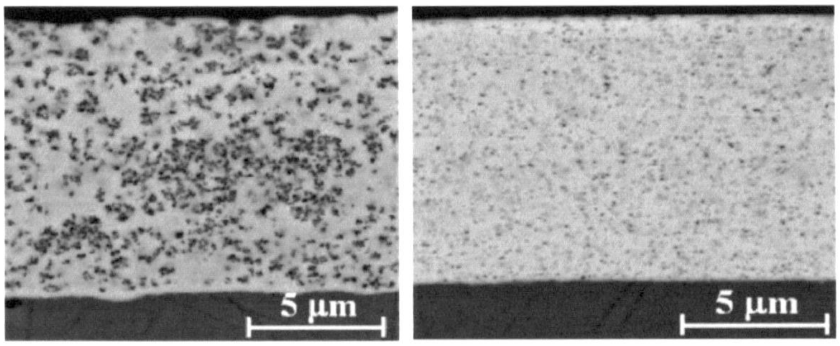

Figure 4. SEM (BSE) image of cross-section of composite Ni-W/Al_2O_3 deposited: A) without US; B) under an ultrasonic field of 35 kHz.

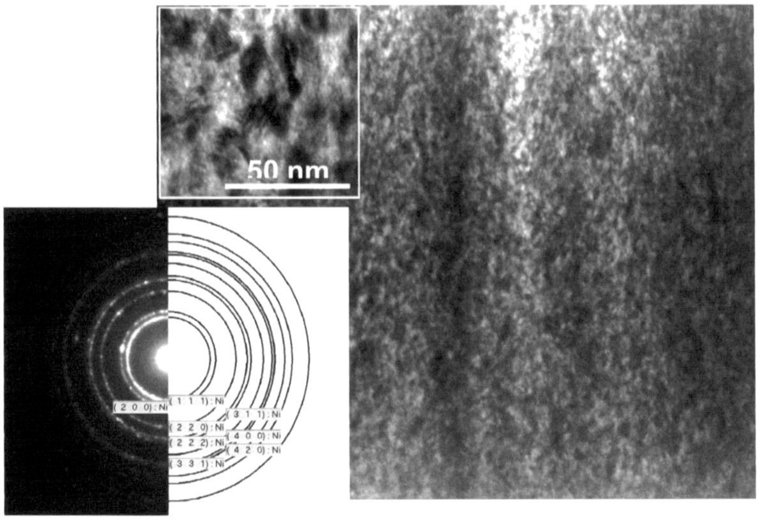

Figure 5. TEM images of Ni-W metallic matrix electrodeposited at 5 A/dm^2 and corresponding electron diffraction pattern (SAED).

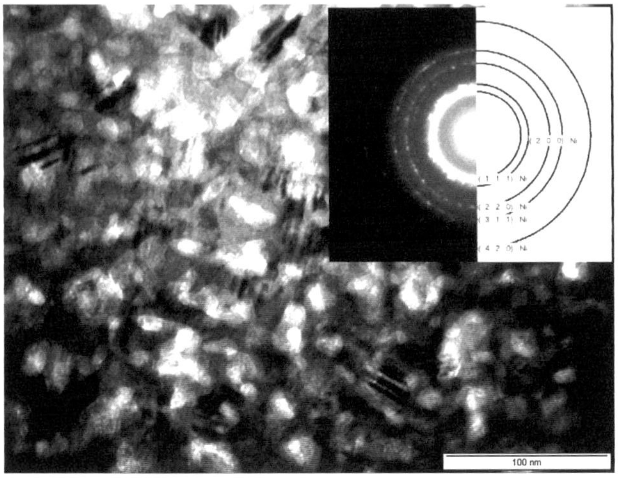

Figure 6. TEM images of an example of Ni-W composite coating (5 wt% Al_2O_3) and corresponding electron diffraction pattern (SAED).

4. Conclusion

- Nanocrystalline Ni-W/Al_2O_3 composites can be produced by co-deposition of nickel, tungsten, and alumina nanoparticles under forced convection conditions and direct current (DC) from aqueous citrate electrolytes in which Al_2O_3 powder is dispersed.
- Application of an ultrasonic field during the electrolysis caused the complete fragmentation of the agglomerates of the ceramic phase, uniformly distributed in the volume of the composite, however, lower frequencies are advisable.
- The difference in the amount and distribution of co-deposited ceramic particles influence the composite properties, especially the hardness and plasticity.

Acknowledgments
The results presented in this paper have been obtained within the project "KomCerMet" (contract no. POIG.01.03.01-14-013/08-00 with the Polish Ministry of Science and Higher Education) in the framework of the Operational Programme Innovative Economy 2007-2013.

References

[1] Kerr C, Barker D and Walsh F 2000 *Trans. Inst. Metal. Finish.* **78** 171
[2] Zhang Z and Chen D L 2006 *Scri. Mater.* **54** 1321
[3] Musiani M 2000 *Electrochim. Acta* **45** 3397
[4] Landolt D 2002 *J. Electrochem. Soc.* **149** S9
[5] Low C T J, Wills R G A and Walsh F C 2006 *Surf. Coat. Tech.* **201** 371
[6] Dong Y S, Lin P H, and Wang H X 2006 *Surf. Coat. Tech.* **200** 3633
[7] Steinbach J and Ferkel H 2001 *Scri. Mater.* **44** 1813
[8] Lampke T, Wielage B, Dietrich D and Leopold A 2006 *Appl. Surf. Sci.* **253** 2399
[9] Shacham-Diamant Y and Lopatin S 1999 *Electrochem. Acta* **44** 3639
[10] Brooman E W 2000 *Met. Finish.* **98** 42
[11] Masalski T B 1990 *Binary alloy phase diagrams.* (USA ASM International)
[12] Lutterotti L and Scardi P 1990 *J. Appl. Cryst.* **23** 246
[13] Oliver W C, Pharr G M 1992 *J. Mat. Research* **7** 1564
[14] Labar J L 2008 *Microsc. Microanal.* **14** 287
[15] Labar J L 2009 *Microsc. Microanal.* **15** 20

IOP Conf. Series: Materials Science and Engineering **32** (2012) 012011 doi:10.1088/1757-899X/32/1/012011

The application of VP-ESEM in microstructure analysis of ceramic macroporous scaffolds for bone tissue engineering

A M Janus[1] **and M Faryna**
Polish Academy of Sciences, Institute of Metallurgy and Materials Science,
25 ul. Reymonta, PL-30059 Krakow, Poland

E-mail: annamariajanus@yahoo.com

Abstract. Macroporous ceramic scaffolds are widely used in the therapy of bone defects. Adaptation of such materials inside the human body depends strongly on their microstructure, particularly on the size of pores and the degree of pore interconnections. In the presented research a novel macroporous ceramic was prepared by means of the polymer sponge method from mixtures of hydroxyapatite of porcine origin and 30 or 50 wt% of biocompatible and bioresorbable phosphate glass. The materials were sintered at 700, 800, and 900 °C and at 700 and 800 °C, respectively. A unique set of VP-ESEM micrographs of ceramics microstructure was recorded. Based on direct measurements taken on these images, histograms representing macropore size distribution and distribution of size of macropore interconnections were plotted for each sample. The obtained data was used to determine average macropore size and average size of macropore interconnections for each material. The average macropore size varied from 532 to 761 μm while the average size of macropore interconnections ranged from 145 to 369 μm.

1. Introduction

A development of materials designed for therapy of bone tissue is of great importance due to the fact that human organism is not able to heal larger bone defects [1], which leads to immobilisation that creates a substantial burden to the society. A large number of materials have been already commercialized and many of them are based on calcium phosphate ceramics [2]. Very often such materials are designed to act as scaffolds, i.e., porous structures of desired geometry which stimulate regeneration of bone, as in this form they can be used to treat load bearing sites which is impossible in case of dense phosphate ceramics [3]. At present, investigations of macroporous ceramics focus on improvement of biocompatibility and bioresorption. Enhancement of biocompatibility can be achieved through the use of materials of biological origin or synthesis of biomimetic materials [4-7] while resorbability can be influenced through the introduction of new phases [8].

Presented research was focused on preparation of resorbable calcium phosphate scaffolds from hydroxyapatite derived from porcine bones. The application of porcine hydroxyapatite was driven by the fact that this material showed better performance in *in vitro* biocompatibility tests compared to stoichiometric synthetic hydroxyapatites [9]. In order to improve resorption of the material, a biocompatibile and resorbable phosphate glass described in [10] was introduced into the material.

[1] To whom any correspondence should be addressed.

As microstructure is one of the key parameters determining the usefulness of macroporous ceramics in bone tissue therapy [11], it was thoroughly examined and characterized using scanning electron microscopy and image analysis techniques.

2. Experimental

2.1. Starting materials

Hydroxyapatite of porcine origin (NAT) was prepared from long porcine bones by means of leaching out of organic matter with 4 M aqueous sodium hydroxide solution. Details of the procedure are given in [9]. Reference materials (synthetic hydroxyapatites) were purchased from Plasma Biotal Ltd. (Captal, UK) and Chema-Elektromet (HA Biocer, PL). As modifier "specular" phosphate was used (composition in mol%: 3 - SiO_2; 45 - P_2O_5; 26 - CaO; 15 - Na_2O; 7 - MgO; 4 - K_2O). Commercially available polyurethane foam was used as the template for replication. Microstructure of the foam, macropore size distribution and distribution of size of macropore interconnections determined in accordance with procedure described in section 2.3 are shown in figure 1.

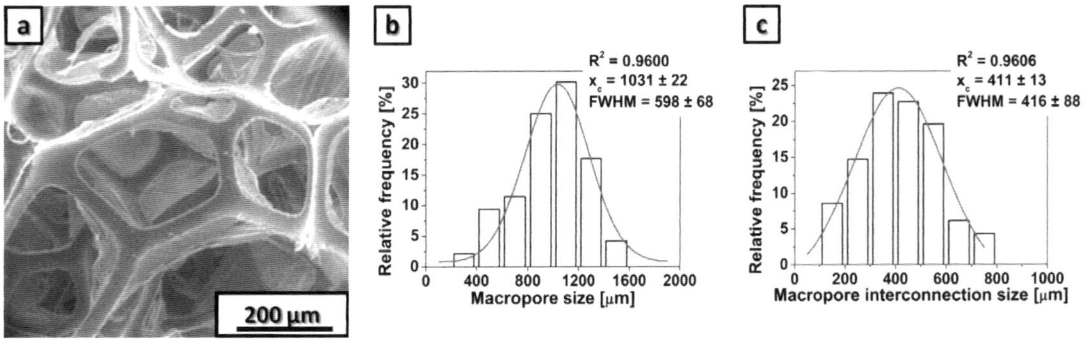

Figure 1. VP-ESEM micrograph (a), pore size distribution (b) and distribution of pore interconnection size (c) of polyurethane sponge template.

2.2. Preparation of macroporous ceramics

Macroporous ceramics were prepared using polymer sponge method. In the first step aqueous slurry of hydroxyapatite, glass and polyvinyl alcohol was prepared through mixing of solid components (36 wt%) with distilled water. In the next steps, shapes of dimensions 1.5 x 1.5 x 1.5 cm cut from polyurethane sponge were soaked with the slurry, the excess of water was removed by means of three-axial pressing and resulting material was dried at ambient temperature in air. The soaking was performed three times. Dried shapes were sintered in air at 700, 800 and 900 °C (ceramics modified with 30 wt% of glass) and 700 and 800 °C (ceramics modified with 50 wt% of glass). Detailed description of preparation of macroporous scaffolds can be found in [12].

2.3. Microscopy examination and image analysis

Microstructure of obtained macroporous scaffolds was examined using FEI scanning electron microscope (SEM) Quanta 3D FEG. Instead of sputter coating the samples were investigated in the environmental mode. Micrographs were recorded under the following operating conditions: accelerating voltage: 20 kV; spot: 4.0; pressure: 90 Pa; working distance (WD): 5.5 to 9.9 mm; detector: low vacuum secondary electron detector (LVSED); chamber gas: water vapour.

Recorded micrographs were used as a source of input data in analysis of size of macropores and size of macropore interconnections. For each material maximum Feret diameter was measured for 150 completely visible macropores (figure 2a - measurement on the left) and 150 macropore interconnections (figure 2a - measurement on the right). The measured Feret diameter was a diameter of the flat object visible in the micrograph and not of the actually existing pore or pore interconnection as it is impossible to take into account various angles at which object are situated with regards to the observation plane. This effect of non normal incidence was considered negligible. Based on data obtained during measurements, histograms representing pore size distribution and distribution of size of pore interconnections were plotted for each sample. Obtained histograms were used to determine average pore size and average size of pore interconnections for each material through fit of normal distribution and subsequent determination of the maximum of the normal distribution.

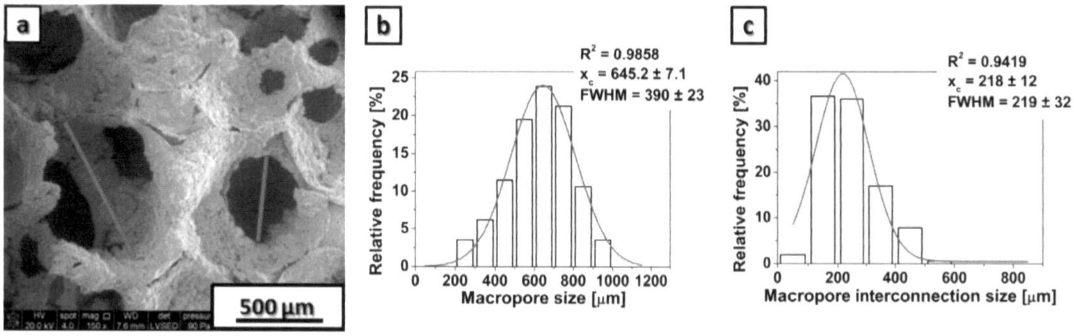

Figure 2. VP-ESEM micrograph (a), macropore size distribution (b) and distribution of macropore interconnection size (c) of NAT30 700 ceramics.

3. Results and discussion

All compositions were successfully sintered apart from the mixture of hydroxyapatite UK and 30 wt% of glass sintered at 700 °C which did not retain structural integrity. Thus, data for material UK30 700 are not presented.

Figures 2 to 4 show representative micrographs, macropore size distributions and distributions of macropore interconnection size of ceramics based on natural origin hydroxyapatite modified with 30 wt% of glass. Figures 5 to 9 show micrographs of the remaining materials (including ones modified with 50 wt% of glass). Table 1 summarizes data on macropore size and macropore interconnection size and their distributions. All produced materials exhibited an internal structure of large number of interconnected macropores. Investigations showed that in obtained ceramics the average macropore size and the average size of macropore interconnections ranged from 532 to 761 µm and 145 to 369 µm, respectively. According to literature the minimum macropore size in macroporous ceramics for bone tissue therapy is 100 µm [13-15] and the optimum macropore size is of several hundreds of µm [14, 15]. The size of macropore interconnections in macroporous ceramics for bone tissue engineering should exceed 50 µm [14, 15]. That means all macroporous ceramics produced in the presented the work showed morphology desired for such materials.

Conclusions

The application of polymer sponge method for preparation of ceramics from porcine origin hydroxyapatite leads to macroporous materials; such materials show desired structure of large number of open and interconnected macropores. From the morphology point of view, macroporous ceramics produced with the use of polymer sponge method from porcine origin hydroxyapatite and phosphate glass can be used as scaffolds in bone tissue engineering. Presented combination of scanning electron

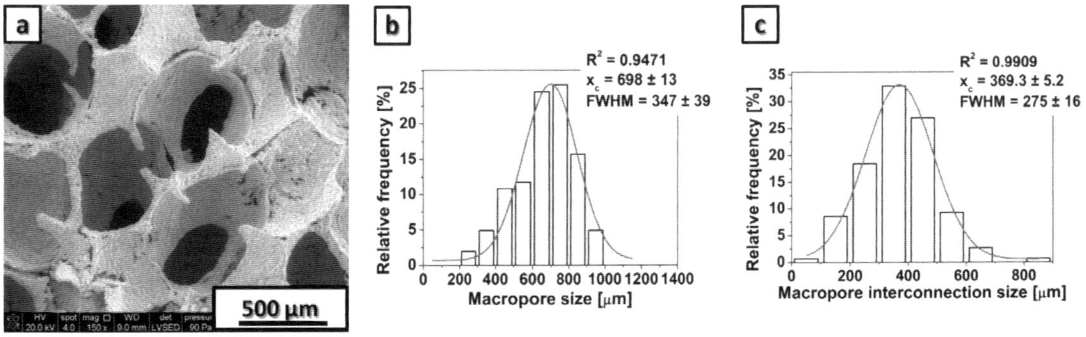

Figure 3. VP-ESEM micrograph (a), macropore size distribution (b) and distribution of macropore interconnection size (c) of NAT30 800 ceramics.

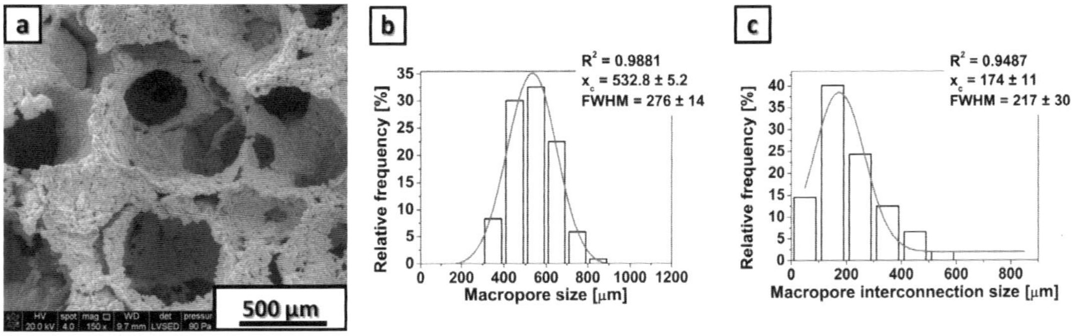

Figure 4. VP-ESEM micrograph (a), macropore size distribution (b) and distribution of macropore interconnection size (c) of NAT30 900 ceramics.

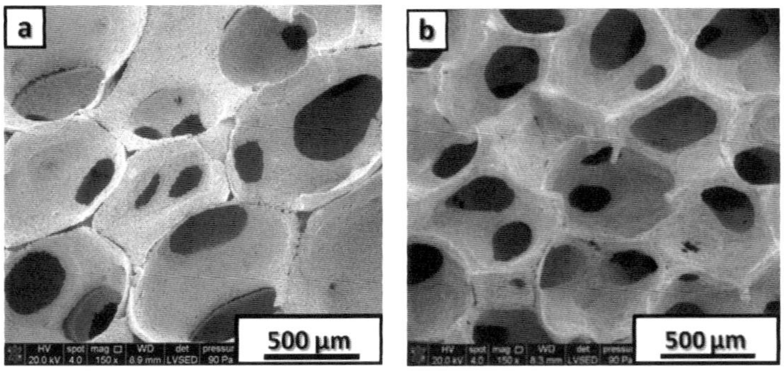

Figure 5. VP-ESEM micrographs of (a) UK30 800 and (b) UK30 900 ceramics.

microscopy and image analysis techniques allows to characterize in detail macroporous structures using quantitative measure. Described approach can be used in analyses of SEM micrographs to characterize any visible morphology feature such as e.g., pores, pore interconnections, grains, crystallites. That makes SEM micrographs a valuable source of quantitative data.

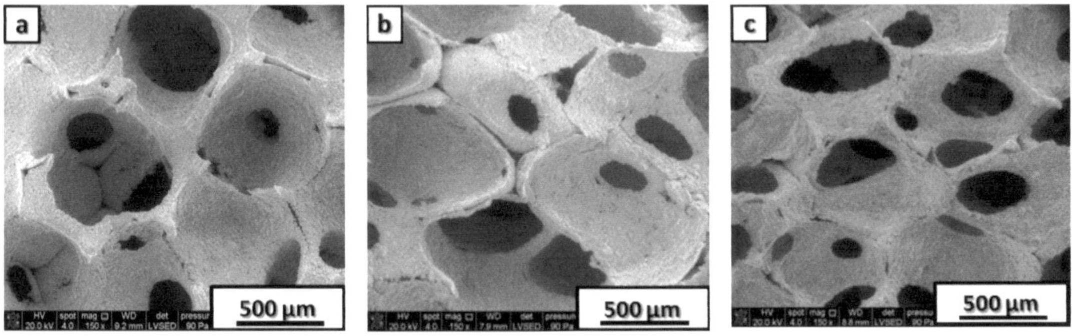

Figure 6. VP-ESEM micrographs of (a) PL30 700, (b) PL30 800, and (c) PL30 900 ceramics.

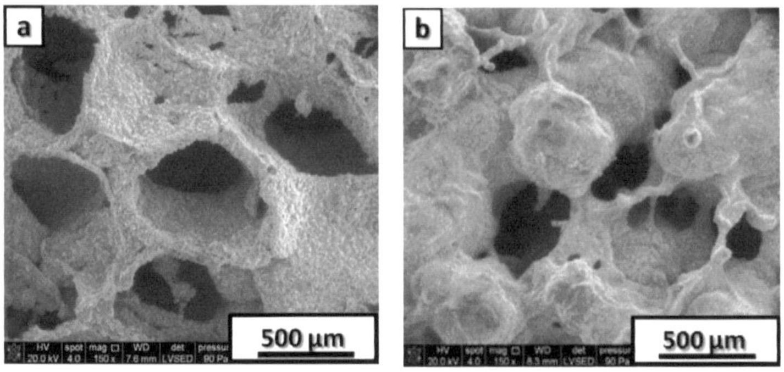

Figure 7. VP-ESEM micrographs of (a) NAT50 700 and (b) NAT50 800 ceramics.

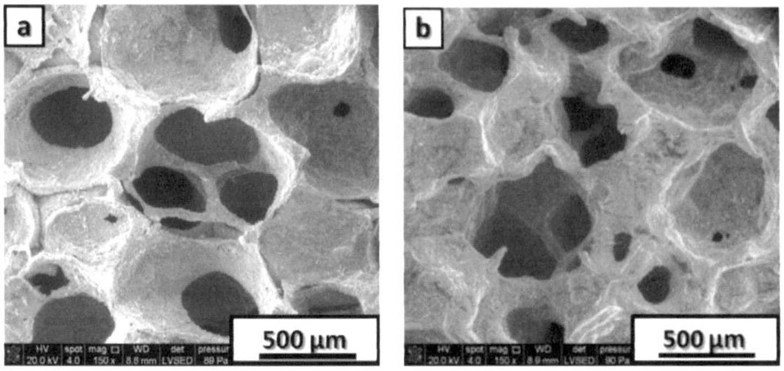

Figure 8. VP-ESEM micrographs of (a) UK50 700 and (b) UK50 800 ceramics.

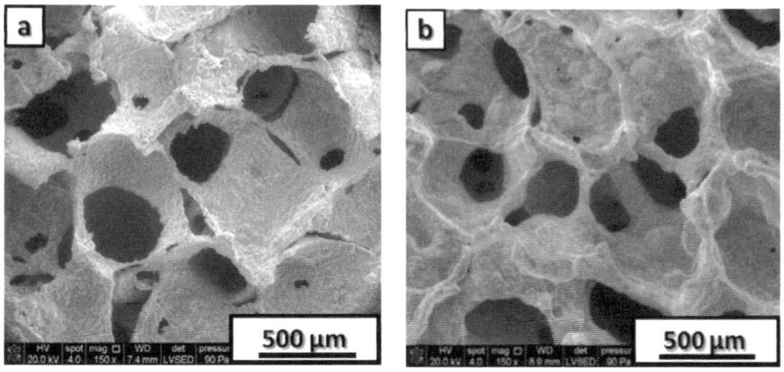

Figure 9. VP-ESEM micrographs of (a) PL50 700 and (b) PL50 800 ceramics.

Table 1. Average macropore and macropore interconnection size and distribution width of these two parameters for hydroxyapatite ceramics modified with 30 and 50 wt% of phosphate glass.

Material	Average macropore size [μm]	Width of macropore size distribution [μm]	Average macropore interconnection size [μm]	Width of macropore interconnection size distribution [μm]
NAT30 700	645.2 ± 7.1	390 ± 23	218 ± 12	219 ± 32
NAT30 800	690 ± 13	347 ± 39	369.3 ± 5.2	275 ± 16
NAT30 900	532.8 ± 5.2	276 ± 14	174 ± 11	217 ± 30
UK30 800	739.7 ± 5.2	383 ± 16	283 ± 17	310 ± 54
UK30 900	627.0 ± 3.7	254.9 ± 9.9	226 ± 13	255 ± 39
PL30 700	704.6 ± 4.0	193 ± 12	248.8 ± 7.4	276 ± 21
PL30 800	761 ± 22	344 ± 64	295 ± 23	388 ± 80
PL30 900	680 ± 13	353 ± 39	256.0 ± 8.1	293 ± 24
NAT50 700	623.2 ± 9.5	383 ± 30	145.3 ± 5.3	161 ± 10
NAT50 800	553.9 ± 4.4	231 ± 11	260.7 ± 5.0	243 ± 14
UK50 700	656 ± 17	409 ± 56	157 ± 25	131 ± 38
UK50 800	153.6 ± 2.0	317.9 ± 5.9	205 ± 11	266 ± 31
PL50 700	619.6 ± 2.6	258.6 ± 7.1	187.7 ± 9.5	266 ± 27
PL50 800	591.2 ± 6.2	337 ± 18	222 ± 21	321 ± 64

Acknowledgements
The research was realized in part at the Materials Science and Chemical Engineering Department of Politecnico di Torino within the framework of a research grant under KMM-NoE programme. The authors would like to thank Ms. Chiara Vitale-Brovarone and Ms. Enrica Verné for their help.

References
[1] Joschek S, Nies B, Krotz R and Göpferich A 2000 Chemical and physicochemical characterization of porous hydroxyapatite ceramics made of natural bone. *Biomaterials* **21** 1645-58
[2] Legeros R Z and Legeros J P 2008 *Bioceramics and their clinical applications.* ed, Kokubo T (Cambridge: Woodhead Publishing Limited) chapters 15-20, pp 326-484
[3] Baino F and Vitale-Brovarone C 2011 Three-dimensional glass-derived scaffolds for bone tissue engineering: current trends and forecasts for the future. *J. Biomed. Mat. Res. A* **97A** 514-35
[4] Ruksudjatir A, Pengpat K, Rujijanagul G and Tunkasiri T 2008 Synthesis and characterization of nanocrystalline hydroxyapatite from natural bovine bone. *Current Appl. Phys.* **8** 270-2
[5] Haberko K, Bućko M M, Mozgawa W, Pyda A, Brzezińska-Miecznik J and Carpentier J 2009 Behaviour of bone origin hydroxyapatite at elevated temperatures and in O_2 and CO_2 atmospheres. *Ceramics Int.* **35** 2537-40
[6] Landi E, Logroscino G, Proietti L, Tampieri A, Sandri M and Sprio S 2008 Biomimetic Mg-substituted hydroxyapatite: from synthesis to in vivo behaviour. *J. Mater. Sci.: Mat. Medic.* **19** 239-47
[7] Komlev V S, Fadeeva I V, Gurin A N, Kovaleva E S, Smirnov V V, Gurin N A and Barinov S M 2009 Effect of the concentration of carbonate groups in a carbonate hydroxyapatite ceramic on its in vivo behavior. *Inorg. Mater.* **45** 329-34
[8] Banerjee S S, Bandyopadhyay A and Bose S 2010 Biphasic resorbable calcium phosphate ceramic for bone implants and local alendronate delivery. *Adv. Engng. Mater.* **12** B148-55

[9] Janus A M, Faryna M, Haberko K, Rakowska A and Panz T 2008 Chemical and microstructural characterization of natural hydroxyapatite derived from pig bones. *Microchim. Acta* **161** 349-53

[10] Vitale-Brovarone C, Baino F, Bretcanu O and Verne E 2009 Foam-like scaffolds for bone tissue engineering based on a novel couple of silicate-phosphate specular glasses: synthesis and properties. *J. Mater. Sci.: Mater. Medic.* **20** 2197-205

[11] Klawitter J J, Bagwell J G, Weinstein A M, Sauer B W and Pruitt J R 1976 An evaluation of bone growth into porous high density polyethylene. *J. Biomed. Mater. Res.* **10** 311-23

[12] Janus A M, Faryna M, Verné E and Vitale-Brovarone C 2007 Preparation of phosphate glass reinforced hydroxyapatite scaffolds for tissue engineering. *Inżynieria Materiałowa* **XXVIII** 265-9

[13] Kim H W, Lee S Y and Bae C J 2003 Porous ZrO_2 bone scaffold coated with hydroxyapatite with fluorapatite intermediate layer. *Biomaterials* **24** 3277-84

[14] Baino F, Vitale-Brovarone C 2011 Three-dimensional glass-derived scaffolds for bone tissue engineering: current trends and forecasts for the future. *J. Biomed. Mater. Res.* **97A** 514-35

[15] Karageorgiou V and Kaplan D 2005 Porosity of 3D biomaterial scaffolds and osteogenesis. *Biomater.* **26** 5474-91

Trace analysis in EPMA

M J Jercinovic[1,3], **M L Williams**[1], **J Allaz**[1] **and J J Donovan**[2]
[1] University of Massachusetts, Department of Geosciences, 611 North Pleasant Street, US-01003 Amherst, Massachusetts, U.S.A.
[2] University of Oregon, CAMCOR, 74 Huestis Hall, US-97403-1241 Eugene, Oregon, U.S.A.

E-mail: mjj@geo.umass.edu

Abstract. Trace element microanalysis has evolved steadily since the early days of EPMA, yet remains an extraordinarily challenging subject. The enhanced capabilities of modern instrumentation, including the use of spectrometers with high X-ray collection efficiencies, high brightness electron sources, and improved stability all contribute to our ability to improve detection limits and analytical spatial resolution. Along with much improved software for data acquisition and analysis, recent progress in EPMA has made the trace realm more accessible than ever. High count precision can be obtained in order to easily bring analytical sensitivity into the single ppm range, but accuracy remains the greatest struggle. With the exception of the calibration, all sources of error encountered in major element analysis are magnified in trace analysis, and other sources become apparent where high spatial resolution is needed. Beam damage and charge effects are difficult problems in high sensitivity, high spatial resolution analysis, particularly in the analysis of insulators. Software can minimize some of the resulting effects on count rates during acquisition in order to improve accuracy, and analysts can empirically evaluate the conditions of analysis (count time, voltage, current, etc.) to try to minimize these effects. Trace analysis is fundamentally an exercise in background characterisation, and the acquisition and evaluation of background is a subject of developing methodology. Background curvature and interferences can result in considerable inaccuracy, but can be dealt with via detailed quantitative wavelength scanning or multi-point spectral acquisitions which allow proper regression of the background shape. In the absence of excellent quality trace element secondary standards of similar matrix to unknowns, blank testing and consistency standards can be used to test at least some aspects of the methods employed. Ultimately, the analyst must rely on accuracy evolving from application of the most rigorous protocols.

1. Introduction

Obtaining trace element concentrations by EPMA (electron probe microanalysis) implies that the spatial context of analysis is crucial. Without this necessity, then bulk techniques employing atomic spectrometry, activation analysis, or X-ray fluorescence, which in some cases obtain limits of detection well below 1 ppb (ultra-trace concentrations), could be used (see [1] for review). It is the balance of spatial resolution and analytical sensitivity that define the critical realm occupied by microbeam techniques. Each has strengths and weaknesses to provide a remarkably powerful set of

[3] To whom any correspondence should be addressed.

tools to address a particular problem. Therefore, analytical sensitivity and spatial resolution remain topical and lively subjects of scientific and technological endeavour.

This paper addresses the essential concepts pertinent to trace analysis by EPMA at high spatial resolution. As part of the discussion of new capabilities, it should be recognized that improvements in counting efficiency and electron beam size to achieve better sensitivity from smaller analytical volumes now necessitates an evaluation of sources of error which may have been negligible under less extreme conditions. We concentrate our examples on lead and actinide analysis in complex, moderately high average atomic number materials as this has been the subject of considerable attention in geologic research in recent years. These samples are particularly pertinent to the discussion of background analysis, especially in materials with varying concentrations of rare earth elements (REEs) and actinides, especially thorium, which can vary in concentration by an order of magnitude even in minerals that are tens of micrometres in size. REE-bearing phosphates also serve as excellent examples of materials which exhibit some degree of beam sensitivity at high electron current density.

1.1. Perspectives

Quantitative analysis at very high spatial resolution has remained an important goal since the inception of materials analysis. Today, a large array of tools exists to characterize trace element concentrations (i.e., below 1.0 wt%) with high spatial resolution analysis, and a number of sources to address the sometimes bewildering assortment of techniques available to researchers (e.g., [2, 3]). In most cases, it can be safely stated that no single technique can address all problems, and that closing in on the most reliable result is usually the consequence of application of a number of techniques. Steady advance in both the theoretical and technological aspects of metrology continue to improve both precision and accuracy in analytical techniques, with the driving forces stemming from both the analytical needs of researchers at one end, and technologic development on the other. Furthermore, the pace of these advancements seems to be progressing at a greater rate than ever.

Trace elements are generally not essential structural constituents of materials, but they are nonetheless of great importance for tracking chemical processes and growth histories. In materials engineering, trace elements are critical for creating essential material properties (band-gap engineering in semiconductors, establishing optical transmissivity in fibre optics, etc.). Investigations of trace element concentrations and spatial variation have great power in the physical sciences and engineering, establishing the relationships between microstructures and the reactions that generate them, also allowing the evaluation of reactions at interfaces. The dilute nature of trace elements is highly advantageous in thermochemical modelling where Henry's Law behaviour can be assumed. Significant applications and advances have been made in igneous and metamorphic petrology, environmental science, paleoclimatology, economic geology, geochronology, cosmochemistry, chemical engineering, and numerous other applications involving materials in the solid state.

Natural and human engineered materials are seldom truly homogeneous, and complex compositional structures commonly provide exciting opportunities to tease out the details of phase relationships in space and time. Today, our ability to image at ever higher resolution has revealed remarkable complexities, and has led to a need for improved micro (and nano) - analytical capability suitable to the task. Furthermore, the need to acquire minor and trace element information from these micro- and nano-domains has added a particular impetus to improve both hardware and software in order to accomplish both the spatial and sensitivity requirements.

1.2 Trace element EPMA

The exciting growth of a number of analytical technologies is exemplified by the new capabilities introduced in EPMA over the past decade, specifically, ultra-high count rate spectrometers, implementation of high brightness cathodes ($REEB_6$, Schottky) and development of sophisticated and powerful software for acquisition (i.e., spectrometer integration, cycled peak-bkg acquisition [4, 5], time dependent acquisition, background acquisition by regression of scan or multi-point spectral data) as well as downstream data reduction (matrix iterated interference corrections, blank correction, etc.). These developments have made the trace element micro-analysis arena more accessible than ever and have substantially revised our concepts of the "limitations" of EPMA. Although highly case-specific, the frontiers of high spatial resolution analysis at exceptional precision and accuracy are being continually pushed to new levels.

There are many splendid contributions addressing trace element EPMA, dating back to the early years of the application of the technique and steadily continuing to the present [6-13]. The precision statistics of analysis have been well characterized, and recent work has concentrated on the improvement of accuracy, particularly with regard to addressing peak interferences [13], background curvature and other features [e.g., 13-15].

Precision and sensitivity in EPMA can always be increased by counting longer and/or increasing amperage without sacrificing spatial resolution. With this in mind however, it is best to consider the following …

"Assuming counting statistics to be the only relevant consideration, the smallest detectable peak may be defined as 3 s.d. of the background count. The probability of the background count-rate exceeding the 3 s.d. limit due to random fluctuations is 0.2 %. The detection limit cannot be reduced indefinitely by accumulating more counts, however, because ***systematic errors in the background correction eventually become significant****."*

S.J.B. Reed [16]

Here, Reed fundamentally addresses the issue of accuracy, the *correctness* of the analysis. Analyses performed by any technique always have an associated (and hopefully reported) estimate of precision, but the complete characterisation of error (how accurate?) may be much larger, and can be far more elusive. There are certainly many sources of error, but the issue of background measurement accuracy (as alluded to above) is among the most fundamental in high spatial resolution trace element EPMA (e.g., [14-19] for discussions relevant to EPMA geochronology). This topic brings a number of matters to the forefront, including precision/sensitivity, and evaluation of accuracy. Both of these are firmly intertwined with the attempted spatial resolution of analysis. In trace analysis, the analyst can be reasonably assured that any difficulties found in major element analysis are magnified in the analysis of very low concentrations, and there are several problems more specific to high spatial resolution trace element analysis.

2. Sensitivity and spatial resolution

The fundamental tradeoff in nearly all micro- and nano-analytical methods is precision versus spatial resolution. Unlike ablation-based micro-analytical methods such as SIMS or LA-ICP-MS which are substantially reliant on the volume of material ablated in order to obtain higher precision, electron beam generated X-ray techniques derive precision from the number of photons counted, a result of beam current, counting time, X-ray production efficiency (fluorescence yields, ionisation cross-sections), and collection efficiency. Counts, and peak/bkg, can be increased by increasing overvoltage (figure 1) which increases electron path lengths in the target and the number of ionisations per unit time. This, of course, directly compromises spatial resolution (figure 2) as the scattering volume increases. Electron beam techniques offer some flexibility such that a number of variables can be adjusted in order to overcome apparent counting limitations in some circumstances (always sample dependent), the typical approach being to substantially increase beam current and/or counting time.

However, the consequential increases in systematic errors due to variations in behaviour of either the instrument or sample must always be factored in (e.g., contamination, beam current drift, compositional change in the sample due to prolonged beam exposure at high current density).

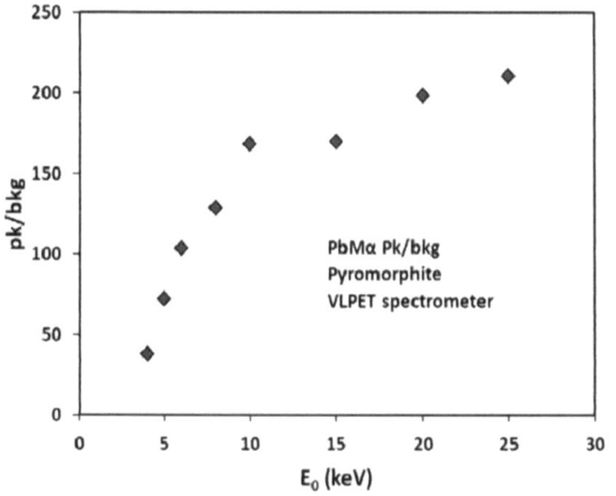

Figure 1. Measured peak to background ratio for Pb Mα from pyromorphite ($Pb_5(PO_4)_3Cl$) as a function of beam accelerating potential (E_0). Measurements made using a Cameca VLPET spectrometer.

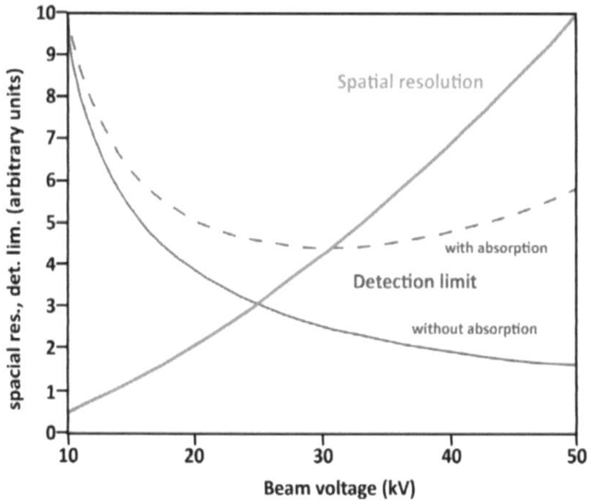

Figure 2. Calculated relative spatial resolution and minimum detectability limit as a function of kV (from Reed [12], Fig. 1). High absorption case illustrates eventual peak intensity decrease as X-ray production volume deepens in specimen.

Attempting quantification of trace concentrations requires an understanding of the basic relationships controlling the precision of analysis. It is instructive, therefore, to revisit the basic relationships for the precision of X-ray emission. The variance of the raw *k*-ratio (net intensity sample/net intensity std.) can be expressed as follows [6, 20]:

$$\sigma_k^2 = k^2\left[\frac{\overline{N}+\overline{N}(B)}{n\left(\overline{N}-\overline{N}(B)\right)^2}+\frac{\overline{N_s}+\overline{N_s}(B)}{n'\left(\overline{N_s}-\overline{N_s}(B)\right)^2}\right] \tag{1}$$

where σ_k is the standard deviation, $\overline{N}$ and $\overline{N}(B)$ are the measured mean counts for peak and background on the sample, $\overline{N_s}$ and $\overline{N_s}(B)$ are the mean counts for peak and background on the standard, and n and n' are the number of measured points on sample and standard, respectively. The variance of the concentration is then:

$$\sigma_C^2 = C^2\left[\frac{\overline{N}+\overline{N}(B)}{n\left(\overline{N}-\overline{N}(B)\right)^2}+\frac{\overline{N_s}+\overline{N_s}(B)}{n'\left(\overline{N_s}-\overline{N_s}(B)\right)^2}\right]*\left[1-\frac{(a-1)C}{a}\right]^2 \tag{2}$$

where σ_c is the concentration standard deviation, C is concentration, and a is the correction factor that relates the *k*-ratio to the concentration. An interesting result of this relationship is that precision for trace element analysis by EPMA is highly dependent on the precision of the sample analysis, and less dependent on the precision of the standardisation, presuming the standardisation is performed on a material with a relatively high concentration of the element of interest (figure 3). This is not to imply that the standardisation is unimportant, but that larger errors in calibration are required in order to substantially affect trace elements compared to major elements. It is also clear that using a standard with a low concentration (and therefore low net intensity) for the element of interest will result in large errors in the final trace element analysis relative to using standards with high concentrations of the element of interest. This should be considered when assessing the errors related to interferences. Interference calibrations will be, in almost all cases, associated with peak tails with low count rates. They can significantly reduce precision when errors are propagated.

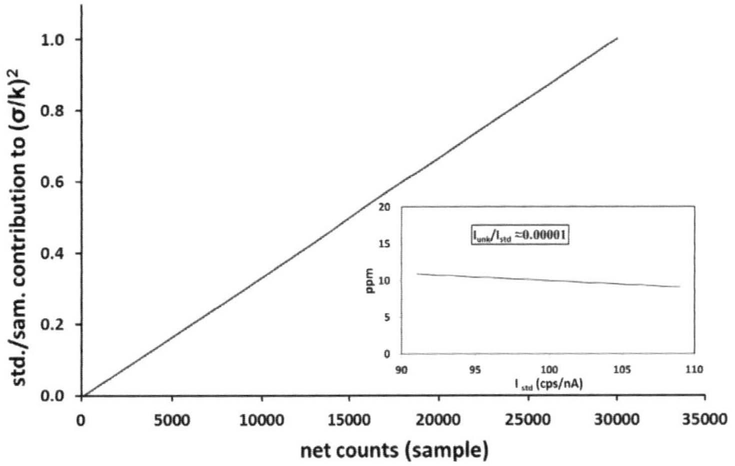

Figure 3. Plot of theoretical variation of contribution of sample and standard counts to the overall precision versus net counts. Inset: theoretical variation in concentration for ~ 10 ppm concentration in sample with ~ 100 cps/nA in the pure element standard, so $I_{unk}/I_{std} \sim 0.00001$.

The accuracy question is similar, as pointed out by Robinson et al. [4]. For major element analysis, the accuracy is dominated by the peak counts for both sample and standard as Pk-Bkg is large for both (so $C \approx f(N/N_S)$), but at trace concentrations, the largest source of inaccuracy is the determination of net counts on the unknown ($C \approx f(N\text{-}N(B))$. A theoretical example is shown in figure 3.

With a standard net intensity of 100 cps/nA, and unknown net intensity of 0.00001 cps/nA (so about 10 ppm), 10 % accuracy error in the standard calibration on known compositions only results in inaccuracies on the order of 1 ppm in the final result.

The sensitivity of analysis can be quantified by calculation of the minimum detectability limit. This has been approached statistically in several different ways, including the simple approximation of Leibhafsky et al. [21]. Based on the Poisson statistics of X-ray emission spectrography, the precision of the counts can be estimated by $N^{1/2}$. Leibhafsky et al. [21] suggest that a peak is detectable from the background at $3\ (N_B)^{1/2}$, where N_B is the number of background counts. The clear implication, and the basis for all trace element analysis, is that detectability is essentially a characterisation of background. The concentration at the limit of detection (C_{DL}) can then be expressed, in simplified fashion, by [20, 22]:

$$C_{DL} = \frac{3\sqrt{\overline{N(B)}}}{\overline{N}_S - \overline{N(B)}} C_{Std} \tag{3}$$

where C_{Std} is the concentration of the element of interest in the standard, $\overline{N(B)}$ is the measured background counts (sample), and $\overline{N_s}$ is the measured peak counts (standard).

Another simplified estimation can be made from the Ziebold [6] expression:

$$C_{DL} > 3.29a/(ntP^*P/B)^{1/2} \tag{4}$$

where a relates composition and intensity, n is the number of measurements, t is the measurement time, P is the count rate (pure element), and P/B is the peak to background ratio (pure element).

More advanced techniques for estimating the minimum detectable concentration also take into account the chosen confidence level, and standard deviation of measurements [7]:

$$C_{DL} = \left[{}^{C_S}\!/_{(\overline{N_S} - \overline{N_{SB}})}\right] \cdot \left[2^{1/2}(t_{n-1}^{1-\alpha})^{S_c}\!/_{n^{1/2}}\right] \tag{5}$$

where C_{Std} is the concentration of the element in the standard, $\overline{N_S}$ and $\overline{N_{SB}}$ are the mean counts for peak and background on the standard, n is the number of repetitions, S_c is the standard deviation of the measurements, and $(t_{n-1}^{1-\alpha})$ is the student factor at the selected confidence level.

Ancey et al. [23] propose a similar approach, including two confidence factors:

$$C_{DL} = \left(\frac{FC_{Std}}{P_{Std} - B_{Std}}\right) * \left(\frac{\lambda(\alpha,\beta)}{2t_{sam}}\right)\sqrt{1 + \frac{4B_{sam}t_{sam}(1 + a_{sam}^{-1})}{\lambda(\alpha,\beta)}} \tag{6}$$

where F is the calculated correction factor, C_{std} is the concentration in the standard, P_{std} is the peak intensity of the standard, B_{std} is the standard background intensity, t_{sam} is the peak counting time of the sample, B_{sam} is the sample background intensity, a_{sam} is a constant relating differences in peak and background acquisition time, and $\lambda(\alpha,\beta)$ is the confidence level for the two risks (α,for the possibility that the conclusion is C > 0, when in fact C = 0; and β, for the possibility that the conclusion is C = 0, when in fact C > 0). The Ancey et al. expression [23], at the 95 % confidence level is used in this paper.

The precision and sensitivity of the analysis depend on the count rate, counting time, and the number of points in the acquisition of the sampled compositional domain. Clearly, the analysis of very small areas limits the number of points that can be taken, especially in beam sensitive materials where several acquisitions from a small region can significantly compromise the integrity of the material, and potentially affect the accuracy of the result. Figure 4 shows the results of acquisitions at different count times comparing 15 to 8 kV for Pb Mα using integrated very large PET spectrometers (Cameca's VLPET). The improvement in sensitivity is most dramatic as several points are accumulated, however, in most cases, only a few ppm of increased sensitivity are gained beyond 5 or 6 points. In fact, the detection limit improves only from 10 ppm to 8 ppm with an additional 45 minutes of acquisition at 15 kV and 200 nA. Detection limits below 10 ppm are attainable under these relatively modest conditions, and can obviously be considerably lower if more spectrometers are integrated on the same element, if higher accelerating potential is used, or if beam current or count time in increased (or, to think of it another way, total Coulombs). Increasing precision in this way (current and time) must be weighed against the potential for instrumental drift and beam damage, which raise questions of accuracy, although software now permits the cycling of peak and background measurements in order to minimize drift and damage effects. Note also that the detection limit achieved per unit time is directly related to beam accelerating potential (overvoltage), but that the use of relatively low voltage (below 10 kV) for high spatial resolution analysis still permits excellent count rates and sensitivity if the requisite detectability limit is in the 10 s of ppm. Performance is dramatically improved by the use of high intensity spectrometers which, in the case of Pb Mα, can yield count rates at 7 kV that are similar to those attained at 15 kV with standard monochromators (figure 5). In situations where the trace elements of interest have low critical excitation potentials, high spatial resolution analysis may be best accomplished with direct acquisition of only the trace components, using a hand-entered major element composition for matrix corrections, or performing the overall analysis with two conditions if appropriate (for example, 7 kV for Pb and actinides, 15 kV for rare earth elements and phosphorous in monazite analysis). Note that the use of high intensity spectrometers are a great advantage in higher spatial resolution trace element analysis, and can also benefit efficiency, particularly to count for less time (or use lower current) in cases where beam damage is a significant issue.

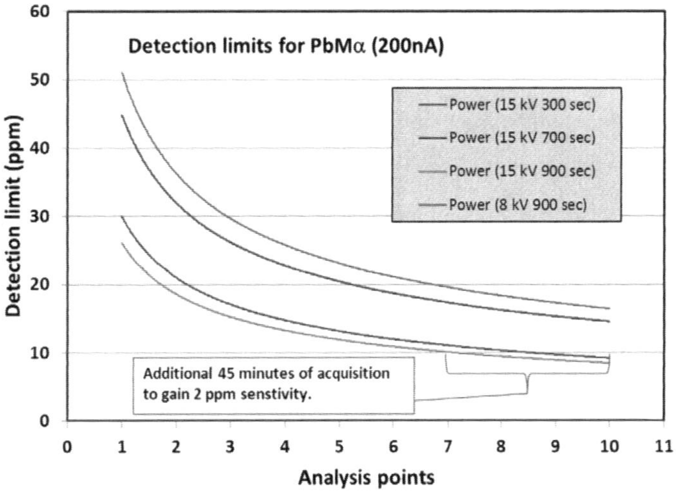

Figure 4. Measured detection limits (95 % conf.) for Pb Mα at different count times as a function of the number of acquired points. 200 nA beam current, 15 and 8 kV accelerating potentials. Counts integrated from two VLPET spectrometers. The measurements were made on Moacyr monazite, a UMass internal consistency standard.

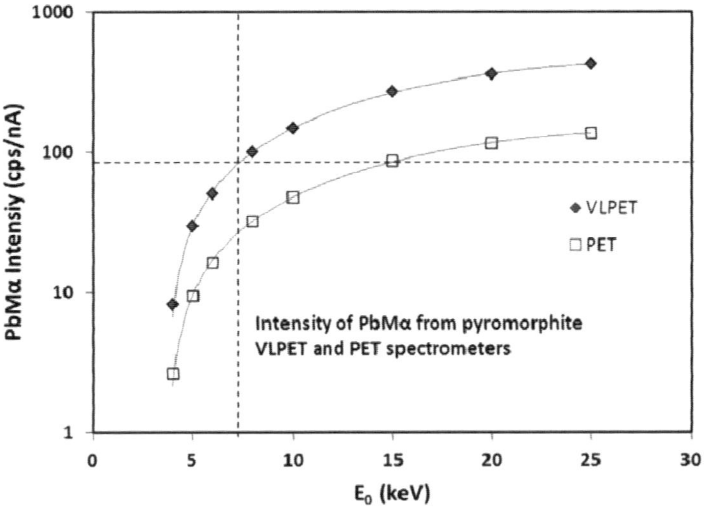

Figure 5. Intensity of Pb Mα as a function of beam accelerating potential comparing conventional PET to VLPET spectrometers.

The relationship between precision and spatial resolution must also be weighed when considering options for the analytical X-ray line to use for the trace element determination. Precision can be increased by increasing overvoltage with the concomitant degradation in spatial resolution. However, higher accelerating potential could also access inner shells in some cases, for example, 20 - 25 kV could be used if U Lα or Pb Lα were to be used for analysis of U and Pb. In such a case, the emission volume for U Lα at 20 kV is actually smaller than the emission volume of U Mα analyzed at 15 kV as the binding energy is much greater for U LIII compared to U MV. In a full analysis, the emission volume dimensions for U Lα are much smaller than that for P Kα (about 0.5 and 1.5 μm depth respectively in monazite), which brings into question the viability of the matrix correction. Precision must be assessed based also on counting efficiency as Pb and actinide L-lines require LIF 200 or preferably LIF 220, sacrificing the high count rates of PET (especially the large and very large monochromators). Also, overlaps must be re-evaluated also, for example Pb Lα is directly overlapped by As Kα, which can occur as a major element in monazite. Absolute generalisations as to choice of analytical line for trace element analysis are not possible without evaluation of all the relevant X-ray generation and measurement parameters.

In situations where high spatial resolution can be somewhat sacrificed, then the beam accelerating potential can be increased, the beam current can be increased (with expected degradation of beam diameter at ultra-high current), and the beam diameter can be increased to mitigate beam damage. Such a situation is illustrated for Ti in quartz or Zr in rutile, where detection limits approach the 1 ppm level without extraordinary operating conditions [13, 24].

The spatial resolution of analysis (or analytical resolution) can be assessed from a theoretical approach which considers the X-ray emission volume, a function of the sample composition, element detected (ionisation potential, emission line energy), excitation volume, beam energy, and beam diameter [15], and can be expressed by:

$$D_{AR} = (D_{beam}^2 + D_{emission}^2)^{1/2} \tag{7}$$

This is the quadrature sum of the error functions representing the beam diameter and the diameter of the emission volume (that includes 99.5 % of the intensity) for the element of interest, and represents a first-order approximation as fluorescence at a distance is not considered. The dimensions of the X-ray emission volume can be obtained from Monte Carlo interaction models (i.e., CASINO [25]) and relationships can be calculated for various beam diameters. For high Z, high density materials, the analytical spatial resolution can be significantly improved by using smaller beam diameters at accelerating potentials below 15 kV (figures 6-8). As beam voltages substantially below 7 - 8 kV are seldom useful for most trace element work (ionisation cross sections too low at low overvoltage), practical trace element nano-analysis by EPMA is limited to the realm of 7 - 15 kV in high Z matrices, and where the beam diameter can be confined to diameters of 100 - 600 nm by use of high brightness cathodes ($REEB_6$ or Schottky). Clearly, maintaining small beam diameters at beam currents useful to perform trace element analysis are necessary, typically 100 - 500 nA in order to attain reasonable precision in 300 to 1000 seconds. The realized analytical resolution, as measured directly in boundary tests, for Th and Pb in monazite (monoclinic LREE, actinide-bearing phosphate) at high current are shown in figure 8. We can conclude, therefore, that trace element analysis at 500 nm spatial resolution is possible in materials of sufficiently high Z and density.

3. Precision and spatial resolution versus accuracy – Beam power density and stability

If one of the goals of investigation is to make exceptionally precise measurements of trace elements at the highest possible analytical spatial resolution, then the other must be to ensure that the results are reasonably accurate. One aspect of such measurements concerns the analysis of insulating, beam-sensitive materials, and what, in fact, constitutes beam sensitivity. The dielectric properties of many materials present problems with accuracy simply from the deposition of charge (see [26-29]). Measurements of elemental concentrations in materials such as glasses, carbonates, apatite ($Ca_5(PO_4)_3(OH,F,Cl)$) and other phosphates, feldspar minerals, and hydrous minerals in general are notoriously difficult, and experience significant beam damage and/or elemental instabilities at even conservative beam conditions. Trace element analysis in such materials suffers also. Other phases

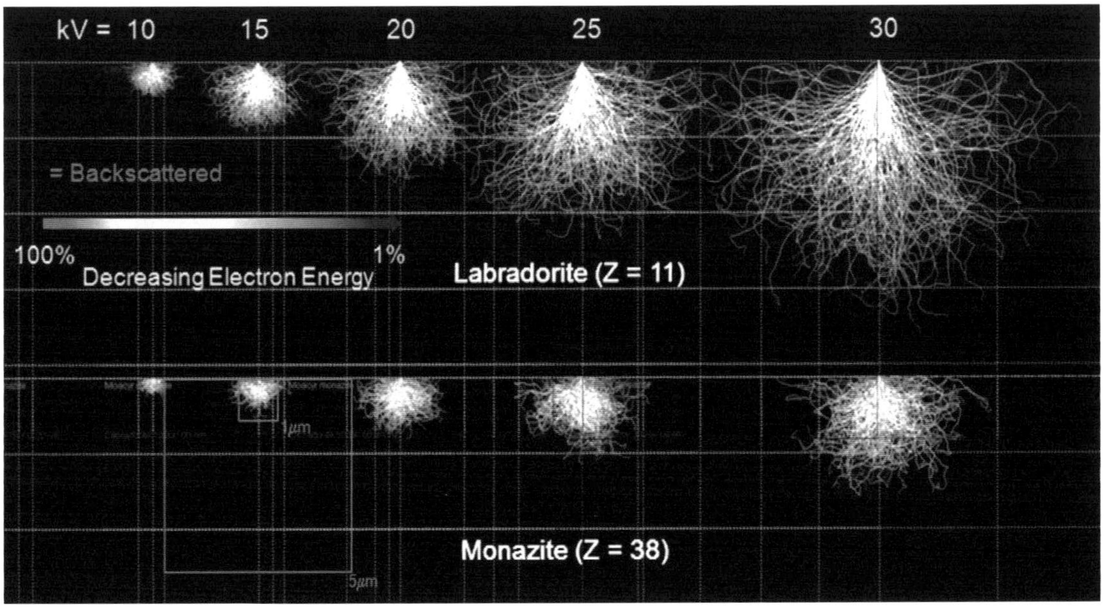

Figure 6. Spatial resolution in monazite ((La,Ce,Nd)PO_4) and labradorite ((Ca,Na)(Si,Al)$_4O_8$), illustrating Monte Carlo simulations of electron scattering as a function of beam voltage. Note red boxes indicating 5 μm and 1 μm dimensions. Z refers to average atomic number. Labradorite and monazite illustrate scattering in relatively low Z (labradorite) and moderately high Z (monazite) matrices.

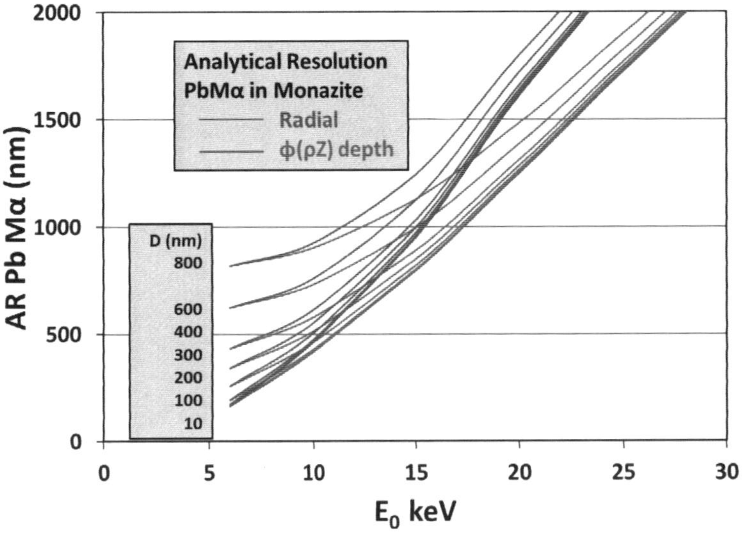

Figure 7. Analytical spatial resolution (AR) in monazite as a function of voltage and beam diameter (D nm). This is the sum of the quadrature of the beam diameter and emission volume containing 99.5 % of the Pb Mα $\phi(\rho z)$ intensity. Both radial and depth ($\phi(\rho z)$) components of analytical resolution are shown.

which may generally behave in a stable fashion during major element analysis may begin to destabilize at the high current densities necessary for high spatial resolution trace element analysis. For example, phosphorous loss is observed in monazite under high beam current density (15 kV,

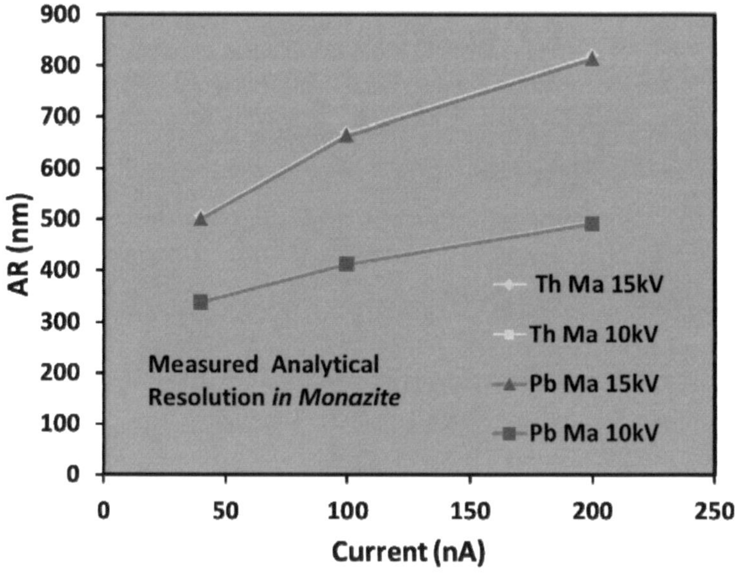

Figure 8. Measured analytical spatial resolution from boundary experiments as a function of beam current and accelerating potential, see [15].

200 nA) over a period of a few hundred seconds [14]. The high analytical spatial resolution permitted by the small electron scattering volumes at lower voltage in high Z, high density phases intensifies the problem. Gunn et al. [30] examined the problem of carbonate analysis, and concluded that adequate stability could only be obtained at values of beam power density (BPD) below 3 μW/μm^2. The estimation of BPD in this case is:

$$BPD = (I*E_0)/A \tag{8}$$

where I is the beam current (nA), E_0 is accelerating potential in (kV), and A is the beam area (μm^2). Here, we will modify this in order to account for the volume of excitation rather than landing area, and also account for the percentage loss of beam electrons emitted from the specimen by backscattering:

$$Simple\ Beam\ Power\ Density = [I-(I*\eta)] * E_0/ V \tag{9}$$

where η is the backscattered electron coefficient, the fraction of beam electrons scattered out of the specimen (determined via CASINO), and V is the estimated interaction volume. Note that this is still an oversimplified model as it does not account for effects such as secondary electron emission, space charge build-up in insulators (internal potential changing the landing energy and deforming the interaction volume), or coating (scattering and grounding of some charge). The interaction volume is calculated as an edge-rounded disk (a shape predicted by Monte Carlo modelling of defocused beam scattering in solids):

$$V = \frac{1}{3}\pi h(3a^2 + 3b^2 + h^2) \tag{10}$$

where a is the radius of the excitation volume, b is the radius of the beam, and h is half the total depth of beam penetration. The radius of the excitation volume is:

$$a = \sqrt{b^2 + d_s^2} \tag{11}$$

where the scattering diameter (d_s) is estimated at beam size = 0.

Calculated results for this simplified beam power density are shown for monazite (Z = 38) in figure 9, and labradorite feldspar (Z = 11) in figure 10. Small beam diameter (below μm) at low voltage increases the electron dose in the specimen tremendously, and has the added effect of shifting the depth of charge deposition closer to the surface, increasing surface heating, charge effects (manifest as carbon deposition near the beam), and interaction with the conductive coating.

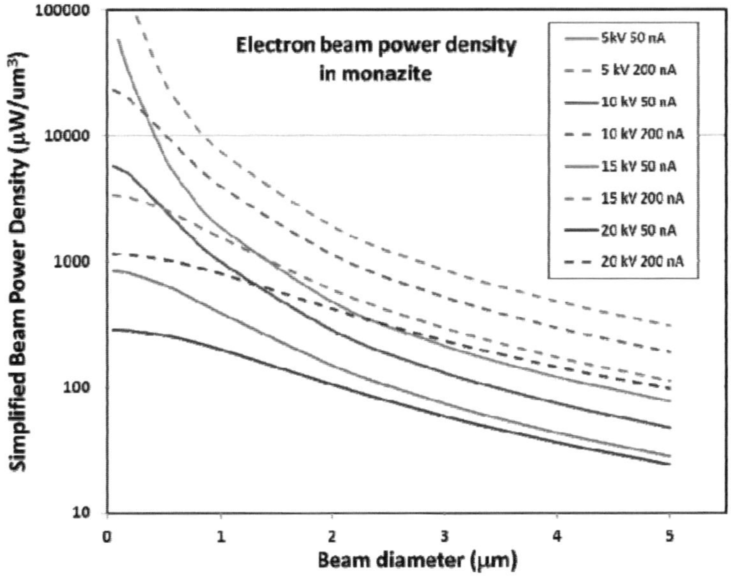

Figure 9. Estimated beam power density for monazite as a function of beam diameter, voltage and current. The influence of coating is not considered.

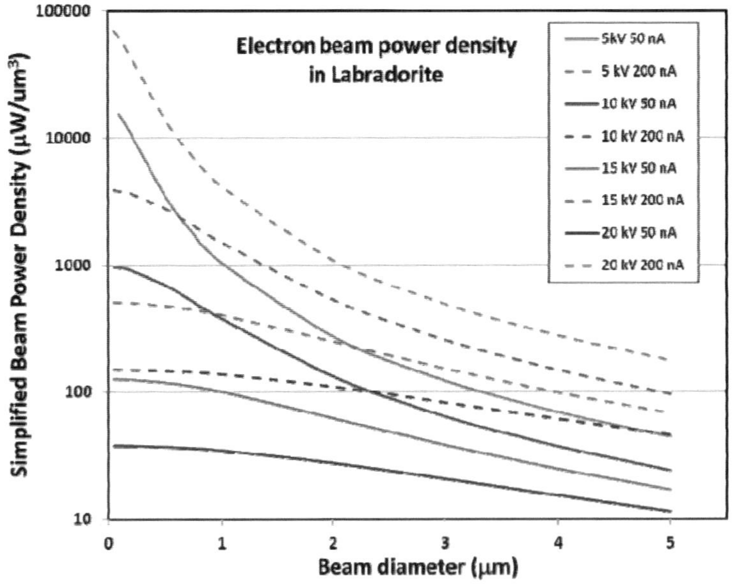

Figure 10. Estimated beam power density for labradorite as a function of beam diameter, voltage and current. The influence of coating is not considered.

The effects of internal charge build-up, heating, and radiation damage have to be assessed for each material to be analyzed. The stability of the measured element as a function of electron beam exposure must be ascertained to avoid inaccurate analysis. Assessing the stability of trace elements, however, is obviously not straightforward (low counting precision in stability tests). The stability of major element emission or absorbed current (figures 11-13) will generally have to serve as proxies, especially if the structural state of the trace elements of interest can be specified, but the effects of surface contamination build-up will also have to be assessed over the time required for a single point analysis. Figure 14 shows the percentage error for the integrated count times for Gd Lα in $GdPO_4$ at different count times and accelerating potential. The beam damage resulting from lower kV at high current is dramatic, leading to nearly 7 % error in the accumulated counts over 900 seconds at 10 kV. Clearly, multiple points with shorter count times are preferable, and the choice of voltage must be assessed based on the spatial requirements and error tolerance. The use of acquisition techniques such as time-dependent evaluation and extrapolation back to time = 0, and cycled peak and background during the course of the analysis [4, 5, 13] are important software developments in improving the accuracy of the analysis of beam sensitive materials.

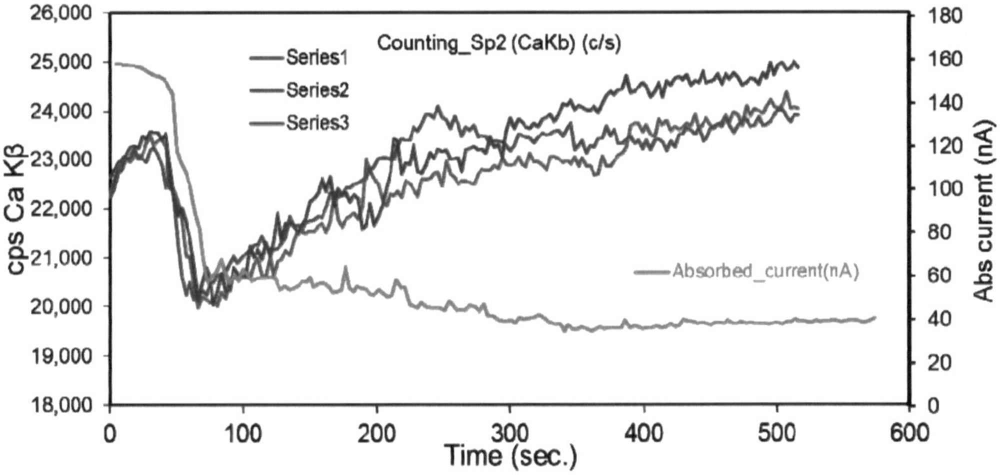

Figure 11. Ca Kα intensity with time in $CaCO_3$ (20 μm beam diameter, 200 nA, 15 kV), and absorbed current measured simultaneously. The resulting inaccuracy during integrated measurement will be large after 1 minute, but lower (compared to time = 0) after 10 minutes. Each series represents a different measurement on the same sample.

Another important variable is the effectiveness of the surface conductive coating. As long as samples and standards are coated simultaneously to ensure constant thickness, either thicker carbon, or metal evaporation offer a means to mitigate charge and thermal effects to some extent. Gold, with exceptionally high electrical and thermal conductivity, is ideal for moderating beam effects [14], but has the drawbacks of high electron and X-ray absorption, as well as producing a complex emission spectrum. The use of silver was proposed by Smith [31] for carbonate analysis, and copper has been used for oxygen analysis in insulators [32]. Plasma cleaning, followed by 2-layer coating first with aluminium (200 Å) then carbon (80 Å) has been proposed for monazite trace element analysis [33]. The use of metals for quantitative EPMA remains interesting and subject to question, and certainly measurement of oxygen through copper has inherent inaccuracy [34]. Results for major element analyses performed on monazite using an aluminium coat are very comparable to those done using a carbon coat.

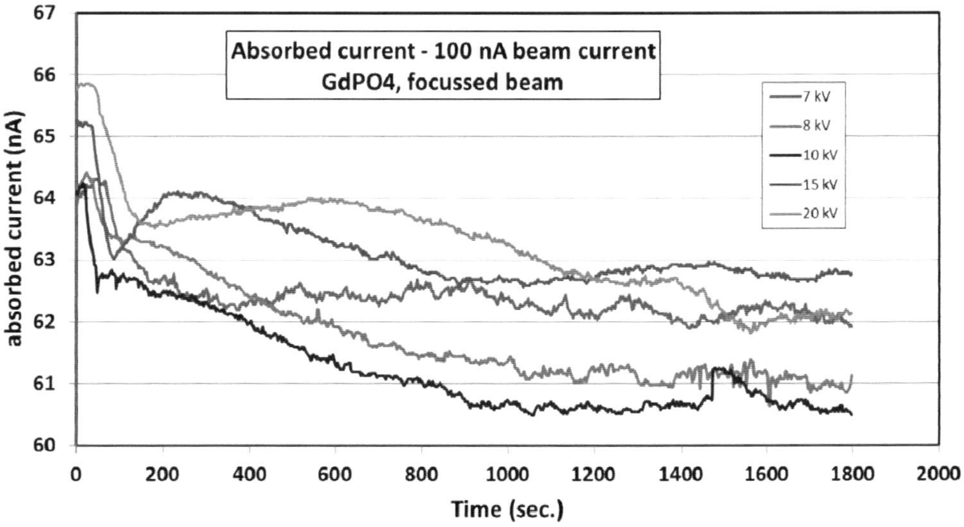

Figure 12. Absorbed current measured in $GdPO_4$ at varying accelerating potential (focussed beam, 100 nA, coating: 200 Å Al and 80 Å C) versus time. Instability in absorbed current is expressed analytically as increased counts over time (figure 13).

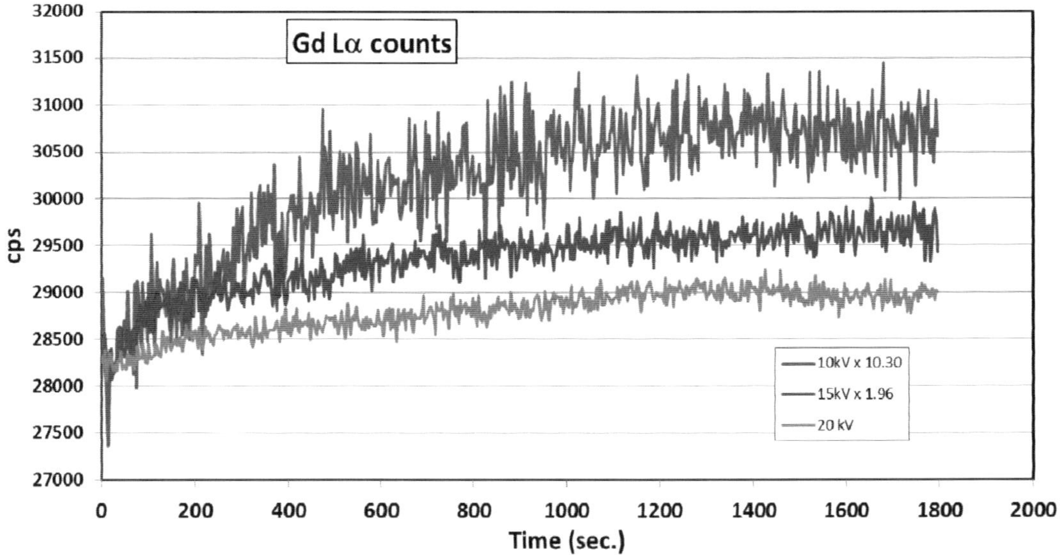

Figure 13. Gd Lα counts from $GdPO_4$ as a function of beam energy and time. Intensities of 15 and 10 kV curves are multiplied artificially to match the initial intensity of the 20 kV example for purposes of comparison.

4. Accuracy, contributions to error, and the blank test

Williams et al. [19] have described the basic contributions to error in trace element analysis applied to EPMA in terms of (1) short-term random errors (counting uncertainty); (2) short-term systematic error (session to session, beyond a single set of analyses: background, sample coating and calibration effects); and (3) long-term systematic errors affecting accuracy (effects stemming from quality of

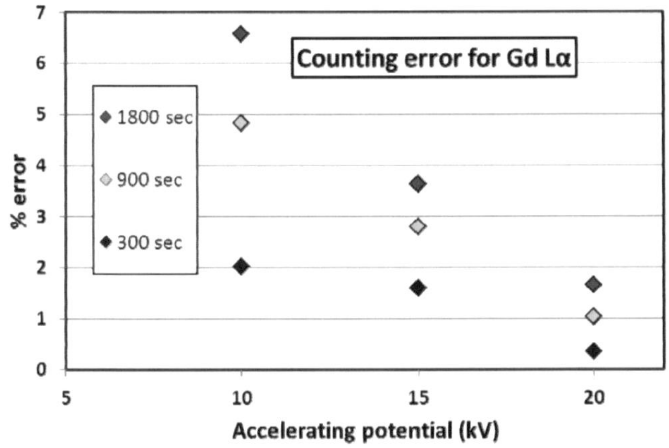

Figure 14. Counting error resulting from prolonged beam exposure at varying kV ($GdPO_4$, 100 nA, focussed beam – see figure 13). Error calculated by integrating all counts for the indicated time intervals, and is relative to the initial 20 kV count rate multiplied by the appropriate time. 10 kV and 15 kV initial values are scaled to be equivalent to the 20 kV values for comparison.

standards, evolving algorithms for interferences, background regression, dead-time corrections, ZAF factors). They stress population sampling as a means of developing a statistical basis for reporting overall error, and also advocate the analysis of trace element "consistency standards", e.g., some material which is analyzed before, during and after sessions in order to assess the cumulative effects of systematic error. This is a critical and simple test which allows the analyst to know that something significant has changed, perhaps very subtle at the major element level, but substantial at the trace level. Assessing the accuracy of analysis based on secondary standards is exceptionally difficult in the trace element realm, primarily due to a lack of materials appropriate to the unknowns and with a reliable, homogeneous trace element concentration of the element or elements of interest. Illustrating accuracy at 2 wt% may not translate at all to accuracy at 200 ppm, and illustrating accuracy for 300 ppm Pb in glass may not translate to accuracy at 300 ppm in a complex phosphate (different interferences, background curvature, etc.). To date, the best approach is to do the analysis by addressing the known sources of error as rigorously as possible. For trace analysis, interferences on peaks and background, and background shape contribute immensely to analytical error (see [15], and figure 15). The key concept is in understanding that the percentage error stems from the error on the net intensity (Pk – Bkg), such that at very low net intensity values, even the smallest perturbation in the background estimate results in very large accuracy errors (figure 16).

Although it remains difficult to create homogeneous, well characterized standards with trace amounts of something, making them with ***none*** of the element of interest can be done more reliably. This gives rise to the blank test [4, 13]. You should be able to analyze an appropriate blank and get a zero concentration result. Any other result, either positive or negative, indicates that there are problems to be addressed before proceeding to unknowns. Some sources of inaccuracy, such as "holes" in the background resulting from secondary diffraction effects in the monochromator can actually be corrected using a blank correction scheme in software [13]. Careful characterisation and analysis of synthetic, Pb-free $GdPO_4$ serves to illustrate several points (figures 17-20), as described below.

Characterisation of background is paramount in trace analysis (as Pk/Bkg approaches 1). In many respects, accurate trace analysis can be considered to be a measurement of background and what can be distinguished from it. Features within the spectral region of interest which can significantly affect the net intensity must be accounted for. This implies that the background spectrum must be evaluated at similar precision to the peak measurement. WDS scanning is essential for this evaluation, to note, in particular, where any interferences may exist, where absorption edges from major elements become significant, and the *shape* of the background, e.g., curvature. Noting where potential interference may exist is an important exercise, but neglects the extent to which unexpected lines can occur as a result of fluorescence at a distance, the presence of elements which should not be in the structure, but in fact

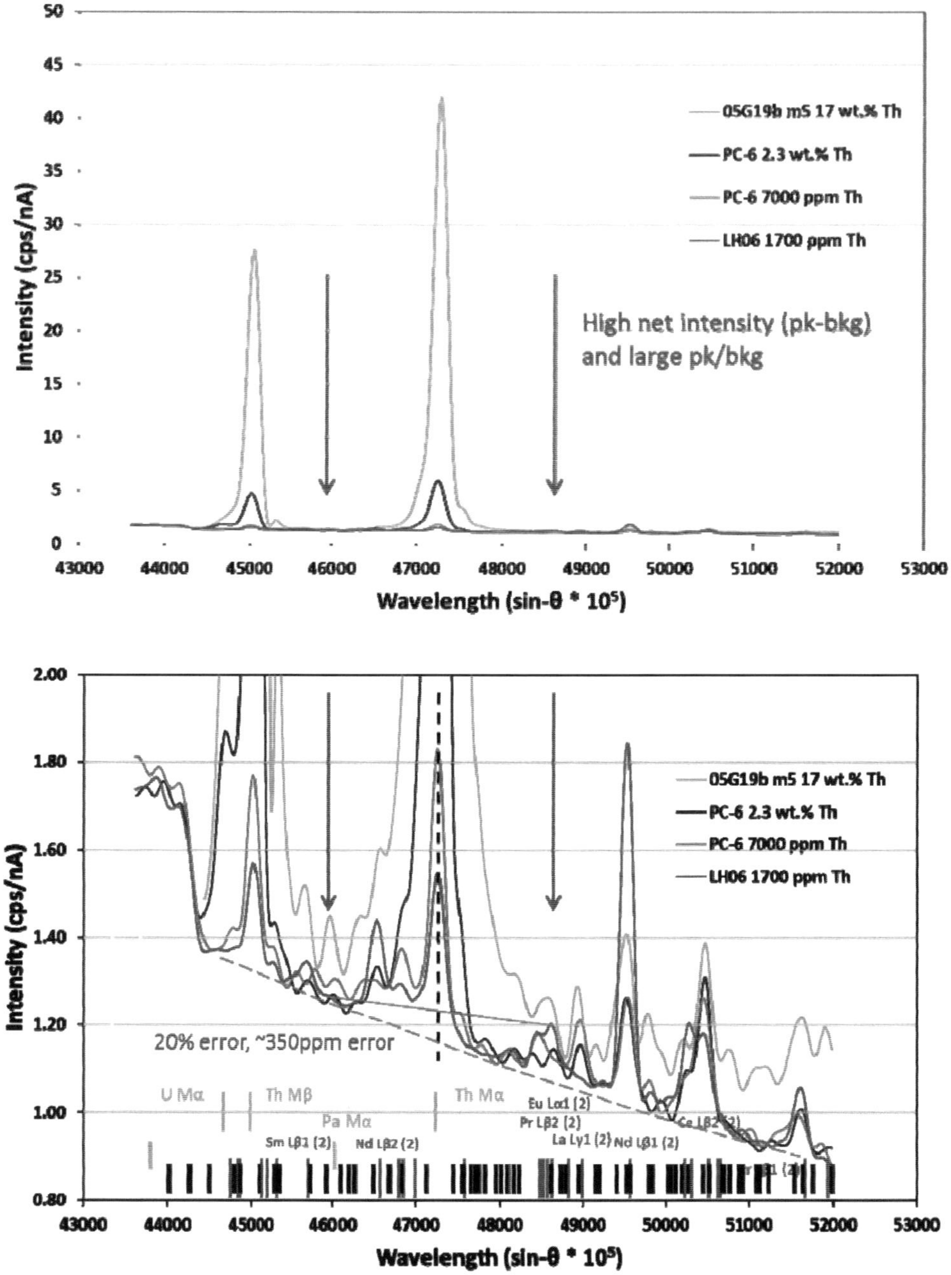

Figure 15. WDS scans of the Th M region of PET in monazites with differing Th concentrations. (a) Thorium as major element, high Pk/Bkg, minor errors on background are inconsequential. (b) Same as above except now scaled appropriate to trace element analysis. If the background positions shown in (a) are used, the error on the net intensity for the LHO6 example (1700 ppm) will be 20 % too low from background interference alone. Red vertical bars are major potential REE interferences; the black vertical bars are minor REE interferences.

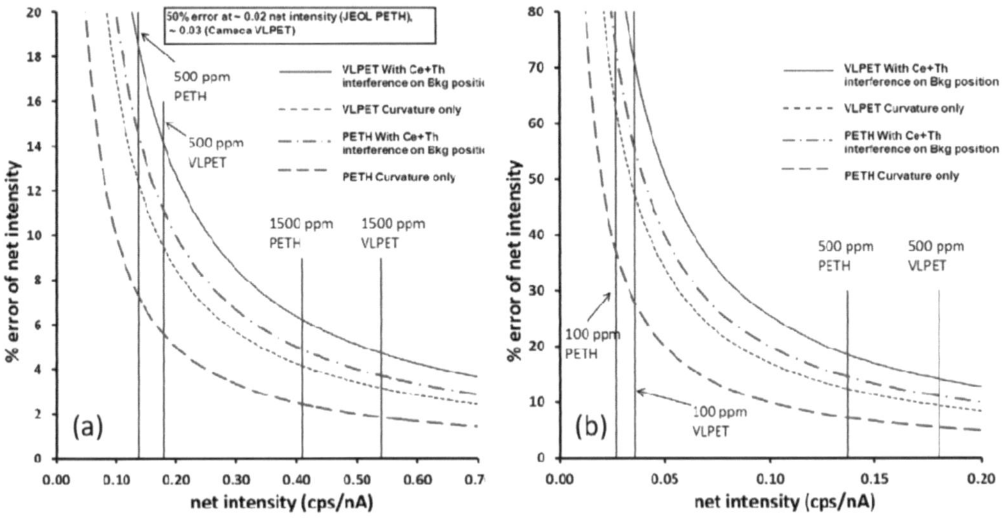

Figure 16. Background error comparing linear interpolation of Pb Mα background intensity using typical high and low λ positions, and exponential background from $GdPO_4$ data. Separate errors can be estimated for the case of curvature only (lower dashed line), and curvature plus interference (upper dashed line). (a) Error (%) on net intensity with decreasing net intensity (peak cps/nA - background cps/nA) for Pb Mα when using linear 2-point background interpolation (neglecting curvature) and adding the observed background interference at lowest intensity region between Pb Mα and Pb Mβ for monazite [15]. The interference portion of the error is dependent on the Th to light rare-earth ratio. (b) Detail of region below net intensity of 0.2, corresponding to concentrations below about 500 ppm for VLPET and PETH.

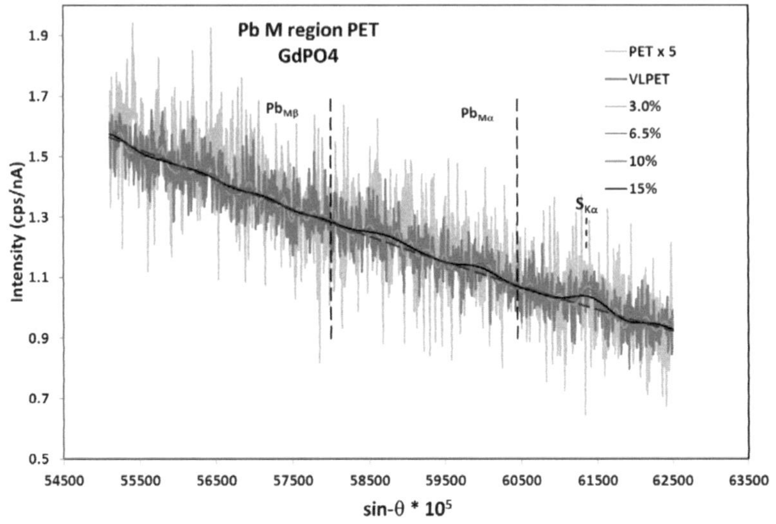

Figure 17. Detailed WDS scan data from $GdPO_4$, Pb Mα - Pb Mβ region of PET. Raw scan data comparing standard PET (scaled 5x intensity) and VLPET. Noise reduction via Savitsky-Golay filter improves accuracy of background and reveals true features. Smoothed curves for 3 %, 6.5 %, 10 % and 15 % data widths (percent of scan in each polynomial fit window) are shown.

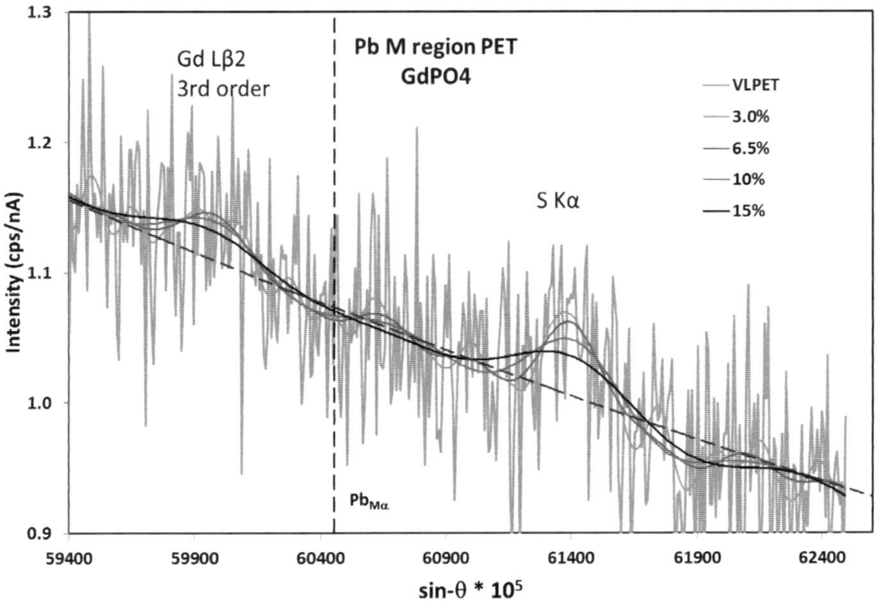

Figure 18. Detail of Pb Mα portion of figure 17 illustrating peak recognition (at .59900 and .61400 sinθ) and effect of varying Savitsky-Golay parameters. 3 % window width includes too much noise, 15 % window width broadening peak tails. Note that peak positions and true background intensities do not change as window size increases.

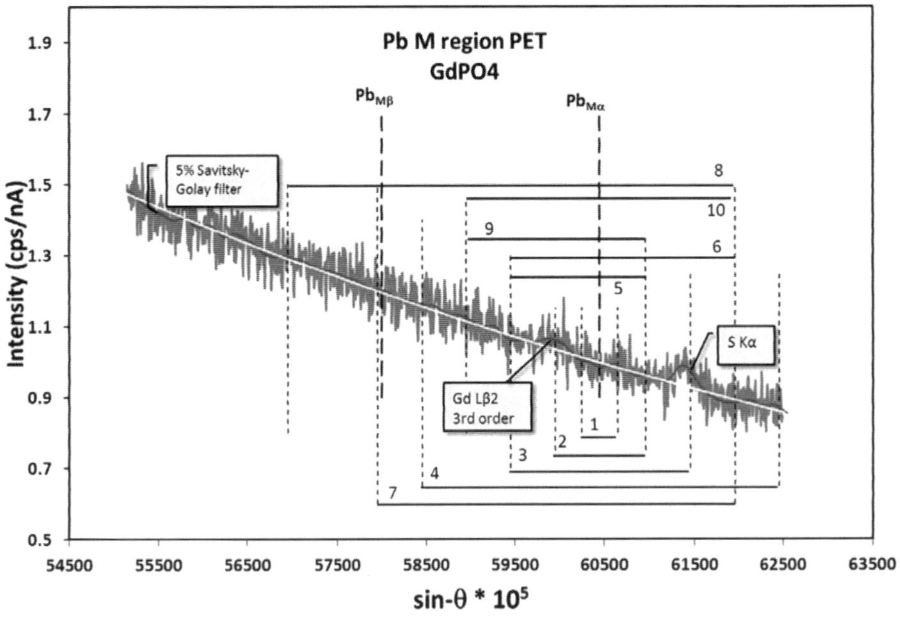

Figure 19. $GdPO_4$ scan showing the background positions (1 to 10) selected for quantitative blank analysis of Pb Mα. Results given in figure 20.

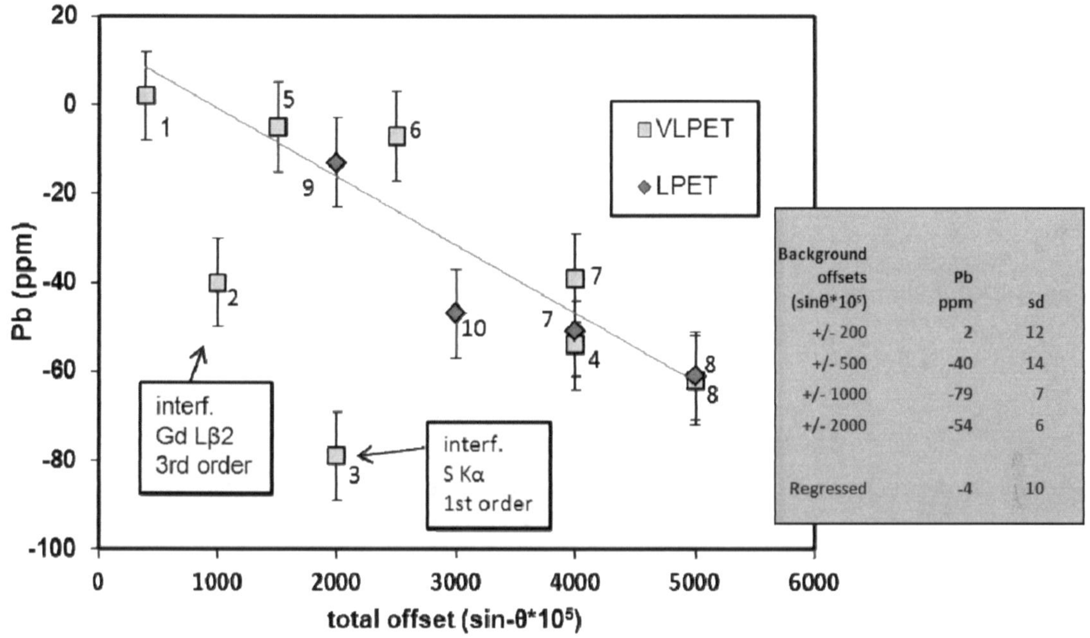

Background offsets (sinθ*10⁵)	Pb ppm	sd
+/- 200	2	12
+/- 500	-40	14
+/- 1000	-79	7
+/- 2000	-54	6
Regressed	-4	10

Figure 20. Blank test for Pb (should be zero) in synthetic $GdPO_4$. Quantitative results of trace element analysis done at 15 kV, 200 nA, for 500 sec. using linear background interpolation of varying positions (see figure 19 for positions). Concentration results versus total wavelength offset for background positions are listed in table to the right.

are, or where catalogued positions may be in error or are incomplete. In addition, high order line positions can vary significantly depending on monochromator d-spacing variation or spectrometer focus – even very slight tuning can measurably shift positions.

Strong absorption edge effects can cause considerable inaccuracy in situations such as Co measurement in Fe, Pb measurement in sulphides, or in a number of measurements of interference corrections such as those for Th on U Mβ, S on Pb Mβ, or Fe on Co Kα. The proper scaling of edge effects with concentration could potentially be corrected in software.

As background curvature is dependent on the Bremsstrahlung emission characteristics of the sample, as well as the collection efficiency and X-ray focusing characteristics of individual spectrometers, the curvature must be established routinely as part of the general protocol. Quantitative WDS scanning and regression analysis is advocated [14, 15] in order to compute the curvature and correctly estimate the background intensity under the peak of interest. In this case, the error on the background is the error of the regression (r.m.s. error). Although such acquisitions are lengthy, this is a rigorous approach which allows features in the background to be accounted for, and for curvature to be calculated directly from the scan. An important step in the process is separating signal (features) from noise (count statistics). This is aided considerably by use of a low pass filter such as the Savitsky-Golay smoothing algorithm [35] which employs local polynomial regression over user-specified data windows. Care must be exercised in setting both the window width (i.e., percent of scan points included for local regression) and the polynomial order such that background features are seen, but noise is still adequately reduced. Regression of the Savitsky-Golay fit data can then be performed to obtain the true background under the peak of interest, and the error can be estimated via evaluation of residuals (r.m.s. error) or by Monte-Carlo methods.

Although background measurement is daunting when examining complex spectra in actinide-REE bearing phases (see figure 15, what part of the scan actually does represent true background?), the quantitative scanning approach can yield good results. $GdPO_4$ is selected as a good test for Pb blank measurement relevant to monazite as there are no major peaks from this synthetic material which interfere with Pb, and the average atomic number and density are quite close to typical monazite ($Z = 38$, ρ = 6 g/cm^3). Pb-, Th-, and light rare-earth element (LREE)-rich monazite will have numerous peaks in this region, so this is an excellent evaluation to determine if the background curve fitting and overall zero (Pb) result are accurate, particularly under the high beam current conditions used for trace element analysis.

As the precision of the scan is increased, certain unexpected features become apparent. Noise filtering reveals two unexpected peaks in the spectrum. One, arising from the 3rd-order Gd Lβ_2 line. At 7.1 keV, this line would be expected to be filtered out with the tight pulse height analysis (PHA) window used in this acquisition as the Pb Mα line is at 2.34 keV. This is the result of the fact that high energy photons still enter the counter, producing an escape peak in the PHA spectrum which extends to much lower energy [17]. The other very minor peak in the spectrum is from S Kα. The presence of sulphur is completely unexpected in this case (although it can exist in natural monazite). This was traced to the carbon coat by analysis of the carbon rods used for coating in this instance. These peaks can be avoided in background analysis if one knows they are there, and the correct background can be regressed from the scan data. If the analyst is unaware, and chooses instead to do 2-point background interpolation, then background positions for Pb Mα analysis may be selected improperly and large errors can be incurred (figures 16 and 20). A background inadvertently obtained on the exceptionally minor (invisible with rapid scan) S Kα peak would ramp out the Pb Mα measured intensity, resulting in a concentration that is 80 ppm underestimated. Likewise, background measurement including the 3rd-order Gd Lβ_2 line would result in a concentration 40 ppm too low. Curvature alone will result in errors if 2-point linear interpolation is used, and that error increases as the background offset increases. Large background offsets result from avoidance of interferences in the background spectrum, a situation that is well illustrated by the measurement of Pb Mα (or Mβ) in monazite [15].

The quantitative WDS scan approach is adequate to the task of background evaluation in the above situation, and has proven to give reasonable results in very complex situations [32, 36, 37]. However, the balancing of scan resolution (precision) with acquisition time remains problematic. Typical scan acquisitions require a considerable time investment in background acquisition in the overall analysis of a single compositional domain (30 minutes or more for the scan). In order to resolve very minor background features, even more beam exposure time or intensity is required, never quite attaining the precision of the peak measurement, especially in the low ppm sensitivity range. An alternative approach, the multi-point background acquisition, has recently been developed (Probe Software, Inc. [38]). This method (figure 21) allows the user to set up to 24 points in the wavelength region around the peak of interest for automatic acquisition, statistical evaluation, and regression (see Allaz et al., this meeting). The advantage is that obvious minor peaks in the background (or holes) will be immediately flagged for exclusion from the background regression and calculation of intensity. Very high background precision is possible with this technique in less time than required for the full, high precision, WDS scan of the same region.

Of particular significance to high spatial resolution trace element EPMA are the effects of fluorescence at a distance. Even in situations where the analyst has taken great care to generate an appropriately small excitation volume at a current and voltage that will still allow an apparently accurate analysis at sufficient sensitivity, critical elements may be fluoresced from boundary phases or compositional domains, which compromise the analysis. There are many possible cases, and one such example is the fluorescence of Th in high-Th material near low-Th monazite (figure 22). The high concentration of light REEs in monazite results in high intensity REE-L emission and fluorescence of Th (or Pb, or U) from the adjacent domain [15]. This is not a hopeless situation, however, as empirical corrections can be applied, or theoretical correction may be possible where the geometry can be modelled, using PENEPMA [39]. An example is the analysis of Ti in zircon near $FeTiO_3$ [40].

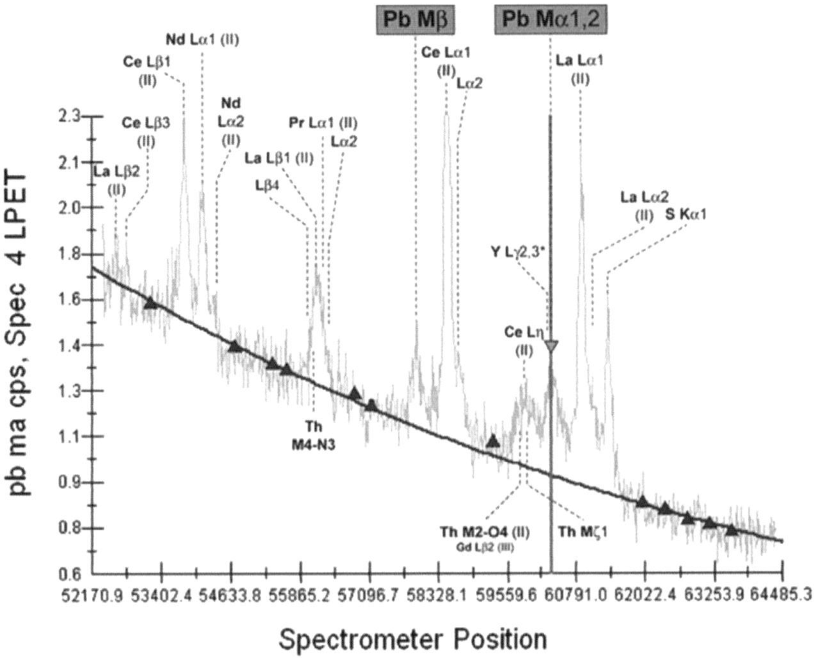

Figure 21. Spectrum (PET) of Pb Mα region in monazite illustrating new multi-point acquisition procedure. User-selected positions are automatically acquired, and resulting data are regressed and optimized to find true background. Points with residual values are eliminated and the line regressed again (for example, the point between Pb Mα and Pb Mβ will be removed). Curvature and interferences are appropriately accounted for.

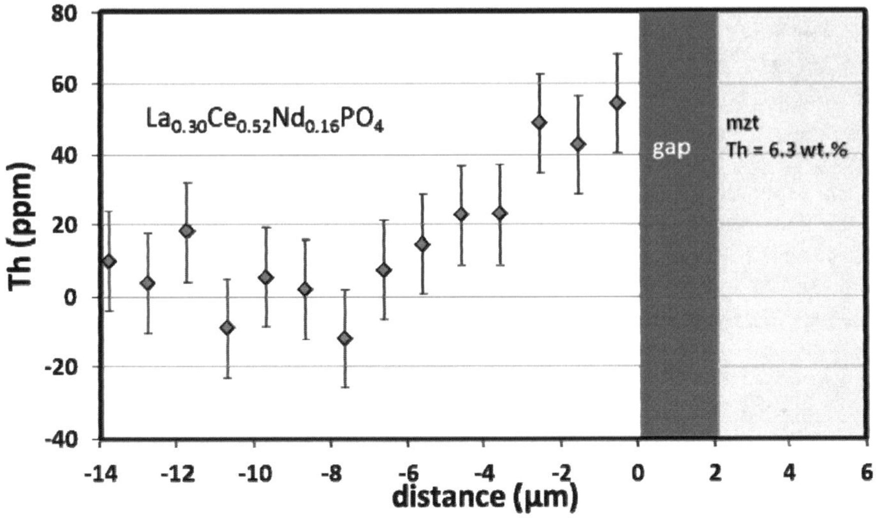

Figure 22. Boundary fluorescence test showing apparent Th concentration in synthetic monazite (Th-free), next to natural Moacyr monazite [15]. Each point represents an average of 6 measurements at the given distance. Individual point analyses were performed at 15 kV, 200 nA (LaB6 cathode), 500 s count time and VLPET spectrometer. The gap is epoxy filled.

Conclusions

Trace analysis by EPMA has many interesting challenges, and new developments in instrumentation and software are resulting in improved potential analytical spatial resolution, detectability limits, and accuracy. Precision, although critical, is not the largest obstacle, especially as modern instruments offer much higher collection efficiency along with stable high beam current output to enable lengthy acquisitions. Accuracy, however, remains the great challenge. With care, the analyst can obtain meaningful results for a number of elements, in some cases approaching the single ppm level in regions below 1 μm in size. Among the most challenging is the situation in which very high spatial resolution, along with high sensitivity are required. Beam damage, internal charge effects, and surface contamination compromise results, and must be considered when setting up the conditions of analysis, and in interpreting results.

The spatial resolution of analysis can extend to well below 1 μm in low voltage (below 15 kV) analysis of high average atomic number phases. However the use of high current and/or lengthy beam exposure time in order to achieve highly sensitive analyses exacerbates damage and charge effects, particularly in insulating materials, bringing accuracy into question. Higher voltage mitigates some of this, and increases count rates, but the obvious tradeoffs are spatial resolution and high matrix corrections. Possible work-arounds involve the use of metallic coatings, and software which cycles peak and background measurements, and can correct for systematic changes in count rates during the analysis.

Trace analysis by EPMA is fundamentally a characterisation and measurement of background, and even minor inaccuracies in the background determination produce extremely large errors in net intensity as increased sensitivity is attempted. Therefore, just extending the protocols for major element analysis into the trace realm is unsatisfactory. Detailed wavelength scanning is critical, and must be performed at a precision determined by the expected net counts at the peak. Such scans can be acquired quantitatively (and accurately smoothed, i.e., Savitsky-Golay) in order to regress background curvature for more accurate background intensity determination. Background curvature and interferences render simple 2-point linear interpolation inadequate (especially in complex materials), potentially resulting in tens to hundreds of ppm errors, but use of quantitative scanning or the new multi-point regression technique offers both high precision and improved accuracy for trace element work.

Blank testing is a useful method for evaluating at least some aspects of the analytical method, and remains one of the few reliable measures in the absence of appropriately low concentration secondary standards. The use of consistency standards offers a method to monitor short term systematic error. Neither of these ensures accuracy in the analysis of unknowns, but they do offer a means to monitor performance and test basic protocols. The existence of interferences on peak and background remain critical as even extremely minor features can result in large errors in the trace realm. For peak acquisition, accurate matrix-iterated interference corrections are absolutely necessary. Careful consideration of analysis near phase or compositional domain boundaries where fluorescence from outside of the primary excitation volume can occur is imperative. These are critical techniques in gaining confidence in trace element analysis.

Acknowledgments

This paper was substantially improved from the comments from two anonymous reviewers, and we greatly appreciate the efforts of coordinating editor János Lábár, and the EMAS 2011 organizing committee.

References

[1] Brown R J C and Milton M J T 2005 *Trends in Analytical Chem.* **24** 266
[2] Martin J W 2002 *The local chemical analysis of materials.* (Oxford: Elsevier)
[3] Alford T L, Feldman L C and Mayer J W 2007 *Fundamentals of nanoscale film analysis.* (New York: Springer)

[4] Robinson B W, Ware N G and Smith D G W 1998 Modern electron-microprobe trace-element analysis in mineralogy. in: *Modern approaches to ore and environmental mineralogy. Cabri, L J and Vaughan D J, eds. Short Course 27.* (Ottawa: Mineralogical Association of Canada) pp. 153-80
[5] Reed S J B 2000 *Microkim. Acta* **132** 145
[6] Ziebold T O 1967 *Anal. Chem.* **39** 858
[7] Goldstein J I 1967 *J. Geophys. Res.* **72** 4689
[8] Bodinier J L, Dupuy C, Dostal, J and Merlet C 1987 *Am. Mineral.* **72** 902
[9] Merlet C and Bodinier J-L 1990 *Chem. Geol.* **83** 55
[10] Donovan J J, Snyder A and Rivers M L 1993 *Microbeam Analysis* **2** 23
[11] Fialin M, Remy H, Richard C and Wagner C 1999 *Am. Mineral.* **84** 70
[12] Reed S J B 2000 *Mikrochim. Acta* **132** 145
[13] Donovan J J, Lowers H A and Rusk B G 2011 *Am. Mineral.* **96** 274
[14] Jercinovic M J and Williams M L 2005 *Am. Mineral.* **90** 526
[15] Jercinovic M J, Williams M L and Lane E 2008 *Chem. Geol.* **254** 197
[16] Reed S J B 1993 *Electron microprobe analysis.* (Cambridge, UK: Cambridge University Press)
[17] Pyle J M, Spear F S, Wark D A, Daniel C G and Storm L C 2005 *Am Mineral.* **90** 547
[18] Williams M L, Jercinovic M J, Goncalves P and Mahan K 2006 *Chem. Geol.* **225** 1
[19] Williams M L, Jercinovic M J and Hetherington C J 2007 *Ann. Rev. Earth Planet. Sci.* **37** 137
[20] Lifshin E, Doganaksoy N, Sirios J and Gauvin R 1999 *Microsc. Microanal.* **4** 598
[21] Liebhafsky H A, Pfeiffer H G and Zemany P D 1960 in: *X-ray microanalysis.* Engstrom A, Cosslett V and Pattee H, eds. (Amsterdam: Elsevier, North-Holland)
[22] Goldstein J I, Newbury D, Joy D, Lyman C, Echlin P, Lifshin E, Sawyer L and Michael J 2003 *Scanning electron microscopy and X-ray microanalysis.* (New York: Kluwer Academic)
[23] Ancey M, Bastenaire F and Tixier R 1978 Applications des methods statistiques en microanalyse. in: *Microanalyse microscopie electronique a balayage.* Maurice F, Meny L, and Tixier R, eds. (Paris: Les Editions de Physique)
[24] Baldwin J A, Powell R, Williams M L and Goncalves P 2007 *J. Metamorphic Geol.* **25** 953
[25] Drouin D, Couture A R, Joly D, Tastet X, Aimez V and Gauvin R 2007 *Scanning* **29** 92
[26] Cazaux J 1996 *X-ray Spect.* **25** 265
[27] Jbara O, Portron, B, Mouze D and Cazaux J 1997 *X-ray Spect.* **26** 291
[28] Boughariou A, Blaise, G, Braga D and Kallel A 2004 *J. Appl. Phys.* **95** 4117
[29] Collected papers in Microscopy and Microanalysis 10, no. 6 Special Issue: The characterization of nonconductive materials, 2004
[30] Gunn J S, Harrowfield I R, Proctor C H and Thresher R E 1992 *J. Exp. Mar. Biol. Ecol.* **158** 1
[31] Smith M P 1986 *J. Sed. Res.* **56** 560
[32] Love G, Cox M G C and Scott V D 1974 *J. Phys. D. Appl. Phys.* **7** 2131
[33] Dumond G, McLean N, Williams M L, Jercinovic M J and Bowring S A 2008 *Chem. Geol.* **25** 175
[34] Bastin G F and Heijligers H J M 2004 *Microsc. Microanal.* **10** 733
[35] Savitsky A and Golay M J E 1964 *Anal. Chem.* **36** 1627
[36] Mahan K H, Goncalves P, Williams M L and Jercinovic M J 2006 *J. Metamorphic Geol.* **24** 193
[37] Mahan K H, Wernicke B P and Jercinovic M J 2010 *Earth Planet. Sci. Lett.* **289** 76
[38] Probe Software, Inc., 885 Crest Dr., US-97405 Eugene, Oegon, U.S.A.
[39] Llovet X and Salvat F 2006 *PENEPMA, A Monte Carlo code for the simulation of X-ray emission spectra using PENELOPE.* (Madison, WI: Workshop Manual)
[40] Fournelle J H 2007 *Microsc. Microanal.* **13** (suppl. 2) 1390 CD

X-ray microanalysis of volcanic ash

S L Kearns[1,3] **and B Buse**[2]
[1] University of Bristol, Department of Earth Sciences, Queen's Road, Bristol BS8 1RJ, Great Britain
[2] now at: Nanyang Technology University, School of Materials Science and Engineering, Singapore

E-mail: stuart.kearns@bris.ac.uk

Abstract. The 2010 eruption of Eyjafjallajökull volcano in Iceland demonstrated the disruptive nature of high-level volcanic ash emissions to the world's air traffic.
The chemistry of volcanic material is complex and varied. Different eruptions yield both compositional and morphological variation. Equally a single eruption, such as that in Iceland will evolve over time and may potentially produce a range of volcanic products of varying composition and morphology. This variability offers the petrologist the opportunity to derive a tracer to the origins both spatially and temporally of a single particle by means of electron microbeam analysis.
EPMA of volcanic ash is now an established technique for this type of analysis as used in tephrachronology. However, airborne particulate material may, as in the case of Eyjafjallajökull, result in a particle size that is too small and too dispersed for preparation of standard EPMA mounts. Consequently SEM-EDS techniques are preferred for this type of quantitative analysis . Results of quantitative SEM-EDS analysis yield data with a larger precision error than EPMA yet sufficient to source the original eruption. Uncoated samples analyzed using variable pressure SEM yield slightly poorer results at modest pressures.

1. Introduction

The Eyjafjallajökull volcano erupted in spring, 2010. The resulting ash cloud caused massive disruption to air traffic through North Western Europe over a period of three weeks. The volcano has an elevation of 1600 m and last erupted discontinuously over a two-year period from 1821-1823. Previously it had erupted in 1612 and before that in about 920 AD [1].

Like many eruptions, the 2010 event was preceded by high levels of seismic activity and ground deformation for a period of about three months. This type of activity is typical of molten magma moving from deeper levels into the magma chamber beneath the volcano resulting in ground deformation. Several thousand earthquakes were measured during this period. The actual eruption started on March 20th in the form of a small-scale fissure vent appearing on the flank of the volcano away from the ice cap. This stage lasted until 12th April and resulted mainly in the effusion of lava rather than emission of ash. These flows, on average about 10 m thickness, travelled up to 3 km from the vent. On 14th April a new stage began which resulted this time in ash emission as well as lava flows. The explosive nature generated an ash column up to 6 km high. As the location was in the main crater underlying the ice cap, large volumes of ice were melted causing floods of mud and

[3] To whom any correspondence should be addressed.

volcanic debris (jokulhaups in Icelandic). The presence of this water turning into steam added to the explosive nature of the eruption and the near instant fragmentation of the ash.

The volcano is a stratovolcano, conical in shape and made up of previously erupted lava, pumice and ash. Other famous stratovolcanoes are Vesuvius, Mt. St. Helens and Mt. Fuji. These are characterized by their propensity for explosive eruption as opposed to a more passive effusion of lava.

2. Nature and chemistry of volcanic products

A typical volcanic ash may consist of three main constituents, the glass, some crystallized minerals and lithic fragments, ripped from the walls of the conduits through which the exploding magma is rapidly rising.

The size and shape of ash particles varies greatly depending on the style of eruption. Grain size is typically less than 2 mm diameter (e.g., figure 1) but a great deal of ash particles may only be 1 μm or smaller.

Figure 1. Backscattered electron (BSE) image of ash from the 1997 eruption of the Soufrière Hills volcano, Montserrat. Note grain size varies from 1 - 50 μm.

Ash is formed as a result of explosive eruption. These particles occur when gases, dissolved in the magma at depth, expand and escape violently resulting in explosion. The explosivity is exacerbated when surface water (or ice in the case of Eyjafjallajökull) becomes involved. These phreatomagmatic eruptions frequently yield shards with a particular angularity, which makes them very abrasive.

The range of compositions from a single volcano or volcanic region can be either very varied or fairly monotonous - best demonstrated by means of compositional variation diagrams such as a TAS (total alkalis – silica) diagram (figure 2). Figure 2 shows the entire range of compositions for the Iceland volcanic region. Some individual volcanoes such as Hekla (red dots) show a very broad range from basaltic through to rhyolitic compositions whereas others such as Katla (green dots lower left) are confined to basaltic composition through both historic and pre-historic eruptions.

Whilst a TAS plot shows clear discrimination of the bulk chemistry of a volcanic rock it simplifies the picture as each rock is potentially made up of several different minerals and frequently an additional glass phase representing the frozen residual melt (e.g., figure 3).

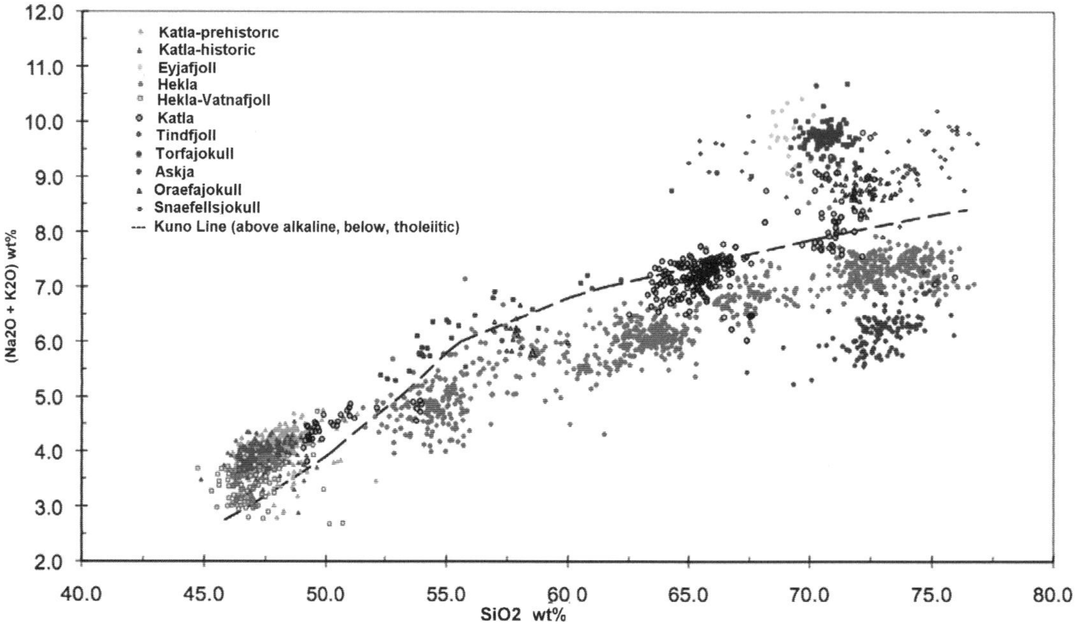

Figure 2. Total alkali versus silica (TAS) diagram for Iceland volcanics. Dashed 'Kuno' line discriminates between alkaline and tholeiitic series (after Le Maitre 1984 [2], Thordarsson pers. comm.).

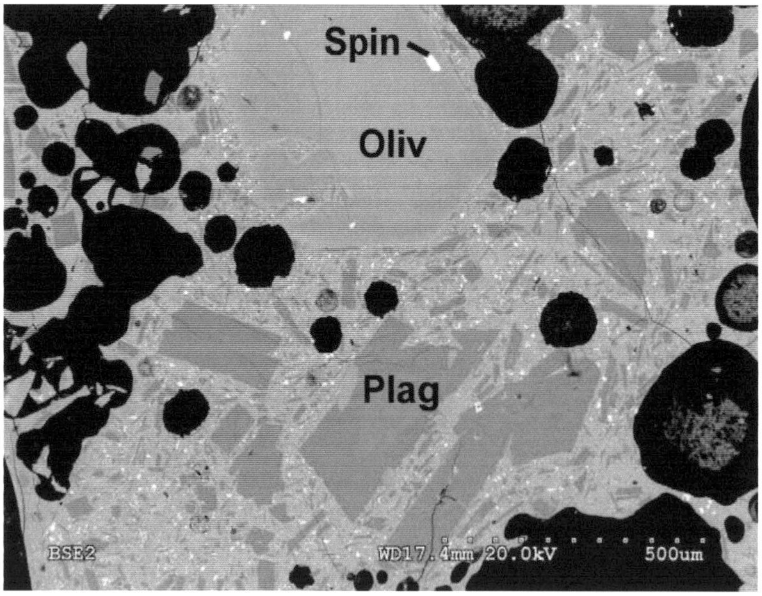

Figure 3. BSE image of ash from March 20th 2011 eruption at Fimmvörðuháls, prior to main phase at Eyjafjallajökull. Note diffuse presence of minerals with interstitial glass (Plag = plagioclase feldspar, Oliv = olivine, Spin = spinel).

3. EPMA of ash

Electron probe microanalysis (EPMA) of ash or more strictly the glass phase found in ash has been the subject of many scientific papers (e.g., [3-5]). A major issue with the analysis is the instability of ash under the electron beam. Alkalis are unstable under the beam due to heating causing diffusion of these cations away from the point of beam impact [6-7]. Lowering the temperature of a sample stabilizes the alkalis for the duration of the analysis (figure 4).

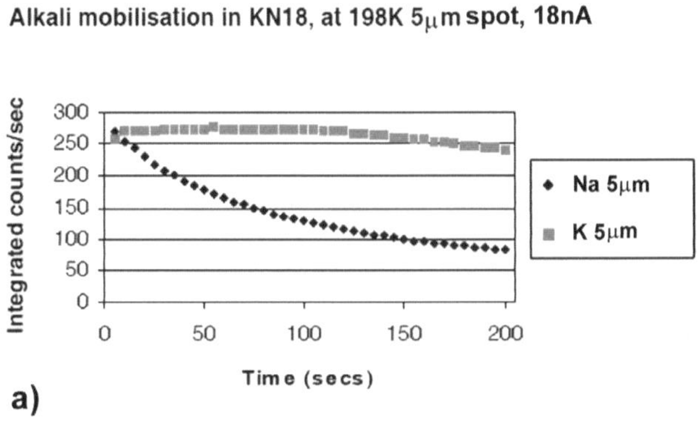

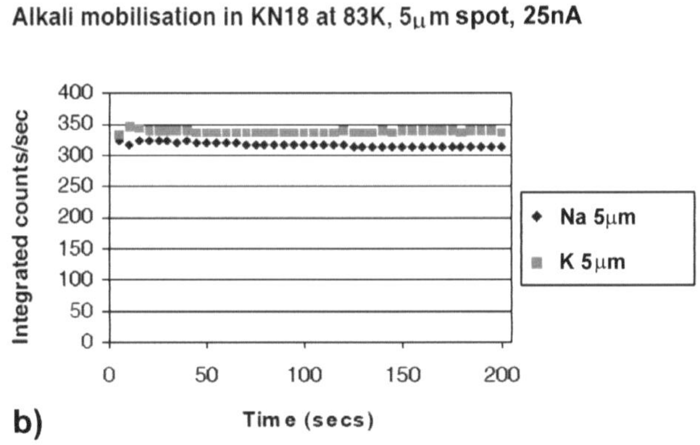

Figure 4. Decay curves from sodium and potassium for KN18 obsidian. a) Decay curve at ambient temperature shows significant sodium migration as soon as the sample is exposed to the electron beam. Potassium only starts to decay after an incubation period of about 100 s of beam exposure. b) Decay curves at liquid nitrogen temperature of 83 K. No alkali migration is observed in 200 s of exposure (Kearns et al. 2002) despite utilising a higher beam current (25 nA)

To mitigate the effect during analysis at ambient temperatures, it is necessary to use low beam currents to reduce the heating and, where possible to defocus the electron beam to an area of a few micrometres. An alternative approach is to let the alkalis diffuse away and use a zero time interpolation to recover the initial alkali concentration ([8]). This can yield good results but can also lead to erroneous data where the sample suffers an incubation period of up to several seconds before the onset of element migration (figure 4a).

There is also evidence from SIMS analysis [3] that anions, in particular hydrogen, as hydroxyl or molecular water can diffuse into the excitation volume – figure 5. Where measurement of water is important it may thus be necessary to perform SIMS analysis prior to EPMA. Nonetheless, glass is routinely and successfully analyzed by EPMA including estimates of water contents [9]. Invariably where a lower beam current is employed then the detection limits are relatively high compared to e.g., anhydrous minerals.

The results of analyzed glass compositions are extremely powerful when trying to provenance source volcanoes of certain deposits and indeed the timing of those events within the evolution of a volcano – many fascinating insights both within and outside earth sciences can be made by this technique such as the following two examples – melt inclusions and tephrachronology.

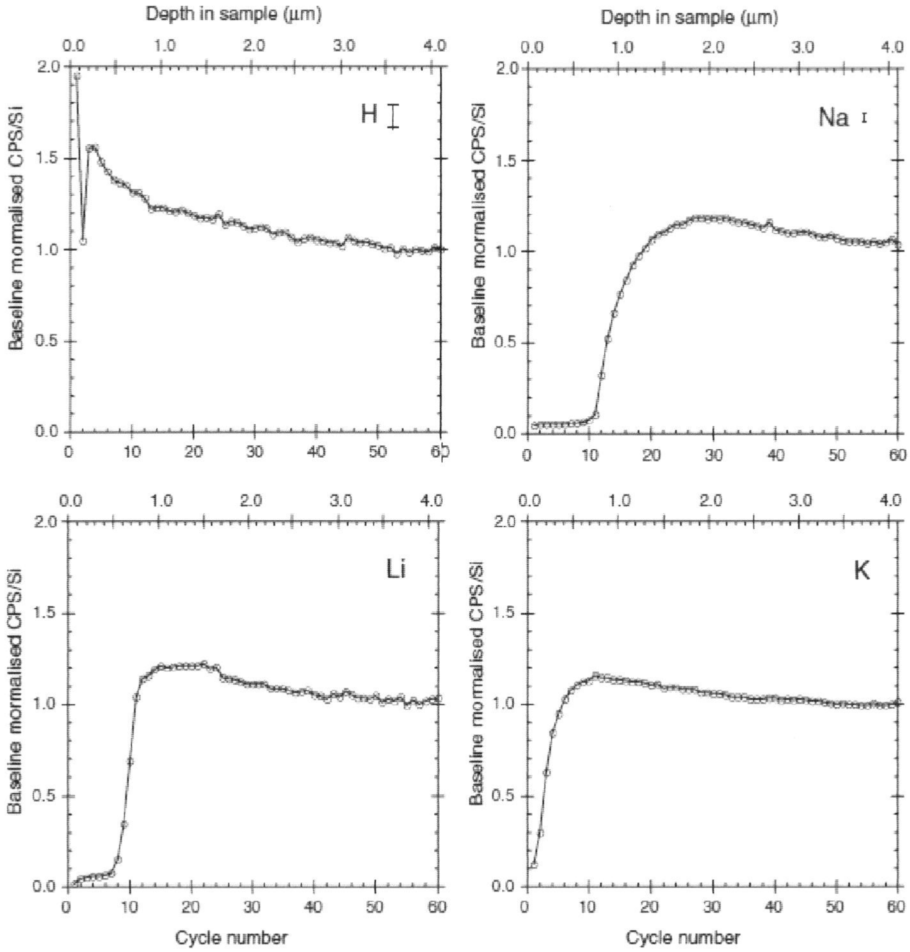

Figure 5. Post irradiation SIMS chemical profiles for H, Li, Na and K in Lipari obsidian glass. Sample irradiated by a 15 kV, 40 nA beam (Humphreys et al. 2005 [3]). Note hydrogen migrates into the excitation volume whereas Na, Li and K all migrate away from the excitation volume. Counting errors are given as vertical bar or are less than size of the marker.

Melt inclusions represent glass remnants of the magma that has been trapped by the growing mineral phase (figure 6). These are often significantly different from the final glass composition, as some of the minerals will have grown earlier, and yield much useful information of the evolution of a magma. Such inclusion analysis together with ion microprobe H_2O estimates enabled Blundy and Cashman [11] to deduce that crystallisation within magma at Mt. St. Helens occurred under rapid isothermal decompression.

4. Tephrachronology

As ash records significantly different composition between some eruptions, layered ash deposits can define a chronological interval in the geological and indeed historical past, e.g., figure 7. This is the basis of tephrachronology.

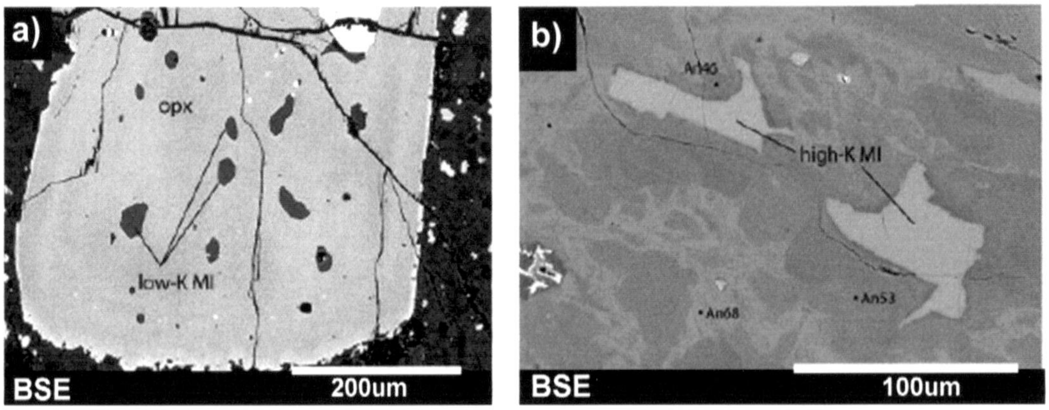

Figure 6. Melt inclusions in a) pyroxene and b) plagioclase from Volcan De Colima, Mexico. Opx = orthopyroxene, K-MI = potassium melt inclusion (Reubi and Blundy 2008 [10]).

Figure 7. Tephra (=ash) layers preserved in SE Iceland. Discreet ash layers are recognised from their composition and traced back to a source volcano. The layer can then be used as a reference marker in the stratigraphy. As ash layers become dispersed with distance from a volcano, so layers become thinner, however use of microanalysis techniques, principally EPMA, enables identification of for example, Icelandic tephras in the British Isles. Image from: http://www.bgs.ac.uk/research/highlights/ashInBritishIsles.html

An example of the application of tephrachronology is detailed in [12] from North Island, New Zealand. The Kaharoa tephra layer dated to the 1314 eruption of Mt. Tarawera records the onset of likely deforestation due to inhabitation by Polynesian settlers as evidenced by an increase in the amount of bracken pollen which post-date the eruption. Interestingly there is also controversial evidence that settlers were present before the eruption based on a rat-nibbled seed found from beneath the tephra. In total 25 tephras have been identified from Okataina caldera volcano in New Zealand over a time period of the last 61000 years. An estimate of the volume of each eruption is determined from the thickness of the layers.

Tephrabase http://www.tephrabase.org/ is a database of all recorded tephra layers maintained at School of Geosciences at the University of Edinburgh where some of the EPMA techniques of tephra analysis were pioneered in the 1980s.

5. Microbeam analysis of airborne particulates

Whilst the analysis of volcanic ash in landfall deposits is relatively routine by EPMA, the chemical characterisation of airborne particulates is much more problematic. The 2010 eruption of Eyjafjallajökull highlights this situation. The widespread disruption to air traffic demanded some very rapid assessment of the extent of ash accumulation on aircraft. The two major controls on ash dispersal from an eruption are eruption style and wind. That ash can travel hundreds to thousands of kilometres downwind from a volcano is recognized in ancient tephra deposits. Significant changes in wind direction and speed result in a complex and changing ash dispersal pattern. At Eyjafjallajökull the ash plume reached 9 km high before dispersal by winds drove the ash cloud southeast across the North Atlantic towards the continental mainland.

As the regulators, scientists and politicians discussed the tolerance on jet engines, commercial airlines were keen to know whether their aircraft had encountered the ash cloud. Sampling of aircraft is not straightforward. Many areas are prone to pick up dust on take-off and landing, other areas accumulate oil deposits and any dust associated in those may have been present for long periods. Seemingly the cleanest surfaces offer the best opportunity to sample dispersed airborne particles; thus surfaces from front of the wing, cowling of engines and windscreens may all yield useful material. Samples are collected on sticky tape peels directly. These are then coated in carbon and analyzed in the SEM.

In order to provenance the ash, to locate its source in both time and space, against known compositions (for example by reference to TAS diagrams) we need to know the accuracy of the quantification from the ED spectrum. Analysing small irregular unpolished material, we would expect a larger scatter of data compared with polished EPMA results.

Figure 8 shows the effect of grain size on the precision of analyses comparing a glass standard – KN18, an obsidian from Kenya, crushed to a fine powder with particles ranging from < 1 μm to 100 μm. The EPMA data, analyzed under conditions chosen to mitigate alkali migration closely matches those obtained by ICP-AES. The SEM-EDS data for the same polished samples shows an offset to lower alkalis and higher silica content. This is due to two factors; firstly the sample is susceptible to alkali migration when using a small spot size (figure 4), and secondly the EDS system uses a standardless phi-rho-z correction and, therefore, produces a normalized total to 100 %. As sodium is lost, other elements become re-normalized and as silica is the most prominent component, it shows a significant offset to higher concentrations. Similarly the system does not take into account non-analyzed elements, e.g., water, which will also adversely increase the silica concentration.

It is clear from figure 8 that the effect of analysing smaller particles is to reduce the precision of the analysis – the envelopes covering all analyses of a particular size range get larger with decreasing particle size such that at 1 μm the relative error is +/- 7 % for silica and +/- 25 % for alkalis (compared to < 2 % for EPMA data).

Monte Carlo simulation of Si Kα X-ray emission for a polished sample of a typical rhyolitic glass suggests electron beam penetration at 15 kV in excess of 3 μm (figure 9) and emitted Si Kα X-rays coming from depths up to 3 μm. When we model the same material as a particle of varying size on a carbon substrate we can observe the effect of the primary beam scatter which causes the loss of precision for the smallest of particles (figure 10).

Many of the ash particles observed are less than three microns in size so we can assume that at 15 kV at least some of the excitation volume will penetrate the substrate – in this case a carbon stub. We know also that the surface of the sample is irregular so the absorption path length will be variable and dependent on the detector geometry and position. Mass and absorption effects of particle analysis are discussed fully in Goldstein et al. - Chapter 9 [13], here it is important to note that these effects are different at different X-ray energies so the net result of irregular shape and variable size with respect to the excitation volume is complex.

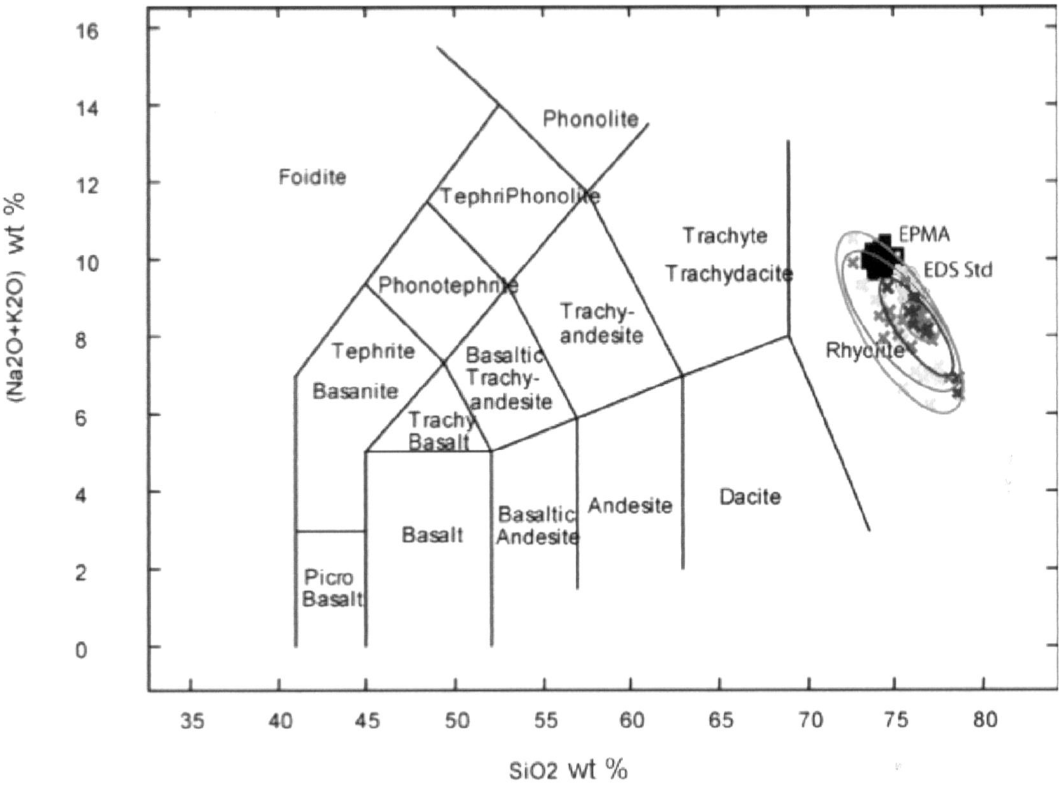

Figure 8. TAS plot of SEM-EDS analyses of obsidian, KN18. Black squares – WDS-EPMA data from polished sample. Yellow squares EDS data from polished sample. Red – 5 μm particle; purple - 4 μm; blue – 3 μm; yellow – 2 μm; green – 1 μm.

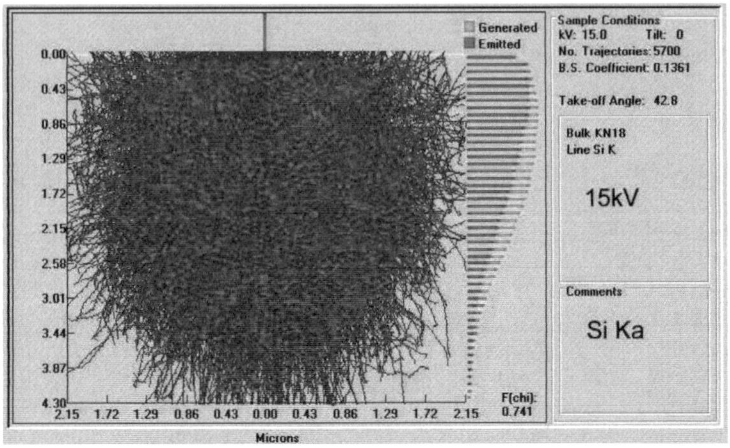

Figure 9. Monte-Carlo simulation of electron beam interaction with KN18 at 15 kV accelerating potential for a polished sample.

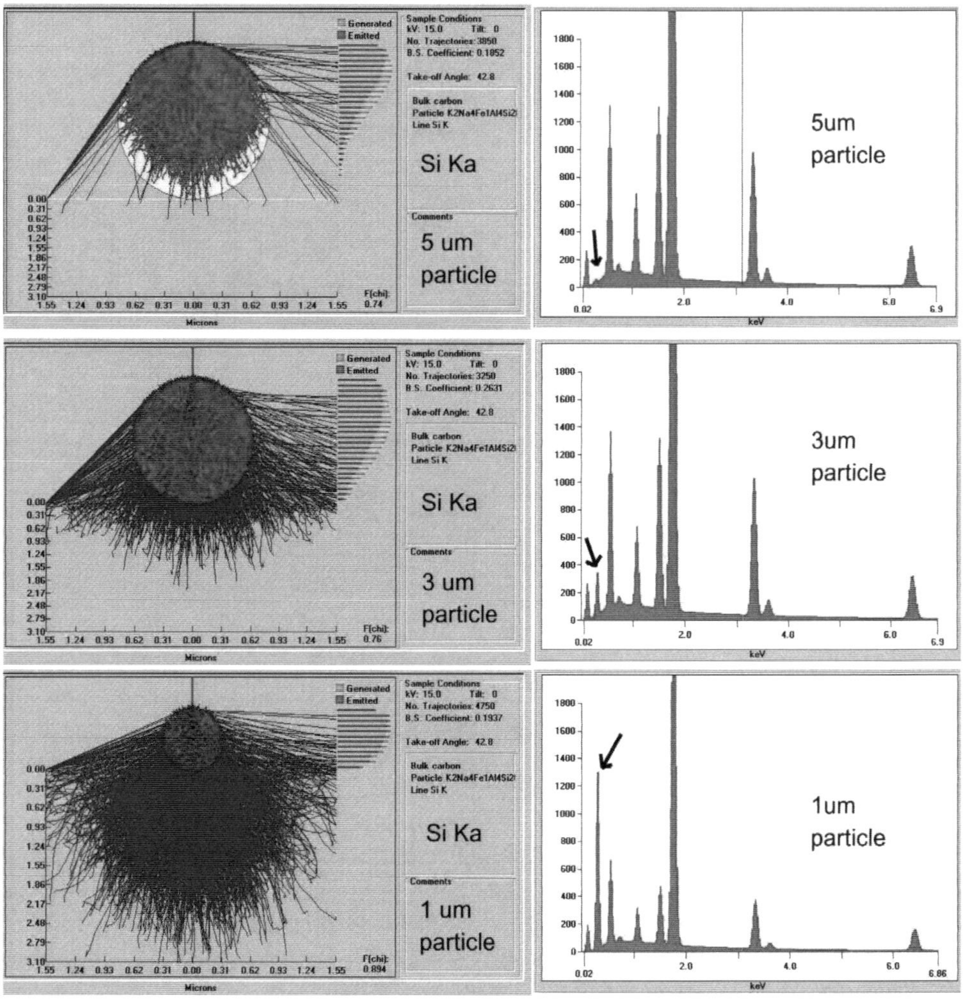

Figure 10. Monte-Carlo simulations of the electron trajectories and X-ray generation for Si Kα and resulting EDS spectra for particles of KN18 at 5, 3, and 1 μm. Carbon Kα X-ray arrowed on EDS spectra.

Our data above show that 5-μm particles yield results close to the known composition for KN18. It is tempting to assume that larger particles are therefore the preferred targets however, when particle size gets significantly larger then geometrical effects become more important. Figure 11 shows an 80-μm particle of KN18 and a series of analyses taken from a variety of surfaces. Resulting data show a significant loss of Na_2O from surfaces where absorption is seriously underestimated.

Analyses of the same material, uncoated under variable pressure conditions of between 5 Pa and 25 Pa shows similar results but error envelopes are slightly larger than those of figure 8. Newbury [14] demonstrates that at high pressures, typical of those used here, primary beam electrons are scattered into a skirt generating emitted X-rays from a significantly larger area than the equivalent focussed beam at high vacuum, thus a larger error would be predicted with analysis under this condition.

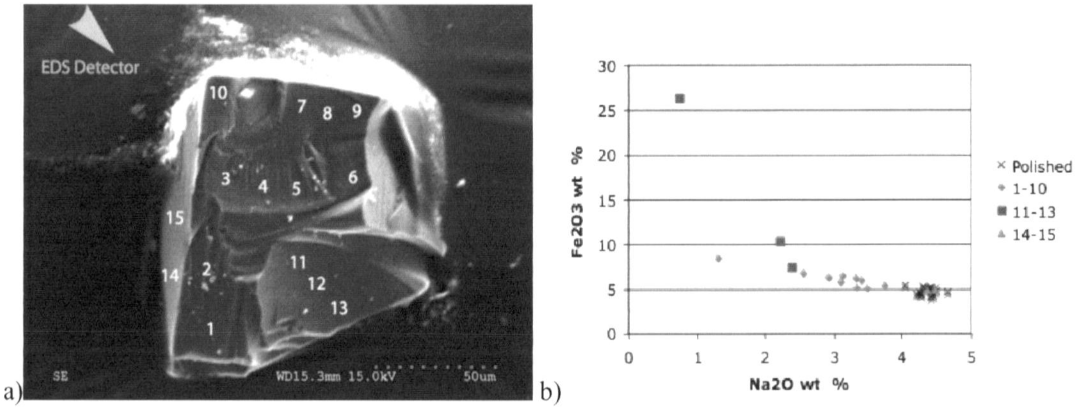

Figure 11. a) SE image of KN18 particle with analysis spots b) data plot of Na_2O (wt%) versus Fe_2O_3 (wt%) of analyses – analysis points are discriminated on basis of orientation to grain faces to the detector (1-10, 11-13, 14-15).

6. Conclusions

The chemistry of volcanic ash is indicative of a particular eruption. Often it is only necessary to look at major element discrimination to be able to separate one ash eruption from another.

Ash clouds may travel hundreds or even thousands of kilometres from source yet still provide a useful source of stratigraphic information via tephrachronology methods. Volcanic glass and ash particles from tephra layers are routinely analyzed by EPMA to yield high precision results where loss of alkalis can be considered negligible or can be corrected for.

EPMA analysis of the Eyjafjallajökull eruption in 2010 yielded first a basaltic component, and then later a trachy-andesite ash, which became entrained at jet stream altitudes and was responsible for severe air traffic disruption.

SEM-EDS analysis of volcanic ash particles results in less precise data than achievable by EPMA-WDS of ash collected at source. This is principally due to the small size (< 3 μm) of the particles and their irregular shape.

References

[1] Thordarson T and Larsen G 2007 Volcanism in Iceland in historical time: volcano types, eruption styles and eruptive history. *J. Geodyn.* **43** 118-52

[2] Le Maitre R W 1984 A proposal by the IUGS subcommission on the systematics of igneous rocks for a chemical classification of volcanic rocks based on the total alkali silica (*TAS*) diagram. *Aust. J. Earth Sci.* **31** 243-55

[3] Humphreys M, Kearns S L and Blundy J 2006 SIMS investigation of electron-beam damage to hydrous, rhyolitic glasses: implications for melt inclusion analysis. *Am. Mineral.* **91** 667-79

[4] Morgan G B and London D 1996 Optimizing the electron microprobe analysis of hydrous alkali aluminosilicate glasses. *Am. Mineral.* **81** 1176-85

[5] Kearns S L, Steen N and Erlund E 2002 Electronprobe microanalysis of volcanic glass at cryogenic temperatures. *Microsc. Microanal.* **8** Suppl. 2, 1562CD

[6] Autefage F and Couderc J-J 1980 Étude du mécanisme de la migration du sodium et du potassium au cours de leur analyse à la microsonde électronique. *Bull. Minéral.* **103** 623-9

[7] Hulinsky V, Jurek K and Gedeon O 1996 Experimental verification of theoretical models simulating the temperature increase in EPMA of glass. *Mikrochim. Acta* **Suppl. 13** 325-32

[8] Nielsen C H and Sigurdsson H 1981 Quantitative methods for electron microprobe analysis of sodium in natural and synthetic glasses. *Am. Mineral.* **66** 547-52

[9] Devine J D, Gardner J E, Brack H P, Layne G D and Rutherford M J 1995 Comparison of microanalytical methods for estimating H_2O contents of silicic volcanic glasses. *Am. Mineral.* **80** 319-32

[10] Reubi O and Blundy J 2008 Assimilation of plutonic roots, formation of high-K 'exotic' melt inclusions and genesis of andesitic magmas at Volcán De Colima, Mexico. *J. Petrology* **49** 2221-43

[11] Blundy J and Cashman K 2005 Rapid decompression-driven crystallization recorded by melt inclusions from Mount St Helens volcano. *Geol.* **33** 793-6

[12] Lowe D 2011 Tephrachronology and its application: a review. *Quat. Geochron.* **6** 107-53

[13] Goldstein J I, Newbury D E, Echlin P, Joy D C, Lyman C E, Lifshin E, Sawyer L and Michael J R 2003 *Scanning electron microscopy and X-ray microanalysis*, 3rd Edition. (New York: Kluwer Academic/ Plenum Publishers)

[14] Newbury D E 2004 Assessing charging effects on spectral quality for X-ray microanalysis in low voltage and variable pressure / environmental scanning electron microscopy. *Microsc. Microanal.* **10** 739-44

An inter-laboratory comparison of EPMA analysis of alloy steel at low voltage

X Llovet[1,8], E Heikinheimo[2], A Núñez Galindo[3], C Merlet[4], J F Almagro Bello[3], S Richter[5], J Fournelle[6] and C J G van Hoek[7]

[1] University of Barcelona, Centres Científics i Tecnològics (CCiT), C/ Lluís Solé i Sabarís 1-3, ES-08028 Barcelona, Spain

[2] Aalto University, Department of Materials Science and Engineering, P.O. Box 16200, FI-00076 Aalto, Espoo, Finland

[3] Acerinox S.A., Departamento I+D+i, Pol. Industrial Palmones s/n, ES-11319 Los Barrios, Spain

[4] Université de Montpellier II, CNRS, Laboratoire Géosciences Montpellier, Place Eugène Bataillon, FR-34095 Montpellier Cedex 5, France

[5] RWTH Aachen University, GFE - Central Facility for Electron Microscopy, Ahornstrasse 55, DE-52074 Aachen, Germany

[6] University of Wisconsin, Department of Geoscience, 1215 West Dayton Street, US-53706 Madison, Wisconsin, U.S.A.

[7] Tata Steel Europe, RD&T, Ceramic Research Centre, P.O. Box 10.000, NL-1970 CA IJmuiden, The Netherlands

E-mail: xavier@ccit.ub.edu

Abstract. We present the results of an inter-laboratory comparison of EPMA analysis at low voltage (5 - 6 kV) of three monophasic alloy steel samples. The aim of the work was to obtain an estimate of the present situation of low-voltage analysis of steel and identify needs for improvement. EPMA analyses of the samples were conducted by seven participant groups using electron microprobes and scanning electron microscopes of different kind, equipped with wavelength- and/or energy-dispersive X-ray spectrometers, and employing their own methodology of analysis. The results using WDS showed essentially an underestimation of the Cr contents, with relative deviations from the reference values ranging from -0.7 % to -17 %, and an overestimation of Fe and Ni, with relative deviations from the reference composition ranging from -4 % to +30 %, and from +14 % to +42 %, respectively. The relative deviations obtained by using EDS showed larger scatter, ranging from -16 % to +54 % for Cr, -0.4 % to +66 % for Fe and from +13 % to +90 % for Ni. Reasons for the differences observed and the scatter of results are discussed.

1. Introduction

Modern electron beam instruments coupled with field-emission electron guns (FEG) have opened new opportunities to perform electron probe microanalysis (EPMA) at low voltages (< 10 kV). At such energies, the penetration range of incident electrons drops from the conventional micrometre scale

[8] To whom any correspondence should be addressed.

down to the sub-micrometre scale, and leads to a significant improvement in the lateral and depth resolution of the technique [1]. This is very useful for the characterisation of materials which contain sub-micrometre phases, such as steel materials that have undergone heat treatment at high temperatures.

However, EPMA at low beam energies poses several difficulties which may affect the accuracy of quantitative results [1-3]. At low voltages, especially below 5 kV, the L-lines of the main metal components such as Fe, Ni and Cr have to be used because the higher energy K-lines cannot be excited but the reliability of EPMA using low energy L-lines is not yet as well advanced as when using the higher energy K-lines. Moreover, at low beam voltages, a surface layer of several nanometres in thickness represents a much larger fraction of the sample and, therefore, the influence of carbon contamination, surface oxidation, or the quality of the sample polish becomes more important [2, 3]. Unfortunately, there is a lack of systematic studies in which the performance of currently available EPMA systems are assessed at low voltage [4].

In this communication, we present the results of an inter-laboratory comparison of EPMA analysis at low voltage of three alloy steel samples. EPMA analyses of the samples were conducted by seven participant groups using electron microprobes and scanning electron microscopes (SEMs) of different manufacturers and models, equipped with wavelength- (WDS) and/or energy-dispersive X-ray systems (EDS), and their own methodology of analysis. The aim of the work was to obtain an estimate of the present situation of low-voltage analysis in the case of steels and identify needs for improvement.

2. Preparation and characterisation of alloy steel samples

2.1. Sample preparation and characterisation

AISI 316, AISI 310S and Incoloy 135 alloy samples were prepared from conventionally produced ingots of the corresponding alloys. The samples were treated so as to obtain a homogeneous structure and a condition equivalent to that of commercial products. Here it should be noted that none of the considered alloys are Ti or Nb stabilized. Thus, to avoid Cr carbide precipitation, the samples were subjected to a solubilisation heat treatment at a temperature of 1200 °C for 90 min, followed by water quenching. The samples finally were embedded in conductive resin and polished in a sequence of steps, finishing with 1 μm diamond on a hard cloth.

The bulk composition of the samples was determined using conventional techniques and procedures used in steel works. In particular, Si, Ti, Cr, Mn, Co, Ni, Cu, Nb, Mo and Sn were analyzed by X-ray fluorescence (XRF) using a Panalytical PW 2606 XRF spectrometer; B, Al, V and Pb were determined by optical emission spectroscopy (OES) using a Spectro 111982 spectrometer; and finally C and S, and N were analyzed by using Leco analysers CS 600 and TC 600. The Fe-content was obtained by difference to 100 %. The bulk composition of the three steel samples is given in table 1. Elements with conventional residual concentrations (< 0.05 wt%) such as B, N, C, Al, S, Ti, and Nb are not listed in the table.

Sample homogeneity was checked by performing 200 EDS analyses on each sample, which yielded standard deviation values (as ± 1σ) less than 2 % for the major elements.

Preliminary EPMA-WDS analyses were performed on a Cameca SX-50 electron microprobe at an accelerating voltage of 20 kV, using Kα-lines of Cr, Mn, Fe, Co, Ni and Cu, and they were found to be in satisfactory agreement with the reference values, with no systematic departures observed. The results of the preliminary analyses are summarized in table 2.

The samples were distributed to the participant laboratories, with no special guideline. The only instruction given was to perform the analyses at an accelerating voltage of 5 or 6 kV, with the procedures and conditions chosen by themselves.

Table 1. Composition of the three alloy samples (in wt%).

Sample	ID code	Si	Cr	Mn	Fe	Co	Ni	Cu	Mo
P11	AISI 316	0.483	17.18	1.483	66.51	0.336	11.24	0.309	2.163
		(0.001)	(0.06)	(0.005)	(0.01)	(0.002)	(0.03)	(0.002)	(0.005)
P13	AISI310S	1.340	25.0	1.610	50.61	0.050	20.02	0.750	0.400
		(0.008)	(0.1)	(0.006)	(0.01)	(0.001)	(0.04)	(0.004)	(0.001)
P15	Incoloy	0.58	28.3	0.74	25.7	0.030	39.7	1.54	3.10
	135	(0.02)	(0.2)	(0.02)	(0.2)	(0.002)	(0.3)	(0.03)	(0.04)

Numbers in parentheses are one standard deviation uncertainties.

Table 2. Results of preliminary WDS analysis at 20 keV.

Sample	Si	Cr	Mn	Fe	Co	Ni	Cu	Mo	Total
P11	0.53	17.8	1.47	66.0	0.32	11.03	0.30	2.27	99.76
	(0.01)	(0.3)	(0.03)	(0.6)	(0.01)	(0.4)	(0.02)	(0.06)	
P13	1.42	25.33	1.58	50.6	0.03	19.5	0.75	0.4	99.67
	(0.04)	(0.3)	(0.05)	(0.7)	(0.01)	(0.3)	(0.03)	(0.03)	
P15	0.60	29.4	0.71	24.6	0.0	39.4	1.6	3.3	99.69
	(0.08)	(0.4)	(0.09)	(0.7)		(0.7)	(0.1)	(0.3)	

3. Experimental conditions of analysis

3.1. Wavelength-dispersive X-ray spectrometry

The equipment and experimental conditions used by the participants are summarized in table 3. All analyses were performed at an accelerating voltage of 6 kV, with the exception of laboratory 6 which used 5 kV. Most of the measurements were performed on electron microprobes (Cameca SX-50, SX-51 and SX-100); only one measurement was performed on a SEM equipped with WDS (Leo 1450). Different dispersing crystals were used, which include TAP, LTAP, PET, LPET, PC0, and PC1, with different pulse-height analyzer settings (in integral and differential mode). All participants used pure reference standards for calibration purposes. The correction procedures used were PAP, XPP and XPHI. Three participants polished the standards and samples before analysis (to remove a possible oxide and contamination layer) and one participant coated samples and standards with a carbon layer. A liquid N_2 trap was used by three laboratories to minimize carbon contamination during the measurements.

All participants obtained the net peak intensity by means of the standard procedure, e.g., by measuring the counting rate at the channel that corresponds to the peak maximum and subtracting the spectral background by linear interpolation of two measurements performed at both sides of the peak. For Cr, multiple background measurements with curved background modelling were performed by participant 6 and participant 1 used parabolic interpolation instead of linear interpolation. Each laboratory performed 10 measurements on each sample, except laboratory 4, which performed 20 measurements.

Table 3. Equipment and experimental conditions used in the WDS analyses.

Lab code	Instrument	Current (nA)	Crystals and Elements	Standards	Correction procedure	Specimen treatment	Remarks
1	Cameca SX-50	100	TAP: Cu, Ni, Fe, Mn, Si PET: Mo PC1: Cr	Pure*	PAP	Polishing	L-N_2 trap
2	Cameca SX-100	200	TAP: Cu, Ni, Fe, Mn, Si LPET: Mo PC1: Cr	Pure*	XPHI	Polishing	L-N_2 trap
3	Leo 1450 Inca Wave	40	PET: Si, Mo TAP: Cu, Ni, Fe, Mn, Cr	Pure	XPP	None	
4	Cameca SX-50	50	PC1: Cr, Fe TAP: Mn, Ni, Co, Cu PET: Si, Mo	Pure*	PAP	Polishing	
5	Cameca SX-100	200	LTAP: Si,Cr,Mn,Fe,Ni,Cu LPET: Mo	Pure	XPHI	None	
6	Cameca SX-51	40	TAP: Si PC1: Cr PC0: Mn, Fe, Ni, Co PET: Mo	Pure	PAP	Carbon coating. No polishing	L-N_2 trap. Multiple backgrounds

* Standards were polished before analysis.

3.2. Energy-dispersive X-ray spectrometry

The equipment and the experimental conditions are summarized in table 4. All analyses were performed at an accelerating voltage of 6 kV, with the exception of laboratory 1 which used 5 kV. Laboratory 1 used two different instruments, which are referred in the table to as 1a and 1b. Participant 4 used simultaneously two SDDs mounted on the same SEM to improve statistics. EDS analyses were performed with and without standards, and were reported as non-normalized and normalized. Only non-normalized results have been considered here. Each laboratory performed 10 measurements on each sample, except laboratory 1, which performed 100 measurements.

4. Results

4.1. Wavelength-dispersive X-ray spectrometry

Representative X-ray spectra recorded at 6 kV around the positions of the Cr Lα-, Fe Lα- and Ni Lα-lines emitted from pure Cr, Fe and Ni reference samples, as well as from samples P11 (Cr, Ni) and P15 (Fe) are shown in figure 1. The spectra from the steel samples have been normalized to the peak maxima from the corresponding pure metal samples to facilitate comparison. The dispersing crystals used are PC1 (for Cr) and TAP (for Fe and Ni).

In the case of Fe and Ni, differences are observed in the high-energy tails of the Lα and Lβ peaks when comparing the spectra recorded from the pure standards with those recorded from the steel samples. These differences are most likely due to different self-absorption, as the latter effect depends on sample composition (the L_3 absorption edges are located at 574 eV for Cr, 707 eV for Fe and 854 eV for Ni). Due to the lower spectral resolution of the PC1 crystal, the Lα- and Lβ-lines of Cr could not be resolved.

Table 4. Equipment and experimental conditions used in the EDS analyses.

Lab Code	Instrument	Current (nA)	EDS	Standards	Correction procedure	Sample treatment	Other remarks
1a	FEG-SEM Leo 982	2.7	SDD INCA	Standardless	XPP	Plasma cleaning	Beam voltage: 5 kV
1b	FEG-SEM Zeiss Ultra 55	1.8	Si(Li) INCA	Standardless	XPP	Plasma cleaning	Beam voltage: 5 kV
2	Cameca SX-50	2	SDD SAMx	Pure*	XPP	Polishing	
3	SEM Leo 1450		SDD INCA	Pure	XPP	None	
4	SEM Jeol 7001F	1.46	2 SDDs Thermo Scientific	Pure	PROZA	None	2 SDDs used simultaneously

*Standards were polished before analysis.

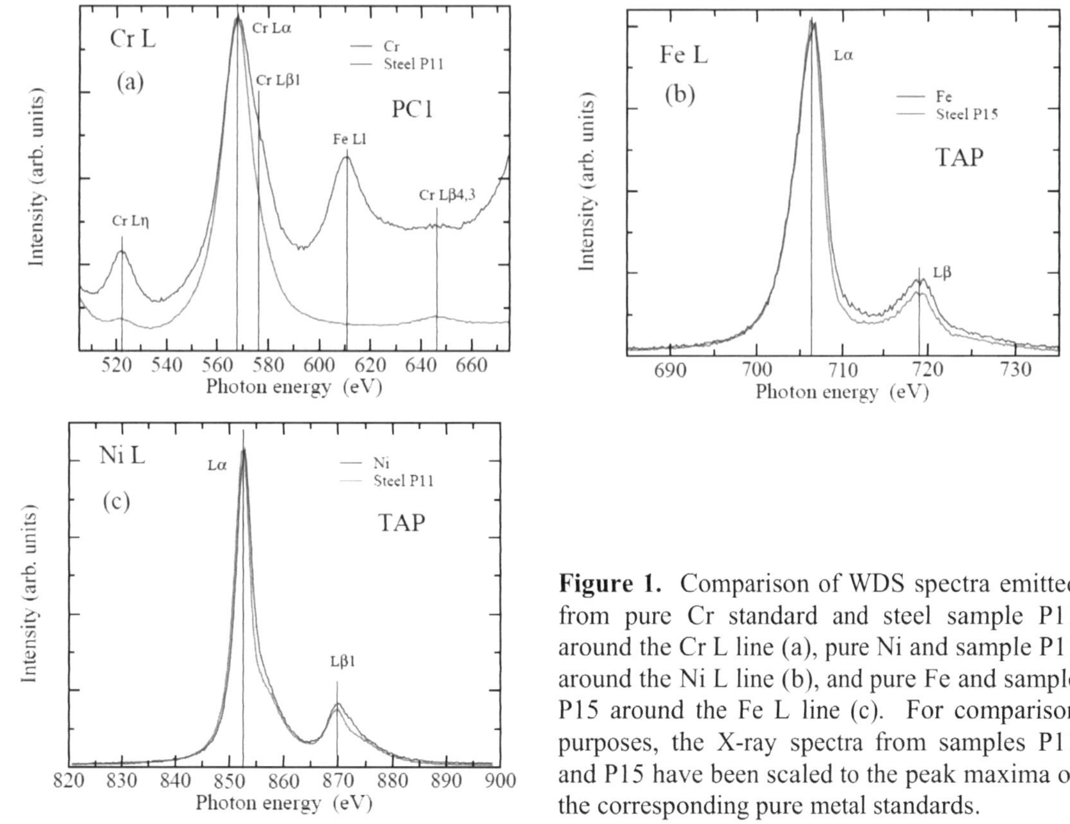

Figure 1. Comparison of WDS spectra emitted from pure Cr standard and steel sample P11 around the Cr L line (a), pure Ni and sample P11 around the Ni L line (b), and pure Fe and sample P15 around the Fe L line (c). For comparison purposes, the X-ray spectra from samples P11 and P15 have been scaled to the peak maxima of the corresponding pure metal standards.

The results of the WDS analyses of the P11, P13 and P15 samples are summarized in tables 5, 6 and 7, along with the standard deviations reported by the participants. The latter were calculated from the number of measurements performed, which is indicated above. The negative Co concentrations reported, which obviously have no physical meaning, are an indication of measurement difficulties (e.g., in background subtraction) found for this element. The sum of the measured concentrations ranged from 100.2 and 124.6 wt%.

Table 5. Results of quantitative WDS analyses of the P11 sample (all values in wt%).

Lab code	Si	Cr	Mn	Fe	Co	Ni	Cu	Mo	Total
Reference composition	0.483	17.18	1.483	66.51	0.336	11.24	0.309	2.163	
1	0.57 (0.02)	14.42 (0.27)	0.88 (0.25)	66.65 (0.7)	0.38 (0.04)	14.65 (0.43)	0.35 (0.03)	2.39 (0.10)	100.29
2	0.52 (0.01)	15.77 (0.14)	1.10 (0.11)	67.86 (0.5)	0.4 (0.05)	14.45 (0.27)	0.35 (0.02)	2.16 (0.04)	102.59
3	0.39 (0.02)	14.33 (0.82)	0.21 (0.28)	86.25 (1.24)	0.27 (0.05)	15.33 (0.53)	0.34 (0.01)	7.54 (0.24)	124.64
4	0.46 (0.06)	15.24 (0.35)	1.49 (0.54)	70.04 (0.3)	0.31 (0.16)	16.0 (0.6)	0.35 (0.06)	2.25 (0.15)	106.14
5	0.50 (0.004)	14.82 (0.77)	1.03 (0.06)	72.97 (0.5)	-0.04 (0.04)	15.13 (1.6)	0.39 (0.06)	2.15 (0.05)	106.99
6	0.51 (0.05)	16.91 (0.64)	0.24 (0.63)	70.36 (0.81)	"Not measured"	14.04 (0.36)	0.59 (0.18)	2.16 (0.35)	104.80

Numbers in parentheses are one standard deviation uncertainties.

Table 6. Results of quantitative WDS analyses of the P13 sample (all values are in wt%).

Lab	Si	Cr	Mn	Fe	Co	Ni	Cu	Mo	Total
Reference composition	1.340	25.0	1.610	50.61	0.050	20.02	0.750	0.400	
1	1.52 (0.05)	21.34 (0.16)	1.03 (0.16)	50.87 (0.73)	0.00	24.97 (0.16)	0.93 (0.04)	0.48 (0.05)	101.15
2	1.38 (0.08)	22.59 (0.35)	1.16 (0.09)	50.8 (1.02)	-0.04 (0.03)	25.11 (0.32)	0.86 (0.04)	0.4 (0.04)	102.3
3	1.15 (0.06)	21.05 (0.60)	0.52 (0.22)	65.02 (1.88)	0.00	25.87 (1.70)	0.79 (0.08)	1.45 (0.18)	115.84
4	1.34 (0.10)	22.99 (0.52)	1.72 (0.63)	56.37 (0.80)	0.06 (0.08)	25.84 (0.65)	0.8 (0.11)	0.47 (0.15)	109.58
5	1.44 (0.06)	20.71 (2.71)	1.26 (0.05)	55.98 (1.04)	-0.14 (0.06)	26.09 (0.35)	0.99 (0.05)	0.43 (0.03)	106.9
6	1.38 (0.1)	23.83 (1.00)	0.0	56.29 (0.81)	"Not measured"	24.57 (0.4)	1.1 (0.13)	0.37 (0.27)	107.5

Numbers in parentheses are one standard deviation uncertainties.

The results of the six participants are in good agreement with each other within a relative standard deviation of 10 % (Si), 7 % (Cr), 9 % (Fe), 3 % (Ni), and 16 % (Cu). For Mn and Mo the agreement is within 60 % and 70 %, respectively. For the latter element, the results from participant 3 are a factor of 4 larger than the other participants and, therefore, a problem with this element is suspected. The results for Co are meaningful only for sample P11; in this case, the results from all participants agree with each other to within 68 %.

Table 7. Results of quantitative WDS analyses of the P15 sample (all values are in wt%).

Lab code	Si	Cr	Mn	Fe	Co	Ni	Cu	Mo	Total
Reference composition	0.58	28.3	0.74	25.7	0.030	39.7	1.54	3.10	
1	0.59	23.91	0.47	24.65	0.00	47.12	1.74	3.37	102.44
	(0.07)	(0.49)	(0.17)	(0.76)		(0.48)	(0.12)	(0.35)	
2	0.49	26.07	0.49	26.06	-0.25	47.77	1.66	2.76	105.29
	(0.05)	(0.39)	(0.14)	(0.41)	(0.05)	(0.26)	(0.10)	(0.21)	
3	0.65	24.46	0.00	29.44	0.00	47.02	1.72	13.01	116.30
	(0.03)	(1.27)		(0.63)		(1.23)	(0.05)	(0.47)	
4	0.57	27.03	0.46	28.62	0.03	46.48	1.69	3.38	108.25
	(0.13)		(0.53)	(1.17)	(0.06)	(0.58)	(0.20)	(0.57)	
5	0.59	23.37	0.51	26.89	0.15	48.41	2.02	3.26	105.20
	(0.13)		(0.12)	(1.32)	(0.05)	(0.80)	(0.22)	(1.57)	
6	0.62	28.12	0.00	27.44	Not	45.48	2.06	3.71	106.60
	(0.10)	(2.10)		(4.9)	measured	(2.28)	(0.41)	(1.28)	

Numbers in parentheses are one standard deviation uncertainties.

Figure 2 shows comparison results in a graphical form for Cr, Fe and Ni, as function of participant code for sample P13. The error bars correspond to the claimed one standard deviation uncertainties. For comparison purposes, the reference values, as well as the average and error bands (at 2σ level) are also displayed. The scattering of results is large as compared to what is usual in EPMA round robins.

The relative variation ΔC of each measurement Cmeas with respect the reference value Cref was calculated as

$$\Delta C = (C_{meas} - C_{ref}) / C_{ref} \times 100, \qquad (1)$$

and the results for Cr, Fe and Ni are shown in table 8. The results show a systematic underestimation of the Cr contents, with relative deviations from the reference values ranging from -0.8 % to -17.5 %, and an overestimation of Fe and Ni, with relative deviations ranging from -4.2 % to +29.6 %, and from +14.4 % to +42.4 %, respectively.

Table 8. Relative deviations (in %) from the reference concentrations for Cr, Ni, Fe for each laboratory. See text for details.

	P11			P13			P15		
Lab. no.	Cr	Fe	Ni	Cr	Fe	Ni	Cr	Fe	Ni
1	-16.1	0.2	30.3	-14.5	0.5	24.7	-15.6	-4.2	18.5
2	-8.2	2.0	28.5	-9.6	0.4	25.4	-8.0	1.3	20.2
3	-16.6	29.7	36.4	-15.7	28.5	29.2	-13.7	14.5	18.3
4	-11.32	5.3	42.4	-8.1	11.4	29.1	-4.6	11.2	17.0
5	-13.8	9.7	34.6	-17.1	10.6	30.3	-17.5	4.5	22.0
6	-1.6	5.8	25.0	-4.6	11.2	22.7	-0.8	6.7	14.4

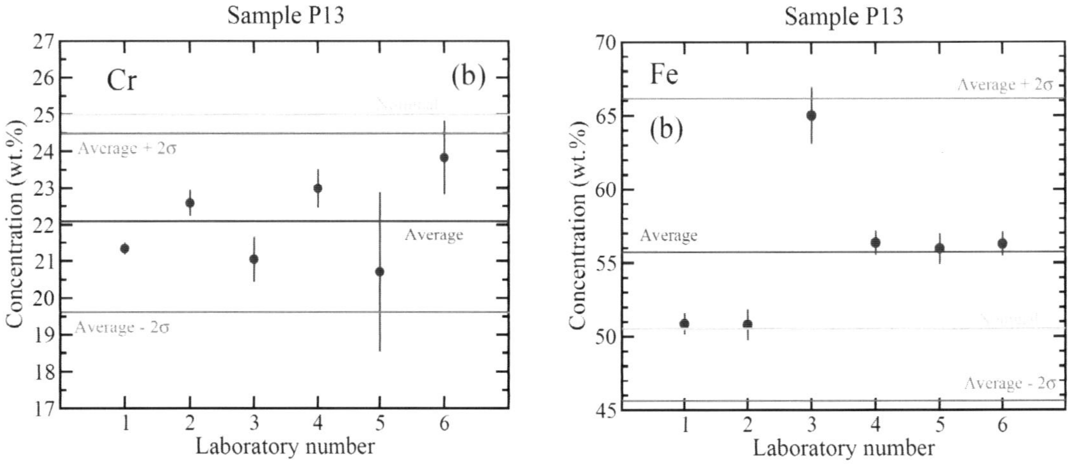

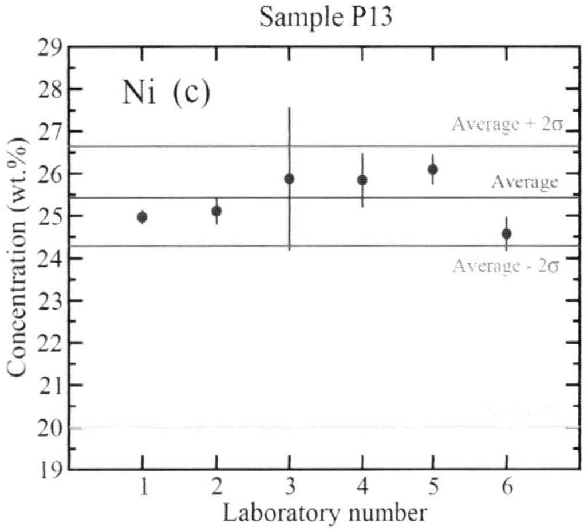

Figure 2. WDS results for Cr (a), Fe (b) and Ni(c) as function of participant code number for sample P13. Error bars correspond to the claimed one standard deviation uncertainties.

4.2. Energy-dispersive X-ray spectrometry

EDS spectra recorded at an accelerating voltage of 6 kV on samples P11 and P15 are shown in figure 3. The spectra show severe peak overlaps for most of the metal L-lines below a photon energy of 1 keV.

The results of the EDS analyses of the P11, P13 and P15 samples are summarized in tables 9, 10 and 11, together with the standard deviations reported by the participants. Although normalized and non-normalized results were reported, only non-normalized values are shown in the tables. In some cases, it was not possible to detect Co and Mn due to the heavy overlap of the major elements. The relative deviations obtained by using EDS, calculated by using Eq. 1, showed larger scatter than for WDS, ranging from -16 % to +54 % for Cr, -0.4 % to +66 % for Fe and from +13 % to +90 % for Ni.

5. Discussion

In the case of the measurements performed by WDS, the participant laboratories used quite different experimental parameters and matrix correction procedures (see table 3). However, similar systematic deviations for the Cr, Ni and Fe concentrations are observed, except in the case of Fe measured by laboratory 3, for which much larger deviations are observed (see table 8). The reasons for such larger deviations are attributed to a deficient Fe standard. The fact that three participants polished samples and standards before analysis and used an anti-contamination device (liquid N_2) suggests that the reason for the systematic deviation might not be due (at least only) to the presence of an oxide layer

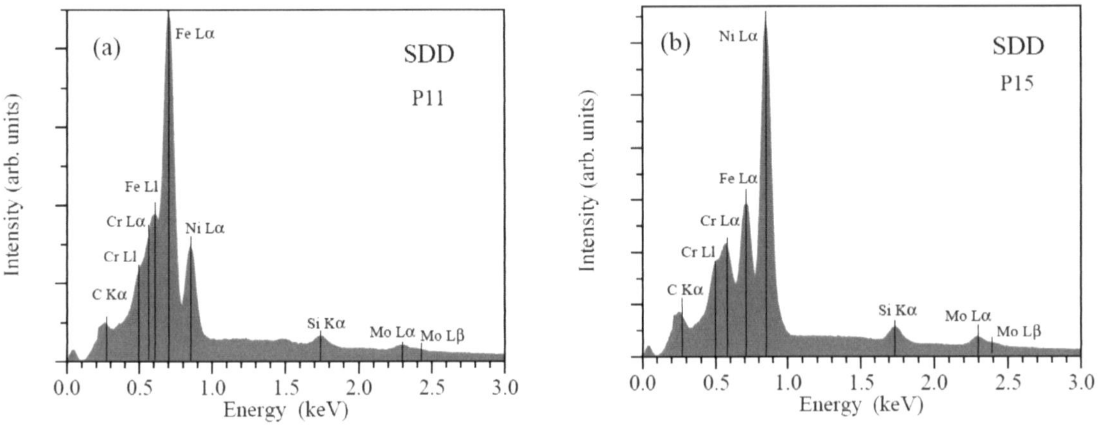

Figure 3. EDS spectra obtained with a SDD detector from samples P11 (a) and P15 (b) at 6 kV.

Table 9. Results of quantitative EDS analyses of the P11 sample (all values are in wt%). Numbers in parentheses are one standard deviation uncertainties.

Lab code	Si	Cr	Mn	Fe	Co	Ni	Cu	Mo	Total
Reference composition	0.483	17.18	1.483	66.51	0.336	11.24	0.309	2.163	
1a	0.67	36.00	-6.21	104.22	-1.87	21.42	0.18	2.82	157.25
	(0.08)	(1.39)	(1.16)	(9.04)	(0.68)	(2.47)	(0.46)	(0.40)	
1b	0.53	28.98	-1.05	100.10	1.02	18.53	0.09	2.83	151.04
	(0.07)	(0.86)	(1.30)	(1.24)	(0.55)	(0.62)	(0.21)	(0.38)	
2	0.44	26.13	0.00	76.98	0.00	13.90	0.39	1.68	119.52
	(0.09)	(0.45)		(0.38)		(0.42)	(0.15)	(0.11)	
3	0.60	23.26	0.47	85.22	2.78	17.26	0.03	5.39	135.00
	(0.04)	(0.69)	(1.00)	(1.07)	(0.48)	(0.35)	(0.10)	(0.27)	
4	0.70	10.25	0.0	73.75	0.0	16.93	0.31	2.54	104.47
	(0.04)	(0.31)		(0.65)		(0.19)	(0.10)	(0.27)	

Numbers in parentheses are one standard deviation uncertainties.

(native oxide layer on the standards and/or passive layer on the steel samples) or to carbon contamination effects. Likewise, it seems plausible not to attribute the observed discrepancies to only background subtraction errors, since very different background "positions" and methods were used by the participants. However, it should be noted that all the participants measured peak intensities and not peak areas in order to extract the X-ray intensities. As a result, possible systematic errors due to peak shape changes between standards and the analyzed steel samples cannot be ruled out. The use of standards with similar composition as those of the analyzed specimens would shift the observed deviations to smaller values, as possible shortcomings of matrix correction procedures and/or measurement errors would be minimized.

In the case of EDS analysis, the results obtained are most likely affected by large systematic errors in the extraction of X-ray intensities due to severe peak overlapping. As a result, only the results obtained by WDS will be considered in the following discussion.

Table 10. Results of quantitative EDS analyses of the P13 sample (all values are in wt%). Numbers in parentheses are one standard deviation uncertainties.

Lab code	Si	Cr	Mn	Fe	Co	Ni	Cu	Mo	Total
Reference composition	1.340	25.0	1.610	50.61	0.050	20.02	0.750	0.400	
1a	1.57	43.92	-7.40	82.19	-3.77	37.96	1.05	0.66	156.17
	(0.11)	(1.23)	(0.80)	(1.51)	(0.47)	(0.65)	(0.20)	(0.29)	
1b	1.44	38.26	-1.39	80.21	0.38	32.66	0.67	0.62	152.86
	(0.12)	(0.97)	(1.00)	(1.63)	(0.58)	(0.68)	(0.24)	(0.32)	
2	1.32	31.54	0.00	63.39	0.00	24.85	0.66	0.25	122.02
	(0.07)	(0.61)		(0.76)		(0.15)	(0.09)	(0.08)	
3	1.48	30.39	1.08	67.77	2.39	28.87	0.59	0.93	133.51
	(0.04)	(0.35)	(1.41)	(0.71)	(0.41)	(0.27)	(0.12)	(0.39)	
4	1.61	18.43	0.01	58.90	0.00	28.91	1.12	0.54	109.50
	(0.06)	(1.19)		(0.50)		(0.47)	(0.09)	(0.25)	

Numbers in parentheses are one standard deviation uncertainties.

Table 11. Results of quantitative EDS analyses of the P15 sample (all values are in wt%). Numbers in parentheses are one standard deviation uncertainties.

Lab code	Si	Cr	Mn	Fe	Co	Ni	Cu	Mo	Total
Reference composition	0.58	28.3	0.74	25.7	0.030	39.7	1.54	3.10	
1a	0.66	43.74	-6.10	41.15	-5.14	70.62	2.38	4.07	151.38
	(0.12)	(1.63)	(0.76)	(2.05)	(0.47)	(1.75)	(0.37)	(1.05)	
1b	0.58	41.51	-1.31	42.84	1.08	62.73	1.56	4.10	153.09
	(0.13)	(1.44)	(0.73)	(2.11)	(0.58)	(1.55)	(0.35)	(0.82)	
2	0.52	30.73	0.00	30.97	0.91	49.23	0.83	2.67	114.95
	(0.04)	(0.21)		(1.16)	(0.01)	(0.41)	(0.12)	(0.19)	
3	0.64	35.05	1.15	37.81	2.74	54.82	1.47	7.70	141.41
	(0.12)	(0.97)	(0.23)	(1.42)	(0.35)	(0.65)	(0.30)	(1.36)	
4	0.76	23.64	0.00	29.92	0.00	52.82	2.33	3.45	112.93
	(0.07)	(1.28)		(0.91)		(0.52)	(0.09)	(0.08)	

Numbers in parentheses are one standard deviation uncertainties.

5.1. Matrix correction procedures

In order to assess the effect of using matrix correction procedures different from those used by the participants, one set of results (laboratory 1, sample P13) was recalculated using different $\phi(\rho z)$-based procedures, namely the PROZA model, and those of Packwood and Armstrong, with the help of the code CALCZAF (details on the tested correction procedures as implemented in CALCZAF can be found in Ref. [5]). For all the tested correction procedures, the same set of MACs was used, namely the FFAST compilation [10]. As shown in table 12, the use of different correction procedures resulted in only small changes of the Cr, Ni and Fe concentrations. Here it should be pointed out that fluorescence effects are almost negligible for the L-lines of the considered elements, owing to their extremely low fluorescence yields.

Table 12. Comparison of WDS results (in wt%) for different correction procedures using the FFAST MACs (*k*-ratios measured on sample P13 by laboratory 1).

Correction procedure/MAC	Si	Cr	Mn	Fe	Ni	Cu	Mo	Total
Reference composition	1.340	25.0	1.610	50.61	20.02	0.750	0.400	
PROZA/FFAST	1.47	22.4	1.07	51.6	25.2	0.87	0.44	103.13
Packwood/FFAST	1.46	22.5	1.09	52.7	26.6	0.93	0.45	105.80
Armstrong/FFAST	1.54	22.7	1.10	52.6	26.0	0.89	0.43	105.28

5.2. Mass absorption coefficients

Although the use of different correction procedures yields small changes in the evaluated concentrations, a problem with a matrix correction parameter such as the mass-attenuation coefficient (MAC) cannot be excluded.

Different MACs are available for the Cr, Fe and Ni Lα X-ray lines in Cr, Fe and Ni respectively, which are summarized in table 13. This table also includes values obtained empirically by Pouchou and Pichoir [12] and by participant 1, using the programme XMAC [3], from measurements on pure Cr, Fe and Ni samples up to 20 kV. For a given X-ray line and element, the mentioned programme determines the MAC by i) considering it as a free parameter in the XPP model, and ii) fitting the predictions of the model to measurements of the X-ray intensity as a function of beam energy. In the case of Cr, the Lα and Lβ peaks appear as a single (Lα+Lβ) line, therefore, the use of the L doublet for analysis will lead to an additional source of uncertainty, as the MAC for the Lα line differs significantly from that of the Lβ line (see e.g., Ref [11]).

Table 13. MACs for Cr, Fe and Ni Lα X-rays in Cr, Fe and Ni, respectively, taken from different compilations (in cm^2/g).

Source	Cr Lα	Fe Lα	Ni Lα
LINEMU: Henke [6], Heinrich [7]	2611	2146	1819
CITZMU: Heinrich [7], Henke and Ebisu [8]	2612	2149	1801
MAC30: Heinrich [9]	2650	2157	1806
FFAST: Chantler et al. [10]	2395	1967	1697
Empirical: Pouchou and Pichoir [12]	3850	3350	3560
Empirical: This work	3612	3559	4277

To assess the effect of using different MACs on the resulting concentrations, the *k*-ratios obtained by laboratory 1 for sample P13 were re-evaluated using the PAP model with the MACs shown in table 13. The results are summarized in table 14 and indicate that the use of empirical MACs gives results closer to the reference values for Fe and Ni, although for the latter element they are still far off. Transition metals such as Fe, Cr and Ni are known to have a narrow and partially empty 3d band and, as a result, the MAC in the vicinity of the L_3 edge changes rapidly with photon energy [13]. These large fluctuations are not accounted for in most available compilations. The use of empirical MACs obtained from EPMA measurements provides a convenient way to overcome, at least partially, this difficulty.

Table 14. Comparison of WDS results (in wt%) for the PAP correction procedure using different MACs (results from sample P13, laboratory 1).

Correction procedure/MAC	Si	Cr	Mn	Fe	Ni	Cu	Mo	Total
Reference composition	1.340	25.0	1.610	50.61	20.02	0.750	0.400	
PAP/LINEMU	1.51	22.7	1.09	52.8	26.3	0.92	0.44	105.91
PAP/CITZMU	1.53	22.8	1.10	52.9	26.5	0.91	0.44	106.27
PAP/MAC30	1.52	22.6	1.09	53.0	26.9	0.94	0.45	106.62
PAP/FFAST	1.50	22.6	1.10	52.7	26.4	0.91	0.44	105.74
PAP/Pouchou	1.51	21.3	1.02	50.6	24.3	0.91	0.44	100.22
PAP/This work	1.51	21.6	1.03	50.5	23.5	0.91	0.45	99.49

5.3. Measurement of k-ratios at different beam energies

To further elucidate the source of the observed discrepancies, *k*-ratio measurements at varying voltages in the range 4 - 20 kV were performed on samples P11, P13 and P15 by laboratory 1. The experimental *k*-ratios are compared in figure 4 with the predictions of the PAP model, using MACs from Ref. [12] (continuous line). For Fe, the agreement between the experimental measurements with the PAP model seems to be satisfactory. Figure 4 also displays the predictions of the PAP model for Ni and Cr re-scaled such that they match the experimental measurements at 4 keV (dashed lines). For the mentioned elements, the differences between the experimental data and the re-normalized PAP results seem to slightly increase with increasing beam energy, which could indicate an absorption anomaly. Although the metals considered here should be in the same chemical state as pure elements (since steel is a solid solution, i.e., with no inter-metallic phases or oxides), we cannot rule out the possibility that the proximity of neighbouring atoms could modify the MACs of the L-lines of these metals. Here it should be pointed out that the modulation of the absorption coefficient by neighbouring atoms is the basis of the well established EXAFS technique.

5.4. Measurements of other alloy steel samples

Further measurements were performed at 6 kV by laboratory 1 on other steel samples of known composition, with Cr, Fe and Ni contents ranging from 1.16 - 28.3 wt%, 25.7 - 96.4 wt% and 0.10 - 39.7 wt%, respectively (see table 15 for reference compositions). For the sake of completeness, table 15 also includes the composition of the P11, P13 and P15 materials.

The results of these measurements for Cr and Ni are displayed in figure 5. As it can be seen, the differences observed between the measured and the reference concentrations increase with increasing concentration, thus confirming the tendencies observed for samples P11, P13 and P15.

Table 15. Concentration of Cr, Fe and Ni (in wt%) of 15 alloy steel reference samples.

	P01	P02	P03	P04	P05	P06	P07	P08	P09	P10	P11	P12	P13	P14	P15
Cr	1.16	4.94	11.4	16.1	17.0	17.3	18.3	18.1	17.4	13.7	17.2	12.4	25.0	23.5	28.3
Fe	96.4	93.6	86.1	65.7	67.3	69.7	70.6	67.6	70.8	72.1	66.5	70.7	50.6	51.5	25.7
Ni	0.10	0.11	0.80	1.52	2.61	4.51	8.13	8.61	9.30	9.92	11.2	14.8	20.0	21.7	39.7

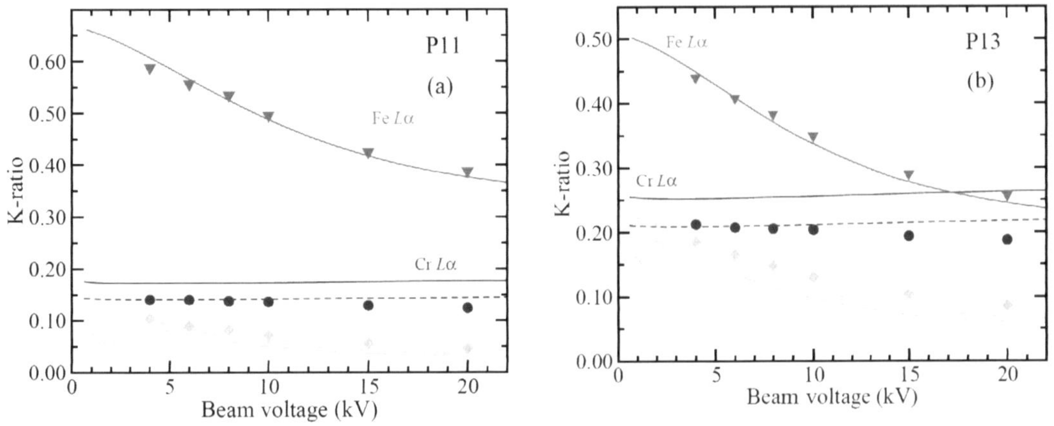

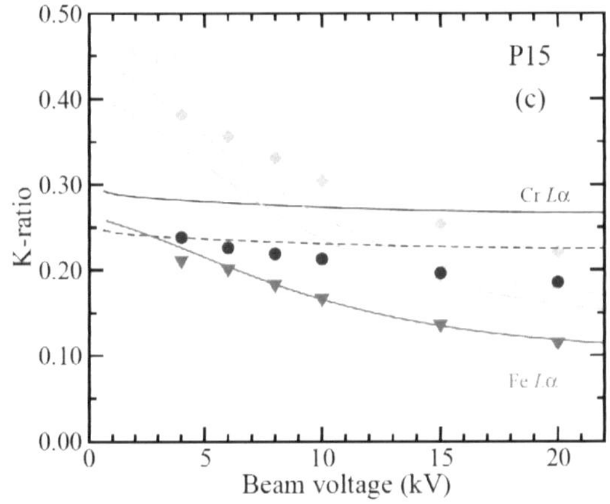

Figure 4. k-ratio as a function of beam energy for Fe, Cr and Ni measured on samples P11 (a), P13 (b) and P15 (c). Symbols are experimental data. Continuous lines are the predictions of the PAP model [12]; dashed lines are the predictions of PAP re-scaled so as to match the experimental measurements at 4 kV.

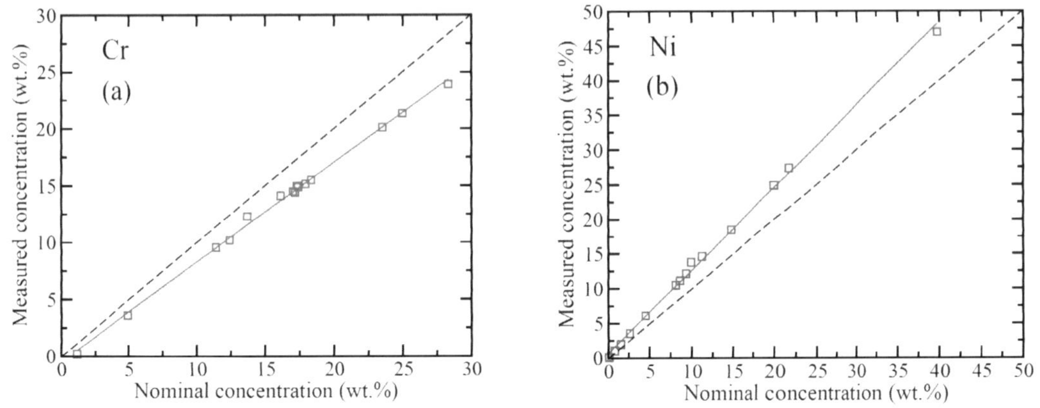

Figure 5. Measured versus reference concentration for Cr (a) and Ni (b) obtained in a series of 15 alloy steel samples with varying Cr and Ni concentrations (see table 15). Symbols are experimental measurements. Continuous lines represent the results of linear regression. Beam voltage is 6 kV.

6. Conclusion

Despite the participant laboratories having used quite different experimental conditions and matrix correction procedures for the WDS analysis, similar systematic deviations in the Cr, Fe and Ni concentrations have been observed. Inaccuracy of the adopted MACs for the L-lines of the metal elements as well as experimental errors in the extraction of X-ray intensities are believed to be the main sources of uncertainty. In the case of EDS analysis, the results obtained are most likely affected by large systematic errors due to severe peak overlapping. Further work is needed for accurate quantitative EPMA analysis of alloy steel at low beam voltage.

References

[1] Newbury D E 2002 *J. Res. Natl. Inst. Stand. Technol.* **107** 605
[2] Willich P and Bethke R 1996 *Mikrochim. Acta* **Suppl. 13** 631
[3] Pouchou J L 1996 *Mikrochim. Acta* **Suppl. 13** 39
[4] Briant M and Balloy D 2008 *Eur. Phys. J. Appl. Phys.* **44** 37
[5] Probe Software Inc., available on-line: http://www.probesoftware.com/Technical.htm
[6] Henke B L, Lee T J, Tanaka R J, Shimabukuro R L and Fijikawa B K 1985 *Atomic Data Nucl. Data Tables* **27** 1
[7] Heinrinch K F J 1966 *The electron microprobe.* McKinley T D, Heinrich K F J and Wittry D B, eds. (New York: John Wiley and Sons) pp. 1035
[8] Henke B L and Ebisu E S 1974 *Adv. X-ray Anal.* **17** 150
[9] Heinrich K F J 1986 *Proc. 11th International Congress on X-ray Optics and Microanalysis.* Brown J D and Packwood R H, eds. (Ontario: University of Western Ontario)
[10] Chantler C T, Olsen K, Dragoset R A, Chang J, Kishore A R, Kotochigova S A and Zucker D S 2005 *X-ray form factors, attenuation and scattering tables (version 2.1).* Available online: http://physics.nist.gov/ffast
[11] Rickerby D G and Wachter N 2000 *Mikrochim. Acta* **132** 157
[12] Pouchou J L and Pichoir F 1991 *Electron probe quantitation.* Heirinch K F J and Newbury D E. (New York: Plenum Press), pp. 31
[13] Chopra D 1970 *Phys. Rev.* A **1** 230

IOP Conf. Series: Materials Science and Engineering **32** (2012) 012015 doi:10.1088/1757-899X/32/1/012015

X-ray microanalysis: the state of the art of SDD detectors and WDS systems on scanning electron microscopes (SEM)

L Maniguet[1,3], F Robaut[1], A Meuris[2], F Roussel-Dherbey[1] and F Charlot[1]
[1] INP Grenoble, Consortium des Moyens Technologiques Communs (CMTC), 1260 rue de la piscine, bât. PHELMA, P.O. Box 75, FR-38402 Saint Martin d'Hères, France
[2] C.E.A., IRFU, Service d'Astrophysique, C.E. Saclay, bât. 709, Orme des Merisiers, FR-91191 Gif sur Yvette, France

E-mail: laurent.maniguet@cmtc.grenoble-inp.fr

Abstract. Over the past decade, the silicon drift detector energy-dispersive X-ray spectrometry systems (SDD-EDS) have emerged as a major breakthrough in X-ray microanalysis on scanning electron microscopes (SEM). This review covers a description of the SDD device principles and the recent technological developments. When high energy resolution or low detection limits are required, WDS detectors remain an effective tool. This review also covers the different WDS systems available on SEMs.

1. Introduction
In the field of X-ray microanalysis, high energy resolution and low energy detection capability are basic requirements. Silicon drift detectors (SDDs) have been known since the concept was invented by Gatti and Rehak in 1984 [1], and they have been available commercially for energy-dispersive X-ray spectrometry (EDS) since 1998 [2]. Over the past decade, SDDs have emerged as a major breakthrough in EDS technology [3]. Originally designed as position sensitive detectors for particle tracking, silicon drift detectors are nowadays used for X-ray spectroscopy, operating close to room temperature and at high count rates. They have opened a new spectrum of experiments in the wide field of X-ray spectroscopy: fluorescent analysis, diffractometry, holography and microanalysis in scanning electron microscopes. SDDs are based on the same physics as Si(Li) detectors, however, the resolution is closer to the theoretical limit 120 eV. In 2000, high resolution cryo-detectors (like microcalorimeters) that operate at very low temperature seemed promising [4, 5]. Unfortunately, they are still being developed and not available for standard X-ray microanalysis applications. When high energy resolution or low detection limits are required, WDS detectors remain an effective tool. Traditional WDS systems are based on curved crystal geometry. Nearly 15 years ago, new parallel-beam wavelength-dispersive X-ray spectrometers were developed.

2. Silicon drift detector (SDD) on SEM

2.1. SDD working principle [6-8]
The silicon drift detector (SDD) consists of a volume of fully depleted high-resistivity silicon, in which an electric field with a strong component parallel to the surface drives electrons generated by

[3] To whom any correspondence should be addressed.

the absorption of ionising radiation towards a small sized collecting anode with a very small capacitance, independent from the total active area of the device. Figure 1 shows the scheme of principle of a single radial SDD.

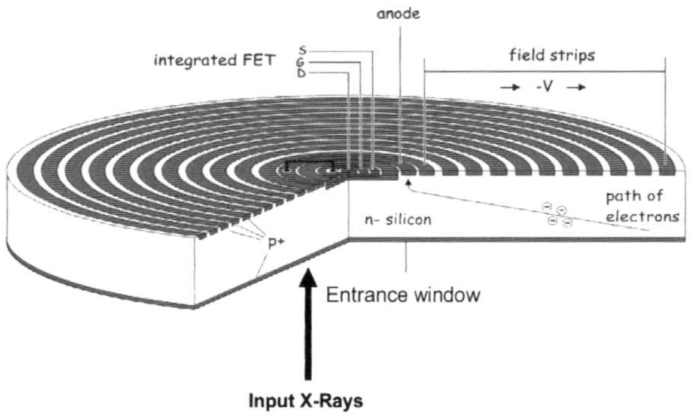

Figure 1. Schematic view of a radial SDD with integrated transistor [6]. Electrons are guided by an electric field toward the small-sized collecting anode in the center of the device.

Semiconductor processing based on a wafer of n-type Si is used to create a pattern on the silicon chip. The symmetry of the detector is radial, with a structure composed of p+ concentric rings implanted on the front side of the device. The purpose of the concentric pattern is to provide a field gradient for the electrons produced by incoming X-rays. Under the influence of this field, the electrons drift towards a small contact that serves as the anode. A continuous and homogeneous p+ junction implanted on the back of the detector acts as a radiation entrance window. Electron-hole pairs are produced in the silicon by the incoming X-ray photons. These electrons produce the signal in relation to the photon energy. They drift toward the anode under the influence of the field, where they are collected, and the charge is transferred to the field-effect transistor (FET). For a 10 mm^2 radial SDD (R = 1.8 mm) with a thickness of 300 µm and a polarisation of 100 V, the maximum drift time is about 40 ns to reach the anode.

2.2. Basic electronics blocks of an EDS system

An EDS system is basically composed of a detector, a signal processor and a digital unit for storing the data and analysing the acquired spectra. For analogue units, the signal processor itself is usually composed of a charge preamplifier with a FET, a pulse shaping amplifier, a fast amplifier to provide the trigger signals for the pile-up rejection unit, a peak sampler and an analogue-to-digital converter. For digital units, the analogue-to-digital converter is placed before the shaping amplifier, which makes the pulse processing digital. The design of the pulse processor, whether digital or analogue, has an effect on the resolution and count rate and must be optimized for SDDs.

2.3. SDD processing technology

Sophisticated semiconductor processing is necessary to produce the device. Only a few companies in the world are able to produce SDD chips [3]. It is interesting to note that these companies have developed a wide range of SDDs for various applications in the field of high-energy physics or X-ray spectroscopy [9-12]. Only part of them can be implemented in a SEM for X-ray microanalysis. EDS manufacturers buy at least the silicon chip from a semiconductor producer. Some EDS companies also buy the cooling system and the associated pulse-processing electronics, while others design downstream components themselves. Consequently, the properties may vary between different manufacturers.

2.4. Resolution [13]

For semiconductor detectors, the energy resolution ΔE depends both on the intrinsic resolution of the detector ΔE_0 and on the electronic noise of the system N:

$$\Delta E^2 = \Delta E_0^2 + N^2$$

The intrinsic resolution of the detector depends on the statistical fluctuations (Poisson statistics) related to the conversion mechanism and sets a lower limit to the achievable resolution. The measured distribution can be described by means of a Gaussian function. For the Gaussian distribution, the full-width-half-max (FWHM) is given by:

$$\Delta E_0 = 2.355\sqrt{F \varepsilon E_0}$$

where F is the Fano factor, approximately 0.12 for silicon and ε the pair creation energy, 3.8 eV for silicon. The theoretical full-width-half-max (FWHM) at the Mn Kα peak is about 122 eV.

The second term is the contribution to the energy resolution of the electronics noise of the detector-amplifier system. This electronics noise is expressed in most of the cases by the equivalent noise charge (ENC):

$$N = 2.355\ \varepsilon\ ENC$$

The squared value of the ENC can be expressed as the sum of three independent contributions:

$$\mathrm{ENC}^2 = \mathrm{ENC}^2_{s} + \mathrm{ENC}^2_{//} + \mathrm{ENC}^2_{1/f}$$

ENC^2_{s} is the white series noise, $\mathrm{ENC}^2_{//}$ is the parallel noise due to the leakage current, $\mathrm{ENC}^2_{1/f}$ is the 1/f series noise. Theses three contributions are given by:

$$\mathrm{ENC}^2_{s} = \mathrm{A\ Ct}^2 / \tau$$
$$\mathrm{ENC}^2_{//} = \mathrm{B\ i_n}^2\ \tau$$
$$\mathrm{ENC}^2_{1/f} = \mathrm{C\ Ct}^2$$

Ct is the value of the total capacitance (detector and preamplifier capacitances), τ is the shaping time used for signal processing, in is the leakage current. Note that the leakage current in depends on the temperature. The coefficients A, B, C are constants which depend on the "shape factors".

Figure 2 shows the schematic variation on the ENC of the detector, as a function of the shaping time. By considering figure 2, we can observe that the contribution of the series noise is inversely proportional to the shaping time, while the contribution of the parallel noise is directly proportional to the shaping time. Note that the contribution of the 1/f noise is independent of τ . As the ENC is dependent on the shaping time, there is an optimum value (τ_{opt}) for this parameter which minimizes the ENC. In nearly all commercial shaping amplifiers, it is possible to choose a shaping time close to the optimum one for the considered application.

In order to minimize both the ENCS and the ENC// contributions and therefore, the ENC value, the detector capacitance should be as low as possible. The reduction of the total capacitance Ct reduces both the τ_{opt} and the ENC (τ_{opt}).

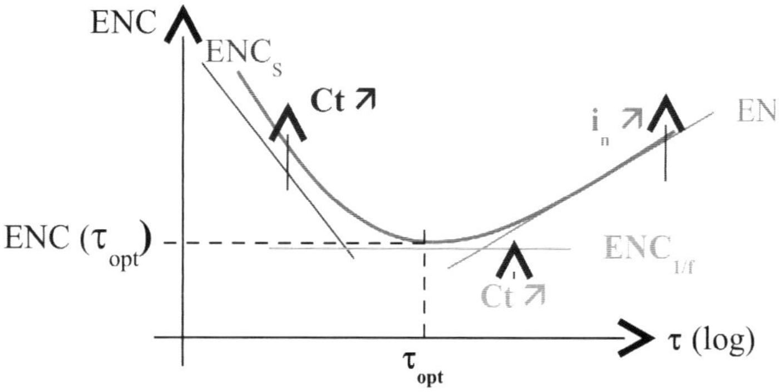

Figure 2. Schematic variation on the ENC of the detector, as a function of the shaping time.

In practice, the detector capacitance is determined by the physical structure of the detector. In a standard diode, the detector capacitance is proportional to the detector sensitive area. Therefore, for high energy resolution, the sensitive area should be kept as small as possible. The main advantage of the SDD is the small value of the collecting anode capacitance which is of the order of 100 fF (1 pF for a Si(Li)). Since this capacitance is mainly given by the area of the anode, reducing this area (a typical anode size is 400 μm_2) helps greatly to minimize the electronic noise contributions of serial white and 1/f noise. Therefore, the capacitance is independent of the active area. Due to its extremely small overall capacitance, the SDD can be operated with very short shaping times (τ_{opt}) of the order of 1 µs (this value is dependent on the operating temperature). Reducing the values of the capacitance and therefore the shaping time is also an advantage because it improves the maximum counting rate of the system. Since the maximum count rate is about $1/\tau$, spectroscopic measurements are possible up to count rates close to 106 cps (τ is the order of 1 µs). In order to improve the achievable energy resolution, the capacitance (detector and preamplifier capacitances) should be minimized. The input capacitance of the front-end electronics should be "matched" to the detector capacitance. The transistor (FET) can be integrated in the central region of the detector (figure 1), with the gate connected to the anode by means of a metal strip. In this way a correct capacitive matching between the detector and the first stage of amplification is achieved and the capacitance is minimized. This solution has also the advantage to reduce the sensitivity of the detector to microphonic noise.

To further improve the performance of the standard SDD with regard to energy resolution and peak-to-background, different types of SDD have been developed such as the SDD droplet called SDD^3 (figure 3) [14]. In this SDD, the anode and the integrated preamplifier field-effect transistor (FET) are no longer centred but moved to the outside margin of the structure where they can be shielded from direct irradiation by the use of a collimator. This device makes it possible to design an anode of a smaller area, thereby reducing the capacitance to a value of 120 fF to be compared with the 200 fF of a conventional SDD. The result is an improvement in energy resolution to about 123 eV, very close to the theoretical limit. Furthermore, incoming X-rays do not generate noise in the charge-sensitive preamplifier, yet the FET remains onboard the chip for fast response.

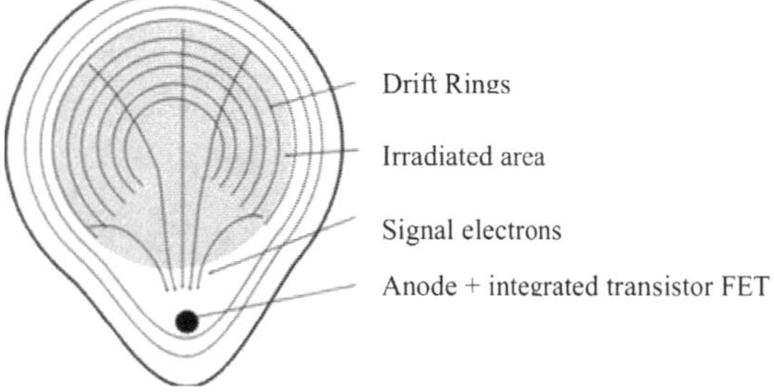

Figure 3. Layout of the "SDD Droplet" detector [14].

This type of SDD is only for small active area. Note that devices with concentric or hexagonal rings are also currently being made with external FETs.

As already mentioned, the leakage current in depends on the temperature. Due to the elaborated process technology used in the SDD fabrication, the leakage current level is so low that the drift detector can be operated with good energy resolution near room temperature (figure 4). Figure 4 shows that under a given temperature, the resolution tends to a constant. Under this temperature, the resolution depends only on the series ENCs and the ENC21/f noise. Decreasing the operating temperature of the device does not improve performance.

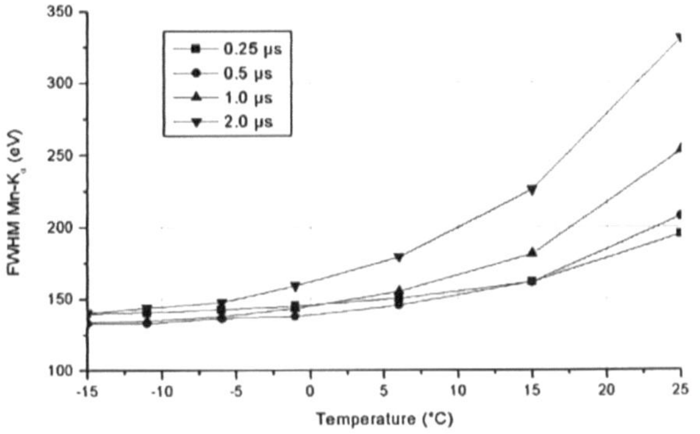

Figure 4. Energy resolution of an SDD as a function of the temperature and of the shaping time [9].

However, the leakage current is mainly proportional to the active area. When a large active area is required, the contribution of the leakage current has to be reduced by cooling. With moderate cooling by stage Peltier elements, the SDD's energy resolution can be compared to that of a Si(Li)- or Ge-detector requiring expensive and inconvenient liquid nitrogen cooling. Thermoelectric (Peltier) cooling brings the device to operating temperature in the range of -10 to -80 °C, often in the time it takes to put the sample in the microscope and pump it down. A second Peltier stage to cool down the first Peltier's hot side can easily be implemented externally. The cooling technology is now firmly established and decreasing the operating temperature of the device does not improve performance.

2.5. Detection efficiency

The detection efficiency is defined as the fraction of the total number of photons emitted by the sample, which interacts in the detector volume and is completely absorbed. The first limitation to the detection efficiency arises from the solid angle available to collect X-rays (collection efficiency). This solid angle depends on the device area and its distance from the point where the electron beam hits the sample. The physical geometry of the detector housing may restrict the location of the detector inside a chamber and may affect the solid angle. The second limitation to the detection efficiency is the quantum efficiency. In the low X-ray energy range (100 eV - 1 keV), the detection efficiency is generally limited by:

- The absorption in the window material in front of the detector. Ultra thin windows are available. Light element detection down to beryllium has been demonstrated.
- The absorption in a dead layer of the detector material itself. Light-element performance depends on silicon crystal processing and on the matching electronics. A careful tailoring of the shallow implantation of the p+ electrode on the back of the SDD ensures good efficiency and avoids incomplete charge collection (ICC): any ion pair generated in the dead layer but not collected results in energy lost to the pulse and produces a broadening of the low energy side of the X-ray peak.

At high energy, the X-ray detection efficiency is limited by the total thickness of the device. SDDs are based on silicon chips from about 300 µm to 500 µm thick. X-rays can penetrate the chip without being detected. The efficiency is higher than 90 % at 10 keV, 30 % at 18 keV and 14 % at 26 keV for a 300 µm thickness. The increase of the detector thickness to 450 µm improves the quantum efficiency to about 45 % at 18 keV and 18 % at 26 keV [13].

2.6. Count rate performance

SDDs have shown remarkable improvements in performance during the last decade, reaching high count rates while maintaining excellent resolution (figure 5). The deviation of the measured output count rate (OCR) versus input count rate (ICR) curves from the ideal curve is a result of the deadtime associated with the processing of pulses. The general shape of these curves is defined by the function OCR = ICR exp(-ICR* t), where t is a time constant which is characteristic for the used pulse former. The quality of the pulse processor has an effect on the resolution and count rates and it must be optimized for the SDD whose properties may vary between different manufacturers. Unfortunately, it is sometimes difficult for users to obtain the curves showing the OCR and the energy resolution as a function of the ICR at several shaping times. In any case, the attainable OCR with SDDs revolutionizes X-ray mapping [15, 16].

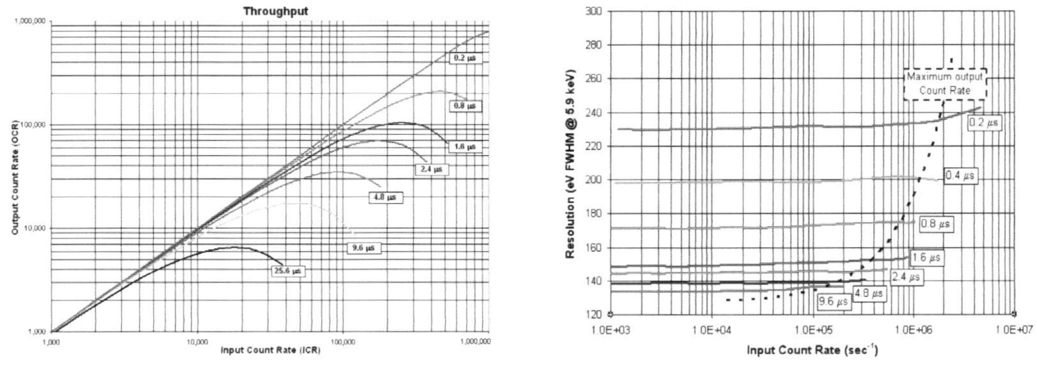

a) b)

Figure 5. Experimental data showing a) the output count rate (OCR), and b) the energy resolution, as a function of the input count rate (ICR) at several shaping times [12].

2.7. Artefacts

Since SDD is constructed from silicon and uses the same detection physics as Si(Li)-EDS, escape peaks may occur in the SDD. Furthermore, pulse pile coincidences can affect the spectrum. Pulse pile-up varies with count rate and is most significant at maximum count rates and with soft X-rays, such as from light elements. Improvements in hardware pile-up rejection and software post processing can minimize the effects of pile-up [17-19]. It is worth noting that, SDD-EDS spectra are subject to significant pulse coincidences due to principal limitations of pulse processors with high count rates, leading to "pile-up peaks" as well as pile-up distortions to the background. Research is currently being performed by the detector community into advanced methods of high-speed pulse processing to correct this problem [16].

2.8. Active area and sensitivity

Although the nature of SDD technology with its associated pulse-processing electronics permits very high count rates, this presupposes that such count rates can be generated by the beam without destroying the sample. Metals, ceramics, and minerals can generally withstand such bombardment. But working with high beam current to get high count rates may induce potentially damage on sensitive sample such as polymers. Furthermore, modern high-resolution SEMs may not be able to generate enough current, and certainly not under high-resolution imaging conditions or at low acceleration voltages. The tendency is to increase the active area in order to increase the sensitivity of the detector even at low beam currents. A larger sensor has the potential to collect more X-rays than a smaller sensor. However, the real factor of X-ray detection is not the size of the detector but its solid-angle. The solid-angle Ω is defined as:

$$\Omega = \frac{S}{d}$$

where S is the active area of the detector and d the sample-to-detector distance. The solid angle is of the order of 0.01 - 0.1sr in a SEM.

Figure 6 shows experimental data of the variation of the OCR as a function of the beam current in a SEM for two different positions of a 80 mm^2 SDD detector: the inserted position and the retracted position. The plots depend strongly on the detector's position and therefore on the solid angle. The solid angle is a key parameter. Unfortunately, it is sometimes difficult for users to obtain this value.

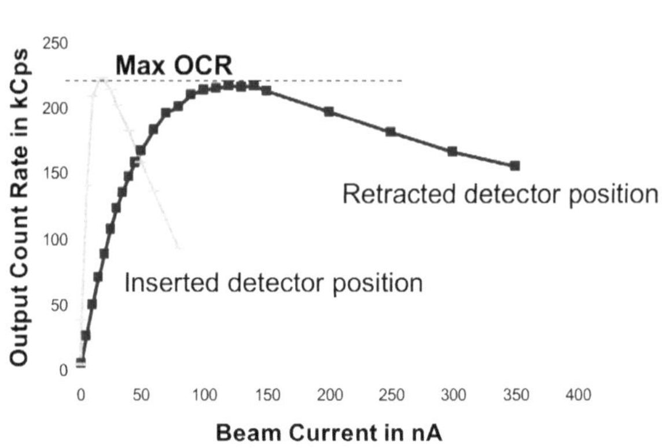

Figure 6. Experimental data showing the output count rate as a function of the beam current in a SEM for two different positions of a 80 mm^2 SDD detector: the inserted position and the retracted position (same fastest shaping time).

In order to increase the active area, different solutions are available:

- Multiple detectors on SEM:
  Several independent SDDs can be installed on the microscope and a single signal processing unit can support these detectors simultaneously. The advantage of this solution is that the total active area is increased while keeping a good solid angle for each detector. It can be also useful for the detection of irregular specimen surfaces: shadow areas on the specimen can be avoided. However, this solution has drawbacks: first, the number of ports available on the specimen chamber is limited; secondly, the cost may be high.
- Large area detectors:
  Single SDDs of large area ranging from a few tens of mm^2 up to few cm^2 (for medical imaging applications) have been produced. But initially, detectors with active areas of only ten square millimetres were installed in SEM-EDS. Larger crystals with areas of thirty square millimetres

were subsequently produced, and they enabled a significant increase in solid angle and hence count rate. It has now become possible to install on SEM-EDS systems larger active area up to 80 mm^2 [20, 21]. An SDD detector of 80 mm^2 area at a distance of 10 mm provides a solid angle Ω of 0.8 steradians. When cooled, these detectors offer an energy resolution similar to the smaller SDD area. A larger sensor size has the potential to collect more X-rays than a smaller sensor size. However, owing to the detectors size, the physical geometry of the detector housing may restrict the location of the detector inside the microscope chamber and the solid angle available may be reduced. This geometric aspect remains to be fully exploited. Furthermore, the count rate capacity remains limited even with a large area detector. Figure 7 shows the variation of the OCR as a function of the beam current for three detectors with different active areas: 10, 50 and 80 mm^2.

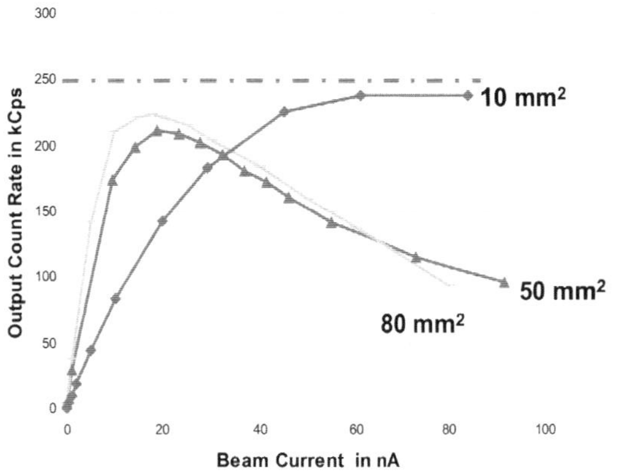

Figure 7. Experimental data showing the output count rate as a function of the beam current in a SEM for three detectors with different active areas: 10, 50 and 80 mm^2 (fastest shaping time).

As can be seen in figure 7, the maximum Ouput Count Rate remains under 250 kCps. The pile up-effect and the pile-up rejection unit induce limitations. Research is currently being performed into advanced methods of high-speed pulse processing to improve the count rate capacity. The improvement can also be achieved using multielement detectors.

- Multielement array detectors:
  To increase detection efficiency, various types of monolithic multielement array detectors have been developed in various application fields [22, 23]. Only a four 10 mm^2 "droplet" elements detector has been commercialized for EDS microanalysis in SEMs [10, 24, 25]. The subdivision of the total active area in several independent detectors allows the increase of the maximum counting rate of the system with respect to a single detector of the same total active area. However, owing to the detector's size, the solid angle available may be reduced. That is why to increase it, a flat prototype detector located directly under the pole piece has been developed [10, 22, 24, 25]. This is a detector with a 4 x 15 mm^2 active detecting area and a middle hole. Because of the flat design, an electron trap (to avoid the influx of electrons) could not be mounted. This detector contains four polymer windows of different thickness which has to be chosen according to the primary excitation otherwise electrons will penetrate the detector and cause spectrum distortion. Therefore, the light element capability is reduced with increasing window thickness. This detector has potential for high count rates (over 106 cps) because of high solid angle of the order of 1 steradian. This prototype is also useful for the detection of irregular specimen surfaces: shadow areas on the specimen can be avoided. However, it still needs to be improved. In the future, it might become possible to incorporate it into the SEM.

3. WDS systems on SEM

3.1. History of the technique

In 1947, in the United States, James Hillier patented the concept of wavelength-dispersive X-ray spectrometry (WDS) for electron microscopes but took the concept no further. The technique was developed in France by Yvette Cauchois and André Guinier, experts in X-ray diffractometers with specially cut and ground quartz crystals. It was used by Raimond Castaing during his Ph.D. thesis supervised by André Guinier in the 1950s at the French Institute for Aeronautical Research (ONERA), to develop the first electron microprobe, now bearing his name. Raimond Castaing also established the theoretical and analytical principles of the technique. WDS spectrometers were introduced as SEM accessories in 1970, these electron microscopes becoming increasingly popular in the late 1960s [26, 27].

3.2. Advantages, limitations and different WDS systems

The excellent spectral resolution of WDS (about 5 eV for the best value) allows better separation of closely spaced X-ray lines compared with EDS. It leads also to high peak-to-background ratio, which controls sensitivity and minimum detection limits. The WDS technique is more time-consuming due to its serial method of X-ray collection [8, 28, 29].

The following paragraphs present the technologies used on WDS systems available commercially, on one hand, the conventional WDS technique, and on the other hand, the parallel beam WDS spectrometers (PBS). Conventional and PBS WDS systems have in common the use of analysing crystals to select X-rays as a function of their wavelength and the detection of the diffracted X-rays by a proportional counter. The X-rays collection efficiency of conventional WDS (figure 8) is limited by the distance between the sample and the crystal (small solid angle of acceptance). The need for greater collection efficiency explained PBS development, these systems combining optics for X-rays collimation. Parallel beam WDS systems were also developed in the middle of the 1990s, to carry out X-ray microanalysis, in particular for low energy photons of low Z elements, taking advantage of the compatibility with high-resolution images obtained on field emission gun microscopes at low accelerating voltages and low beam current.

Figure 8. A conventional inclined WDS spectrometer. Note the position of the turret and the proportional counter (Oxford Instruments).

3.3. Conventional WDS systems

Following the Bragg's law for X-ray diffraction, the incidence angle of emitted X-rays on the analysing crystal (placed at an angle θ) equals the angle of reflection (detector placed at 2θ). Therefore, this WDS detector is a fully focusing spectrometer, where the fixed X-ray point source on the specimen, the analysing crystal and the proportional counter move on the same circle (the focussing Rowland circle) for different wavelengths, to maximise the intensity of the measured characteristic signal (figure 9a) [8, 28, 29]. The circle radius is relatively large (210 mm) in order to accommodate various SEMs. On the latest FEG-SEM, the working distance for WDS analysis with this system is ranging between 8.5 and 15 mm.

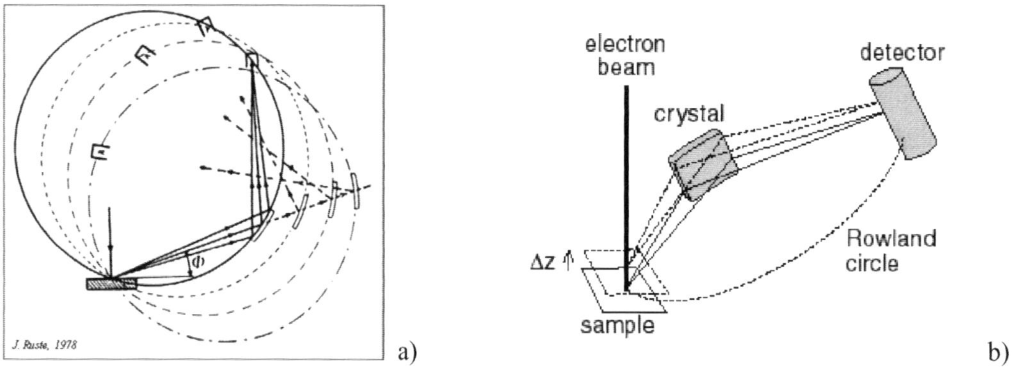

Figure 9. a) Movement of the fully focusing spectrometer. Φ is the X-ray take-off angle. b) A schematic of the inclined spectrometer offering less z-axis sensitivity.

The spectrometer is mounted with the Rowland circle inclined, so its plane is no longer parallel to the electron beam, thus reducing the sensitivity to z sample positioning and defocusing (figure 9b). The inclined orientation of the spectrometer, rather than the actually horizontal one, provides an elevation angle and improves the constant X-ray take-off angle value on untilted samples (between 30° - 35°).

To partially compensate for the loss of intensity due to the large distance between X-ray point source and crystal, quite large crystals are used (4.1 cm x 1.6 cm) as they allow to increase the solid angle of X-ray collection, of some unities of millisteradians. Hence, Johansson bent and ground PET, TAP, LiF crystals are used, with the advantage of being completely tangent to the focusing Rowland circle, in order to maximize collection efficiency. Light elements analysing crystals consisting of bent synthetic multi-layers (Johann 'crystals'), which cannot be ground without destruction, are used as well. Up to six crystals can be mounted on the spectrometer turret to analyse the entire spectrum. The 2θ-Bragg angle is ranging between 33° and 135°. So, this conventional system covers a large range of photon energy from about 100 eV to 15 keV.

A variable slit is placed in front of the counter to eliminate background from scattered radiation. It affects measured intensity and peak-to-background intensity ratio.

For low Bragg angles, the analysing crystal moves close to the sample, so a specific WDS port is needed to mount the system on the column. Due to its size and geometrical requirements for installation, only one conventional WDS spectrometer is generally mounted on the SEM column [20].

3.4. Parallel beam WDS systems

A completely different design of spectrometer combines collimating optics with wavelength-dispersive crystals (figure 10). Thanks to the collimating optic placed close (~ 2 cm) to the sample surface, isotropic X-rays emitted from the sample are re-directed into a parallel beam on the analysing crystal and then into a proportional counter, with a good collection efficiency (figure 11) [30-32]. Since collimated or nearly parallel beam of X-ray is obtained, bent crystals are replaced by large flat crystals (typically, 5 cm x 2 cm), with less aberration and the benefit of improving diffraction efficiency. Generally, up to 6 various flat crystals can be fixed on the turret.

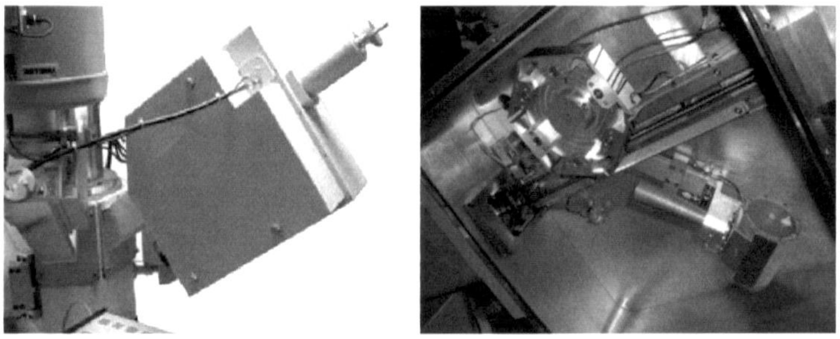

Figure 10. A parallel beam WDS spectrometer. Note the hexagonal-shaped turret and the position of the proportional counter (Parallax Research).

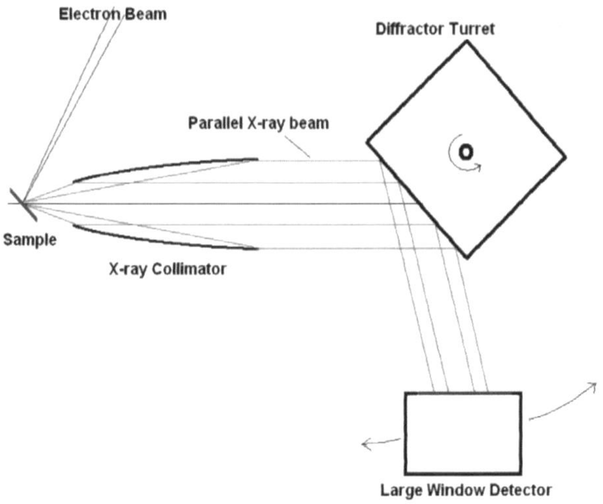

Figure 11. Schematic of the parallel beam WDS spectrometer (Parallax Research).

The analysing crystal remains at a fixed distance from the sample for all wavelengths and simply rotates to present varying incident Bragg angles. Depending on the SEMs, the distance between the entrance of the X-rays collimator (the collimator nose) and the crystal is generally about 20 cm. The distance from the crystal to the counter is about 10 cm. So, these systems present a compact design. The fixed position of the crystal in the spectrometer housing, far from the inside SEM chamber, enables the system to be mounted even on very small ports; in most cases, EDS ports are suitable.

3.4.1. The collimating optic. This collimating lens uses the total X-ray reflection phenomena at small incident angles. Total reflection of X-rays occurs if the incident angle on the smooth surface optic is less than a critical angle which depends on the mirror material and the X-ray energy. The following paragraphs present the different optics used for X-ray collimation.

3.4.1.1. X-ray concentric mirrors for low energy X-ray microanalysis. The low-energy grazing incidence optics consists of 3 nested tubular metallic mirrors, composed of an Al alloy covered by a 20 µm thick Ni layer. Each mirror presents a paraboloid shape in the longitudinal section and is truncated at the entrance aperture to provide a two-reflection optic for better collimation efficiency (figure 12) [33, 36, 37]. The largest grazing angles occur at the entrance aperture while the smallest ones occur at the exit aperture (figure 13). Only one paraboloid mirror would concentrate the collimated X-rays into a narrow ring, leaving uncollimated X-rays inside the ring. These are collimated by a smaller nested mirror, the maximum grazing angle of which corresponds to the minimum grazing angle of the outer mirror. As the critical angle becomes smaller with increasing energy, the outermost mirror – ring 1 – collimates photons with lower energies and the innermost mirror – ring 3 – the photons with higher energies (on figure 12b: ring 1: E < 500 eV, ring 2: E [500 - 1500 eV], ring 3: E [1500 - 2400 eV]). A stop exit aperture blocks uncollimated X-rays from getting out of the device. This device is about 2 cm in diameter, 10 cm in length and the diameter of the input aperture is some unities of mm.

a)

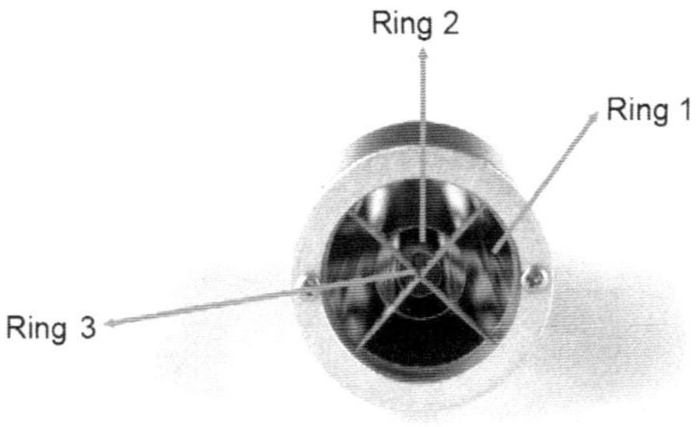

b)

Figure 12. Three nested X-ray mirrors: view of a) the entrance aperture, and b) the exit aperture (Parallax Research, EDAX).

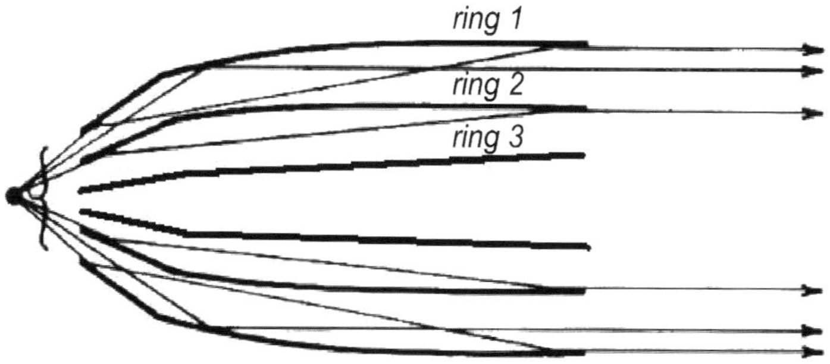

Figure 13. Schematic of the longitudinal cross section of three nested X-ray mirrors and collimated X-rays [36, 37].

For photon energies below 2.5 keV, the grazing angles are sufficiently high (1° - 6°) that this optic can accept a large solid angle of X-ray collection. It is roughly 150 msr for photon energy of 200 eV (typically, B Kα line), 80 msr at 525 eV (O Kα line) and 3 msr at 2.5 keV.

This range of operation from 80 eV to 2.5 keV, is suited for light elements analysis from Be to S (B, C, N, O especially) with an energy resolution < 20 eV (6 eV at Si Kα). Some systems with improved nested optics based on the principle described previously, allow extending the analysing energy range to 3.3 keV. These also associate, between the crystal and the counter, a conical optic that preferentially directs the collimated X-rays into the detector while removing the non-parallel scattered X-rays. This has the advantage to improve the signal-to-background intensity ratio [34].

3.4.1.2. Polycapillary collimation optics. Reflection efficiency of paraboloid mirrors deteriorates quickly at energies above 2 - 3 keV (figure 14). To cover a higher range of photon energy (typically from 150 eV to 10 keV), polycapillary optics are used (figures 14 and 15). These collimating lenses are made up of hundreds of thousands of small hollow glass tubes fused into a single bundle, individual channels having inside diameters typically in the range of 8 to 50 µm [38, 39]. The order of the reflection angle is of some tenths of degrees for X-rays in the 3 - 10 keV energy range and the solid angle of X-ray collection about 10 msr for energies above 2 keV.

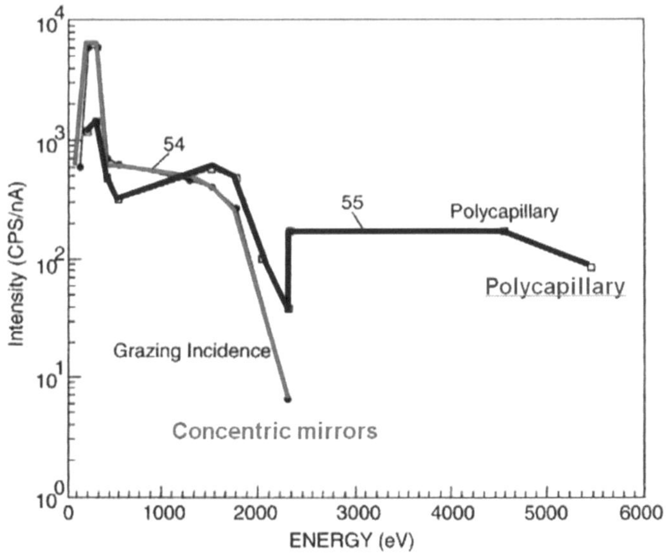

Figure 14. Plot of X-ray intensity as a function of energy of X-rays passed through different PBS optics: the concentric mirrors have higher reflection efficiency than the polycapillary optic for lower energies, whereas the polycapillary optic is more effective for higher energies [41].

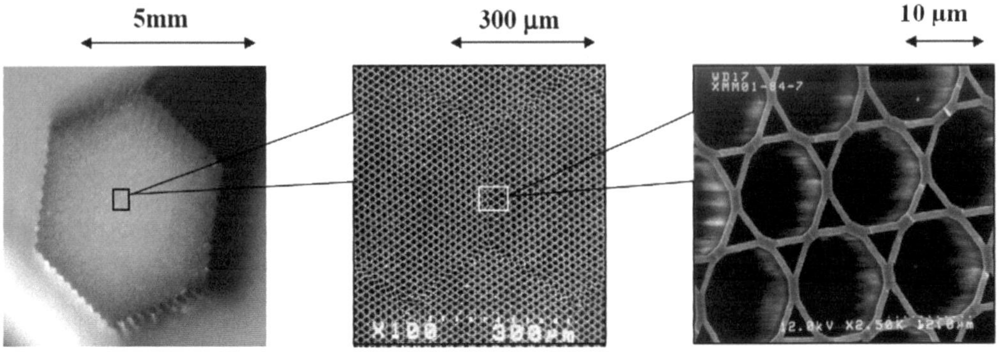

Figure 15. Polycapillary optics (X-ray Optical Systems [40]).

X-rays are conducted along the inside surface of the capillary optic by total external reflection and are re-directed into a parallel beam on a flat analysing crystal.

3.4.1.3. Hybrid optics. To benefit from previously presented optics specificities, some systems use an hybrid optic that employs a polycapillary optic (more dedicated to high photon energies, up to ~ 10 keV) nested inside paraboloid mirrors (for low energy X-ray microanalysis) [41]. The hybrid optic provides solid angles of collection of 184 msr. below 2 keV and of 12 msr. above 2 keV.

3.5. Sample and optics positioning, analysis set-up

For the different WDS systems are presented the aspects of sample positioning and also optics positioning regarding the focal optic of the PBS device [33-35].

3.5.1. Conventional WDS systems. There is no accurate adjustment of the sample height position, as far as the inclined orientation of the spectrometer reduces the sensitivity to z sample positioning and defocusing (figure 16a and b). The sample is simply placed at the recommended working distance, varying between 8.5 and 15 mm. according to the position of the system port on the SEM column.

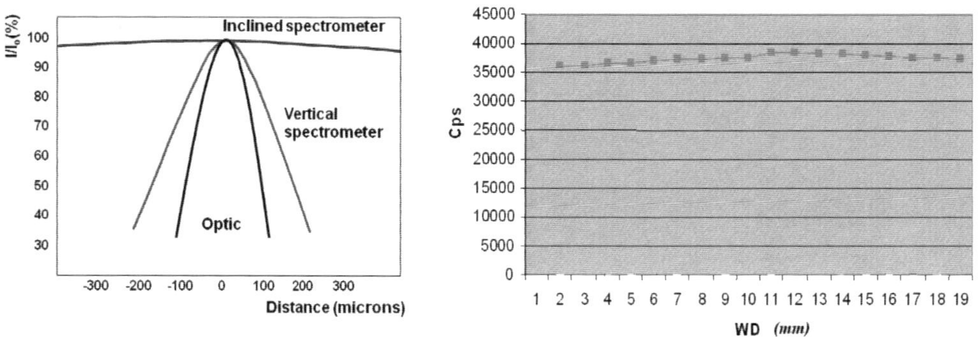

a) b)

Figure 16. a) For different spectrometer orientations, intensity loss as a function of the deviation to the optimum sample position. b) For the inclined spectrometer, intensity variations as a function of the working distance (Oxford Instruments).

After peak search is carried out by moving in a coupled way the monochromator and the counter, the variable width of the slit, placed in front of the counter, is adjusted to obtain a compromise between background intensity reduction and maximum peak intensity (the order of the slit width is 1 mm.).

3.5.2. Parallel beam WDS systems. These systems are sensitive to positioning as the point source of X-rays on the sample must be at the focal point of the collimating optics. The input focal distance is 18 mm for both three-nested mirrors and polycapillary optics (figure 17a and b). The input focal spot size corresponds to the size of the x-ray source area that can be well collimated by the optic (figure 17c). It is determined by the following factors: the input focal distance, the optic design and the X-ray source. The input focal spot size changes with the critical angle that means it is dependent on the X-ray energy. For both three-nested mirrors and polycapillary optics, X-rays of lower energy, corresponding to higher critical angles, can be reflected by the optics from a larger area than photons of higher energies. As an example, with the nested mirrors, the input focal spot size is 200 μm at 100 eV (Be Kα), about 100 μm for C Kα, N Kα, O Kα and 25 μm at 1.75 keV (Si Kα). The output divergent angle of the collimated X-rays is about 0.1° across the entire energy spectrum whatever optic device is considered.

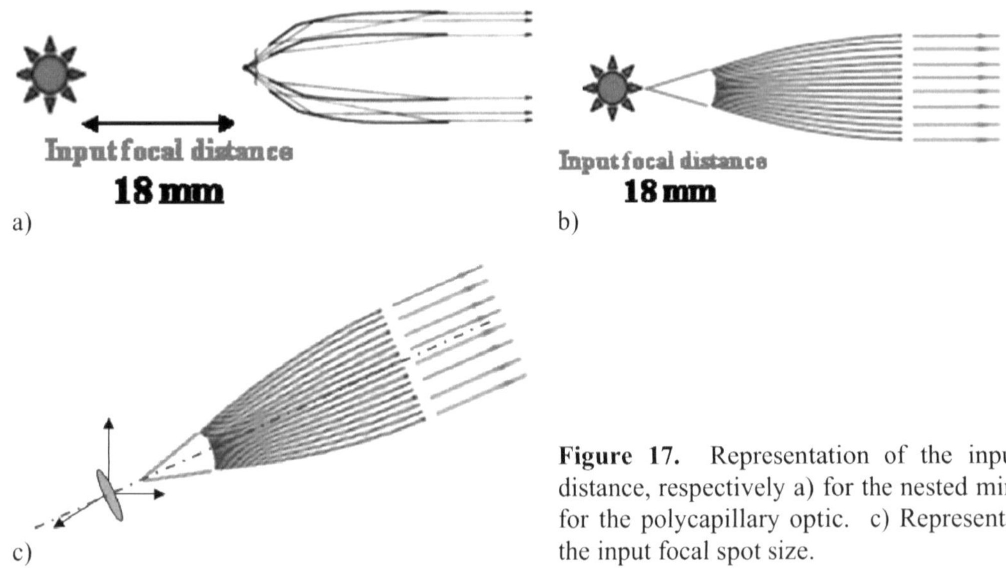

Figure 17. Representation of the input focal distance, respectively a) for the nested mirrors, b) for the polycapillary optic. c) Representation of the input focal spot size.

Considering a characteristic peak and so setting the crystal and the proportional counter to the corresponding Bragg angle value, the sample is positioned at the correct working distance by moving it up and down until the maximum intensity of the selected line is found. An automatic alignment procedure of the sample is generally integrated as a WDS system facility. Then, the spectrometer moves from side to side to maximize peak counts. Finally, by using software control of the optic motors, the collimator nose moves in and out along its axis to obtain optimal spectrometer intensity. The procedure can be repeated for another characteristic peak.

3.6. Examples of WDS spectra

WDS spectra were acquired on olivine sample at 20 keV, for different crystals, on a Cameca SX50 microprobe corresponding to a conventional system, and on a PBS system (figure 18). These acquisitions show similar results on the point of view of spectral resolution.

4. Conclusion

In the field of X-ray microanalysis, SDD detectors on scanning electron microscopes are developing extremely fast. These detectors, which are easy to use and efficient, have replaced standard diodes. SDD technology is now firmly established and further improvements are expected. WDS systems are still used to provide excellent spectral resolution, accurate measurements for quantitative analysis, especially for minor, trace and also light elements. In both EDS and WDS spectrometry, an extensive range of systems/equipments is now available on SEMs, each including specific features depending on the field of application.

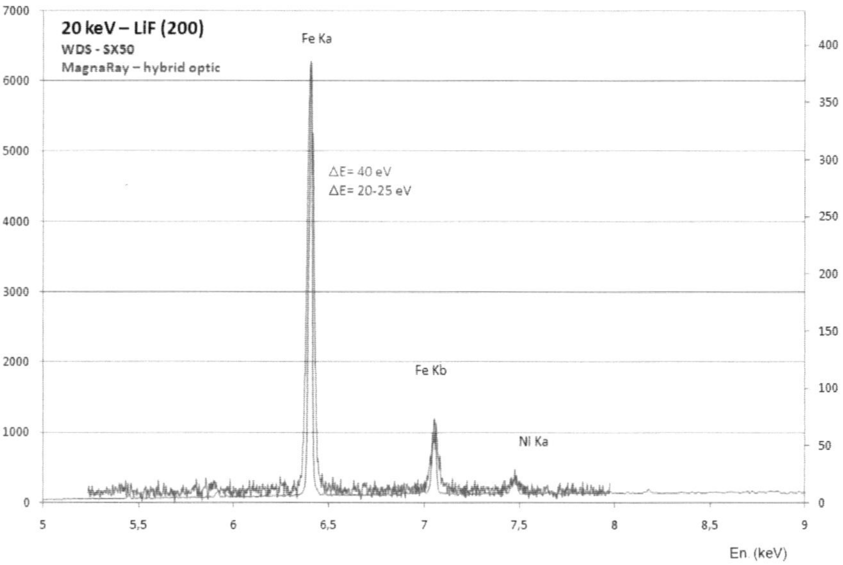
20 keV – LiF (200)
WDS - SX50
MagnaRay – hybrid optic
Fe Ka
ΔE= 40 eV
ΔE= 20-25 eV
Fe Kb
Ni Ka
En (keV)

a)

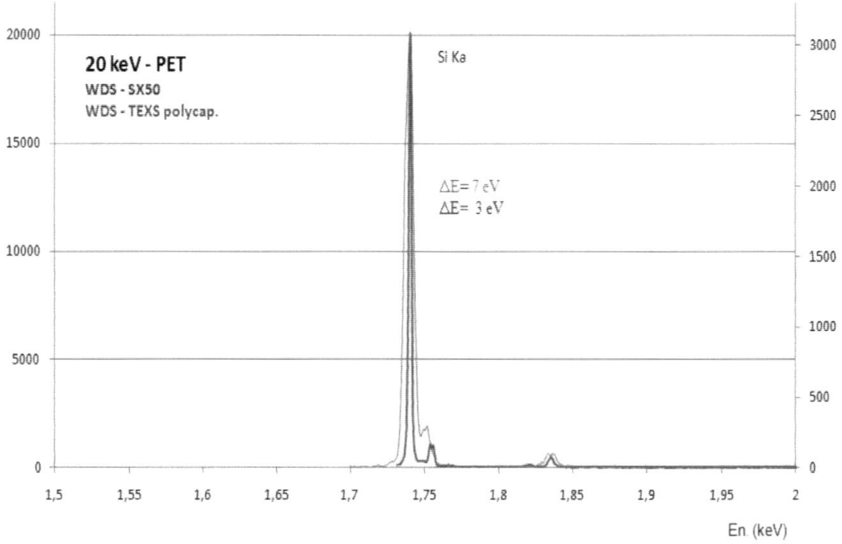
20 keV - PET
WDS - SX50
WDS - TEXS polycap.
Si Ka
ΔE= 7 eV
ΔE= 3 eV
En (keV)

b)

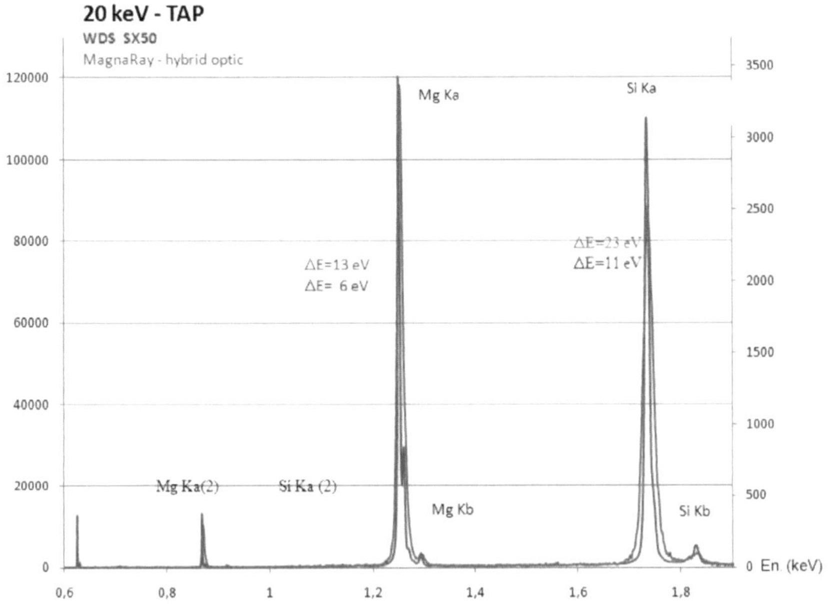

c)

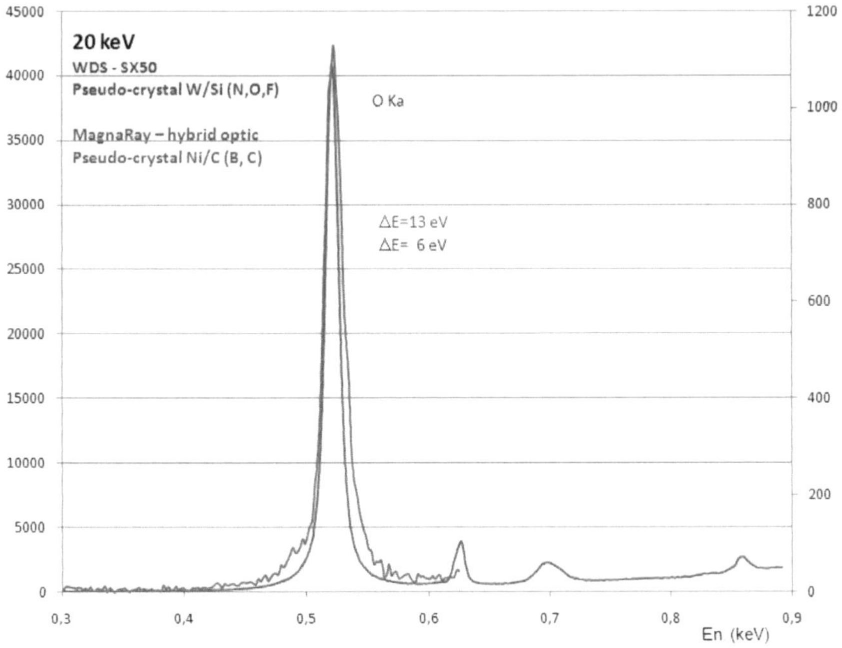

d)

Figure 18. WDS spectra acquired at 20 keV on a Cameca SX50 microprobe and on a PBS system, using respectively, a) LiF (200); b) PET; c) TAP crystals; and d) synthetic multi-layers.

Acknowledgments
The authors thank the following people for providing information for this document: Dale E. Newbury (NIST), Claudie Josse (Université de Bourgogne), Frederic Christien (Polytech'Nantes), Yann Borjon-Piron (Polytech'Nantes), François-Xavier Lefevre (CNRS CRISMAT, Caen), Raynald Gauvin (Université McGill, Montréal), Guillaume Wille (BRGM, Orléans), Philippe Lasson (Synergie4), Ralf Terborg (Bruker), Peter Statham (Oxford Instruments), Eric Pasquier (Oxford Instruments), Jean-François Thiot (SAMx), David Barresi (EDAX), David O'Hara (Parallax Research), Thierry Grenut (Elexience), Beat Gade (Thermo Fisher Scientific), and Ning Gao (XOS).

Disclaimer
Certain commercial equipment, instruments, or materials are identified in this paper to foster understanding. Such identification does not imply recommendation or endorsement by the Grenoble Institute of Technology, nor does it imply that the materials or equipment identified are necessarily the best available for the purpose.

References
[1] Gatti E and Rehak P 1984 *Nucl. Instr. Meth. A* **225** 608-14
[2] Strüder L, Meidinger N, Stötter D, Kemmer J, Lechner P, Leutenegger P, Soltau H, Eggert F, Rohde M and Schülein T 1998 *Microsc. Microanal.* **4** 622
[3] Friel F 2008 Breakthrough in EDS technology. *Advanced materials and processes* 41-43
[4] Kenik E A, Joy D C and Redfern D 2004 *Microchim. Acta* **145** 81-5
[5] Demers H, Gauvin R, Anderson I M, Joy D C and Kenik E A 2005 *Microsc. Microanal.* **11** (suppl 2)
[6] Kemmer J and Lutz G 1987 *Nucl. Instr. and Meth. A* **253** 365-77
[7] Struder L, Fiorini C, Gatti E, Hartmann R, Holl P, Krause N, Lechner P, Longoni A, Lutz G, Kemmer J, Meidinger N, Popp M, Soltau H and von Zanthier C 1998 *Mikrochim. Acta* **suppl. 15** 11-9
[8] Matthews M B 2010 *Book of Tutorials and Abstracts of 9th EMAS Regional Workshop* (Antwerp: EMAS)
[9] Product information by PnSensor GmbH, info@pnsensor.de
[10] Niculae A, Soltau H, Lechner P, Liebl A, Lutz A, Strüder L, Eckhard R, Schaller G and Schopper F 2007 *Microsc. Microanal.* **13** (Suppl. 2)
[11] Product information by Ketek, info@ketek.net
[12] Product information by Amptek Inc., www.amptek.com
[13] Longoni A and Fiorini C 2005 X-ray detectors and signal processing. in: *Handbook of practical X-ray fluorescence analysis.* pp. 203-259.
[14] Lechner P, Pahlke A and Soltau H 2004 *X-ray Spectrom* **33** 256-61
[15] Newbury D E 2005 *Scanning* **27** 227-39
[16] Newbury D E 2009 *Spectroscopy*
[17] Statham P J 2007 *Microsc. Microanal.* **13** 1428-9
[18] Elam T, Anderhalt R, Sandborg A, Nicolosi J and Redfern D 2008 *Microsc Microanal.* **14** (Suppl. 2) 1260
[19] Mott R B 2007 *Microsc. Microanal.* **13** (suppl. 2) 1434
[20] Product information by Oxford Instruments, www.oxinst.com
[21] Statham P J 2010 *Microsc. Microanal.* **16** 1304-5
[22] Alberti R, Buzzetti S, Fiorini C, Guazzoni C, Klatka T, Lechner P, Longoni A and Strüder L 2006 Advanced monolithic arrays of silicon drift detectors for elemental mapping applications. in: *Proc. of the Int. Symp. on Detector Development for Particle, Astroparticle and Synchrotron Radiation Experiments, SNIC Symp. (Stanford, California, 3-6 April)*
[23] Lechner P et al. 2000 *Proc SPIE* **4012** 592
[24] Product information by Bruker nano, www.bruker-nano.com

[25] Terborg R and Rohde M 2008 New developments in state of the art silicon drift detectors (SDD) and multiple element SDD. in: *EMC 2008, vol. 1: Instrumentation and Methods, (14th European Microscopy Congress, Aachen, Germany, 1-5 September)*, pp. 633-4
[26] Philibert J 2001 *Microsc. Microanal.* **7** 94-9
[27] Wittry D B and Barbi N C 2001 *Microsc. Microanal.* **7** 124-41
[28] Goldstein J and al. 2003 *Scanning electron microscopy and X-ray Microanalysis,* 3rd edition. (New York: Kluwer-Plenum)
[29] Crisci A and Ruste J 2008 *Microscopie électronique à balayage et microanalyses.* GN MEBA and Brisset F, eds. (Paris: EDP Sciences)
[30] Love G 2002 *Mikrochim. Acta* **138** 115-24
[31] Van Hoek C and Koolwsijk M 2008 *Microchim Acta* **161** 287-93
[32] ThermoFischer Scientific, notes
[33] Product information by Edax, www.edax.com
[34] Product information by Parallax Research, www.parallaxray.com
[35] Product information by ThermoFisher Scientific, www.thermofisher.com
[36] O'Hara D B 1997 US Patent 5,682,415
[37] O'Hara D B 1998 US Patent 5,768,339
[38] Gibson D and Gibson W 2002 *Adv. X-ray Anal.* **45** 286-97
[39] Gao N and Janssens K 2004 *Polycapillary X-ray optics, X-ray spectrometry: recent technological advances.* Tsuji K, Injuk J and Van Grieken R, eds. (New York: John Wiley & Sons Ltd.)
[40] Product information by XOS, www.xos.com
[41] Mc Carthy J J et al. 1999 US Patent 5,926,522

Uncertainty and capability of quantitative EPMA at low voltage – A review

C Merlet[1,3] and X Llovet[2]
[1] Université de Montpellier II, CNRS, Laboratoire Géosciences Montpellier, Place Eugène Bataillon, FR-34095 Montpellier Cedex 5, France
[2] University of Barcelona, Centres Científics i Tecnològics (CCiT), C/ Lluís Solé i Sabarís 1-3, ES-08028 Barcelona, Spain

E-mail: merlet@gm.univ-montp2.fr

Abstract. The use of electron probe microanalysis (EPMA) at low voltage allows improvements in the lateral resolution and surface sensitivity of the technique down to the sub-micrometre scale, and it is also useful in minimizing X-ray absorption (for the analysis of ultra-light elements), charging effects and secondary fluorescence. However, EPMA at low voltage is accompanied by a number of problems which may affect the accuracy of quantitative results. A large number of these problems arise from the need of using low-energy X-ray lines such as L- and M-lines, which have low fluorescence yields, large uncertainties in the mass attenuation coefficients, and are affected by several spectroscopic difficulties (e.g., peak shifts and overlaps). Other factors that play an important role at low voltage are carbon contamination, surface oxidation, the quality of sample polishing, the influence of a conductive coating and the limitations of matrix correction procedures. In this paper, we illustrate the capabilities of EPMA at low voltage and we examine in detail the different sources of uncertainty that may affect the accuracy of quantitative results.

1. Introduction

The incorporation of the field emission gun (FEG) in the scanning electron microscope (SEM) and, more recently, in the electron microprobe, has stimulated the operation of these instruments at low voltage. Using FEG electron sources, it is possible to obtain a finely focussed (< 100 nm) electron beam with relatively high current (< 100 nA) even at low beam energies (< 5 keV). At such energies, the penetration range of incident electrons drops from the conventional micron scale down to the sub-micrometre scale, and leads to a significant improvement in the lateral resolution and surface sensitivity of electron probe microanalysis (EPMA) [1]. This opens a wide range of new possibilities for the characterisation of materials that are heterogeneous on a sub-μm scale. The operation of EPMA at low voltage is also useful for the analysis of ultra-light elements, as it largely reduces X-ray absorption (the main source of uncertainty for such elements), for a number of insulating materials, as charge compensation effects allows to analyze them without any conductive coating [2], and the analysis near phase boundaries, as it reduces secondary fluorescence effects.

However, EPMA at low-beam energies is accompanied by experimental and analytical problems which may affect the accuracy of quantitative results [3]. At low-beam voltages, low-energy X-ray

[3] To whom any correspondence should be addressed.

lines such as L- and M-lines have to be used, and these lines are often affected by spectroscopic difficulties (peak shifts and overlaps) and larger uncertainties in the associated mass attenuation coefficients (MACs). Moreover, the fluorescence yields for the L- and M-lines are generally lower than for the conventional K-lines. As a result, peak-to-background ratios are also lower and the detection limit worsens. At low voltage, the overvoltage ratio, defined as $U = E_0/E_c$, where E_0 is the beam energy and E_c is the critical excitation energy of the considered line, may also be relatively low (< 2), decreasing significantly the X-ray intensity. Also, the approximations underlying matrix corrections procedures (e.g., $\phi(\rho z)$ models) at low voltage are not well established. For instance, the validity of expressions used in conventional correction procedures for the calculation of quantities such as the ionisation cross-section or the stopping power at low voltages (or close to the ionisation threshold) may be questioned. Finally, at low beam voltages, a surface layer of several nanometres in thickness represents a much larger fraction of the sample and, therefore, the influence of carbon contamination, surface oxidation, the quality of the sample polish or a conductive coating becomes more significant. Here, measurements at multiple voltages may become useful to separately characterize a sample substrate under a layer or multiple layers, using suitable thin film programmes [4].

In this paper, we illustrate the advantages of the use of EPMA at low voltage by means of several examples and we examine the difficulties that may affect the accuracy of quantitative results.

2. Analytical resolution of EPMA

Figure 1 shows the trajectories of electrons in Si, computed by using Monte Carlo simulations, for different electron incident beam energies. We can see that the penetration range of electrons reduces as the beam energy decreases.

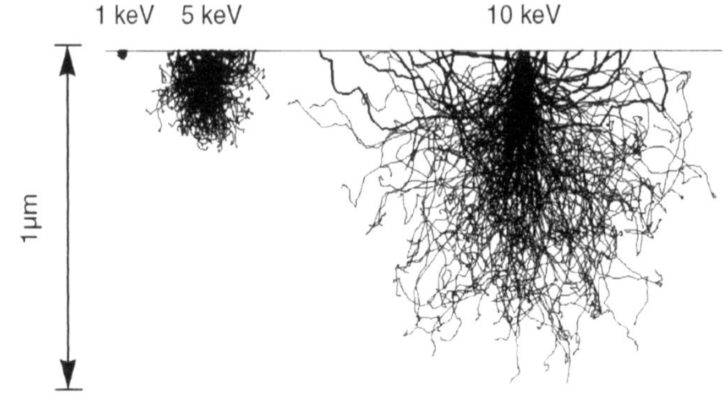

Figure 1. Electron trajectories in Si with E = 1, 5 and 10 keV, computed using Monte Carlo simulations.

The spatial resolution of EPMA can be estimated from the effective ionisation range of electrons R_x, or X-ray range, defined as the average path length that electrons travel before slowing down to an energy equal to the critical excitation energy of the considered electron shell. The ionisation range can be estimated by means of a simple relationship [5]:

$$R_X(\text{nm}) = \frac{33.A}{\rho.Z}\left(E_0^{1.7} - E_c^{1.7}\right) \quad (1)$$

where E_0 and E_c are expressed in keV, A is the atomic weight (in g/mol), Z is the atomic number, and ρ is the density (in g/cm^3). Here it should be noted that the beam diameter needs also to be taken into account to estimate the spatial resolution, especially at low beam energies.

Table 1 lists X-ray ranges calculated by using Eq. (1) for Fe Kα (E_c = 7.11 keV) and Fe Lα (E_c = 0.707 keV) X-rays emitted from C, Si, Fe and Au, and for incident electron energies in the range 2.5 - 30 keV. Notice that Fe is considered here as a ``trace'' element immersed within C, Si, Fe and Au matrices. At 20 keV, the effective ionisation range for Fe Kα ranges from 3.9 μm for C to 470 nm for Au, which would correspond to the conventional μm-scale spatial resolution of EPMA. However, at 2.5 keV, the effective ionisation range for Fe Lα drops substantially and goes from 123 nm for C down to 18 nm for Au. The improvement in spatial resolution at 2.5 keV is worth noticing. We also note that, the overvoltage ratio for the Fe Kα-line is above 2 for conventional beam energies (> 15 keV), but it drops below 2 for beam energies lower than 12 keV.

Table 1. X-ray range (in μm) calculated from Eq. (1) for Fe Kα (E_c = 7.11 keV) and Fe Lα (E_c = 0.707 keV) in C, Si, Fe and Au matrices and beam energies in the range 2.5 - 30 keV.

		X-ray range							
		Fe Kα-line (E_c = 7.11 keV)					Fe Lα-line (E_c = 0.707 keV)		
Matrix	keV	E_0 = 30	E_0 = 25	E_0 = 20	E_0 = 15	E_0 = 10	E_0 = 10	E_0 = 5	E_0 = 2.5
	Density (g/cm^3)	U_0=4.2	U_0=3.5	U_0=2.8	U_0=2.1	U_0=1.4	U_0=14	U_0=7	U_0=3.5
C	2.26	8.66	6.13	3.94	2.10	0.645	1.45	0.435	0.123
Si	2.32	8.46	5.98	3.84	2.04	0.629	1.41	0.424	0.120
Fe	7.86	2.67	1.76	1.13	0.603	0.199	0.447	0.134	0.038
Au	18.85	1.29	0.74	0.47	0.252	0.096	0.216	0.065	0.018

3. Advantages of EPMA at low voltage

3.1. Quantification of thin films and multilayers

The first benefit of using EPMA at low voltage is the improvement in surface sensitivity, which is useful for the analysis of very thin over-layers.

EPMA was originally developed for the analysis of bulk samples (i.e., those samples that are homogeneous both laterally and in depth at the μm-scale) but it was later extended so as to allow the analysis (composition and thickness) of thin layers with thicknesses below 1 - 2 μm [4]. Such improved depth resolution is achieved mainly because the depth distribution of X-ray production can be predicted within a reasonable accuracy by using an analytical $\phi(\rho z)$ model. Analysis of thin layers by EPMA requires in many cases the measurement of X-ray intensities (*k*-ratios) at different electron incident energies and the use of a thin film programme which determines the layer thickness and composition by fitting the predictions of the $\phi(\rho z)$ model to the measured *k*-ratios (see figure 2).

The optimal choice of accelerating voltage and X-ray line for the analysis of thin layers and multilayers depends strongly on the dimension of the multilayered structure: the lower the incident electron energy, the shallower the electrons penetrate into the specimen and, as a result, the *k*-ratio becomes more sensitive to resolve structures on the sub-μm scale.

This is illustrated in figure 2, which shows the variation of the *k*-ratio for Ti and Si Kα X-rays, as a function of beam energy, emitted from a Si 17 μg/cm^2 / Ti 18 μg/cm^2 bi-layer on Si. In this example, *k*-ratio measurements at 3 and 4 keV are essential to infer the correct near-surface configuration of the sample. If measurements are only carried out above 5 keV, it becomes almost impossible to predict that a Ti layer is buried under Si [6]. This example illustrates that measurements at low voltage are essential to determine the near-surface structure of stratified samples.

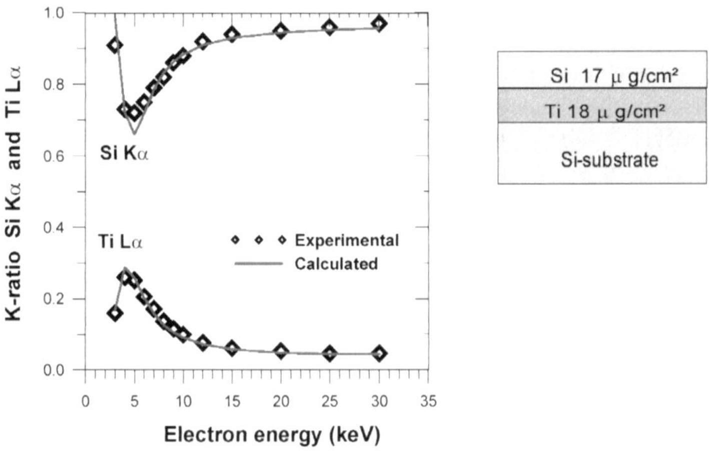

Figure 2. Comparison between calculated and measured *k*-ratios for Si Kα and Ti Kα as a function of beam energy from a Si 17 μg/cm^2 / Ti 18 μg/cm^2 bi-layer on Si. Symbols are experimental data measured by Bastin et al. [6]. Continuous lines are the predictions of X-FILM [4].

3.2. Quantification of ultra-light elements

As mentioned previously, EPMA at low voltage reduces the effective ionisation range of electrons and, as a consequence, X-ray absorption is also reduced. This potentially improves the accuracy of quantitative analysis for ultra-light elements, for which the uncertainties in the MACs are much more important relatively to more energetic lines (in some cases these uncertainties may be of 1000 %).

Figure 3 illustrates the effect of the uncertainty in the MAC of O Kα in TiO_2 on the O-concentration as calculated with the X-PHI model [7]. For the considered MAC uncertainties (from 0 % up to 10 %), the maximum error in the O-concentration is obtained at a beam energy of approximately 15 keV. Above 15 keV, the error in the concentration becomes relatively constant, with a value similar to the MAC uncertainty. However, below 15 keV, the effect of the MAC uncertainty on the O-concentration is significantly reduced. For example, at 5 keV, the error in the O-concentration drops to 4 %, even if the MAC uncertainty is 10 %. Notice the MAC of O Kα in Ti adopted in X-PHI (20000 cm^2/g) is slightly lower than the value reported in Henke's tabulation (22100 cm^2/g) [8]. This difference is most likely due to the effect of chemical bonding (see below) and experimental uncertainties affecting Henke's database. Similar discrepancies are also found in other MAC tabulations obtained from EPMA measurements.

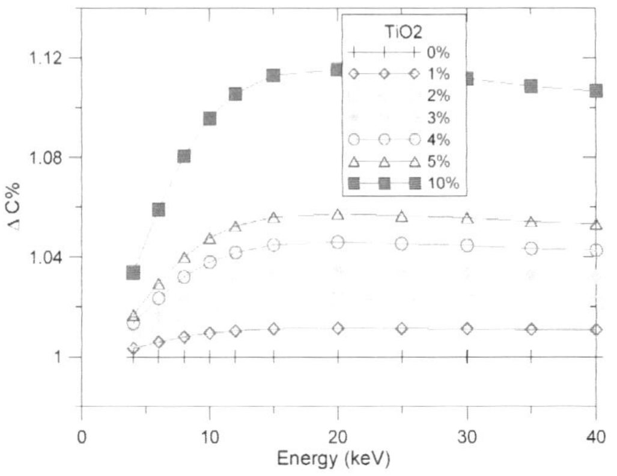

Figure 3. Relative error in the O-concentration (ΔC%) in TiO_2 for different uncertainties in the MAC of O Kα, ranging from 0 % up to 10 %, as a function of beam energy.

3.3. Improvement in spatial resolution

The development of the FEG-SEM and, more recently, the FEG electron microprobe, has made it possible to achieve finely focused (< 100 nm) electron beams with a relatively high currents (< 100 nA) even at low beam energies (< 5 keV). At such energies, the X-ray range drops from the conventional μm-scale down to the sub-μm range permitting the acquisition of -ray maps and point analyses with sub-micron resolution. FEG-EPMA fills the gap existing between conventional EPMA (with μm resolution) and X-ray analysis in the transmission electron microscope (with nanometre resolution).

To illustrate the improvement in lateral resolution that can be achieved at low voltage, X-ray maps were acquired on a sample consisting of hematite-ilmenite-rutile using a Jeol 8500F FEG electron microprobe and were compared to the results obtained with a conventional Cameca SX-100 electron microprobe [9]. Figure 4 shows Nb Lα, Fe Kα, Ti Kα X-ray maps, as well as the backscattered (BSE) electron image, of an area approximately 15x15 μm^2 in size acquired at 15 keV using the conventional microprobe and Nb Lα, Fe Kα, Ti Kα maps of approximately the same region obtained at 6 keV using the FEG electron microprobe. The improvement in lateral resolution is worth noticing, especially for Nb. This is also illustrated in table 2, which tabulates the X-ray range and the lateral resolution, in ilmenite calculated by using Eqs. (1) and (2) for the different X-ray lines used in the maps, at $E_0 = 15$ keV and $E_0 = 6$ keV. The mass density of ilmenite is in the range of 4 - 5 g/cm^3. The X-ray range is also computed for Zr Lα. The lateral resolution, R_S, can be obtained using the simple formula

$$R_S = \sqrt{4(R_x - Z_m)^2 + d_0^2}\,, \qquad (2)$$

which requires knowledge of the X-ray range (R_x), the depth of the maximum of the X-ray depth distribution (Z_m) and the beam diameter (d_0). A beam diameter of 100 nm was used in table 2.

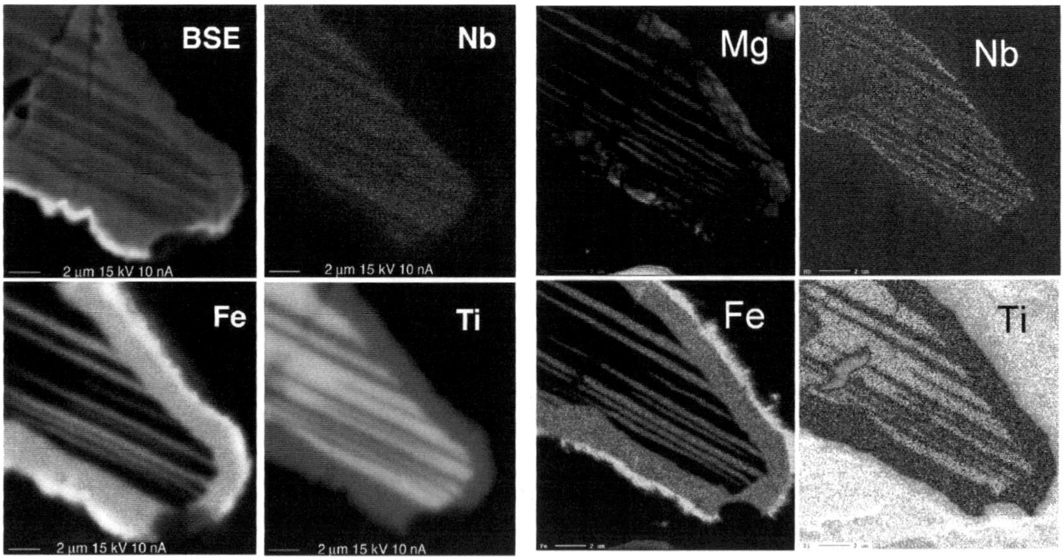

a) b)

Figure 4. X-ray maps for the distribution of Mg, Nb, Fe and Ti in a sample consisting of hematite-ilmenite-rutile. a) Results obtained with a conventional Cameca SX-100 electron microprobe at 15 keV [9] (the BSE image is also shown). b) Results obtained with a Jeol 8500F FEG electron microprobe at 6 keV.

Table 2. X-ray range (in μm) and lateral resolution (R_S) for Mg Kα, Ti Kα, Fe Kα and Lα, Zr Lα and Nb Lα X-rays from ilmenite at 15 keV and 6 keV, based on Eqs. (1) and (2).

		E_0 = 15 keV					E_0 = 6 keV		
X-ray line	E_c keV	U_o (E_0/E_c)	Range	R_S	X-ray line	E_c keV	U_o (E_0/E_c)	Range	R_S
Mg Kα	1.3	11.5	1.30	1.95	Mg Kα	1.3	4.6	0.257	0.399
Ti Kα	4.97	3.0	1.12	1.68	Ti Kα	4.97	1.2	0.077	0.171
Fe Kα	7.11	2.1	0.95	1.43	Fe Lα	0.71	8.5	0.271	0.418
Zr Lα	2.22	6.8	1.27	1.90	Zr Lα	2.22	2.7	0.227	0.354
Nb Lα	2.37	6.3	1.26	1.90	Nb Lα	2.37	2.5	0.221	0.346

In order to examine the resolution achieved at low voltage, X-ray line scans were further acquired across the hematite-ilmenite-rutile at an energy of 6 keV and beam current of 15 nA, using the FEG electron microprobe. For comparison purposes, similar X-ray line profiles were obtained at 15 keV using a conventional electron microprobe. Figure 5 shows the X-ray line scans, of approximately 14 μm in length, corresponding to Mg and Ti Kα, and Zr, Fe and Nb Lα X-rays, as a function of the distance along the line. The location of the line profiles (white line superimposed on the Fe map) is also shown.

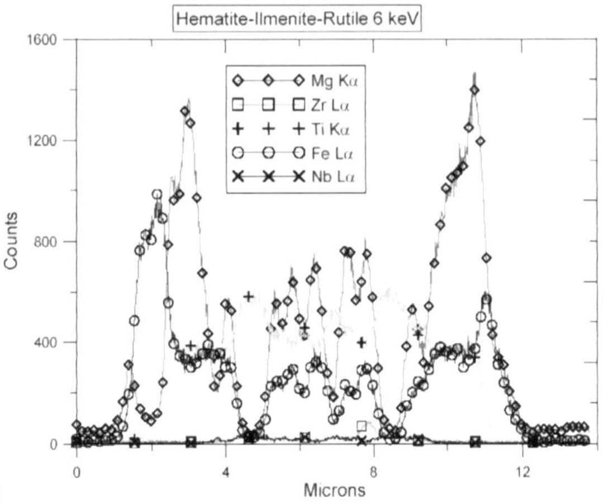

(a)

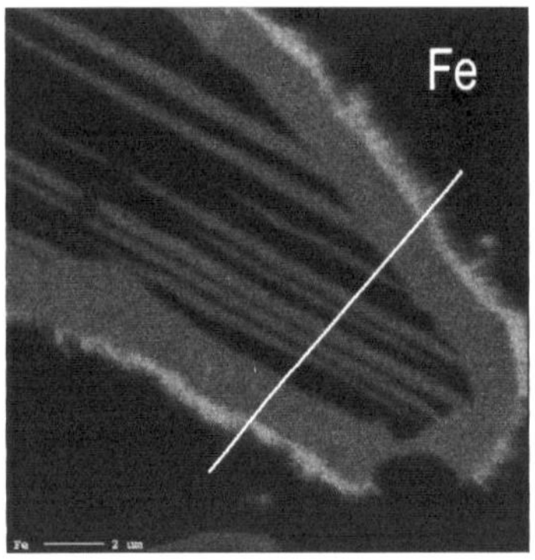

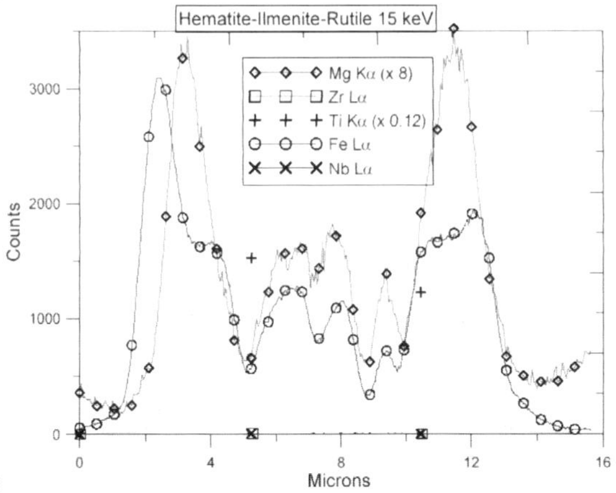

(b)

Figure 5. X-ray line scans across the hematite-ilmenite-rutile sample obtained with a FEG electron microprobe at 6 keV (a), and with a conventional electron microprobe at 15 keV (b). To facilitate comparison, the Ti and Mg profiles obtained at 15 keV have been multiplied by a factor of 0.12 and 8, respectively. The location of the line scan is shown on the Fe map.

It should be noted that the X-ray line scans shown in figure 5 are qualitative in nature, and therefore, counting rates are not proportional to element concentration. For instance, the counting rates measured on the ilmenite phase for Ti, Mg and Fe shown in figure 6a (400, 1300 and 400 counts/s, respectively) do not actually reflect the differences in concentration of these elements (TiO_2 62 wt%; MgO 20 wt%; FeO 18 wt%) [9]. This is so because X-ray counting rates depend not only on sample composition but also on other factors such as the overvoltage ratio or spectrometer efficiency.

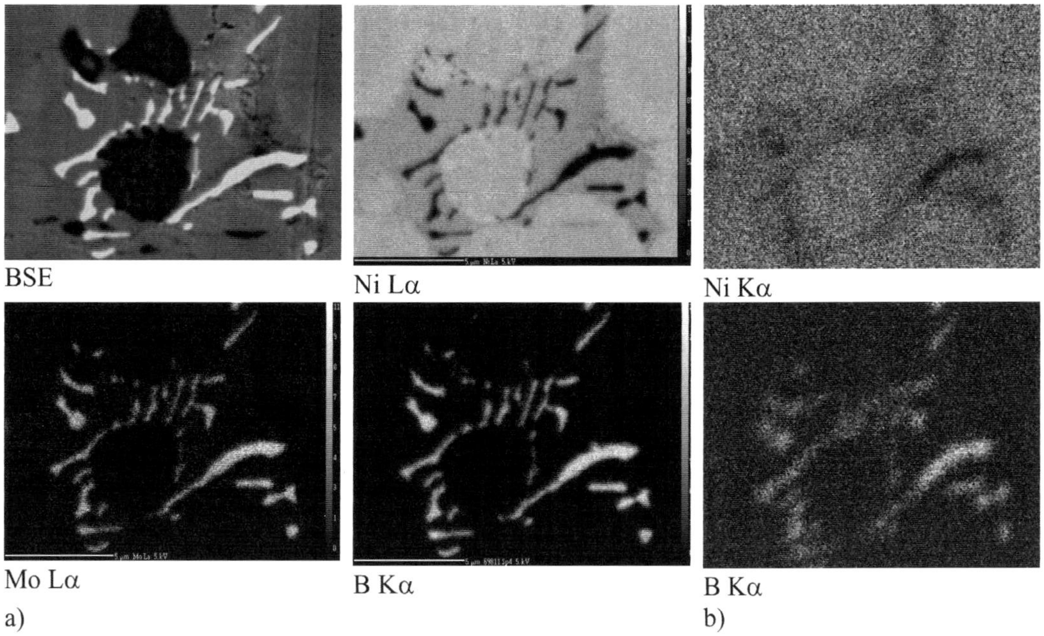

Figure 6. X-ray maps for the distribution of Ni, Mo and B in a Ni-base superalloy sample. a) Results obtained with a Cameca SX-Five FEG electron microprobe at 5 keV. b) Results obtained with a conventional Cameca SX-100 microprobe at 10 keV [10].

The width from 10 % to 90 % of the maximal intensity of a line scan across a sharp interface can be used as a measure of the lateral resolution. For the hematite-ilmenite interface, the lateral resolution of the Ti Kα- and Fe Lα-lines is 290 nm and 390 nm, respectively. Notice that, in this case, the resolution of the Ti Kα-line is slightly lower than that of the Fe Lα radiation mainly because at 6 keV, the Ti K-shell is excited at a very low over-voltage ratio (U = 1.2), which makes the X-ray range to drop drastically (see table 2). In comparison, the Ti Kα intensity measured at 15 keV (figure 5b) is 20 times higher. Notice that at 15 keV the rutile phase cannot be distinguished from the ilmenite phase.

A second example concerns the analysis of sub-micron borides occurring in Ni-base superalloys. In this case, X-ray maps of Ni, Mo and B were acquired at 5 keV and 40 nA using a Cameca SX-Five FEG electron microprobe, and the results were compared to those obtained earlier with a conventional Cameca SX-100 electron microprobe [10]. In the latter case, the maps had been obtained with a beam voltage of 10 kV and a beam current of 1 nA, in order to achieve the best possible resolution. The area scanned was approximately 14×14 μm^2 in size. In the case of the FEG microprobe, the analyzed X-ray lines were Ni Lα, Mo Lα and B Kα, while for the conventional microprobe, Ni Kα- and B Kα-lines had been used. The improvement in image resolution is demonstrated, especially for boron. The use of FEG microprobes opens new possibilities for X-ray mapping at high spatial resolution.

3.4. Reduction of fluorescence effects
Another benefit of using EPMA at low-voltage is the reduction of secondary fluorescence near the boundary between two phases. Secondary fluorescence may lead not only to erroneous quantitative results but also to qualitative misinterpretation [11], especially in trace element analysis (and the two adjacent phases have similar atomic number). In our hematite-ilmenite-rutile example, EPMA analyses for Ti performed at 15 keV at the hematite phase shows systematically small Ti-contents. However, if the analyses are performed at 6 keV, no Ti is detected. This is so because at 6 keV, the main source that produces secondary fluorescence, namely Fe Kα X-rays, cannot be excited.

3.5. Quantification of insulating materials
In EPMA, analysis of insulators is typically performed by coating the sample with a conductive layer, generally C or Au, to prevent charging effects. However, in some materials, charge self-compensation can be achieved by suitably reducing the beam voltage. The principle is similar to that used for imaging insulators in the FEG-SEM without conductive coating. A beam voltage (usually between 2 - 3 keV) can be tuned such that the injected charge from the electron beam is compensated by the secondary and backscattered electrons leaving the specimen. To do that, knowledge of the potential at the sample surface is required, which can be determined by measuring the Duane-Hunt limit when charge compensation is reached [2].

Figure 7 shows the variation of O and Al *k*-ratio, as a function of measuring time, from a non-coated sapphire sample (Al_2O_3) with a beam voltage of 2.5 kV. The measurements were performed at the same sample position for almost 2 h, with an electron beam of 500 nA of current and 20 μm in diameter. The stability with time of the measured Al and O *k*-ratios is worth noticing. It would not be possible to achieve such stability at higher beam energy due to charging effects. The sample was plasma etched before the measurements and an O_2-jet was used to minimize C contamination during measurements. In spite of this, a decrease in X-ray intensity at a t = 80 min is observed which corresponds to the effect of C contamination after a beam shift (of approximately 5 μm) most likely due to charge implantation in the sample (second-order backscattered electrons). By measuring C Kα X-rays simultaneously with Al and O, the loss of X-ray intensity could be estimated to correspond to a 1.4 nm thick C layer (see section 4.2.4). We note that the (uncorrected) *k*-ratios are close to the nominal concentrations (O 47 wt%; Al 53 wt%), despite the high MAC of O Kα in Al (6700 cm^2/g). We also note that the overestimation of the Al *k*-ratio could be due to the effect of a native oxide layer on the Al standard and C contamination on the O standard (Fe_2O_3) (see section 4.2.5).

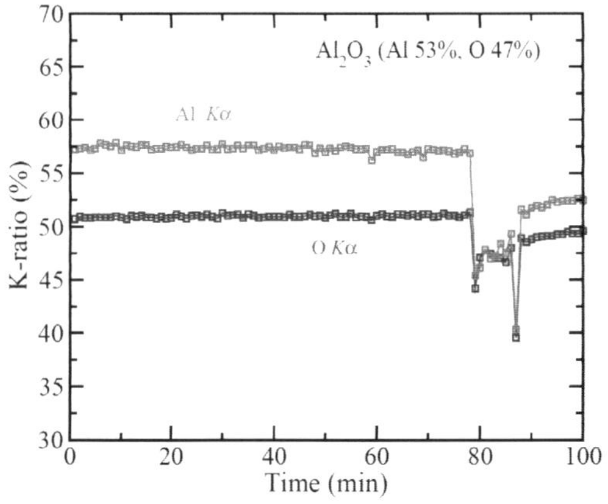

Figure 7. *k*-ratio for Al and O, as a function of irradiation time, measured on a Al_2O_3 insulator sample. Beam energy and beam current are 2.5 keV and 500 nA, respectively.

4. Factors affecting the accuracy of low-voltage EPMA

In this section, the experimental and analytical difficulties that may affect the accuracy of quantitative EPMA analysis at low voltage and the steps taken to overcome them are discussed. These include difficulties associated with the measurement of X-rays, sample characteristics, matrix correction procedures and chemical bonding effects [12].

4.1. Spectroscopic difficulties

In conventional EPMA, different X-ray lines can be selected to analyze most elements. For example, the first series of transition metals can be analyzed using their K- or L-lines. In general, K-lines are preferred because the K X-ray yield is typically one order of magnitude higher than the L X-ray yield, and because the MACs are lower. However, at low voltage, the available X-ray lines reduce to the less intense L- or M-lines. These limitations affect the accuracy of quantitative results.

Figure 8a compares portions of X-ray spectra emitted from the rutile phase of the hematite-ilmenite-rutile sample, acquired at 2.5 keV, using and not using an O_2-jet. In both cases, the C coating was removed before the measurements by in-situ etching (see section 4.1.4), in the attempt to minimize X-ray absorption (at 2.5 keV, a 25 nm thick C coating would reduce the emitted X-ray intensity by 50 %). Due to their low fluorescence yields, the intensity of the two visible Ti L-lines is even lower than that of O Kα, in spite of the higher concentration of Ti in rutile (~ 60 wt%) relatively to O (~ 40 wt%). The spectrum acquired with an O_2-jet provides better peak-to-background ratios and reduces the spectral overlap caused by the C K-line (from contamination layer). Yet, the M-lines of Zr and Nb are not observed, even at a Nb-concentration of 2 wt%.

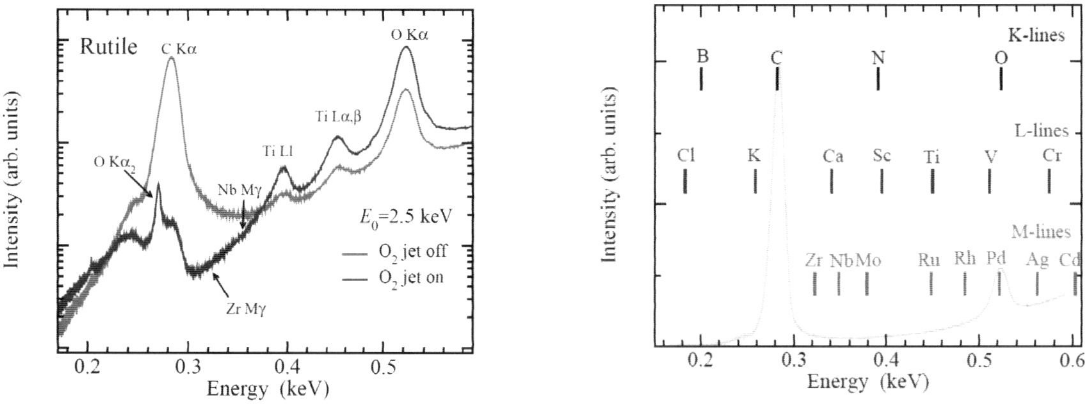

Figure 8. X-ray spectra emitted from a rutile sample (a) and from a freshly polished Zr standard (b). In the latter case, the different K-, L-, and M-lines available in the considered photon energy range are also displayed. The spectra were obtained on a Cameca SX-100 electron microprobe at an incident electron beam of 2.5 keV, using a Ni/C multilayer crystal (2d = 9.5 nm) and a 90 % Ar - 10 % CH_4 flow proportional counter, equipped with a polypropylene window, and operated at 1 bar of pressure.

A portion of the X-ray spectrum emitted from a freshly polished Zr standard obtained at 2.5 keV is shown in figure 8b, together with the energy positions of all K-, L- and M-lines that are available in the considered energy range (0.2 - 0.6 keV). Notice that, despite the use of a wavelength-dispersive (WD) X-ray spectrometer, the Zr M-lines are not detected due to the combined effect of high absorption plus extremely low fluorescence yield. Only two lines are visible in the spectrum, which correspond to C and O Kα X-rays from C contamination and oxidation layers (no anti-contamination devices were used during the acquisition). Obviously, the presence of these two lines compromises the sensitivity of the technique. In general, since the available X-ray lines will be located at a narrow photon energy range, low-voltage EPMA will require the use of high resolution X-ray spectrometers.

At high voltages ($E_0 > 15$ keV), the peak-to-background ratio (P/B) ratio is relatively high and allows the determination of trace elements (with concentration less than 0.001 wt%) within reasonable counting times (t < 500 s). However, at low voltages, the P/B drops significantly and leads to worse detection limits. This is illustrated in figure 9, which shows the variation of the peak intensity and the P/B ratio for Fe Lα X-rays emitted from Fe as a function of the over-voltage ratio.

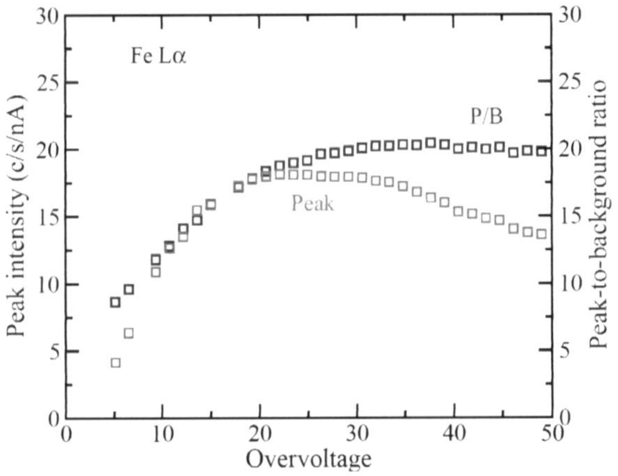

Figure 9. Variation of the peak and peak-to-background ratio (P/B) as a function of over-voltage ratio, for Fe Lα X-rays emitted from Fe.

4.2. Sample characteristics

At low voltage, a surface layer of several nanometres in thickness represents a large fraction of the sample and, therefore, the influence of hydrocarbon contamination, surface oxidation, and the quality of the sample polish becomes more significant than at higher voltages. A surface contamination layer may grow relatively fast on a freshly polished specimen at a vacuum of 10^{-5} Pa. As shown in figure 8b, the interfering effect of the C and O X-ray peaks, which correspond to contamination layers, is not negligible. Other sample characteristics such as roughness, insulator character and crystallographic orientation may also be important, as discussed below.

4.2.1. Hydrocarbon contamination. Carbon contamination occurs when hydrocarbon molecules present in the chamber are polymerized by the electron beam at the sample surface. The resulting polymer grows during electron irradiation because of its low vapour pressure and low surface mobility. The hydrocarbon molecules may originate from different instrumental sources (pump oil, vacuum grease, O-ring seals, etc.), the specimen itself, or they can also be introduced during sample preparation, storage or transportation in the air. Specimen hydrocarbon contamination typically diffuses towards the edge of the irradiated area and forms a "ring" around the beam point of impact. Beam-induced contamination is difficult to avoid even by improving the chamber vacuum (in terms of lower chamber pressure and lower hydrocarbon contents). Several techniques have been proposed to reduce carbon contamination, both at the specimen preparation stage and during measurements [13]:

1. Clean the specimen with detergent or carbon dioxide snow.
2. Heat the specimen with an electric lamp in air or in vacuum in order to desorb hydrocarbons from its surface.
3. Sputter away the surface layer in plasma cleaner.
4. Nitrogen purging of the specimen holder and specimen chamber.
5. Flood the surrounding area with electrons by defocusing the electron beam or by scanning at lower magnification in order to polymerize the surface hydrocarbons and prevent them to diffuse around the focused probe.

6. Flood the area with a gas (O_2, Ar, N_2) jet device so as to desorb hydrocarbons from its surface and keep a low hydrocarbon concentration during the measurement. This is the principle exploited by electron beam induced etching (EBIE), to etch sub-µm structures (see below), and electron-beam-induced deposition (EBID), to build sub-µm structures [14].
7. Use a N_2 cold finger in the chamber to condense hydrocarbons out of the specimen.

Figure 10 shows an example of the use of EBIE to improve the accuracy of EPMA at low voltage. In this example, an O_2-jet at a pressure of 4.10^{-4} Pa was used, together with a 2.5 keV electron beam of 500 nA and 50 µm in diameter, to etch a 25 nm thick C layer deposited on an hematite sample (Fe_2O_3). The etching rate was approximately 0.16 nm/min for a surface of 2000 μm^2. EBIE is based on the interaction of the incident and the emitted secondary electrons with the surface atoms in the presence of a precursor gas. The etching rate depends on different factors such as beam energy, current and size, temperature of the substrate material, chamber pressure, and geometry of the gas injector. EBIE is more efficient at low beam energies (~ 2 - 3 keV) because of the high yield of secondary electrons, whose contribution is dominant. As shown in figure 10, EBIE is time consuming for a large area, but the efficiency increases for smaller areas.

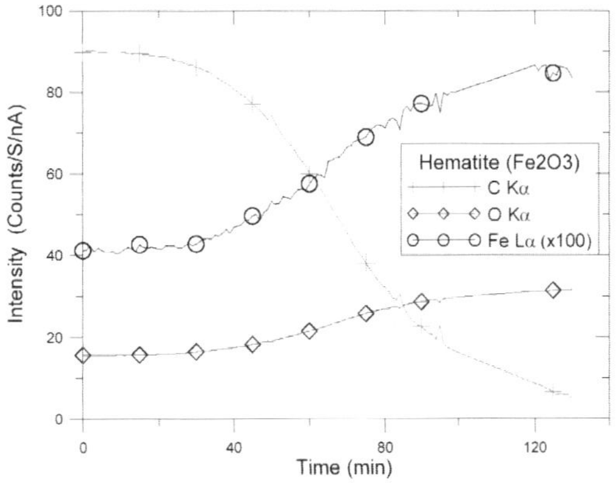

Figure 10. Electron-beam-induced etching (EBIE) of a C coated, hematite sample (Fe_2O_3) using an O_2-jet on an electron microprobe. The thickness of the C coating is 25 nm. Beam conditions: energy 2.5 keV, current 500 nA and diameter 50 µm.

Even if the EBIE technique is applied, it is useful to measure the C Kα X-ray line simultaneously with the X-ray lines of interest in order to take into account the effect of any residual C layer by using a thin-film programme.

4.2.2. Surface oxidation. Sample surface oxidation occurs in non-oxide samples. A variety of techniques can be used to reduce this effect, both at the specimen preparation and measurement stages:
1. Use freshly polished samples and standards (whenever possible).
2. Sputter away the surface layer with Ar or N_2 plasma cleaner.
3. Flood the area with a gas (N_2, Ar) jet device simultaneously with a low beam energy and high beam current in order to etch the oxidation layer using the EBIE principle.

Here it is useful to measure the O Kα X-rays simultaneously and use a thin film programme to take into account the effect of any residual oxidation layer.

4.2.3. Roughness. As discussed by Sorbier et al. [15], the uncertainties arising from sample roughness become more important at low voltage. By means of Monte Carlo simulations, these authors showed that a roughness of 10 % of the X-ray range leads to an error of 1% in the resulting concentration.

4.2.4. Insulator character. In conventional EPMA, insulating samples are routinely coated with a conductive layer [22] in order to neutralize the injected charge from the beam, with almost no influence on the evaluated composition. However, at low voltage, the energy loss of electrons in the conductive layer can be relatively important, especially when the overvoltage is low, and it can even hinder the detection of some elements. Measurements at multiple voltages may provide the information required to separately characterize an insulator sample from the conductive coating layer.

4.2.5. Crystallographic orientation. The influence of the crystallographic orientation on the X-ray yield is usually disregarded in conventional analysis conditions. However, Meisenkothen et al. [16] have shown that at low overvoltage ratios, electron channelling may produce X-ray variations as large as 25 %, which will obviously affect the accuracy of quantitative analysis.

4.3. Limitations of quantitative procedures

Quantitative EPMA at low voltage follows the same methodology as conventional EPMA. The starting point of quantitative methods is to establish a relation between the concentration of the element of interest C_A and the measured X-ray intensity I_e. This relationship can be written as:

$$I_e = C_A \cdot \frac{N^o}{A} .n. \frac{\Omega}{4\pi} \varepsilon.\omega_j P_{nl} .(1 + T_{CK}). \frac{1}{\cos\alpha} .Q_l^A(E_o).\left(\int \phi(\rho z).\exp(\chi \rho z).d\rho z\right)\left(1 + \sum f_c + f_{FC}\right) \quad (3)$$

where $N°$ is the Avogadro number, A is the atomic weight, n is the number of incident electrons, Ω is the solid angle of detection, ε is the detector efficiency, ω_j is the fluorescence yield, P_{nl} is the weight of the line, $(1+T_{CK})$ is a factor that takes into account Coster-Kronig transitions, α is the sample tilting angle, $Q_l^A(E_o)$ is ionisation cross-section, E_0 is the accelerating voltage, E_j is the excitation energy, $\phi(\rho z) = f\left(E_o, E_j, \alpha, \sum C_i\right)$ is the depth-distribution of X-rays, $\chi = {}^{\mu}\!/_{\rho} \operatorname{cosec}(\theta)$, where ${}^{\mu}\!/_{\rho}$ is the MAC and θ the take-off angle, and $\left(1 + \sum f_c + f_{FC}\right)$ is a factor that corrects for characteristic and continuum fluorescence effects.

In practice, a number of analytical, approximate expressions have been proposed to parameterize the $\phi(\rho z)$ function. These parameterisations employ quadrilateral, Gaussian, parabolic or exponential functions, which can be easily integrated to obtain C_A from I_e. The analytical $\phi(\rho z)$ expressions are defined by a few parameters (such as its maximum or its value at the surface $\phi(0)$), which depend on E_0, Z, E_i, α, and sample composition, and are usually computed from physical quantities and/or fits to experimental or Monte Carlo simulation data. To improve the accuracy of quantitative analysis, some $\phi(\rho z)$ models have been refined by imposing the qualification that the integral of $\phi(0)$ should be equal to the number of ionisations. This condition can be written as:

$$I_0 = Cst.C_A \frac{N_o}{A}.\frac{R}{S} = Cst.C_A \frac{N_o}{A} Q_l^A(E_o).\int \phi(\rho z).d\rho z\,, \quad (4)$$

where R and S are the backscattering and stopping power factors, respectively. Figure 11 shows the $\phi(\rho z)$ function for an Al target at 15 keV, for different excitation energies. We can see that the shape of this distribution changes rapidly with excitation energy (over-voltage ratio).

$\phi(\rho z)$ models adopt simple expressions for the calculation of R, S and $Q\ (E_0)$. For example, S is generally computed by using the Bethe formula and $Q\ (E_0)$ is also determined by means of the Bethe ionisation cross-section (see e.g., [18]). Although these formulas have been often tuned [23] to provide the best match between the predicted X-ray intensities and selected experimental measurements, their validity at low incident electron energies is not firmly established. This is illustrated in figure 12, which compares different calculations and experimental measurements of the cross section for the production of Au Mα X-rays and the stopping power for Cu. In the case of the ionisation cross-section (figure 12a), the experimental measurements are only reproduced by the most

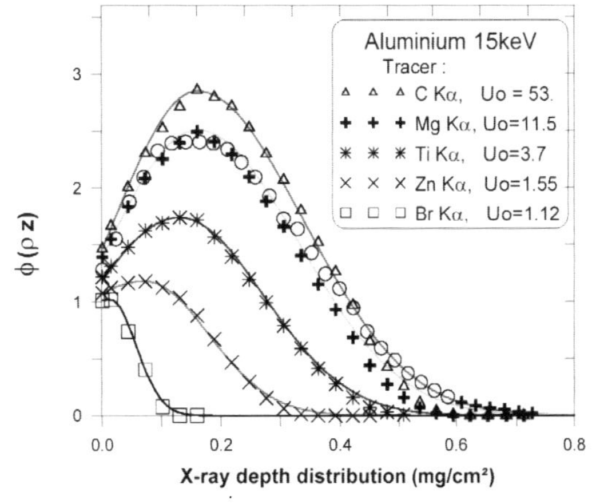

Figure 11. Depth-distribution of X-ray production for various X-ray lines (tracer elements) in Al at 15 keV. Circles are experimental measurements of Castaing and Hénoc [17]; other symbols are Monte Carlo simulation results; continuous lines are the predictions of the X-PHI model [7].

sophisticated calculation based on the distorted-wave Born approximation (DWBA) [19], whereas the Bethe-Powell formula, as well as other predictive formulas widely used in EPMA, systematically underestimate the experimental data especially near the ionisation threshold. As for the stopping power, this can be evaluated by using the Bethe formula only for electrons with energy larger than the excitation energy of the most tightly bound shell (figure 12b).

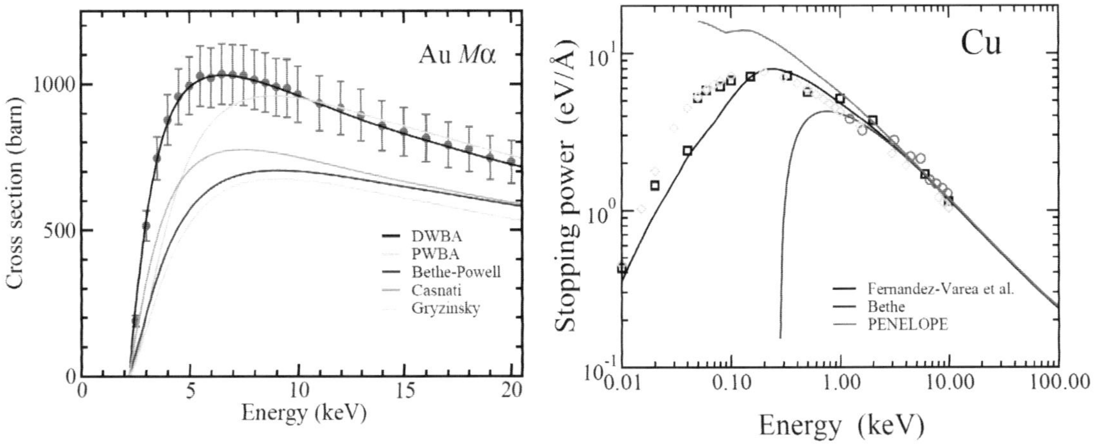

Figure 12. Cross-section for the production of Au Mα X-rays (a) and stopping power of electrons in Cu (b). Continuous lines are the results of different calculations and predictive formulas, some of them widely used in EPMA. Symbols are experimental data taken from the literature.

Recent measurements of the surface ionisation φ(0) for L- and M-shells [20] have shown the limitations of currently available EPMA models for such lines at low energies. While the more sophisticated calculations performed with the Monte Carlo simulation code PENELOPE [20], which adopts ionisation cross-sections computed from the DWBA, were found to reproduce the measured φ(0) values reasonably well, none of current available φ(0) parameterisations was found to reproduce consistently the experimental data [20].

4.4. Chemical bonding

X-ray emission lines that correspond to electron transitions involving outer valence electrons are known to be sensitive to chemical bond [24]. As a result, a shift in peak position as well as a change in peak distribution is observed when comparing X-ray lines of the same element in different chemical environments. For some low-energy lines, such as the L-lines of the first series of transition elements, the chemical state of the transition atom may also be responsible for changes in the X-ray distribution as well as in the self-absorption. The use of empirical MACs obtained from EPMA measurements at variable voltage may be useful to overcome these difficulties.

Conclusions

Low-voltage EPMA offers advantages for the characterisation of thin films and multilayers, ultra-light elements and insulators, but more importantly, it opens a wide range of new possibilities for the characterisation of materials that are heterogeneous on a sub-micrometre scale. However, low-voltage EPMA is affected by experimental and analytical difficulties that deteriorate the accuracy of quantitative results and requires a more careful approach than conventional EPMA.

The recent improvements in EPMA instrumentation represent a major step towards the application of EPMA at low voltage on a routine basis. Nevertheless, further studies must still be made in order to improve the reliability of quantification models at such energy range. It is to be hoped that these studies will provide improved sets of fundamental parameters, such as MACs, ionisation cross-sections, fluorescence and Coster-Kronig yields and stopping powers, consistent with EPMA models and suitable for low-voltage applications.

Acknowledgements

We would like to thank Fernanda Guimarães and Michel Outrequin for performing the X-ray maps on the FEG EPMA instruments.

References

[1] Newbury D E 2002 *J. Res. Natl. Inst. Stand. Technol.* **107** 605-19
[2] Cazaux J 2010 *J. Electron Spectros. Rel. Phenomena* 176 58-79
[3] Willich P and Bethke R 1996 *Mikrochim. Acta* Suppl. **13** 631-8
[4] Llovet X and Merlet C 2010 *Microsc. Microanal.* **16** 21-32
[5] Castaing R 1952 Ph.D-thesis. Publication ONERA no. 55
[6] Bastin G F, Dijkstra J M, Heijligers H J M and Klepper D 1993 *Microbeam Anal.* **2** 29
[7] Merlet C 1994 *Mikrochim. Acta* **114/115** 363
[8] Henke B L, Gullikson E M and Davis J C 1993 *Atom. Data Nucl. Data Tables* **54** 181
[9] Kalfoun F, Ionov D and Merlet C 2002 *Earth Planet. Sci. Lett.* **199** 49-65
[10] Pascal C, Merlet C, Marin-Ayral R M, Tédenac J C and Boyer B 2004 *Mikrochim. Acta* **145** 147-51
[11] Llovet X and Galan G 2003 *Am. Mineral.* **88** 121
[12] Llovet X, Heikinheimo E, Núñez A, Merlet C, Almagro J F, Richter S, Fournelle J, van Hoek C J G 2012 *IOP Conf. Ser.: Mater. Sci. Eng.* **32** 012014
[13] Egerton R F, Li P and Malac M 2004 *Micron* **35** 399-409
[14] Randolph S J, Fowlkes J D and Rack P D 2006 *Crit. Rev. Solid State Mater. Sci.* **31** 55-89
[15] Sorbier L, Rosenberg E and Merlet C 2004 *Micros. Microanal.* **10** 745-52
[16] Meisenkothen F, Wheeler R and Uchic M D 2009 *Microsc. Microanal.* **15** 83-92
[17] Castaing R and Hénoc J 1965 *Proc.* 4th *IXCOM.* Castaing R, Deschamps P and Philibert J, eds. (Paris: Hermann) pp. 120
[18] Salvat F, Llovet X and Fernández-Varea J M 2004 *Microchim. Acta* **145** 193-202
[19] Bote D and Salvat F 2010 *Phys. Rev. A* **77** 042701
[20] Merlet C and Llovet X 2011 *X-ray Spectro.* **40** 47-54

[21] Salvat F, Fernández-Varea J M and Sempau J 2009 *PENELOPE-2008: A code system for Monte Carlo simulation of electron and photon transport.* (Issy-les-Moulineaux, France: OECD/NEA Data Bank)
[22] Cazaux J 2004 *Microsc. Microanal.* **10** 670-84
[23] Joy D 1998 *J. Micros.* **191** 74-82
[24] Remond G, Gilles C, Fialin M, Rouer O, Marinenko R, Myklebust R and Newbury D 1996 *Mikrochim. Acta* Suppl. **13** 61-86

IOP Conf. Series: Materials Science and Engineering **32** (2012) 012017 doi:10.1088/1757-899X/32/1/012017

Experimental check of the use of unconventional reference materials for EDS analysis in a TEM by extrapolation method based on pure elements

M Nacucchi[1,4], M Alvisi[1], D Altamura[2], V Pfister[1], D Valerini[1], D Mello[3] and C Giannini[2]
[1] ENEA, Materials Technology Unit, Research Centre of Brindisi, S.S. 7 Appia – km 706.00, IT-72100 Brindisi, Italy
[2] Consiglio Nazionale delle Ricerche (CNR), Institute of Crystallography, Via G. Amendola 122/O, IT-70126 Bari, Italy
[3] ST Microelectronics, QA and Phys. Lab., Stradale Primosole 50, IT-95100 Catania, Italy

E-mail: michele.nacucchi@enea.it

Abstract. In this work, the use of unconventional reference materials to determine experimentally the Cliff-Lorimer factor for EDS quantitative analysis with a TEM is checked by means of an alternative experimental procedure. The *k*-factor is determined by the extrapolation method based on pure elements, by measuring the normalized X-ray intensities emitted by thin films of pure gold and pure silver of different thickness, accurately measured by X-ray reflectivity. The goal of this work is to confirm the value of the *k*-factor previously obtained by the use of unconventional reference materials consisting of a bi-layer of pure gold on pure silver. The current result is in accordance with the previous one when considering their error bars, however, their relative difference is about 13 %, probably due to some uncertainties in mass thickness measurements. The mass thickness measurement of the layers of pure elements needs to be performed by different methods in order to reduce its uncertainty.

1. Introduction

The ever increasing needs of nanotechnologies to control the matter and fabricate devices on the nanoscale increase the demand of microstructural and microanalytical characterisation of materials. Knowledge of the local composition and chemistry is crucial in order to optimize performance. The analytical electron microscopy (AEM) performed in the (scanning) transmission electron microscope (TEM) is the only technique which allows both atomic resolution imaging and elemental analysis of areas as small as some nm^2. Although energy-dispersive X-ray spectrometry (EDS) is commonly employed in transmission electron microscopy to chemically characterize the sample on the nanometre scale, limitations in the use of this technique for quantitative analysis persist. The main reason for this is the difficulty to prepare reference materials (RMs) that are transparent to the electron beam, have an extremely high spatial homogeneity, and are stable under electron bombardment. Only very few systems, like the ones showing complete solid solubility, are suitable to this purpose. The most popular method to quantify elemental concentrations in thin film X-ray microanalysis is the one

[4] To whom any correspondence should be addressed.

established to Cliff and Lorimer [1]. In the simplest case, the concentration of both elements A and B of an unknown binary alloy is related to the ratio of their X-ray intensities through the equation:

$$\frac{C_A}{C_B} = k_{AB} \frac{I_A}{I_B} \tag{1}$$

($C_A + C_B = 1$) where the proportionality constant k_{AB} is the so called Cliff-Lorimer factor, or *k*-factor. The *k*-factor depends on spectrometer characteristics that are not well known or stable for a long time, such as the quantum efficiency of the detector; hence, even if the *k*-factor can be theoretically determined, it is better to derive it from the direct measurement of samples of known composition. Traditionally, the experimental determination of *k*-factors requires a number of binary alloys as reference materials, using one solvent and different solutes. Silicon and iron are generally used as solvents in mineralogy and metallurgy, respectively. Because of the difficulty to find or prepare reference materials, the feasibility to perform quantitative X-ray microanalysis in a TEM with no need of RMs has been explored since more than twenty years. For example, it was possible to determine both the thickness and composition of thin self-supporting film, by performing measurements under different specimen tilt angle, with the aid of a Monte Carlo code [2, 3]. These methods are more complex than that of Cliff and Lorimer and they do not work in all the cases.

Results obtained in a recent work clearly demonstrated the equivalence of a bi-layer of pure gold on pure silver with respect to a binary alloy foil of the same elements [4]. Two different bi-layer films, made of well characterized and homogeneous single layers with known mass thickness, were used to determine independently the Cliff-Lorimer factor in the ratio eq. (1), giving the same value. The bi-layer RMs were prepared by radio frequency magnetron sputtering (RF-sputtering). This physical deposition technique is suitable to prepare such reference materials due to the high reproducibility of the coating thickness and lateral thickness uniformity. Furthermore, RF-sputtering allows obtaining sharp interfaces, even thinner than 1 nm, and amorphous layers or polycrystals with a grain size in the nanometre range. As a consequence of this new approach to the quantitative analysis, if the operator had a calibrated set of such RMs, he could measure directly the *k*-factors for a large number of elements and then apply them in successive analysis whichever the optical conditions and the beam current are. In the cited work [4], the experimental *k*-factor was compared to the theoretical value computed by using confidential data of the electron microscope manufacturer. The agreement was very good, but the theoretical computation has to be considered just as an estimate, because the obtained value depends on the compilations used for the ionisation cross-section, the X-ray fluorescence yield and partial yield. However, it should be noted that the absolute values in the data sets are much more different than the relative values for a selected element-pair within any single data set.

In this work an alternative experimental procedure is employed to validate the use of RMs consisting of a bi-layer of pure gold on pure silver. In this case, the *k*-factor was determined by the extrapolation method based on pure elements, by measuring the normalized X-ray intensities emitted by thin films of pure gold and pure silver of different thickness, the latter accurately measured by X-ray specular reflectivity (XSR) [5]. The goal of this work is to confirm the value of the *k*-factor previously obtained by the use of the unconventional reference material. The test samples are gold-aluminium and silver-aluminium bi-layers grown on a support film of amorphous carbon on a copper TEM grid. The bottom aluminium layer was the same for all the samples and its presence was used only for normalisation data purposes.

2. Materials and methods

2.1. Specimen preparation

Au-Al and Ag-Al bilayers were deposited by RF sputtering at room temperature simultaneously on two kinds of substrates: commercial holey carbon film grid (300 mesh Quantifoil) for TEM analyses, and silicon (100) for the measurement of film thickness and mass density through profilometry and X-ray reflectivity. The vacuum chamber was evacuated down to a background pressure of $\sim 5\times10^{-7}$ mbar, which is mainly constituted by water vapour, as determined by in-situ mass spectrometry. The pressure of the Ar sputtering gas was set at $\sim 2\times10^{-2}$ mbar. Before the film deposition, each target (99.99 % purity) was shuttered and pre-sputtered in order to remove surface contamination.

A first 30 nm Al bottom layer was deposited at RF power of 50 W. This layer served as a reference in the EDS analyses, as the Al Kα line intensity was used as a normalisation factor in the EDS analyses of the bilayers. Au and Ag films were deposited on the Al layer at a constant power of 25 W and 15 W respectively, and four nominal layer thicknesses were realized: 10, 15, 20 and 25 nm for the Au films (in the Au-Al bi-layers), and 8, 12, 16 and 20 nm for the Ag films (in the Ag-Al structures). These thicknesses were chosen in order to minimize the mass absorption. Also, Al single layers were deposited on silicon substrates for the characterisation of the Al bottom layer.

2.2. Specimen characterisation

2.2.1. X-ray reflectivity. The two Al single layers were first characterized as reference layers in order to use the obtained Al parameters as starting values for the simulation of the XSR patterns of the Au/Al and Ag/Al bi-layers, respectively. Afterwards, Au or Ag parameters were derived, and Al parameters were refined by simulating the XSR patterns for each bi-layer. As an example, in figures 1a and 1b are reported the experimental and simulated XSR patterns of Au/Al and Ag/Al bi-layers, together with the experimental and simulated XSR reference patterns of the relative Al single layer.

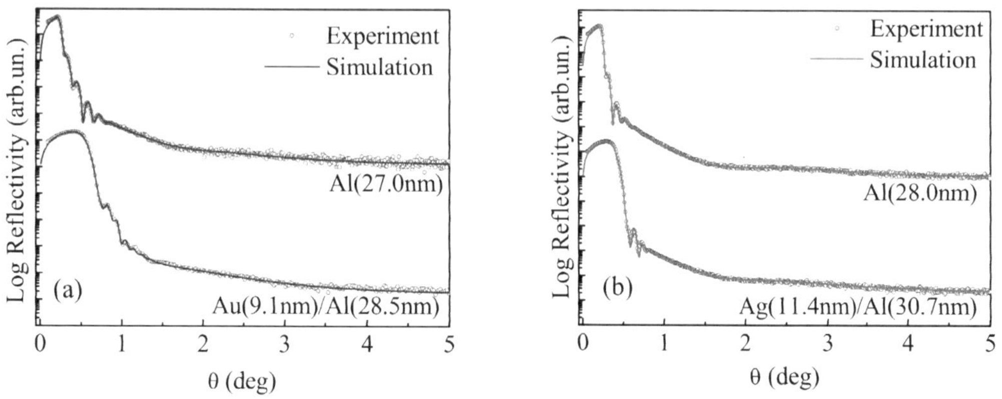

Figure 1. Experimental (symbols) and simulated (line) X-ray specular reflectivity patterns of Au/Al (a) and Ag/Al (b) bi-layers, together with relative Al single layer reference pattern.

The actual Al thickness was found to vary between ± 7 % of the mean value in the case of Au/Al bi-layers, and ± 4 % of the mean value in the case of Ag/Al bi-layers. Such variations are compatible with those expected due to the different position of each specimen on the sample holder in the sputtering deposition chamber, and with the error associated with the XSR simulation for Al thickness in the bi-layer, estimated as ± 2 %. Al density values were between 2.5 and 2.7 g/cm^3, therefore, close to the density value expected for bulk Al (2.7 g/cm^3). The Au and Ag parameters derived for each bi-layer are reported in tables 1 and 2, respectively. Au density values are also found close the value expected for bulk Au (19.3 g/cm^3) within 4 %, which is within the accuracy of the XSR technique [6]. On the other hand larger differences are observed in Ag density, in particular in batches 1 and 2 (table 2) with respect to the bulk density value (10.5 g/cm^3), which can be ascribed to partial oxidation of the Ag layer, due to the residual water vapour inside the vacuum chamber during the deposition and to the exposure of the films to the outside atmosphere. As expected, this effect is more evident in Ag films than in Au ones, due to easier oxidation of Ag with respect to Au.

Table 1. Parameters of the Au/Al bi-layers and Al reference layer from XSR analysis.

Sample	Thickness (nm)		Mass density (g/cm^3)		Mass thickness (μg/cm^2)	
	Au	Al	Au	Al	Au	Al
AuAl-01	9.1	28.5	18.7	2.5	17.0	7.1
AuAl-02	13.3	30.8	18.8	2.7	24.9	8.3
AuAl-03	17.6	29.2	19.1	2.5	33.5	7.3
AuAl-04	26.3	30.5	19.1	2.5	50.1	7.6
Al reference sample	-	27.0	-	2.7	-	7.3

Table 2. Parameters of the Ag/Al bi-layers and Al reference layer from XSR analysis.

Sample	Thickness (nm)		Mass density (g/cm^3)		Mass thickness (μg/cm^2)	
	Ag	Al	Ag	Al	Ag	Al
AgAl-01	8.9	29.2	8.1	2.6	7.2	7.7
AgAl-02	11.4	30.7	8.8	2.7	10.0	8.3
AgAl-03	13.9	31.2	10.1	2.5	14.0	8.1
AgAl-04	17.6	32.0	9.7	2.5	17.0	8.1
Al reference sample	-	28.0	-	2.7	-	7.6

2.2.2. TEM. The goal of the TEM-EDS characterisation was to acquire the X-ray spectra to determine experimentally the Cliff-Lorimer factor of the elements, by using the extrapolation method. The instrument was a FEI Tecnai G2 F30 equipped with a Schottky emitter with an Edax Si(Li) X-ray detector, LN_2 cooled, protected by an ultra-thin window. The accelerating voltage was 300 kV. The specimen was positioned at the eucentric height and during the analysis it was tilted to the manufacturer's recommended value of 20° toward the X-ray detector in order to optimize the X-ray collection. As the elevation angle of the detector is also equal to 20 deg, the take off angle of emerging X-rays was 40°.

For each pure film of gold or silver, three X-ray analysis points were sampled directly in TEM mode, in a central area of the grid. All the four samples of each element were analyzed the same day in order to reduce the differences in the experimental conditions. Following the successful procedure already used in the previous work concerning the Au/Ag bi-layers, a large spot of about 0.4 μm diameter, obtained by defocusing the condenser system, was used, in order to avoid possible short range nm fluctuations [4]. In this way also the risk of sample damage and carbon contamination is reduced, because defocusing the spot reduces the local beam energy density on the sample surface. A cold trap around the specimen stage and short counting times equal to or less than 200 s also contributed to minimize the carbon contamination. X-ray counts were collected along a line of three points. The average distance between points was a few micrometres.

The selected X-ray lines for the analysis were Au Lα and Ag Kα. As these lines are free from interferences, a continuous background curve under the peak was obtained by connecting linearly two points from both sides of the peak. EDS spectra confirmed the hypothesis of a slight oxygen contamination, but a correct quantitative evaluation of the oxygen content was not possible due to the low oxygen signal. Nevertheless, at a first approximation the results reported in this work are not influenced by the presence of this small oxygen contamination because the computations are based on the mass-thickness (a lighter and thicker oxidized layer is equivalent to a thinner and heavier unoxidized layer).

The presence of the bottom aluminium layer in all the samples is just for normalisation purposes. As the intensity of the Al Kα-line is proportional to the electron beam current, if the ratio between the Au Lα- or Ag Kα-counts and the Al Kα-counts in the same spectrum is calculated for each analysis point, possible fluctuations due to variations of the beam emission current are automatically cancelled. This normalisation factor must be the same for all the samples concerning the same element, gold or silver. Furthermore, it should be the same for both the pure elements, in a sense that it must be corrected for any possible difference between one element and the other. The normalisation factor is proportional to the thickness of the aluminium layer; for this reason the Al layer was deposited in the same run for all the samples of a same element. Experimental results for Au Lα and Ag Kα intensities are summarized in tables 3 and 4 respectively.

The previous considerations are true if the absorption of the Al Kα-line by the top layer does not vary appreciably among the samples. According to Tixier and Philibert [7], this is checked by computing the product:

$$\chi(\rho t) = (\mu / \rho)_{AlK\alpha}^{Au/Ag} \cos ec\psi \cdot (\rho t)_{Au/Ag} \qquad (2)$$

where $(\mu/\rho)_{AlK\alpha}$ is the mass absorption coefficient (MAC) for Al Kα X-ray line in the top layer of gold or silver having a mass thickness ρt and ψ is the take-off angle of emerging x-rays, equal to 40° as already noted. The sample is considered so thin that the X-ray absorption is negligible if $\chi/(\rho t) < 0.1$. The energy of the Al Kα X-ray line is 1.49 keV; its MAC in gold is 2.140 E+03 cm^2/g, whereas in silver is 2.875 E+03 cm^2/g [8]. From these values, it can be easily verified that the thin film criterion is largely satisfied by all the layers of both gold and silver: $\chi/(\rho t)$ is equal to 0.069 or 0.031 for the thicker layer of gold or silver respectively, so it is not necessary to take into account the absorption of the Al Kα-line by the top layer.

Table 3. Experimental Au Lα intensities for the various films of pure gold.

		AuAl-01	AuAl-02	AuAl-03	AuAl-04
Spectrum 1	I(Au Lα)	15040	22409	28221	36225
	I(Al Kα)	23867	22822	22016	22301
	I(Au Lα)/I(Al Kα)	0.6302	0.9819	1.2819	1.6243
Spectrum 2	I(Au Lα)	15000	22521	28361	36451
	I(Al Kα)	25240	22775	23090	22378
	I(Au Lα)/I(Al Kα)	0.5943	0.9889	1.2283	1.6289
Spectrum 3	I(Au Lα)	15213	22497	28356	36586
	I(Al Kα)	24722	22613	22690	22672
	I(Au Lα)/I(Al Kα)	0.6154	0.9949	1.2497	1.6137
Normalized intensity average		0.61 ± 0.02	0.989 ± 0.006	1.25 ± 0.03	1.622 ± 0.008

Table 4. Experimental Ag Kα intensities for the various films of pure silver.

		AgAl-01	AgAl-02	AgAl-03	AgAl-04
Spectrum 1	I(Ag Kα)	10569	17661	23409	29929
	I(Al Kα)	49666	52041	51290	48865
	I(Ag Kα)/I(Al Kα)	0.2128	0.3394	0.4564	0.6125
Spectrum 2	I(Ag Kα)	10897	17481	24063	30016
	I(Al Kα)	50446	52644	53015	48087
	I(Ag Kα)/I(Al Kα)	0.2160	0.3321	0.4539	0.6242
Spectrum 3	I(Ag Kα)	10393	18076	24179	29908
	I(Al Kα)	50704	52289	53597	48728
	I(Ag Kα)/I(Al Kα)	0.2050	0.3457	0.4511	0.6138
Normalized intensity average		0.211 ± 0.006	0.339 ± 0.007	0.454 ± 0.003	0.617 ± 0.006

3. Results and discussion

3.1. Experimental determination of Cliff-Lorimer factor by extrapolation based on pure elements

A way of determining the Cliff-Lorimer factor is to exploit the linear relationship existing between the normalized X-ray line intensity and the layer thickness [9]. Indeed, the intensity of a given net peak of an element A is given by:

$$I_A = J(Q\omega a)_A \rho(N_0 / A_A)(\Omega / 4\pi)\varepsilon_A C_A t \tag{3}$$

where J is the number of electrons striking the sample during the live counting time, Q is the electron ionisation cross-section for the particular shell of interest, ω is the X-ray fluorescence yield, a is the partial yield of the line, ρ and t are the mass density and the thickness of the sample, N_0 is the Avogadro number, A_A is the atomic weight of the element A, $\Omega/4\pi$ is the fractional collection solid angle, ε_A is the detector quantum efficiency for X-rays of element A at the selected peak energy, and C_A is the weight concentration of the same element. If one measures, on thin films of pure A and pure B, the ratio I_A/J and I_B/J, respectively, and plots these values as a function of the thickness-related parameter $(\Omega/4\pi)N_0\rho t$, a linear plot is obtained, passing through the origin of axes. The slope of the two straight lines are $(Q\omega a\varepsilon/A)_A$ and $(Q\omega a\varepsilon/A)_B$, wherefrom the k-factor

$$k_{AB} = \frac{(Q\omega a\varepsilon / A)_B}{(Q\omega a\varepsilon / A)_A} \tag{4}$$

can be deduced. As the k-factor is a ratio, it does not matter the metric unit on the horizontal axis, i.e., the mass thickness of the films can be reported directly. The thickness and the mass density of the various points in the current case were determined by XSR. Furthermore, as the number of the emitted Al Kα photons is proportional to the electrons striking the sample, the X-ray intensities data on the vertical axis were normalized by the corresponding Al Kα peak area, measured in the same spectrum.

As small differences in the mass thickness of the bottom aluminium layer exist among the samples grown in the same deposition run, the normalized intensity has to be corrected for them. Thickness calibration tests allowed us to measure a thickness non uniformity of about 5 % within a distance of 2.5 cm from the centre of the sample. As a consequence, a thickness non-uniformity lower than 5 % is expected, due to the smaller dimensions of our samples. Because of the normalisation operation, the average intensity of Au Lα-line emitted from a sample having a thicker aluminium layer underlying the gold one should be divided by a greater normalisation factor with respect to a sample with a thinner aluminium layer at the bottom. If no corrective action was taken, the normalisation operation could be not neutral. Let's assume the thickness of the aluminium layer of the thinner gold sample (AuAl-01) as reference value. Then, the normalized intensity average of Au Lα-line of any other gold sample is corrected by multiplying it by the ratio of the mass thickness of the aluminium layer of the current sample divided by that one of the reference sample (AuAl-01). In a similar way, the average normalized Ag Kα intensities were corrected for the small differences existing in the normalisation factor, by assuming the thickness of the aluminium layer of the thinner silver sample (AgAl-01) as reference value. Table 5 reports the coordinates of the plot points for both the series of samples.

Thickness dependence of the Au Lα X-ray production for pure gold is plotted in figure 2. The slope of the interpolation line, determined by using the (weighted) least square approximation is equal to:

$$m_{AuL\alpha} = 0.0376 \pm 0.0026 \ (\mathrm{cm^2/\mu g}) \tag{5}$$

In a similar manner the thickness dependence of the Ag Kα X-ray production for pure silver was deduced. Thickness dependence of the Ag Kα X-ray production for pure silver is plotted in figure 3. The slope of the interpolation line, determined by using the least square approximation, is equal to:

$$m_{AgK\alpha} = 0.0346 \pm 0.0024 \ (\mathrm{cm^2/\mu g}) \tag{6}$$

Table 5. Coordinates of the points of the pure gold and pure silver lines.

Plot point	Sample	Mass thickness (μg/cm^2)	Corrected Normalized I(Au Lα)	Weight of the point[a]	Sample	Mass thickness (μg/cm^2)	Corrected Normalized I(Ag Kα)	Weight of the point[a]
0	fictious	0	0		fictious	0	0	
1	AuAl01	17.0	0.613	3.10E+03	AgAl01	7.2	0.211	3.29E+04
2	AuAl02	24.9	1.153	2.37E+04	AgAl02	10.0	0.363	2.15E+04
3	AuAl03	33.5	1.283	1.39E+03	AgAl03	14.0	0.473	1.43E+05
4	AuAl04	50.1	1.736	1.74E+04	AgAl04	17.0	0.648	2.91E+04

[a] The weight is equal to $1/\sigma^2$, where σ is the deviation on the normalized line intensity reported in table 4.

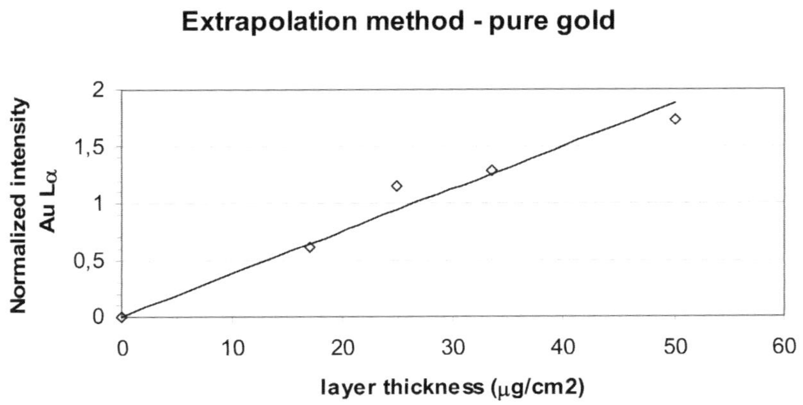

Figure 2. Normalized Au Lα intensity as a function of the mass thickness of pure gold layers. Full rhomb symbols represent the experimental points. The best fit line is computed by the least square approximation.

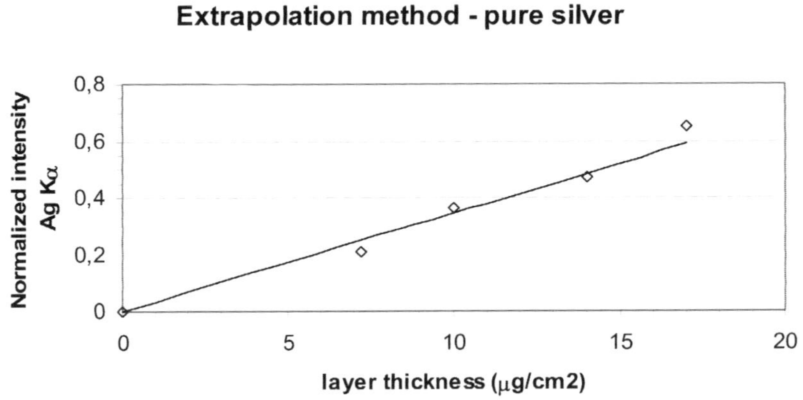

Figure 3. Normalized Ag Kα intensity as a function of the mass thickness of pure silver layers. Full rhomb symbols represent the experimental points. The best fit line is computed by the least square approximation.

Note the linearity of both the plots: the correlation coefficient is equal to 0.90 and 0.95 for the gold and silver line respectively; such high values of the correlation coefficient confirm the existence of a homogeneous linear relation between normalized X-ray intensity and mass thickness of the layer. This result is not surprising since the thin film criterion is fulfilled in the whole range of measurements: the self absorption of the Au Lα-line in the gold layer, as well as the Ag Kα-line in the silver one, is absolutely negligible. Then, the *k*-factor is obtained by computing the ratio between the two slopes:

$$k_{AuAg} = \frac{m_{AgK\alpha}}{m_{AuL\alpha}} = 0.92 \pm 0.09 \qquad (7)$$

This value has to be corrected for the difference of the normalisation factor used for the gold series and the silver one. In other words, the obtained *k*-factor must be divided by the ratio of thicknesses of the aluminium layer used in the silver deposition and in the gold one. The XSR measures on the single Al layers were used for this purpose, so finally:

$$k_{AuAg,corr} = k_{AuAg} \frac{(\rho t)_{Al}^{AgSeries}}{(\rho t)_{Al}^{AuSeries}} = 0.96 \pm 0.09 \qquad (8)$$

The large uncertainty affecting the results is a consequence of both the uncertainty on the layer mass thickness determination and the successive normalisation operations by means of factors each affected by its own measurement error. For comparison, the value of the *k*-factor obtained by using Au/Ag bi-layers RMs and reported in reference [4] was equal to 0.84 ± 0.08. The error bars overlap but the relative difference is about 13 %. Such difference cannot be attributed to a modification of physical properties of the detector because the interval between the two acquisitions was about one year and no significant problems occurred to the instrument in the meantime. The main issue is that the extrapolation method requires an accurate thickness determination of all the samples of both the series but, in this case, the presence of the bottom aluminium layer used for normalisation purposes made XSR measurements insufficient. Conclusions reported in reference [4] already noted that the mass thickness determination needs to be accomplished by means of different techniques, e.g., also by electron microprobe. A preliminary feasibility test has been performed by Rutherford backscattering with a focussed beam and proton induced X-ray emission. Both the techniques are able to measure directly the mass thickness in the range of values useful for the characterisation of the bi-layered RMs for TEM [10, 11]. The former technique offers the advantage that it can be used both on samples grown on silicon substrate and on TEM grid. Anyway, the current result is not in contrast with the previous one.

3.2. Uncertainties

The uncertainty on the k_{AuAg}-factor was computed through the quadratic sum of the relative errors on the slopes of the best fit lines, obtained by the least square approximation. In this case, if the uncertainty on the slope m was deduced by using the standard errors propagation formula it should result underestimated. For this reason it was estimated by the following considerations. The uncertainty on the slope *m* is a consequence of the statistical counting error affecting the normalized intensity average of Au Lα-or Ag Kα-lines and of the uncertainty in the mass thickness of the layers. The latter is due to the experimental error on thickness and mass density determined by XSR. These two quantities are not independent variables and the uncertainty on the mass density dominates the one related to the thickness. The relative error on mass density can be as greater as 4 % [6]; so, the relative uncertainty of the mass thickness was assumed equal to 4 %. As a consequence, both the relative errors on $m_{AuL\alpha}$ and $m_{AgK\alpha}$ are about 7 %. Finally, the reported uncertainty of the *k*-factor is obtained by computing the square root of the sum of the squared relative errors on the slopes and by multiplying the result by the value of *k*.

Conclusions

An experimental check of the use of unconventional reference materials recently introduced to determine the Cliff-Lorimer factor for EDS quantitative analysis with a TEM has been performed by means of the extrapolation method based on pure elements. The obtained result is compatible with the previous one, but an unambiguous conclusion is still not possible due to the large uncertainty affecting the obtained *k*-factor value. Nevertheless, this new approach to the quantitative analysis based on bi-layered RMs could give advantages in the usual laboratory procedure. Indeed, if the operator had a calibrated set of such RMs, he could directly measure the *k*-factors for a large number of elements and then apply them in successive analyses whatever the optical conditions and the beam current are. However, in order to use bi-layers of pure elements instead of binary alloys as reference materials for TEM-EDS analysis, it is necessary to accurately measure the mass thickness of each layer and to reduce the relative uncertainty of the measure down to a few percent, otherwise the combined error on the *k*-factor easily mounts to 10 % or more, as the current case has shown. Experimental techniques that allow determining the mass thickness directly and accurately, such as the electron microprobe, are suitable for the purpose. For example, the Pouchou and Pichoir thin film analysis method applied to bi-layers of known concentration and without common elements (e.g., pure elements of different atomic number) can allow a mass thickness determination as accurate as the measurement of the major element concentrations in a bulk sample by the usual ZAF method. Then, the characterisation of these bi-layered RMs needs the use of more than one technique for thickness measurements such as XSR, electron microprobe or Rutherford backscattering. For the moment, the advantage of the proposed unconventional RMs is that, in principle, it is possible to produce a large number of different bi-layers in order to cover a great number of elements in the periodic table. Next scheduled step will be the preparation of bi-layered RMs suitable for the study of diffusion problems in electronic devices. Further improvements to their characterisation techniques will be adopted.

Acknowledgements

The authors whish to thank Dr. Aldo Armigliato and Prof. Rodolfo Rosa for their useful discussion. The authors are also grateful to Prof. Lucio Calcagnile and Dr. Gianluca Quarta for their preliminary tests by Rutherford backscattering with a focussed beam and proton induced X-ray emission. They also thank Mr. Martino Palmisano for TEM sample preparation. This work has been partially supported by the SEED project "X-ray synchrotron class rotating anode microsource for the structural micro imaging of nanomaterials and engineered biotissues (XMI-LAB)"- IIT protocol no. 21537 of 23/12/2009.

References

[1] Cliff G and Lorimer G W 1975 *J. Microsc.* **103** 203-07
[2] Rosa R and Armigliato A 1989 *X-ray Spectr.* **18** 19-23
[3] Armigliato A and Rosa R 1990 *Ultramicroscopy* **32** 127-36
[4] Nacucchi M, Alvisi M, Altamura D, Pfister V, Re M, Signore M A and Vittori Antisari M 2010 *IOP Conf. Series: Mat. Science and Engng.* **7** 012020
[5] Pietsch U, Holy V and Baumbach T 2004 *High resolution X-ray scattering: from thin films to lateral nanostructures.* (New York: Springer Verlag) Chapters 7 and 10
[6] Bergese P, Bontempi E and Depero L E 2006 *Appl. Surf. Sci.* **253** 28
[7] Tixier R and Philibert J 1969 *Proc. 5th Int. Congr. on X-ray Optics and Microanalysis.* Mollenstedt G and Glaukler K H , eds. (Berlin: Springer Verlag) pp. 180
[8] http://www.nist.gov/pml/data/xraycoef/index.cfm, Table 3, Data elemental media

IOP Conf. Series: Materials Science and Engineering **32** (2012) 012017 doi:10.1088/1757-899X/32/1/012017

[9] Armigliato A 1992 *Proc. of the Int. School on Electron Microscopy in Materials Science (Brindisi, IT, 7 -19 October 1991)* Merli P G and Vittori Antisari M, eds. (Singapore: World Scientific Publishing) pp. 431-56
[10] Calcagnile L, Quarta G, D'Elia M, Muscogiuri D, Maruccio L, Butalag K, Gianfrate G, Sanapo C and Toma U 2005 *Nuclear Instr. Meth. Phys. Res. B*, **240** 22-5
[11] Demortier G, Quarta G, Butalag K, D'Elia M and Calcagnile L 2008 *X-ray Spectrom.* **37** 178-83

IOP Conf. Series: Materials Science and Engineering **32** (2012) 012018 doi:10.1088/1757-899X/32/1/012018

EPMA and EBSD analysis of the chemical and structural changes in 16 wt% chromium stainless steel during deformation

A Núñez Galindo[1,3], X Llovet[2] and J F Almagro Bello[1]
[1] Acerinox S.A., Departamento I+D+i, Pol. Industrial Palmones s/n, ES-11319 Los Barrios, Spain
[2] University of Barcelona, Centres Científics i Tecnològics (CCiT), C/ Lluís Solé i Sabarís 1-3, ES-08028 Barcelona, Spain

E-mail: andres.nunez@acerinox.com

Abstract. In this study, four AISI 430 (EN 1.4016) steel samples of different initial composition and annealing treatment were submitted to uni-axial tension deformation. The as-supplied and deformed materials were characterized by electron channelling contrast imaging, electron backscatter diffraction and electron probe microanalysis. The results show a clear correlation between chemical composition, annealing treatment and strain level obtained until fracture.

1. Introduction

AISI 430 (EN 1.4016) is the most spread ferritic stainless steel grade and one of the simplest steel, containing only 16 wt% of chromium as alloying element. The use of this steel has grown because of its lack of nickel, which makes its price relatively low and stable. This steel grade is typically used in construction, automotive and food industry. Demanding applications of this steel are also related to conformation, usually stretching and deep drawing, for household appliances parts (such as grills, pans, pressure cookers, dishwashers, washing machines, etc.). In addition, AISI 430 (EN 1.4016) ferritic stainless steels can be easily conformed in cold, being the most common conforming processes those of stretch-forming and deep drawing.

Failures in ferritic stainless steel conformation are known to be related to problems of structural homogeneity caused by the remaining primary solidification structure in the final product [1-4]. However, recent investigations have revealed that such failures could also be due to chemical segregation [5, 6]. Unfortunately, knowledge about the level of segregation, in terms of different operating practices [7] and evolution with the forming process is at present quite limited. Furthermore, the information provided by structural characterisation techniques such as electron backscatter diffraction (EBSD) and electron channelling contrast imaging (ECCI) have not been adequately exploited.

In this context, it is necessary to improve the basic knowledge about the deformation mechanisms of these materials, as well as the effect of their microstructure and segregations on the dynamic behaviour during deformation. It is also necessary to establish the deformation mechanisms of this

[3] To whom any correspondence should be addressed.

steel grade, and evaluate the effect of chemical compositional changes and process variables on the resulting deformation mechanisms.

In this study, four AISI 430 (EN 1.4016) steel samples, of different composition, were subjected to uni- and bi-axial tension deformation. The resulting samples were first analyzed by using conventional techniques available in a stainless steel work in order to determine bulk composition, mechanical properties and structure and grain size. In a second step, textures and secondary phases were characterized by EBSD [8-11], and crystallographic orientation and defects were studied by ECCI [12]. Finally, electron probe microanalysis (EPMA) was used to identify chemical segregations. The correlation between chemical composition, annealing treatment and strain level obtained until fracture is discussed.

2. Experimental procedure

Four EN 1.4016 steel samples of thickness 0.4 mm with different initial compositions (basic and modified) were taken from the daily production of Acerinox, SA (www.acerinox.es). The samples were subjected to normal and high annealing treatments. In the case of normal treatment, the materials are annealed at a speed of 53.7 m/sec, which is a standard speed for a thickness of 0.4 mm, and the surface of the steel band reaches a maximum temperature of 940 °C. In the high annealing treatment, the processing speed of the materials is 43.0 m/sec, which corresponds to an increase of 10 % in the carry-in time into the furnace as compared with the normal treatment. After the annealing treatment, the samples were subjected to uni-axial tension deformation. The deformation tests were stopped after 1 %, 10 % and 20 % of the deformation (denoted to as -T1, -T10 and -T20 in the sample identification).

The chemical composition of the samples was analyzed using conventional techniques and instruments available in a stainless steel factory. Specifically Si, P, Ti, Cr, Mn, Co, Ni, Cu, Nb, Mo and Sn were analyzed by X-ray fluorescence (XRF) using a Panalytical PW2606 XRF spectrometer. B, Al, V and Pb were analyzed by spark optical emission (S-OES) using a Spectro 111982 spectrometer. Finally, C and S, and N were analyzed using Leco CS 600 and TC 600 analyzers, respectively.

The mechanical tests were carried out using different equipment, depending on the parameters to be measured. For the study of high-tensile strength, yield strength, elongation, coefficient of plastic anisotropy and coefficient of acrimony, a RK100 Roell+Korthaus Universal Testing Machine was used. To obtain the Erichssen, KWI and LDR parameters, a Zwick Roell BUP600 drawing was used.

Grain size analysis was performed after polishing and etching the samples (Vilella's reagent), for 60 to 75 s. The observation and measurement of grain size were performed with an Olympus GX71 optical microscope equipped with an "AnalySIS Inclusion Inspector" image analysis system.

The characterisation by EBSD and ECCI was performed on a Zeiss ULTRA 55 FEG-SEM using polished longitudinal sections of the samples. The sample sections were prepared by conventional metallographic procedures, with a final polish with colloidal silica suspension OPS. The FEG-SEM used is equipped with an HKL Channel 5 EBSD system, from Oxford Instruments. Backscattered electron (BSE) images were obtained at an accelerating voltage of 20 kV and a working distance of 1 - 2 mm; EBSD maps were acquired at 20 kV and a working distance of 10 mm.

The quantitative imaging of element distribution was performed on a Cameca SX-50 electron microprobe using wavelength-dispersive X-ray spectrometers. X-ray maps were acquired at an accelerating voltage of 20 kV and beam current of 100 nA, with a dwell time of 50 ms/pixel and they were converted into quantitative maps using the software packages HiMax and XMAS (SAMx, France). Each map covered an area of approximately 16 x 16 μm^2.

3. Results and discussion

3.1. As-supplied materials

3.1.1. Composition, structure and mechanical properties. The materials selected for this study are summarized in table 1, together with their global chemical composition and type of annealing treatment applied. The composition of samples 500-0A-S and 500-0B-S is referred to as "basic" composition, while that of samples 500-1A-S and 500-1C-S is a "modified" composition, with lower contents of interstitial elements such as C and N.

Optical microscope images showing the structure of samples 500-0A-S and 500-1C-S are displayed in figures 1 and 2. The structure of material 500-1C-S can be seen to show a better re-crystallisation than that of material 500-0A-S, as the grains and distribution of carbides within the grains is better differentiated.

Table 1. Composition, annealing treatment and metallurgical properties of the as-supplied materials.

Sample identification	Typical Composition (wt%)	Annealing	Mechanical tests			Grain Size	
			r_m	Δr	KWI	OM	EBSD
500-0A-S	C: 0.050; N: 0.035 Si: 0.35; Cr: 16.3	Normal	0.93	-0.40	24.91	10.5-11	12.3
500-1A-S	C: 0.025; N: 0.025 Si: 0.45; Cr: 16.7	Normal	1.14	-0.21	33.42	10	12.2
500-0B-S	C: 0.050; N: 0.035 Si: 0.35; Cr: 16.3	High	0.95	-0.21	45.11	10	11.9
500-1C-S	C: 0.025; N: 0.025 Si: 0.45; Cr: 16.7	High	1.15	-0.19	58.80	9.3	10.2

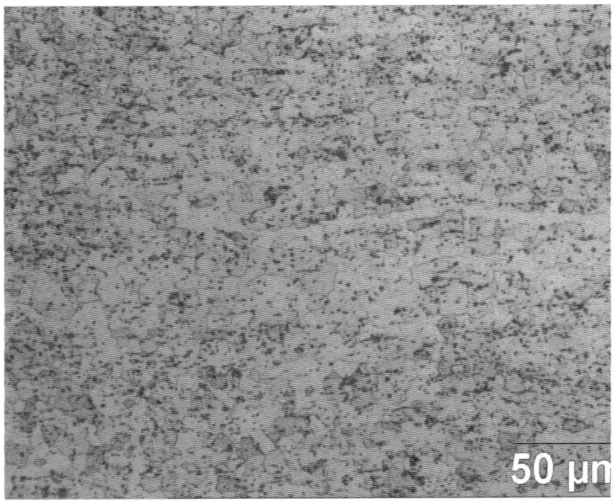

Figure 1. Optical microscope image showing the structure of material 500-0A-S.

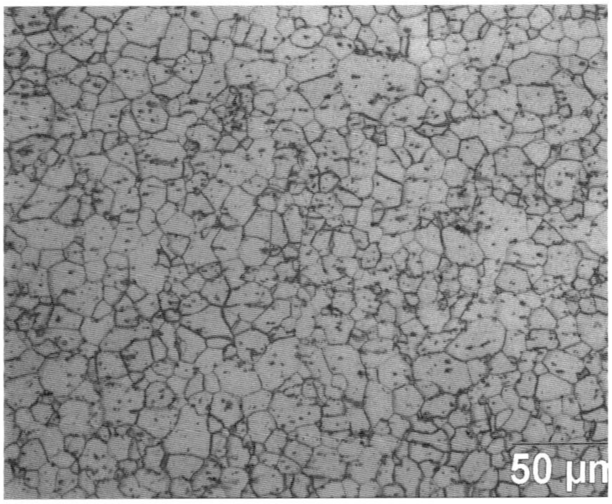

Figure 2. Optical microscope image showing the structure of material 500-1C-S.

The results of the mechanical tests performed on the as-supplied samples are listed in table 1, which include the average anisotropy, r_m, the in-plane anisotropy, Δr [13], and the bore expanding test (KWI test). The results obtained for parameters r_m and Δr indicate that the materials with modified composition are better suited for drawing than those with basic composition are. In addition, the KWI parameter, which gives valuable information on drawing, indicates that the 500-1C-S material is better suited for deep drawing than the rest of the tested materials.

The results of grain size measurements obtained by conventional optical microscopy (OM) and EBSD are listed in table 1. EBSD measurements respond only to variations in the relative misorientation between adjacent areas, and they are not affected by the action of chemical reagents as in conventional analysis. Grain sizes obtained by OM and EBSD agree satisfactorily with each other.

3.1.2. Electron channelling contrast imaging (ECCI). ECCI provides information on crystallographic orientation and defects in the material under study [12]. ECCI images are usually obtained by tilting the sample, which increases the BSE yield and enhances electron channelling. High-angle BSE information is used to locate areas of tension, dislocations, or other crystalline defects [14, 15].

Here it should be noted that the high accelerating voltage (20 kV) and the short working distance (1 - 2 mm) used allow the observation of fine structures in grain boundaries, which largely coincide with the dislocations that can be observed by transmission electron microscopy (TEM). In these conditions, tilting the sample is not required.

Figures 3 to 6 show general and detailed BSE micrographs corresponding to the samples 500-0A-S (figures 3 and 4) and 500-1C-S (figures 5 and 6). The contrast observed is the result of the different crystallographic orientation of the grains with respect to the impinging electron beam (different zone axis). For all the samples, a crystalline structure of equi-axed grains randomly oriented can be observed. The scratches on the polished surface are a consequence of the presence of numerous chromium carbide precipitates.

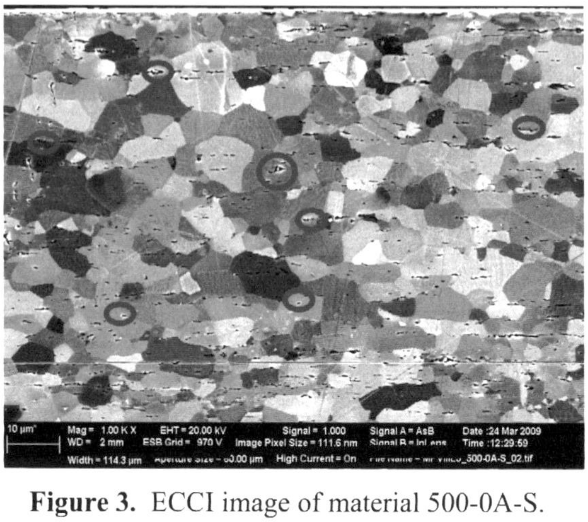

Figure 3. ECCI image of material 500-0A-S.

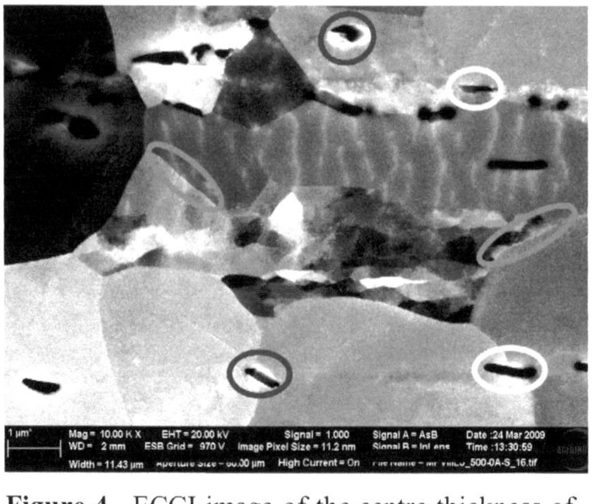

Figure 4. ECCI image of the centre thickness of material 500-0A-S.

Both the 500-0A-S and 500-1C-S materials show carbo-nitride precipitates (shown in purple ovals) which reveal crystal inconsistency between inclusions and the matrix. The number of precipitates is larger for the 500-0A-S material. In addition, the 500-0A-S material shows some dislocations (orange ovals), which are likely to be due to accumulated tension at the grain boundaries [16-19].

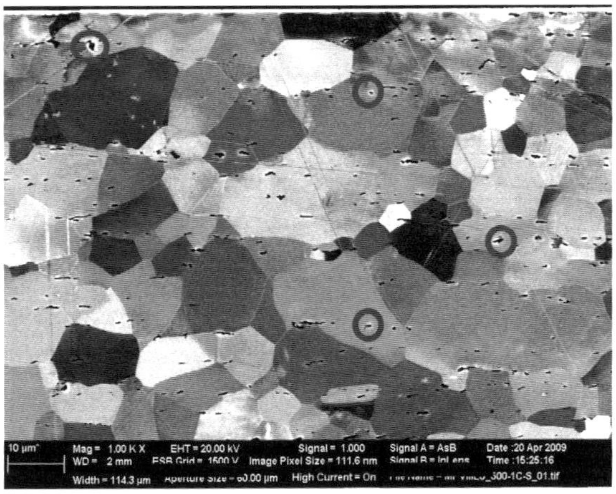

Figure 5. ECCI image of material 500-1C-S.

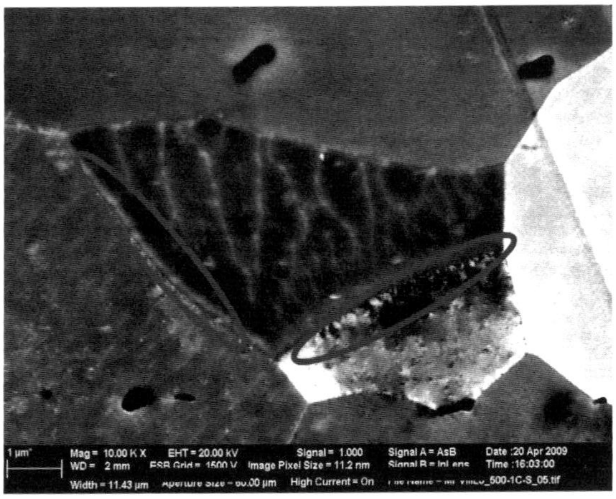

Figure 6. ECCI image of the centre thickness of material 500-1C-S.

Some grains "take advantage" of the presence of a precipitate at the grain boundaries in order to reduce tensions in the material (shown in yellow ovals).

For the 500-1C-S material (figure 6), grain boundaries with complex structures can be observed (shown in blue ovals). These kinds of boundaries are the result of the high density of dislocations. A large number of boundaries with dislocation networks are observed in the 500-1C-S material.

3.1.3. EBSD and EPMA. EBSD analysis allowed obtaining information of the whole sample thickness for the main phase, as the observation plane was ND-LD (ND and LD stands for normal direction and line, or rolling, direction). The FEG-SEM characterisation was carried out using a polished sample in longitudinal section (ND-LD plane). After the analysis, a cleaning procedure was applied, which consisted of removing EBSDs that could not be indexed and isolating points that were incorrectly indexed and appeared as Wild Spikes. The removed points were filled in using copies of their neighbouring points. If a given point had more than the minimum number of indexed neighbours (5 - 6), then it was replaced by the most common neighbouring orientation.

Figure 7 shows grain, sub-grain, and coincidence site lattice (CSL) maps corresponding to the 500-0A-S material. The grain and sub-grain maps are obtained from misorientation angles (sub-grains, 1 - 10º, in light grey and grain boundaries, > 10º, in black). The CSL maps are obtained from a specific coincidence of the crystal lattices across the boundary. This information is needed to define more accurately the grain size by EBSD (table 1) because CSL are not considered as grain boundaries but special boundaries (i.e., Σ3 is the well known twin boundary in fcc materials). The lines with different colours correspond to CSLs with different Σ values (Σ is the reciprocal number density of lattice points that are coincident in two adjacent grains), axis and angle. Figure 8 shows the inverse pole figure (IPF) map corresponding to the same material. In this latter map, the colour code refers to the crystallographic direction parallel to the rolling direction.

In the 500-0A-S material, areas with high density of sub-grain boundaries are observed. These areas also represent grain areas with preferred orientation.

Figure 9 shows grain, sub-grain and CSL; and figure 10 the ODF function, corresponding both to the 500-1C-S material. The ODF figure indicates a predominance of gamma fibre and an absence of alpha fibre. The presence of the {311} <136> texture component [20, 21] is typical of a good transition from hot rolling to cold rolling-annealing texture in ferritic steels. As a result, this material presents a better re-crystallisation, and it can be advanced that it will present the best drawability performance.

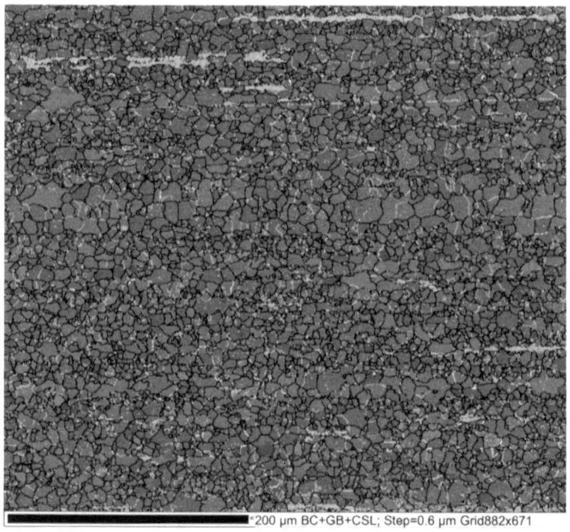

Figure 7. Grain, sub-grain and CSL maps of material 500-0A-S. BC+GB+CSL; grid 882x671 pixels; step 0.6 µm.

Figure 8. Orientation IPF map of material 500-0A-S (the colour code refers to the crystallographic direction parallel to RD). BC+IPF X1; grid 882x671 pixels; step 0.6 µm.

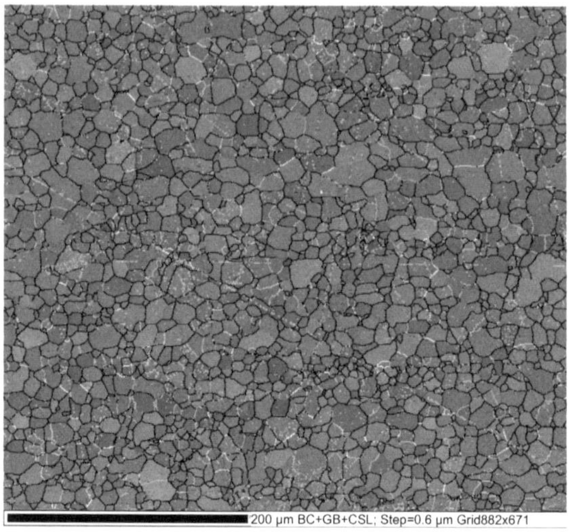

Figure 9. Grain, sub-grain and CSL maps of material 500-1C-S. BC+GB+CSL; grid 882x671 pixels; step 0.6 µm.

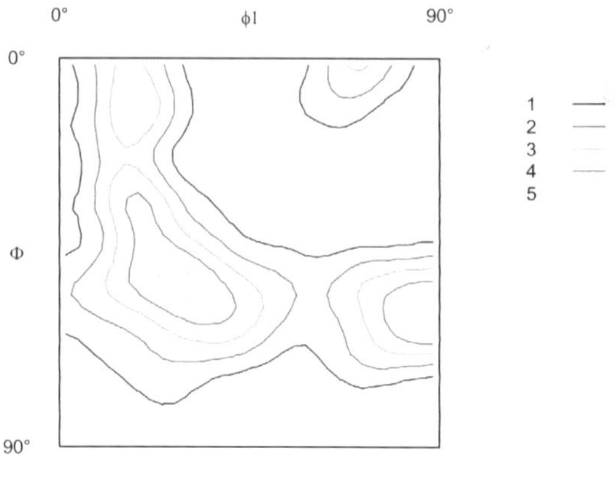

Figure 10. ODF function of material 500-1C-S. 5, green line corresponding to the peak, means 5.51; and 1, purple line corresponding to the valley, means 1.1.

Figures 11 and 12 shows false colour quantitative EPMA maps for the Cr distribution in each studied material, 500-0A-S and 500-1C-S, respectively. The colour level coding corresponds to mass%. The maps show compositional variations as horizontal bands, parallel to the rolling direction. Similar behaviour can be observed for Fe, while for Si and V, only segregations of these elements can be observed (for the sake of simplicity the maps corresponding to Fe, Si and V are not shown here). The number of precipitates seems to be larger for the 500-0A-S material (basic composition).

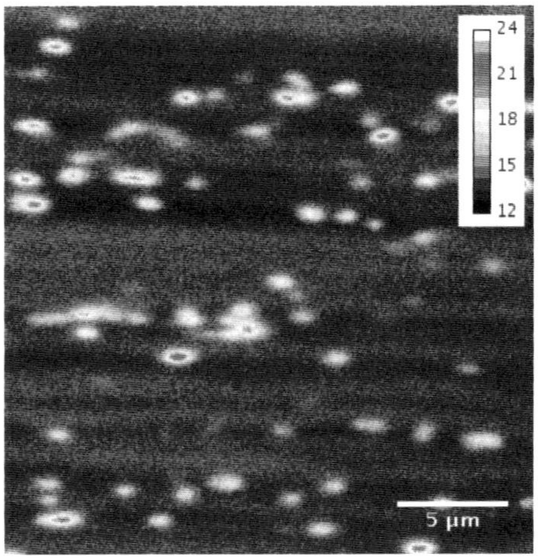

Figure 11. EPMA map for Cr corresponding to sample 500-0A-S.

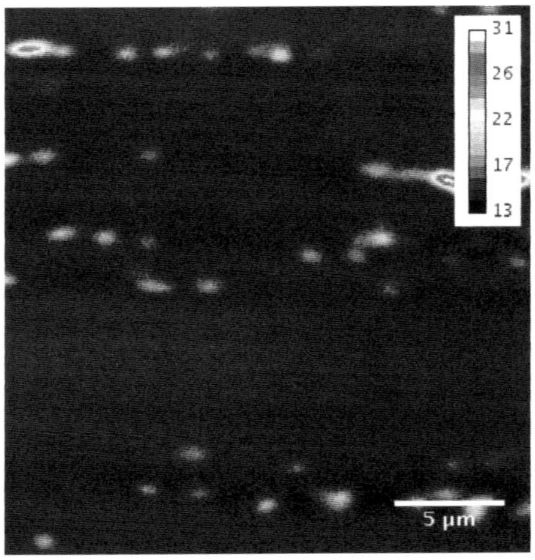

Figure 12. EPMA map for Cr corresponding to sample 500-1C-S.

3.2. Uni-axial tension deformation

3.2.1. Electron channelling contrast imaging. Figures 13, 14 and 16 show ECCI images corresponding to the material 500-1C after 1 %, 10 % and 20 % tension deformation. With 1 % of deformation, the crystalline structure of equi-axed grains randomly oriented remains. Dislocation networks typically originated by accumulated tension in grain boundaries (shown in orange ovals) can be observed. With 10 % of deformation, the grains present changing contrast (shown with green arrows) due to stress associated with the presence of dislocation networks. At 20 % of tension deformation, in addition to the grain stress contrast (green arrows), some line defects can be observed (yellow arrows). These line defects have resulted in dislocations networks defining sub-grains. Therefore, the density of crystalline defects increases with plastic deformation in the grains, especially dislocations, i.e., "extra" planes of atoms of the material inserted into the crystal structure. This results in a complex sub-grain structure, which is generated because of deformation.

Figures 15 and 16 compare ECCI images corresponding to materials 500-0A and 500-1C, respectively, after deformation of 20 %. Both materials show stress and dislocation areas (green arrows). For the 500-0A material, bright areas of inter- and intra-granular stress can be observed. These areas are not visible in the 500-0A material. Line defects (yellow arrows) can be mainly seen in the 500-1C material, which have resulted in dislocations networks defining sub-grains.

3.2.2. EBSD and EPMA. Figures 17 and 18 show EBSD maps corresponding to the 500-0A-T20 material. Similar maps for the 500-1C-T20 material are shown in figures 19 to 20.

As for the corresponding as-supplied material, the ODF function of material 500-1C-T20 (figure 20), the gamma fibre predominates and no alpha fibre is observed.

Figures 21 and 22 show EPMA maps for Cr corresponding to the 500-0A-T20 and 500-1C-T20 samples, respectively. The maps show compositional variations of Cr as horizontal bands, parallel to the rolling direction. A similar behaviour is observed for Fe (not shown here).

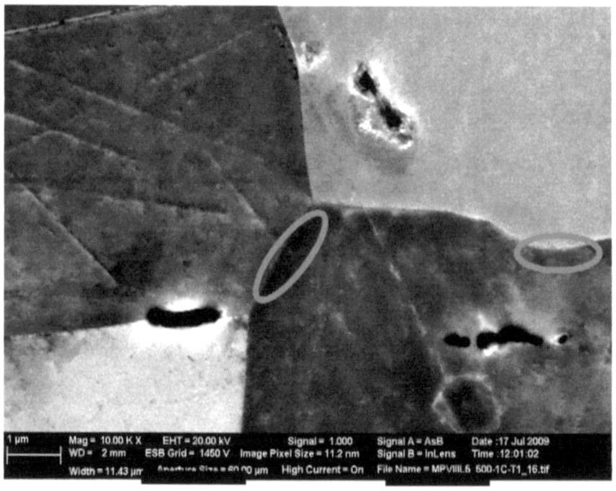

Figure 13. ECCI image of material 500-1C deformed by tension of 1 % (500-1C-T1).

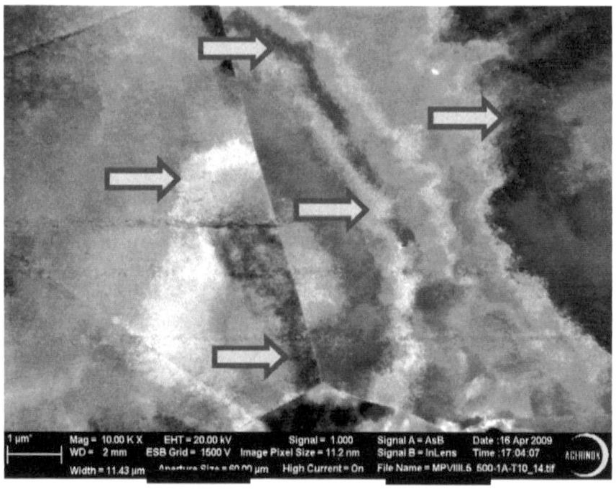

Figure 14. ECCI image of material 500-1C deformed by tension of 10 % (500-1C-T10).

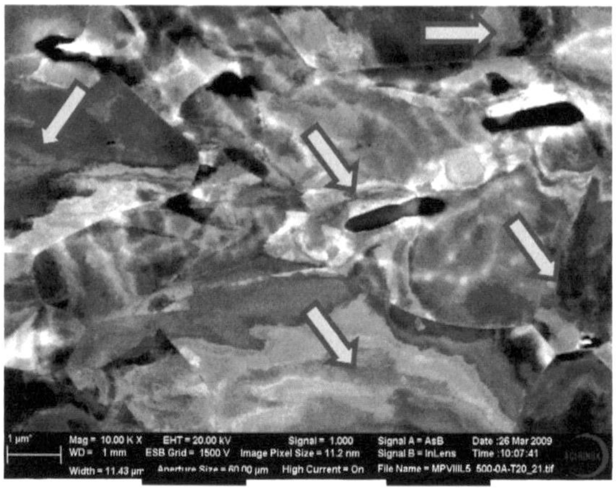

Figure 15. ECCI image of material 500-0A deformed by tension of 20 % (500-0A-T20).

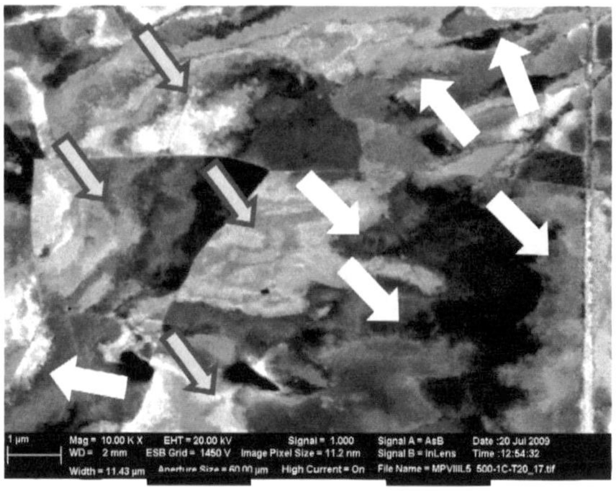

Figure 16. ECCI image of material 500-1C deformed by tension of 20 % (500-1C-T20).

Figure 23 compares average Cr concentration profiles, for samples 500-0A-S and 500-0A-T20, extracted from the corresponding EPMA maps of figures 11 and 21 respectively. The x-axis represents distance along the vertical direction on the maps. The y-axis is the horizontally average Cr-concentration, which is obtained by averaging pixel values of all pixels corresponding to the same image raw. Similar comparison for the 500-1C-S and 500-1C-T20 samples is presented in figure 24. In the case of the 500-1C material, the Cr bands in the deformed material appear to be slightly thicker than those observed in the as-supplied sample. Conversely, for the 500-0A material, the Cr bands in the deformed sample are thinner than those in the as-supplied sample.

In theory, any of the studied materials should show a contraction of the lateral area after the deformation test, which should cause a reduction of the material section. While the 500-1C material (with modified composition) seems to follow the expected behaviour after deformation, the 500-0A material (basic composition) does not follow it. This abnormal behaviour could be due to a migration

Figure 17. Grain, subgrain and CSL map of material 500-0A-T20. BC+GB+CSL; grid 882x648 pixels; step 0.6 µm.

Figure 18. Orientation IPF map of material 500-0A-T20. BC+IPF X1; grid 882x648 pixels; step 0.6 µm.

Figure 19. Grain, subgrain and CSL map of material 500-1C-T20. BC+GB+CSL; grid 882x671 pixels; step 0.6 µm.

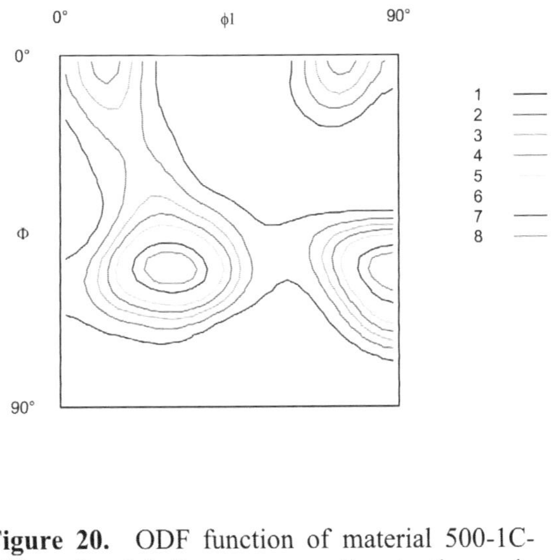

Figure 20. ODF function of material 500-1C-T20. 8, pink line corresponding to the peak, means 8.97; and 1, purple line corresponding to the valley, means 1.12.

of the alloy elements to non-preferred orientation places, originating new precipitates, and making it difficult for the dislocations produced during the tension deformation to move.

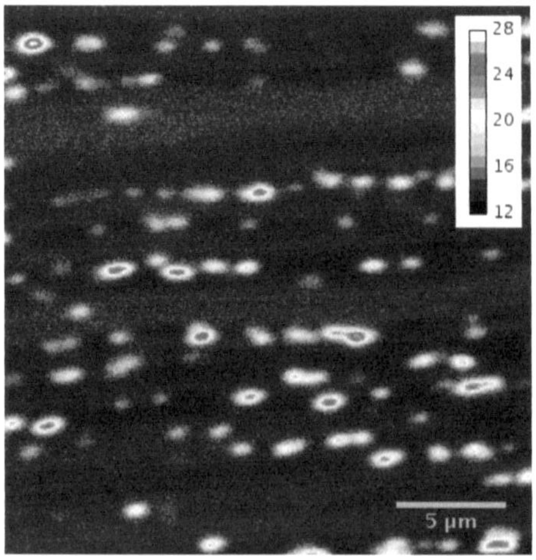

Figure 21. EPMA map for Cr corresponding to the sample 500-0A-T20.

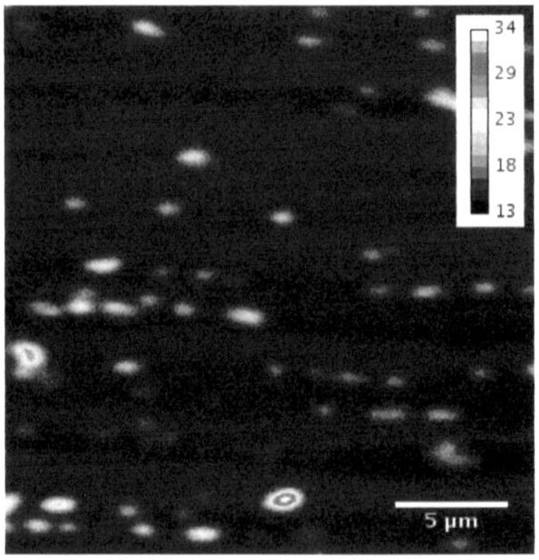

Figure 22. EPMA map for Cr corresponding to the sample 500-1C-T20.

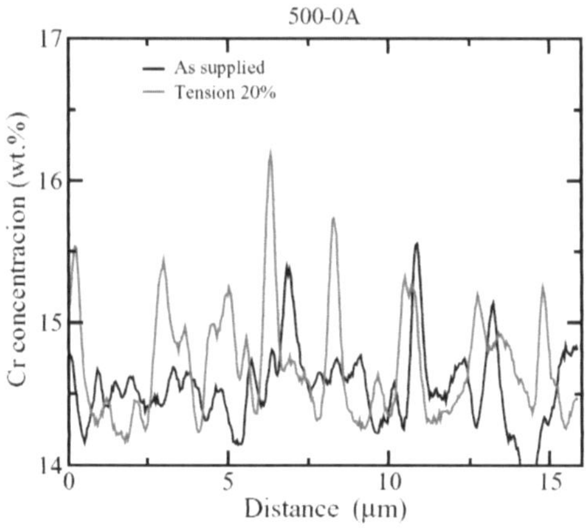

Figure 23. Average Cr profile along the vertical direction for samples 500-0A-S (as supplied) and 500-0A-T20 (after 20 % tension deformation). See text for details.

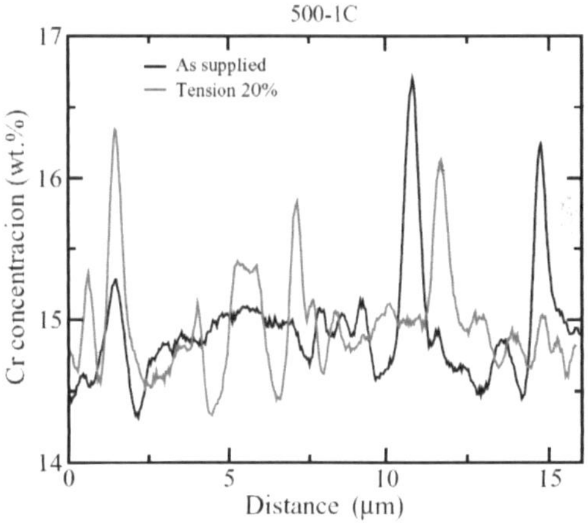

Figure 24. Average Cr profile along the vertical direction for samples 500-1C-S (as supplied) and 500-1C-T20 (after 20 % tension deformation). See text for details.

4. Conclusions

The study of the structure, mechanical properties, as well as the ECCI and EBSD analysis of the as-supplied materials, show that the material with the modified chemical composition and high annealing treatment has a better re-crystallized structure and drawability behaviour than that with basic composition and normal annealing treatment.

The materials subjected to uni-axial tension deformation show a correlation between chemical composition, annealing treatment and strain level obtained until fracture as follows:

- The materials with modified chemical composition and higher annealing process show almost no segregation areas. This is so because these materials develop a dense sub-grain network and, as a result, the external distortion generates a strong dislocation grid that seems to inhibit any segregation phenomena.
- In the materials with the basic composition and normal annealing treatment, the higher content of interstitial elements seems to prevent the formation of sub-grain networks, which helps to absorb the deformation energy. This facilitates the reorganisation of the elements, and allows the segregation and precipitation of alloying elements such as Cr.

References

[1] Shin H J, An J K, Park S H and Lee D N 2003 *Acta Material.* **51** 4693-706
[2] Park S H, Kim K Y, Lee Y D and Park D J 2002 *ISIJ Intern.* **42** 100-5
[3] Sinclair C W, Robaut F, Maniguet L, Mithieux J D, Schmitt J H and Brechet Y 2003 *Adv. Eng. Mater.* **5** 570
[4] Benthem K, Krämer S, Sigle W and Rühle M 2002 *Mikrochim. Acta* **138** 181-93
[5] Presslinger H, Mayr M, Tragl E and Bernhard C 2002 *Steel Research Int.* **77** 107-15
[6] Almagro J F, Llovet X, Heredia M A, Luna C and Sánchez R 2008 *Mikrochim. Acta* **161** 323-7
[7] Huh M Y and Engler O 2001 *Mat. Sci. Eng.* A **308** 74-87
[8] Ferreira H, Santana S, Neto P, Pires R, Nogueira V, de Oliveira P and Souto S 2007 *Mat. Res.* **10** 359-66
[9] da Costa C S, Candido F S and Pinto A L 2005 *Mat. Sci. Forum* **495-497** 173-8
[10] Sato Y S, Nelson T W and Sterling C J 2005 *Acta Mater.* **53** 637-45
[11] Shimada M, Kokawa H, Wang Z J, Sato Y S and Karibe I 2002 *Acta Mater.* **50** 2331-41
[12] Trager-Cowan C, Sweeney F, Trimby P W, Day A P, Gholinia A, Schmidt N-H, Parbrook P J, Wilkinson A J and Watson I M 2007 *Phys. Rev.* B **75** 085301
[13] ASTM International 2007 Designation: A 370-07a. V. 01.03, ISBN 978-0-8031-4284-8
[14] Jaksch H, Steigerwald M and Drexel V 2005 *Microsc. Microanal* **11** Suppl 2, 758-9
[15] Yasuda M, Kawata H and Murata K 1996 *Phys. Stat. Sol. (a)* **153** 133-44
[16] Karato S 1987 *Physics Chem. Minerals* **14** 245-8
[17] Kamaladasa R J and Picard Y N 2010 *Microscopy: Sci., Techn., Appl. and Educ.* **3** 1583-90
[18] Kuwano N, Itakura M, Nagatomo Y and Tachibana S 2010 *J. Elec. Microsc.* **3** 1-7
[19] Epishin A, Link T and Nolze G 2007 *J. Microsc.* **228**(pt. 2) 110-7
[20] Gobernado P 2004 *Structure and properties of grain boundaries in electrical steels.* Ph.D.-thesis. (Ghent: Ghent University) ISBN 978-90-8578-292-6
[21] Gobernado P, Petrov R, Ruiz D, Leunis E and Kestens L A I 2010 *Adv. Eng. Mat.* **12** 1077-81

Effect of TiN nano-coating on the interface microstructure of carbon fibres-AZ91 alloy composite

A Olszówka-Myalska[1] and A Botor-Probierz
Silesian University of Technology, Faculty of Materials Engineering and Metallurgy, ul. Krasińskiego 8b, PL-40019 Katowice, Poland

E-mail: anita.olszowka-myalska@polsl.pl

Abstract. Magnesium matrix composites reinforced with carbon fibres, C_f,without surface modification and coated with TiN nanolayer, $(C_f)_{TiN}$, were investigated. AZ91 magnesium alloy and carbon fibres of T300B (Toray) PAN type were chosen as components. In the experiment infiltration in vacuum as a method of component consolidation and chemical vapour deposition (CVD) for carbon fibres surface modification were applied. Structural investigations were performed in the Hitachi 3200S field-emission scanning electron microscope (FE-SEM) and on a FEI Tecnai G2 FEG high-resolution transmission electron microscope (TEM) equipped with energy-dispersive X-ray spectrometer (EDS) and high angle annular dark field (HAADF) detectors.
SEM observations revealed that at the interface of AZ91-C_f composite destructive phases were formed. An application of TiN nano-coating as a surface modification ensured proper wettability of carbon fibres by liquid metal and protection against the destructive products formation. The microstructure of AZ91-$(C_f)_{TiN}$ composite interface was multilayered. TEM investigation revealed a continuous layer of mixed alumina and magnesia just at the carbon surface, followed by a layer with a dominant concentration of titanium, enriched with Al, O, Mg and C and subsequently a layer of nano-sized $Al_{12}Mg_{17}$ crystals in the magnesium matrix. Therefore, an applied TiN nano-coating can be classified as an active barrier in AZ91-C_f system.

1. Introduction

Light weight metallic materials, such as Mg and its alloys, are getting popular in the designing of magnesium matrix composites, especially reinforced with carbon materials. The greatest advantage of this system is the comparable density of both components ~ 1.8 g/cm^3. The carbon-magnesium system offers unique properties in comparison to conventional metals and alloys. The system is characterized as a material with high specific strength, good thermal conductivity and low density [1-4]. It is necessary to solve some problems for composite processing, such as proper wetting and reactivity of carbon fibres in contact with liquid metal [5-6].

Because of their technological properties magnesium alloys containing aluminium (i.e., AZ91) have been used as composite matrix and described in several papers [7-9]. However, hydrophilic carbides (Al_4C_3, Al_2MgC_2) form at the fibre-matrix interface. Therefore, modification of magnesium alloy composition and application of technological coatings in order to eliminate the formation of hydrophilic carbides were proposed in some papers [10-12].

[1] To whom any correspondence should be addressed.

For instance, TiN nano-coating was proposed by Popowska et al. [12]. They revealed an improvement of wettability and a decrease of carbides formation intensity but the microstructure of the interface and stability of TiN was not characterized in detail. The latter problem was analyzed for the SiO_2 nano-coating application in the AZ91-C_f system [13] where it has been shown that the transformation of the initial SiO_2 into a mixed oxide layer containing silicon and aluminium took place.

The aim of present paper is to characterize the interface microstructure in AZ91-$(C_f)_{TiN}$ composite with SEM and TEM methods and to analyze the role of TiN in the interface formation.

2. Experiment

2.1. Composite processing

2.1.1. Surface modification of carbon fibres. Carbon fibres of FT300B (Toray) type in roving form were applied in experiments. Chemical vapour deposition (CVD) was used for fibre surface modification. During CVD at the temperature of 980 °C the mixture of N_2 and $TiCl_4$ gases was applied.

2.1.2. Components consolidation. Carbon fibres uncoated and coated with TiN were applied as a reinforcement of AZ91 alloy (9 % Al and 0.6 % Zn). Composite samples were obtained in the same conditions by infiltration process in Degussa press. The process was performed in vacuum at a temperature of 670 °C for 20 minutes followed by cooling under the loading. Composite samples with a 20 mm diameter each were obtained.

2.2. Methods of materials microstructure investigation

The microstructure was preliminarily investigated using a scanning electron microscope (FE-SEM Hitachi 4200S, Hitachi Ltd., Tokyo, Japan) at an accelerating voltage of 15 kV, equipped with an energy-dispersive X-ray spectrometer (EDS) (Noran 3500, Thermo Electron Corporation, Waltham, MA, Si-Li detector, 130 eV resolution). The Kα X-ray series (for all elements) were used for elemental mapping and line-scans. X-ray diffraction measurements (XRD) were made with a Jeol JDX-7S diffractometer (equipped with plate graphite monochromator, using Cu Kα radiation at 40 kV and 20 mA). X-ray diffraction patterns of fibres were recorded in the range of diffraction angles 2θ from 10° up to 90° with a step size of 0.02°. More detailed structural investigations of the fibre-matrix interface were performed on a FEI Tecnai G2 field-emission high-resolution transmission electron microscope equipped with a HR CCD Gatan UltraScan camera, high angle annular dark field (HAADF) detector, at an accelerating voltage of 200 kV equipped with EDS (Edax Phoenix Si(Li)-detector with an UTW ultra-thin window for light elements analysis).

3. Results

3.1. Composite processing

As a result of fibre surface modification, a thin coating of TiN with a thickness of approximately 100 nm was formed. It covered the fibres very well and was uniform. It was composed of very fine and thickly packed grains and it had the golden colour typical for TiN (figures 1a and 1b). X-ray diffraction measurements (XRD) confirmed the presence of crystalline TiN (figure 2).

After components hot-pressing consolidation the composite samples of 20 mm diameter each were obtained for subsequent structural examination.

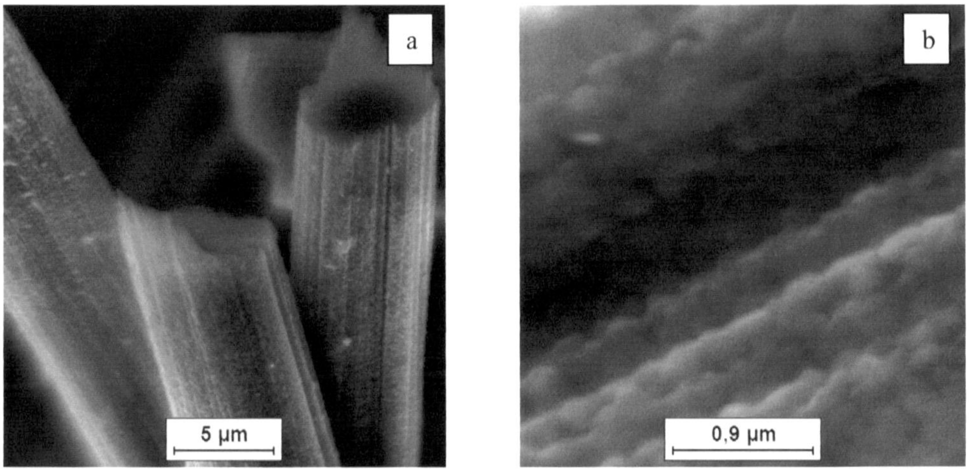

Figure 1. SEM micrographs of TiN coating deposited on T300B carbon fibres by the CVD method.

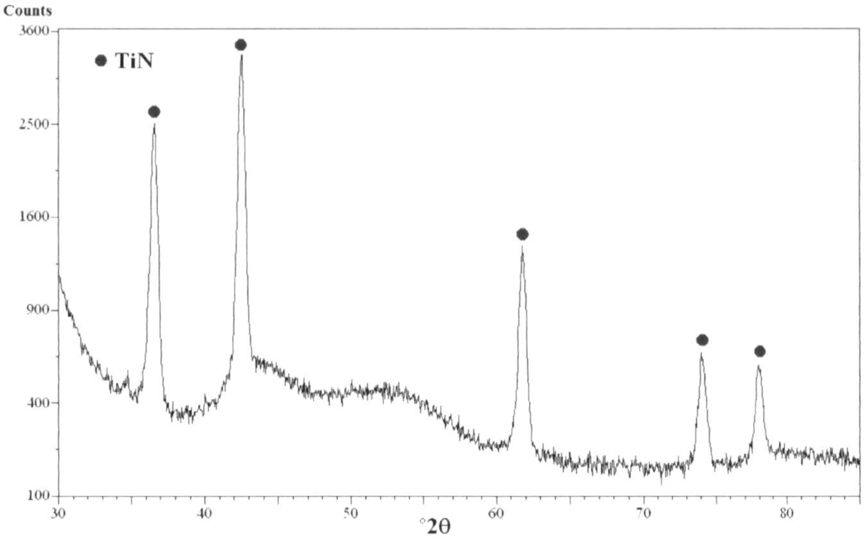

Figure 2. XRD patterns of carbon fibres roving coated with TiN by the CVD method.

3.2. Composites microstructure investigation

3.2.1. SEM observation. The observation was performed on polished and fractured cross-sections (figures 3-6). The reinforcement was properly surrounded with nonporous metal matrix. However, in the case of the composite made with uncoated fibres micro-pores were formed at the interface on the polished cross-sections (figure 3a and 3b) as a result of contact with water, while that effect was not recorded in the AZ91-$(C_f)_{TiN}$ composite (figure 4). Moreover, in the AZ91-C_f system very thin needle-like phases oriented from fibre into the matrix were observed (figure 3b). This type of interface suggests the presence of hydrophilic carbide. Additionally, in case of the AZ91-$(C_f)_{TiN}$

composite a characteristic thin and regular layer was detected around the fibres (figure 4). On the fractured cross-sections of both composites (figures 3c and 5) the effect of carbon fibres pulling-out was not observed.

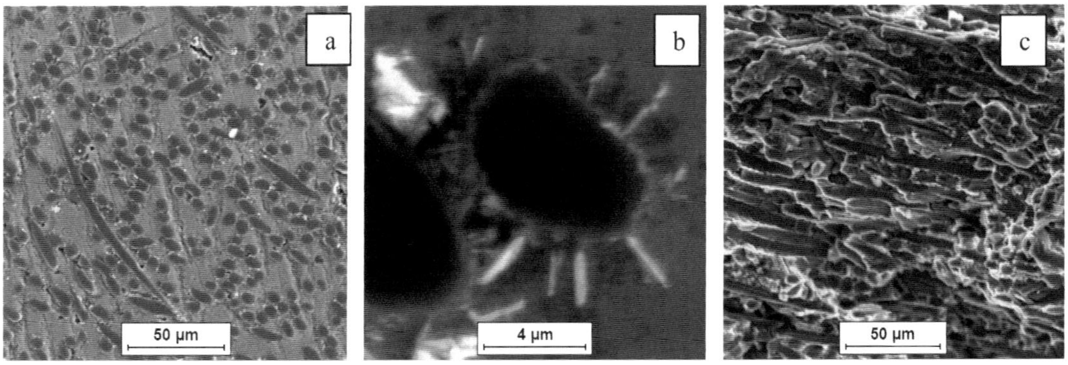

Figure 3. Microstructure of the AZ91-C_f composite without surface modification of the fibres. a & b) polished cross-section; and c) fractured cross-section (SEM).

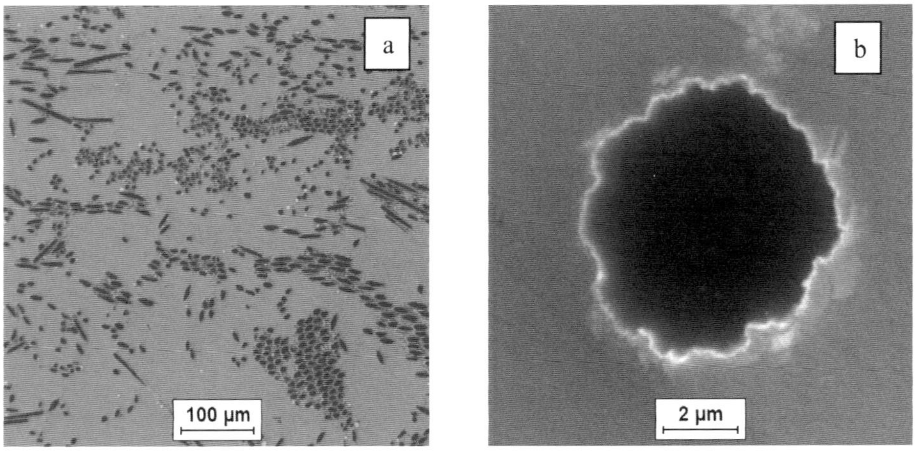

Figure 4. SEM micrograph showing the microstructure of a polished cross-section of a AZ91-$(C_f)_{TiN}$ composite.

As a results of X-ray microanalysis (figures 6 and 7) it was found that in the AZ91-$(C_f)_{TiN}$ composite the interface layer was enriched with aluminium, titanium and oxygen.

3.2.2. TEM observation. The observation of the fibre-matrix interface by TEM (figure 8) indicated the presence of a regular zone between fibre and matrix with irregular plate-shaped crystals located near its vicinity. Phase identification of the continuous zone by selected-area diffraction pattern (SADP) was impossible, but irregular crystals around that zone were characterized as $Al_{12}Mg_{17}$-phase, visibly alike to the other plate shaped crystals visible in the matrix. The HAADF image with X-ray maps (figure 9) showed that the continuous zone consisted of two layers. One of them located directly at the

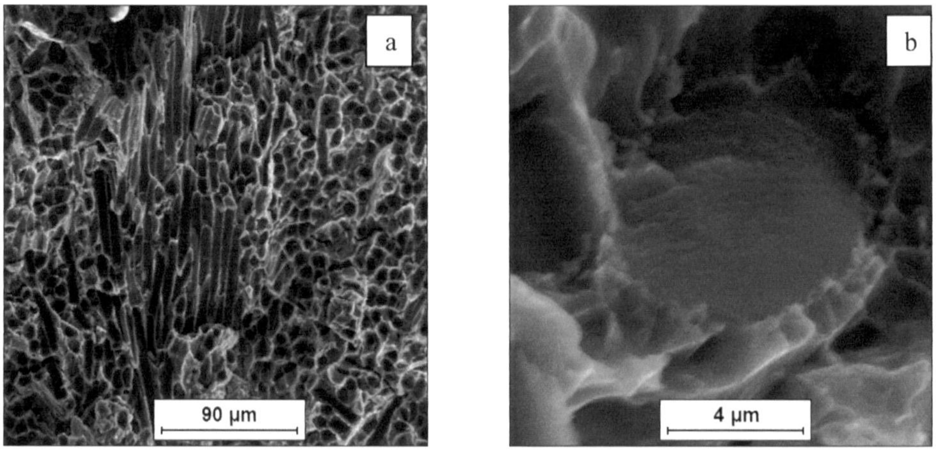

Figure 5. SEM micrograph showing the microstructure of fractured cross-section of a AZ91-$(C_f)_{TiN}$ composite.

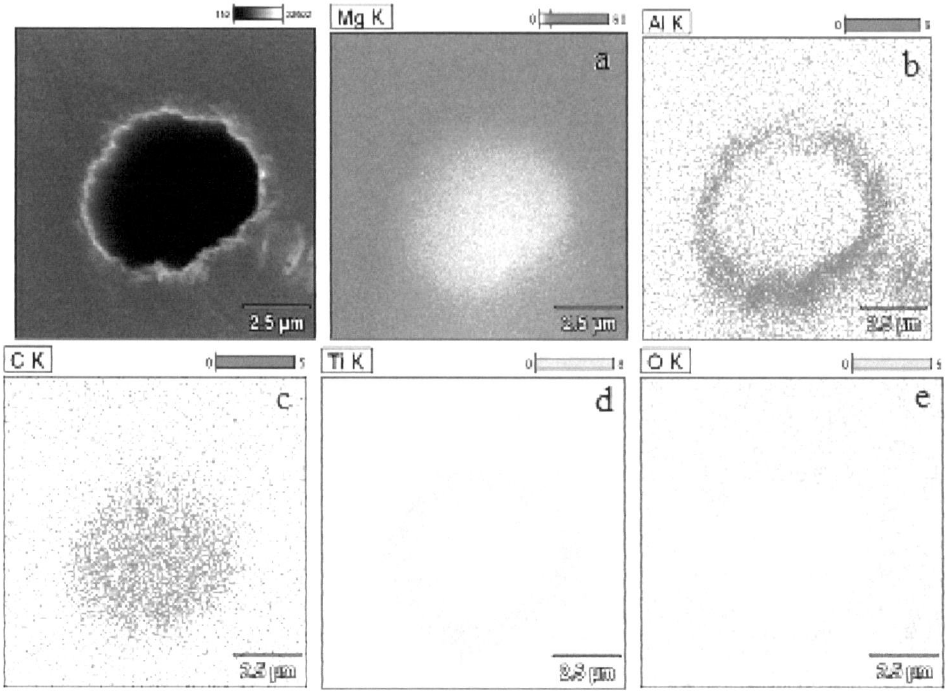

Figure 6. Interface of the AZ91-$(C_f)_{TiN}$ composite. X-ray mapping of: a) Mg; b) Al; c) C; d) Ti; and e) O.

carbon surface consisted of oxygen, aluminium and magnesium, while in the next layer titanium was detected as the dominant element and additionally oxygen, carbon, aluminium and magnesium were observed. The presence of nitrogen at the interface was not revealed.

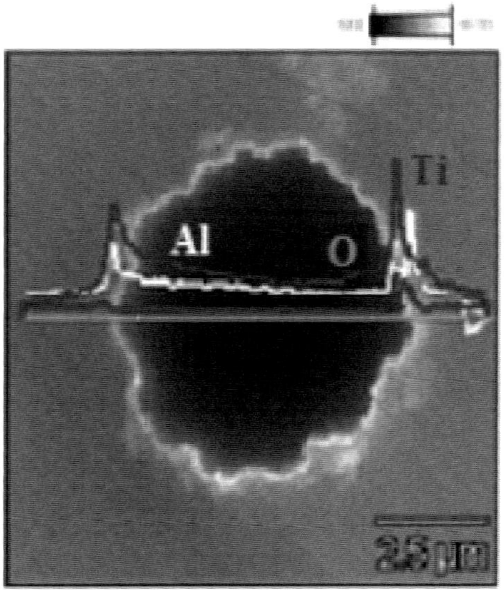

Figure 7. X-ray line scans along the interface of the AZ91-$(C_f)_{TiN}$ composite (SEM+EDS).

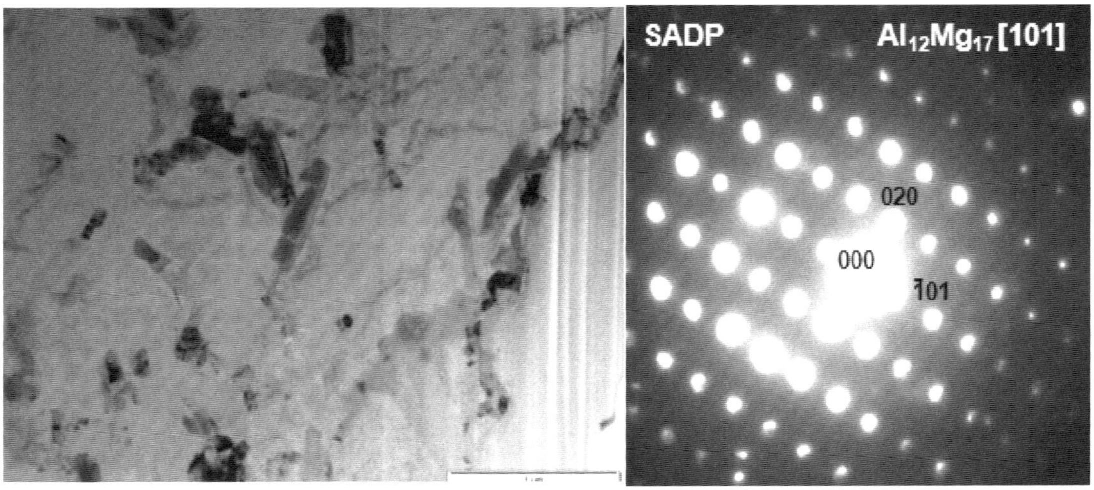

Figure 8. Interface of the AZ91-$(C_f)_{TiN}$ composite. Bright field image (BFI) with selected-area electron diffraction pattern (SADP) of the $Al_{12}Mg_{17}$-phase.

4. Discussion

Results of SEM observations of both composites indicate that the parameters applied in the process of components consolidation are proper from the point of view of good infiltration of the carbon fibres and the close continuous contact of the lamellar-shaped fibres with the metal. The absence of pulling-out effect in the composites indicates that chemical bonding was formed in both composite with and without carbon fibres' surface modification (figures 3c and 5). Generally, the results of the performed investigations suggest that fibre surface modification with nano-coating is desirable for a composite with an AZ91 matrix with respect to the TiN-coating protection against formation of hydrophilic carbide, which appeared at the interface in AZ91-C_f composite without modification (figure 3) and led to partial degradation of AZ91-C_f composite interface. The latter result has been confirmed in the literature data for similar experiments on C-Mg alloy systems [12].

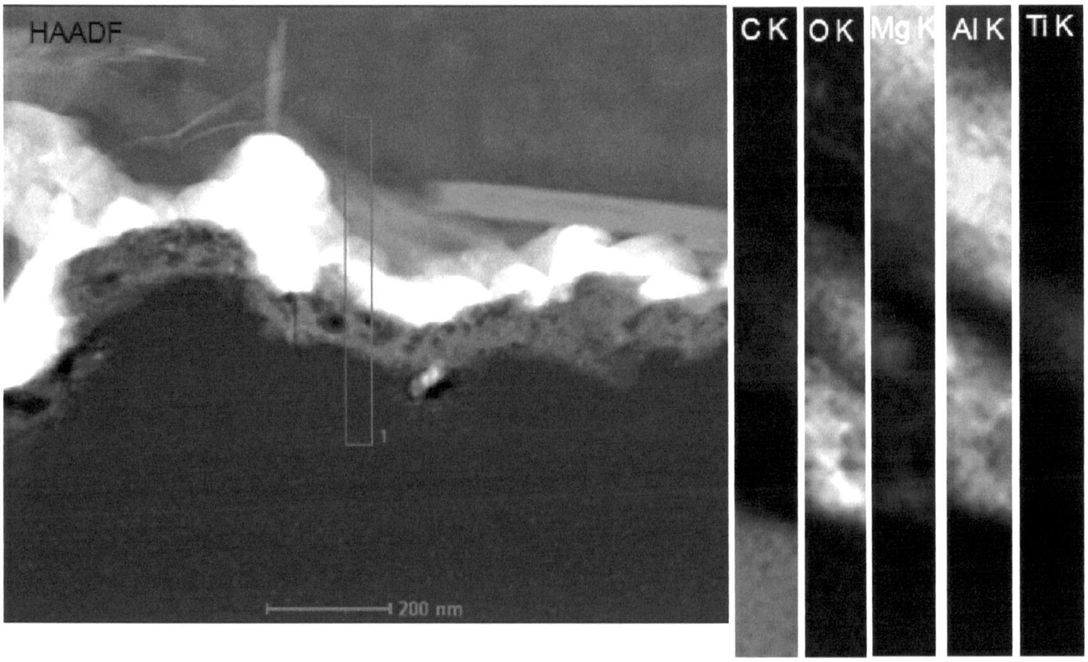

Figure 9. Interface of the AZ91- $(C_f)_{TiN}$ composite. HAADF (z-contrast and X-ray elemental mapping).

Results of SEM/EDS investigations of AZ91-$(C_f)_{TiN}$ composite revealed only that the interface is enriched with aluminium, titanium and oxygen (figures 5 and 6). Not all light element X-ray lines had sufficient intensity. The nitrogen Kα intensity was not measured because of an overlap with the Ti Lα-line. The oxygen Kα intensity is distinctly higher in the transition region than in the matrix (which can be considered as a background (bremsstrahlung)). Carbon Kα mapping gave only the visible difference in the intensities in the carbon fibre and magnesium alloy matrix but its content in the transition region has to be examined by TEM/EDS for samples prepared by FIB technique.

For a more accurate characterisation of the interface microstructure, TEM examination was necessary in order to reveal its multilayered nature (figure 7). EDS measurements show the presence of Ti, Al, Mg, O and C in the zone composed of a uniform double layer and of an additional region of nano-sized plates distributed in the magnesium matrix. A double layer of new phases instead of the initial TiN nano-layer was formed during components consolidation by hot pressing. Their precise phase identification requires further investigations but the X-ray maps suggest the appearance of some unexpected effects. The oxide layer formed exactly on the carbon fibres is the result of interaction between oxygen absorbed by the fibres during their manufacturing and atoms of metals (Al and Mg), which have diffused through the nano-grained TiN technological coating. Moreover, an initial TiN nano-layer has been subjected to transformation into a layer of similar geometry but with different chemical composition. It consisted of titanium as dominant element, with additional oxygen and carbon, and was doped with some amount of aluminium and magnesium. That may suggest the presence of titanium oxide and titanium carbide molecules in the layer. However, nitrogen was not detected by SEM/EDS and TEM /EDS but its presence cannot be excluded. The electron SADP of nano-sized plates in interface region showed $Al_{12}Mg_{17}$-phase formation. Generally, it is the phase typical for AZ91 magnesium alloy and its crystallisation around carbon fibres was observed also in other composites [14-16].

Investigations of the AZ91-$(C_f)_{TiN}$ interface microstructure showed that the applied TiN nano-layer does not eliminate chemical interaction between carbon reinforcement and AZ91 matrix, therefore, the TiN is only an active barrier, suitable for diffusion process control at the interface. It explains the results of some experiments [12] where the protective role of TiN was insufficient.

A very important and a new result of the presented experiment is the detection of an evident oxide layer around the fibres. That effect was observed in composites without any coating and one may expect that the presence of the oxides is the result of oxygen presence in the magnesium alloy. An application of a TiN barrier showed that oxygen absorbed by carbon reinforcement participated in oxide layer formation. This observation allows to explain the process of interface microstructure formation in the case of carbon fibre modification with a SiO_2 nano-layer, where SiO_2 transformation into complex oxide took place [13], which was not evident upon structural investigations of the AZ91-$(C_f)_{SiO2}$ system.

5. Summary

Some conclusions of the performed investigations can be drawn:

1. Proper infiltration of Ti coated carbon fibres by AZ91 magnesium alloy was obtained and fibres pulling-out effect was not observed what suggests that a chemical bonding was formed in the AZ91-$(C_f)_{TiN}$ composite.
2. The TiN nanolayer is however the effective protection in the AZ91-C_f system against a formation of hydrophilic phases in applied conditions of components consolidation but the nanolayer microstructure evolves as a result of diffusion processes. Therefore, the role of TiN can be characterized as a reactive diffusion barrier.
3. A region of several nanolayers was formed at the interface of AZ91-$(C_f)_{TiN}$ composite which consisted of continuous complex oxide and complex Ti containing layers, as well as the zone of nano-plates of $Al_{12}Mg_{17}$ phase surrounded by Mg matrix.
4. Presence of complex oxide nanolayer between fibre and Ti containing nanolayer suggests that oxygen absorbed by carbon reinforcement affects final interface microstructure in C-Mg system independently on applied carbon surface modification.

Acknowledgements

This work was supported by the Polish Ministry of Science and Higher Education (project no. 4689/B/T02/2007/33).

References

[1] Hufenbach W, Andrich M, Langkamp A and Czulak A 2006 *J. Mater. Proc. Tech.* **175** 218-24
[2] Lindroos V K and Talvitie M J 1995 *J. Mater. Proc. Tech.* **53** 273-84
[3] Ye H Z and Liu X Y 2004 *J. Mater. Sci.* **39** 6153-71
[4] Mordike B L and Ebert T 2001 *Mater. Sci. Engng. A* **302** 37-45
[5] Bouix J, Berthet M P, Bosselet F, Fawe R, Peronnet M, Viala J C, Vincent C and Vincent H 1997 *J. Phys. IV France* **7** 191-205
[6] Feldhoff A, Pippel E and Woltersdorf J 1997 *J. Microsc. Oxford* **185** 122-31
[7] Korner C, Schaff W, Ottmuller M and Singer R F 2000 *Adv. Engng. Mater.* **6** 327-37
[8] Oakley R, Cochrane R F and Stevens R 1995 *Key Engng. Mater.* **104-107** 387-416
[9] Kleine A, Hemptenmacher H, Dudek H J, Kainer K U and Krüger G 1995 *J. Mater. Sci. Lett.* **14** 358-60
[10] Pei Z L, Li K, Gong J, Shi N L, Elangovan E and Sun C 2009 *J. Mater. Sci.* **44** 1124-31
[11] Feldhoff A, Pippel E and Woltersdorf J 2000 *Adv. Engng. Mater.* **8** 471-80
[12] Popovska N, Gerhard H, Wum D, Poscher S, Emig G and Singer R F 1997 *Mater. Design* **18** 239-42
[13] Olszówka-Myalska A and Botor-Probierz A 2010 *IOP Conf. Series: Mat. Sci. Engng.* **7** 012022

[14] Faraji G and Asadi P 2011 *Mater. Sci. Engng. A* **528** 2431-40
[15] Braszczyńska-Malik K N 2008 *Kompozyty* **3** 242-6
[16] Song M, Wu G, Yang W, Jia W, Xiu Z and Chen G 2008 *J. Mater. Sci. Engng.* **26** 931-5

IOP Conf. Series: Materials Science and Engineering **32** (2012) 012020 doi:10.1088/1757-899X/32/1/012020

A geochemical investigation of trace elements in well RN-17 at Reykjanes geothermal system, SW-Iceland

L P Ottolini[1,6], **N Raffone**[2], **G Ó Fridleifsson**[3], **S Tonarini**[4], **M D'Orazio**[4,5] **and G Gianelli**[4]

[1] Consiglio Nazionale delle Ricerche (CNR), Istituto di Geoscienze e Georisorse (IGG), Unità di Pavia, via A. Ferrata 1, IT-27100 Pavia, Italy

[2] Università di Pavia, Dipartimento di Scienze della Terra e dell'Ambiente, via A. Ferrata, 1, IT-27100 Pavia, Italy

[3] HS Orka hf, Brekkustigur 36, IS-260 Reykjanesbaer, Iceland

[4] Consiglio Nazionale delle Ricerche (CNR), Istituto di Geoscienze e Georisorse (IGG), Unità di Pisa, via G. Moruzzi 1, IT-56124 Pisa, Italy

[5] Università di Pisa, Dipartimento di Scienze della Terra, via S. Maria 53, IT-56126 Pisa, Italy

E-mail: ottolini@crystal.unipv.it

Abstract. Here we present the results of an investigation of well RN-17 at Reykjanes geothermal system (Reykjanes Peninsula, SW-Iceland). We show that the adoption of micro-spot secondary ion mass spectrometry (SIMS) together with bulk-rock techniques, such as X-ray fluorescence (XRF), inductively coupled plasma mass spectrometry (ICP-MS), and thermal ionisation mass spectrometry (TIMS) carried out on 10 basaltic drill-cuttings over a depth of 3000 m gives us insights into the geochemical characteristics of the well. Whole rock and mineral phases, i.e., plagioclase, clinopyroxene, epidote and amphibole, indicate that the rocks pertaining to the cuttings were affected by hydrothermal alteration that has been able to modify significantly the Sr isotope ratios and the concentration of K, Rb, Cs, Ba and B. Seawater/rock ratio, evaluated for $^{87}Sr/^{86}Sr$, achieved its maximum for the shallow cutting RN-17-650, decreasing although not systematically with depth, with the minimum value for the deep cutting RN-17-2800. Moreover, the occurrence of Na-rich plagioclase, epidote and amphibole down to the well bottom are evidence of the persistence of fluid/rock interaction at depth.

1. Introduction

Hydrothermal processes in the oceanic crust produce supercritical fluids that form a major pathway connecting the earth's mantle with the oceans and the atmosphere. These fluids thus have a profound impact on the dynamics of mid-oceanic-ridges and global geochemical cycles [1]. The hydrothermal processes may be investigated by: i) direct sampling of fluids on and off the ridge–axis; ii) petrological studies of rocks from deep ocean drilling and ophiolite exposures [2, 3]. In this context, the position of Iceland as a sub-aerial sector of the Mid-Atlantic Ridge, with a wealth of geothermal activity, makes it a particularly favourable location for these studies since the repeated seismicity and volcanic activity within this active rift create high permeability and high temperatures at drillable depths.

[6] To whom any correspondence should be addressed.

Significant geothermal activity occurs along the Reykjanes Peninsula with specific sites controlled by the complex local tectonics and the recent and ongoing volcanic activity. The Reykjanes field is associated with the southernmost of five volcanic fissure swarms, which are located within the active tholeiite volcanic zone of the peninsula [4-7]. The stratigraphy of the Reykjanes field is divided into two formations, the uppermost 1000 m of the thermal area is characterized by hyaloclastite tuff, breccias, tuffaceous and marine sediments; the deeper part is dominated by submarine basaltic pillow lavas and intrusive rocks [8]. The rock permeability is the key physical parameter controlling the mechanisms of water/rock interaction and the evolution of submarine alteration. Permeability results from the combination of primary volcanic features (explosivity, brecciation, porosity), fracturing upon thermal contraction of the basaltic rocks due to cooling, and subsequent tectonic fracturing and faulting in the seismically active dilation zone of the plate boundary. Recent studies have shown that the intensity of alteration and secondary mineralisation is directly related to the texture of the protolith [9, 10]. In well RN-17 a new high-temperature amphibole zone was defined below 2400 m depth, transitional into amphibolite-grade alteration, and $^{87}Sr/^{86}Sr$ ratios within alteration minerals were observed to significantly shift towards seawater values with increasing depth, and, therefore, confirming deep penetration of seawater into the Reykjanes system [9, 10]. The peak hydrothermal alteration at shallow depth occurred during late Pleistocene, when an ice-sheet covered the peninsula. In late glacial time, hydrostatic pressures (P) and temperatures (T) were elevated at shallow depth resulting in higher temperature alteration than current P-T conditions allow [8, 11-13]. Fresh water fluid inclusions observed in quartz [8] were interpreted to imply that meteoric water had dominated the geothermal system in Pleistocene time, confirmed by isotopic constraints on ice age fluids by [14].

In the present work we have used bulk-rock (XRF, ICP-MS, and TIMS) analyses and SIMS investigations on 10 drill cuttings (depth from 400 to 3000 m) from the well RN-17. The possibility to measure at a small scale the concentration of trace elements with high precision and accuracy, minimum material consumption, limited sample preparation and virtually null contamination is a formidable tool offered by SIMS in the study of mineral phases of the well RN-17, i.e., plagioclase, clinopyroxene, epidote and amphibole. They were separated from the 10 drill cuttings because no core was available. For comparative purposes, we have included some additional samples: a surface lava (GIOV-2), lacking any sign of hydrothermal alteration, collected at the top of well RN-17, and a core-sample (2246 m dolerite dyke) from well RN-19, which is located 1.1 km NE of well RN-17 [13]. The dolerite was selected since it was the only core sample available from Reykjanes at the time this research was carried out. The results of the present investigation on the geochemical characteristics of the well RN-17 follow those obtained previously by SIMS on light (Li, Be and B) and volatile elements (F, Cl) in the most abundant mineral phases of the same drill cuttings of the well [15, 16]. As a result of this research, SIMS appears a fundamental complement to bulk-rock techniques in the study of complex geological areas.

2. Analytical methods for trace and isotope analysis

An overview of all analyzed rock samples is shown in table 1. All samples are of basaltic composition and olivine-free except the dolerite intrusion from RN-19. As seen in the table 1, half of the samples are of submarine extrusive origin (tuffs, pillow lavas or breccias and erosion products of the same origin); the other half of the samples are intrusive dykes, ranging in granularity from fine-grained to doleritic. Mineralogical assemblage consists of magmatic (plagioclase, clinopyroxene and opaque) and hydrothermal phases (amphibole, quartz, epidote, calcite and clay minerals). Plagioclase is the most abundant phenocryst, but additionally clinopyroxene is also found as phenocryst in the coarser grained samples below 1100 m depth. Epidote always occurs from 650 m down to 3000 m. The sample list and information summary about the analytical techniques used, the mineral phases analyzed for the different cuttings are shown in table 2.

As said before, since the rock thin sections were not available for this study – except for the dolerite dyke from well RN-19 –, fragments of cuttings were embedded in epoxy-resin mounts, and analyzed by EPMA for major/minor elements. Mineralogical observations made by [17] on drill

Table 1. Analyzed samples, rock types and rock origin for the wells RN-17 and RN-19 at Reykjanes geothermal field.

Sample		Rock type	Origin
RN-17	400	Marine sediment	Eroded hyaloclastite
RN-17	650	Basaltic tuff	Submarine tuff
RN-17	750	Basaltic pillows	Submarine pillow basalt
RN-17	1100	Basaltic pillows	Submarine pillow basalt
RN-17	1450	Basaltic intrusion – fine-medium grained	Dyke
RN-17	1550	Basaltic intrusion – coarse grained	Dyke
RN-17	1700	Basaltic intrusion – coarse grained	Dyke
RN-17	2150	Dolerite intrusion	Dyke
RN-17	2800	Basaltic intrusion – coarse grained	Dyke
RN-17	3000	Pillow basalt	Submarine pillow basalt
RN-19	2246 (core)	Dolerite intrusion	Dyke or sill

Table 2. Sample list and information summary for the cuttings from well RN-17 (Reykjanes geothermal system, Reykjanes Peninsula, SW-Iceland).

Sample	Rock type	Analytical methods	Analyzed minerals	$^{87}Sr/^{86}Sr$	$^{143}Nd/^{144}Nd$
Drill-cuttings					
RN-17-400	Basalt	XRF, EPMA, ICP-MS, TIMS, SIMS	Plg, Cpx	WR	WR
RN-17-650	Basalt	XRF, EPMA, ICP-MS, TIMS, SIMS	Plg, Cpx, Ep	WR	WR
RN-17-750	Basalt	XRF, EPMA, ICP-MS, TIMS, SIMS	Plg, Cpx, Ep	WR	WR
RN-17-1100	Basalt	XRF, EPMA, ICP-MS, TIMS, SIMS	Plg, Cpx, Ep	WR, Plg, Cpx, Ep	WR
RN-17-1450	Basalt	XRF, EPMA, ICP-MS, TIMS, SIMS	Plg, Cpx, Ep, Amph	WR	WR
RN-17-1550	Basalt	XRF, EPMA, ICP-MS, TIMS, SIMS	Plg, Cpx, Ep	WR	WR
RN-17-1700	Basalt	XRF, EPMA, ICP-MS, TIMS, SIMS	Plg, Cpx, Ep	WR	WR
RN-17-2150	Basalt	XRF, EPMA, ICP-MS, SIMS, TIMS	Plg, Cpx, Ep, Amph	WR, Plg, Cpx, Ep	WR
RN-17-2800	Basalt	XRF, EPMA, ICP-MS, TIMS, SIMS	Plg, Cpx, Ep, Amph	WR	WR
RN-17-3000	Basalt	XRF, EPMA, ICP-MS, TIMS, SIMS	Plg, Cpx, Ep, Amph	WR, Plg, Cpx	WR
Unaltered lava					
RN-17-GIOV-2	Basalt	XRF, EPMA, ICP-MS, TIMS, SIMS	Plg, Cpx	WR, Plg, Cpx	WR
Core-sample					
RN-19-2246	Dolerite	XRF, EPMA, ICP-MS, TIMS, SIMS	Plg, Cpx, Amph	WR, Plg, Cpx	WR

RN = Reykjanes; XRF = X-ray fluorescence; EPMA = electron probe microanalysis; ICP-MS = inductively coupled plasma mass spectrometry; TIMS = thermal ionisation mass spectrometry; SIMS = secondary ion mass spectrometry; WR = whole rock; Plg = plagioclase; Cpx = clinopyroxene; Ep = epidote; Amph = amphibole.

cuttings of wells RN-9, RN-10 and RN-17 showed that they typically range in size from <0.5 to 2 mm at most (and generally decrease in size with increasing drilling depth), thus precluding detailed evaluation of paragenetic sequences of hydrothermal minerals, characterisation of coexisting phase relations and analysis of the time sequence of mineral filled cross cutting fracture systems (vein).

Our samples were ran by Cameca SX-50 electron microprobe (CNR-IGG, Padova) and Jeol 8200 SuperProbe (Dipartimento di Scienze della Terra, Università di Milano). The crystals were analyzed with a beam current of 15 nA and acceleration voltage of 15 kV, using the following standards: albite (Na); orthoclase (K); wollastonite or diopside for plagioclase analysis (Si and Ca); $MnTiO_3$ (Mn and Ti); Al_2O_3 (Al); Fe_2O_3 (Fe^{3+}); Cr_2O_3 (Cr); and MgO (Mg) for the Cameca SX-50 microprobe; omphacite (Na); orthoclase (K); wollastonite (Si and Ca); rodonite (Mn); ilmenite (Ti); anorthite (Al); Fe_2O_3 (Fe^{3+}); pure chromium (Cr); and olivine (Mg and Fe^{2+}) for the Jeol 8200 SuperProbe.

Trace elements were investigated by SIMS in selected clinopyroxene, plagioclase, epidote and amphibole minerals, using a Cameca IMS 4f ion microprobe installed at CNR-IGG, Pavia. Due to a significant small-scale intergrowth (in some cases at a few-micrometres scale) of the mineral assemblage, we took special care to get data representative of the trace-element concentration in single mineral phases. Moreover, the different thickness in the sample from which the X-rays (EPMA) and the secondary ions (SIMS) are produced during electron and ion microprobe analysis, and the different beam size adopted for major and trace elements (typically, 1 - 2 μm Ø for the electron beam and 10 μm Ø for the ion beam) sometimes caused a mismatch between major- and trace-element compositions. When necessary, we carried out our SIMS analysis with a very small $^{16}O^-$ ion beam (down to ~ 5 μm Ø), monitoring the ion signals for Ti in clinopyroxene, Na and/or K in plagioclase, and comparing the SIMS data of these test elements with those obtained by EMPA for the same elements at the same microspot areas. For all the ion probe measurements, the primary current intensity ranged between 4 and 20 nA, with a beam spot of ~ 5 - 25 μm in diameter. The width of the energy slit was 50 eV and the voltage offset applied to the sample accelerating voltage (+ 4500 V) was -100 V. This corresponded to the analysis of secondary ions with emission kinetic energies in the range: ~ 75 - 125 eV. The filtered positive secondary ions were extracted and focussed under an ion image field of 25 μm. We used the largest contrast diaphragm and field aperture (400 and 1800 μm ∅, respectively). We analyzed rare earth elements (REE) with a procedure similar to that developed by [18] for silicates. La, Ce, Nd, Sm, Er and Yb were analyzed by measuring the signals from one isotope, i.e., 139La, 140Ce, 146Nd, 149Sm, 167Er and 174Yb. For Eu we used a deconvolution procedure to remove BaO interferences from Eu isotopes. Gd isotopes were monitored at mass 154, 160 and 162 in order to discriminate the Gd (and Dy) signal from CeO and NdO interferences. Dy was monitored also through the isotope 163. Moreover, in the case of Er, we improved the reliability of analysis by evaluating the $^{151}Eu^{16}O$ contribution at mass 167 from the Eu^+ signal and from the oxidation ratio (i.e., EuO^+/Eu^+) as determined on suitable standards under the adopted analytical conditions. Similar procedure was followed to correct Yb from GdO interference at mass 174. For the other trace elements, we detected signals from the following isotopes ^{45}Sc, ^{47}Ti, ^{51}V, ^{52}Cr, ^{85}Rb, ^{88}Sr, ^{89}Y, ^{90}Zr, ^{93}Nb, ^{133}Cs, ^{138}Ba, ^{178}Hf and ^{232}Th (similar analytical procedures were recently used by L. Ottolini in [19]). All of them were monitored in the same analytical run together with the REE ion signals. $^{45}Sc^+$ and $^{52}Cr^+$ were further corrected for $^{29}Si^{16}O^+$ and $^{24}Mg^{28}Si^+$ interferences at mass 45 and 52, respectively. $^{30}Si^+$ was selected as the isotope of the reference element (Si) for these matrices. International reference materials, i.e., NIST SRM 610 and R315 [20], in addition to CNR-IGG standards, i.e., kaersutite Soda Springs (KSS) and alkali-olivine-basalt BB (in-house standard of the Geochemical Institute, Göttingen University) were used in the calibration procedure. Accuracy for REE and other trace elements is estimated to be of the order of 10 % rel. at ppm (wt) level. Below 1 ppm, precision is constrained by counting statistics in the range 10 - 30 %. Accuracy is typically of the order of 15 % for Th; 20 - 25 % for Cs and Rb.

Whole-rock major and trace element analyses were performed at the Earth Sciences Department of Pisa University. Major elements were determined by X-ray fluorescence on an ARL 9400 XP+ spectrometer using $Li_2B_4O_7$ glass disks (sample/flux ratio = 1:7). Estimated precision (relative

IOP Conf. Series: Materials Science and Engineering **32** (2012) 012020 doi:10.1088/1757-899X/32/1/012020

standard deviation, RSD) is about 1 % for SiO_2 and about 2 % for the other major elements except those with low concentrations (~ < 0.50 wt%), for which the standard deviation is about ± 0.05 wt%. Loss on ignition (L.O.I.) was determined by gravimetry at 1000 °C after pre-heating at 110 °C. The XRF results are shown in table 3.

Table 3. Whole-rock major element concentrations (wt%) of the studied cuttings by XRF.

sample	**RN-17-GIOV-2**	**RN-17-400**	**RN-17-650**	**RN-17-750**	**RN-17-1100**	**RN-17-1450**	**RN-17-1550**	**RN-17-1700**	**RN-17-2150**	**RN-17-2800**	**RN-17-3000**	**RN-19-2246**
SiO_2	49.76	51.25	48.42	45.57	48.99	48.87	49.47	48.23	45.95	48.42	48.29	48.77
TiO_2	1.18	1.40	1.52	2.15	1.46	0.65	2.37	1.63	1.91	1.49	1.24	0.57
Al_2O_3	14.22	13.38	14.06	12.91	13.78	15.00	12.89	14.93	13.07	13.28	13.48	19.31
FeO_T	12.02	10.017	9.35	12.22	11.46	8.07	13.40	10.45	12.92	11.00	10.91	8.11
MnO	0.20	0.16	0.18	0.21	0.19	0.15	0.23	0.19	0.22	0.20	0.26	0.14
MgO	7.26	7.15	8.50	9.36	6.91	9.65	5.93	6.99	10.91	8.61	7.75	6.32
CaO	11.65	6.83	8.02	10.75	11.99	13.45	10.03	12.52	10.64	12.01	11.66	14.30
Na_2O	1.91	2.27	1.97	1.42	2.17	2.05	2.46	1.88	1.68	1.88	1.87	1.70
K_2O	0.11	1.46	1.20	0.09	0.04	0.04	0.30	0.04	0.09	0.10		
P_2O_5	0.11	0.15	0.16	0.26	0.15	0.06	0.27	0.16	0.13	0.17	0.13	0.05
Total	98.42	94.07	93.38	94.94	97.14	97.99	97.35	97.02	97.52	97.16	95.59	99.27
L.O.I.	-0.18	3.98	4.53	3.47	1.12	1.65	1.13	1.35	1.58	1.19	2.20	0.48

The concentrations of the bulk-rock trace elements were determined by ICP-MS (VG PQ II Plus). Sample powders were dissolved in PFA vessels on a hot plate using HNO_3 + HF. The sample solutions, spiked with Rh, Re and Bi as internal standards, were measured by external calibration using international reference materials of basaltic composition. Analytical precision, assessed by repeated analysis of the in-house standard HE-1 (Mt Etna hawaiite, n = 94), is between 2 and 5 % RSD, except for Gd, Tm, Be, Sc, Pb (6 - 8 % RSD). Detection limits, corresponding to 3σ of the blank concentrations, vary between 2 and 60 ppb for all elements, except Cr, Zr and Sc (400 - 1000 ppb). All the experimental data are listed in table 4.

Sr and Nd isotope ratios were determined by TIMS at IGG-CNR, Pisa, using a Finnigan MAT 262 multicollector mass spectrometer running in dynamic mode after conventional procedures for Sr and Nd separation from the matrix. Measured $^{87}Sr/^{86}Sr$ ratios were normalized to $^{86}Sr/^{88}Sr$ = 0.1194; $^{143}Nd/^{144}Nd$ ratios to $^{146}Nd/^{144}Nd$ = 0.7219. During the collection of the isotopic data for this study, the mean measured value of $^{87}Sr/^{86}Sr$ for the standard NIST-SRM 987 was 0.710242 ± 0.000013 (2σ, n = 25) and that of $^{143}Nd/^{144}Nd$ for the standard La Jolla was 0.511847 ± 0.000008 (2σ, n = 25). The JNdi-1 standard [21] was additionally analyzed giving a $^{143}Nd/^{144}Nd$ value of 0.512100 ± 0.000010 (2σ, n = 25). The Sr and Nd blank levels were about 0.3 ng during the period of chemistry processing. The analyses were carried out on about 100 mg of whole-rock powders washed ultrasonically with deionized "milli-Q" water and on 5 to 20 mg of mineral handpicked under a binocular microscope and washed with deionized "milli-Q" water. The pertaining results for bulk rock-analysis and mineral separates are summarized in table 5.

Table 4. Whole rock trace element concentrations (ppm) by ICP-MS.

sample	RN-17-GIOV-2	RN-17-400	RN-17-650	RN-17-750	RN-17-1100	RN-17-1450	RN-17-1550	RN-17-1700	RN-17-2150	RN-17-2800	RN-17-3000	RN-19-2246
La	3.8	6.4	6.8	9.7	6.2	2.42	12.4	7.1	7.0	7.2	4.3	1.06
Ce	9.9	15.6	16.5	23.7	15.5	6.1	30.0	17.6	17.0	17.7	11.0	3.00
Pr	1.52	2.26	2.42	3.5	2.23	0.90	4.2	2.59	2.44	2.56	1.65	0.52
Nd	7.6	10.8	11.6	16.0	10.6	4.5	19.8	12.3	11.7	11.9	8.0	2.78
Sm	2.45	3.04	3.12	4.4	3.03	1.45	5.0	3.4	3.33	3.30	2.48	1.05
Eu	0.86	1.10	1.23	1.55	1.07	0.61	1.60	1.26	1.20	1.10	0.84	0.47
Gd	3.07	3.44	3.5	4.7	3.5	1.81	5.1	3.6	3.7	3.4	3.09	1.47
Tb	0.59	0.64	0.62	0.80	0.64	0.35	0.90	0.61	0.64	0.60	0.57	0.30
Dy	4.0	4.2	3.8	5.1	4.1	2.48	5.7	4.0	4.0	3.9	3.8	1.99
Ho	0.90	0.91	0.82	1.09	0.91	0.62	1.21	0.85	0.84	0.88	0.83	0.47
Er	2.48	2.56	2.23	2.95	2.55	1.66	3.4	2.39	2.32	2.27	2.37	1.39
Tm	0.39	0.38	0.32	0.44	0.40	0.23	0.48	0.32	0.34	0.34	0.36	0.21
Yb	2.47	2.54	2.02	2.77	2.42	1.57	2.96	2.09	2.09	2.11	2.22	1.25
Lu	0.37	0.33	0.29	0.37	0.38	0.23	0.41	0.28	0.31	0.30	0.31	0.20
Sc	41	36.5	31	37	38	44	41	38	40	41	43	38
V	345	312	282	352	346	256	476	303	270	314	286	229
Cr	138	190	275	363	144	467	92	236	392	336	286	108
Rb	1.92	14.9	8.1	1.41	0.43	0.67	4.7	0.70	1.91	2.28	0.29	0.64
Sr	100	110	152	144	152	104	161	167	170	147	129	96
Y	25.3	25.0	23.2	29.6	26.8	16.4	33.0	23.3	23.9	23.8	23.2	13.3
Zr	54	72	76	108	75	23.3	103	51	59	62	51	19.9
Nb	4.8	8.8	8.6	14.9	8.1	2.54	18.4	10.3	8.8	10.7	6.4	1.02
Cs	0.04	0.04	0.06	0.15	0.01	0.02	0.03	0.03	0.07	0.03	0.01	0.02
Ba	28.3	920	347	63	34	26.5	64	33.1	44	52	20.2	10.9
Hf	1.57	1.99	2.05	2.80	2.02	0.75	2.89	1.61	1.74	1.87	1.55	0.61
Ta	0.32	0.60	0.56	1.00	0.59	0.18	1.24	0.71	0.57	0.71	0.42	0.07
Pb	0.44	3.4	2.23	1.85	1.47	0.89	0.45	0.81	1.52	1.65	1.31	0.68
Th	0.27	0.46	0.34	0.38	0.45	0.14	0.73	0.31	0.36	0.40	0.30	0.10
U	0.08	0.24	0.21	0.35	0.25	0.06	0.27	0.11	0.17	0.16	0.09	0.03

3. Bulk-rock geochemistry

The cuttings exhibit a large range in REE, Cs, Rb, Th, U, Ba, Nb and Ta contents, as well as in the light-to-heavy rare earth elements (LREE/HREE) fractionation. The C1-chondrite normalized REE concentrations are plotted in figure 1, from which we can see that all samples fall in the range of Icelandic tholeiites [22-31]. The recent work by [9] on RN-17 has revealed the cuttings to consist of hydrothermally altered basalt with a composition that is transitional between Mid-Ocean Ridge Basalt (MORB) and Ocean Island Basalt (OIB); the basalts are primarily tholeiites and are in good agreement with the range of published Icelandic and Reykjanes Ridge data.

As for the other trace elements analyzed in the present study (whose C1-chondrite plot is not reported here), the concentration of some of them increase from depth to surface: 0.13 - 6.5 for Rb_N, 0.05 - 0.8 for Cs_N, 8.6 - 27.4 for Ba_N from the deeper to intermediate cuttings, and up to 150 and 390 (Ba_N) in the shallower ones (RN-17-650 and RN-17-400). Also, the U/Th fractionations show significant increase, i.e., 0.30 - 0.62 from the bottom to the surface.

Differently from the cuttings, the dolerite dyke (RN-19-2246) is LREE-depleted (La_N/Yb_N = 0.59) and is characterized by the lowest REE (figure 1) and other trace element concentrations.

Table 5. Sr and Nd isotope data by TIMS.

Sample	$^{87}Sr/^{86}Sr$		$2\sigma_{mean}$	$^{143}Nd/^{144}Nd$		$2\sigma_{mean}$
RN-17-400 whole rock	0.704333		0.000008	0.513028		0.000009
RN-17-650 whole rock	0.70465	±	0.000008	0.513037	±	0.000009
RN-17-750 whole rock	0.704075		0.000011	0.513006		0.000008
RN-17-1100 whole rock	0.703830	±	0.000010	0.513059	±	0.000008
RN-17-1100 plg	0.704293	±	0.000011			
RN-17-1100 epidote	0.704778	±	0.000008			
RN-17-1100 cpx	0.703495	±	0.000007			
RN-17-1450 whole rock	0.703997		0.000008	0.513073		0.000007
RN-17-1550 whole rock	0.703267		0.000008	0.513017		0.000007
RN-17-1700 whole rock	0.703507		0.000008	0.513013		0.00001
RN-17-2150 whole rock	0.703567	±	0.000010	0.513026	±	0.000010
RN-17-2150 plg	0.703209	±	0.000009			
RN-17-2150 epidote	0.703728	±	0.000014			
RN-17-2150 cpx	0.703371	±	0.000007			
RN-17-2800 whole rock	0.703253		0.00001	0.513008		0.000009
RN-17-3000 whole rock	0.703590	±	0.000010	0.513051	±	0.000009
RN-17-3000 plg	0.703465	±	0.000010			
RN-17-3000 cpx	0.703251	±	0.000010			
RN-19-2246 whole rock	0.703101	±	0.000010	0.513084	±	0.000008
RN-19-2246 plg	0.703019	±	0.000011			
RN-19-2246 cpx	0.703161	±	0.000009			
RN-17-GIOV-2 whole rock	0.703198	±	0.000009	0.513073	±	0.000008
RN-17-GIOV-2 plg	0.703216	±	0.000010			
RN-17-GIOV-2 cpx	0.703398	±	0.000010			

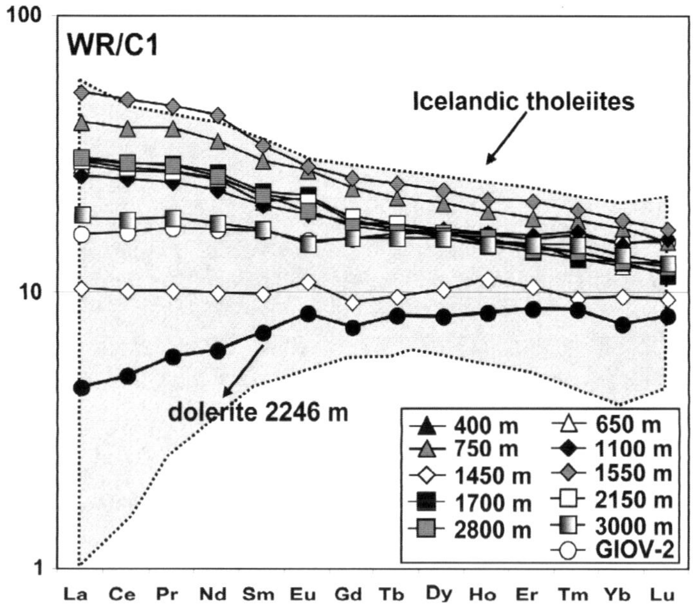

Figure 1. C1-chondrite-normalized REE patterns for selected whole-rock (WR) cuttings and core-sample from well RN-17 and RN-19 (Reykjanes geothermal system), respectively. Normalisation values from [59]. All the REE data for our samples fall in the range of Icelandic tholeiites (grey field, from literature).

The whole rock concentration of the highly-mobile elements Li and B investigated by [15] show a wide range of B concentrations (2.44 - 12.38 ppm), whereas the variability of Li is narrower (1.85 - 5.66 ppm): in particular, the highest B values occur in the cuttings from RN-17-2150 while the highest Li concentration was measured in the unaltered surface lava RN-17-GIOV-2 and in the K-rich sample RN-17-650 (figure 2a). Li contents are generally lower than those of young basalts from Mid Ocean Ridges (3.6 - 29.9 ppm) and Ocean Islands (2.8 - 15.7 ppm) ([32] and references therein). On the contrary, B abundances are within the literature values for young basalts ([33] and references therein). In order to avoid compositional differences related to the evolution degree of the samples, we have compared B/Ce ratios versus Ce (figure 2b), and Li/Yb ratios versus Yb (figure 2c) of our samples with those from the literature. Figure 2b shows that the cuttings are enriched in B (B/Ce = 0.39 - 0.84) with respect to young basalts from MOR (B/Ce = 0.05 - 0.15: [34-36]) and OIB (B/Ce = 0.01 - 0.16: [33, 36]) whereas Li/Yb ratios (1 - 2.7) are similar or slightly lower than MORBs (Li/Yb = 1.6 - 3: [32, 34-35, 37]) and OIBs (Li/Yb = 2.3 - 4.5: [32, 38]).

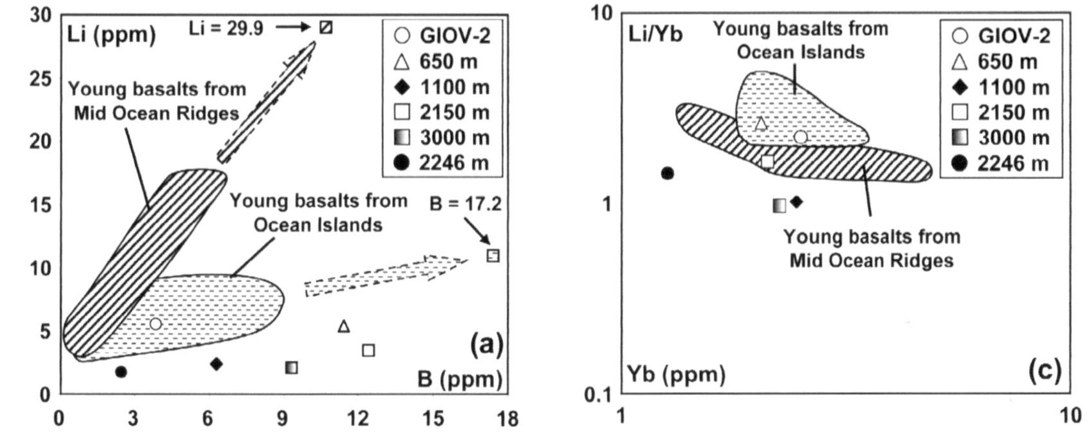

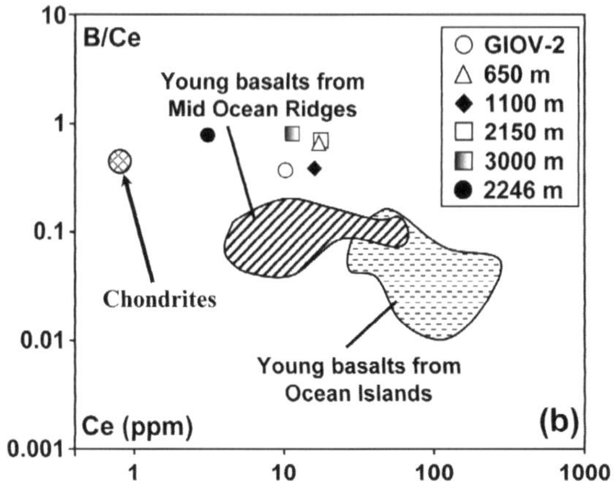

Figure 2. Plots of whole-rock data for: (a) Li versus B (ppm); (b) B/Ce versus Ce (ppm); and (c) Li/Yb versus Yb (ppm) of selected cuttings from Reykjanes geothermal field (data from [15]). The partial fields for Mid Ocean Ridges and Ocean Islands young basalts [32, 33] are shown for comparison. The extrapolated trend for Li (29.9 ppm) and B (17.2 ppm) in Mid Ocean Ridge and Ocean Island basalts (from [33]) is also reported for comparison.

Measured $^{87}Sr/^{86}Sr$ ratios of bulk rocks vary between 0.70325 and 0.70465 in the cuttings, whereas lower values characterize the unaltered lava and the RN-19-2246 dolerite dyke ($^{87}Sr/^{86}Sr$ = 0.70320 and 0.70310, respectively) (see table 5). The Sr isotope ratio is generally higher for the more K-enriched and shallower cuttings (e.g., RN-17-400, RN-17-650), and progressively decreases with depth (figure 3). In contrast, the $^{143}Nd/^{144}Nd$ values vary within a narrower range (0.51301 - 0.51308) without any correlation with depth (see table 5). Thus, the $^{143}Nd/^{144}Nd$ variations of the studied cuttings are probably due to magmatic processes since hydrothermal alteration is expected to have no effect on $^{143}Nd/^{144}Nd$ [39]. The first process investigated is the increasing of ^{143}Nd due to radioactive decay of ^{147}Sm. The $^{143}Nd/^{144}Nd$ versus Sm/Nd plot of figure 4 shows that our samples fall close to a linear array, formed by Icelandic rocks, with a slope corresponding to an age of about 170 Ma [28]. Since the crust beneath the Reykjanes Peninsula has near-zero age, it has not had sufficient time to develop a distinctive Nd isotope ratio [28, 40]. On the other hand, this linear array suggests a mixing process between melts derived from a depleted MORB-like source with melts derived from a more enriched mantle source. In this frame, the negative correlations between $^{143}Nd/^{144}Nd$ ratios and some of the compositional parameters which are relatively insensitive to low-pressure crystal fractionation, such as La/Yb (figure 5a) and Nb/Zr (figure 5b) [41] also require at least two different mantle sources: one depleted (D) and one enriched (E). The D component, which includes cuttings RN-17-1450, the unaltered lava RN-17-GIOV-2 and the dolerite dyke RN-19-2246, has $^{143}Nd/^{144}Nd$ (0.51307 - 0.51308), LREE/HREE and Nb/Zr ratios (La/Yb = 0.85 - 1.5; Nb/Zr = 0.05 - 0.11) comparable to those of the source of depleted Reykjanes Peninsula lavas [39]; the E component, which includes cuttings RN-17-400, -650, -750, -1100, -1700, -2150, -2800 and -3000, has $^{143}Nd/^{144}Nd$ (0.51301 - 0.51307), LREE/HREE and Nb/Zr ratios (La/Yb = 1.9 - 3.5; Nb/Zr = 0.11 - 0.20) that approach those of the source of enriched Reykjanes lavas [39]. The high La/Yb ratio (4.2) of cuttings RN-17-1550 may require the existence of a third source, similar to that of Iceland mildly alkaline lavas [42]. However, the Nb/Zr range of mildly alkaline lavas (figure 5b) does not seem to confirm such a hypothesis. As an alternative, the geochemical features of sample RN-17-1550 could be ascribed to a slightly lower degree of partial melting of the enriched source.

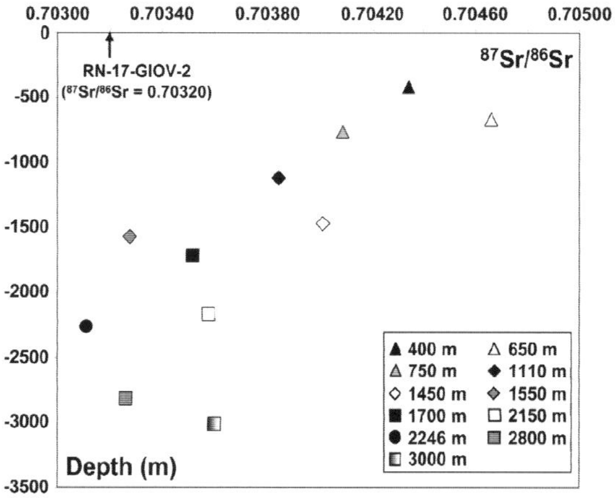

Figure 3. Whole-rock $^{87}Sr/^{86}Sr$ versus depth (m) for the studied cuttings.

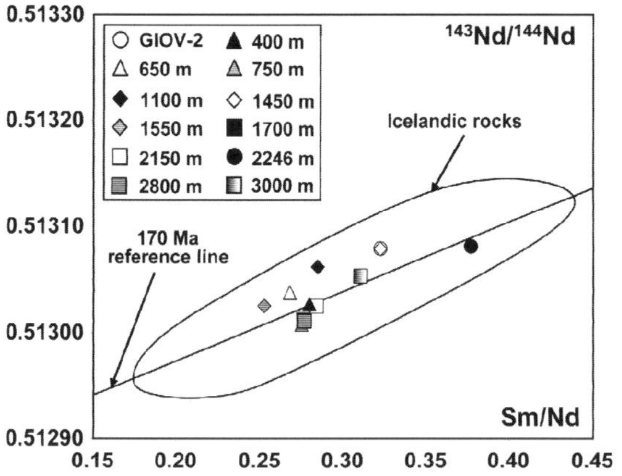

Figure 4. Plots of $^{143}Nd/^{144}Nd$ versus Sm/Nd for the studied cuttings. The field of Icelandic rocks, along with a linear array with a slope corresponding to an age of about 170 Ma, is reported for comparison [28].

4. Mineral geochemistry

EPMA and SIMS data for all analyzed plagioclase are listed in tables 6 and 7, respectively. The occurrence of both igneous and secondary plagioclase is confirmed by the diagrams of K_2O (wt%) and FeO_T (wt%) versus the anorthite (An) index (see for instance that for FeO_T in figure 6). Feldspar with An < 30 % (albite or oligoclase) occurs as replacing magmatic plagioclase [13]. Plagioclase with An > 30 % from intermediate to deep cuttings (e.g., RN-17-1550, -1700, -2150 and -3000) and the

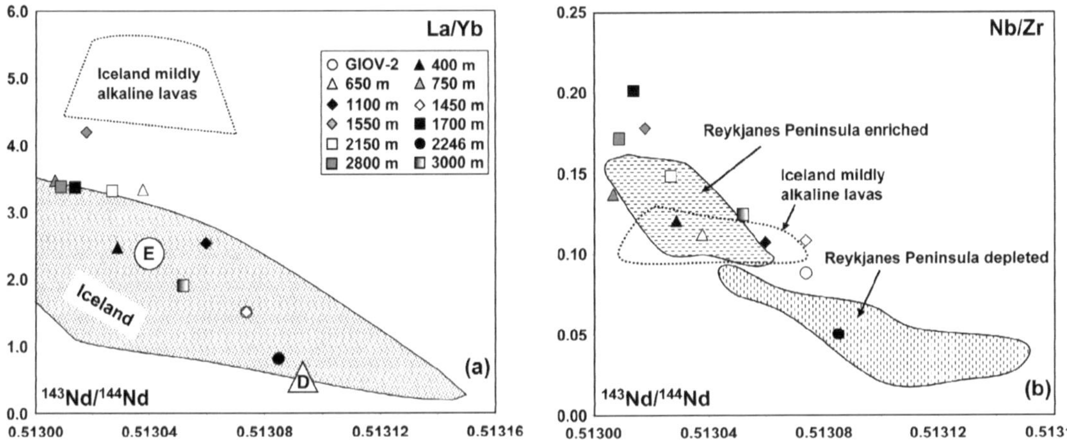

Figure 5. Plots of $^{143}Nd/^{144}Nd$ versus (a) La/Yb and (b) Nb/Zr for the studied samples. The negative correlations require at least two different sources in the mantle beneath Reykjanes geothermal field: enriched (E) and depleted (D) sources. The fields for Iceland rocks [41], Iceland mildly alkaline lavas [42], Reykjanes Peninsula depleted and enriched lavas [39] are shown for comparison.

unaltered lava (RN-17-GIOV-2) are characterized by higher LREE and lower HREE concentrations (LaN/YbN = 21 - 33) than those from the shallower ones (e.g., RN-17-400, -750, -1100 and -1450: LaN/YbN = 5 - 12). The chondrite-normalized REE concentrations, obtained by SIMS, are shown in figure 7. All plg crystals have marked positive Eu anomaly (Eu_N/Eu^*_N = 3 - 17, where $Eu^*_N = (Sm_N \times Gd_N)^{1/2}$) (figure 7) as well as positive Sr anomalies (Sr_N/Sr^*_N = 14 - 44, where $Sr^*_N = (Ce_N \times Nd_N)^{1/2}$) together with negative Zr anomaly (Zr_N/Zr^*_N = 0.02 - 0.08), where $Zr^*_N = (Nd_N \times Sm_N)^{1/2}$. In contrast, secondary feldspar (An < 30 %) from RN-17-650 cutting shows very low LREE (La_N/Yb_N = 0.92) contents, and marked negative Ce (Ce_N/Ce^*_N = 0.2) (figure 7) and Ti (Ti_N/Ti^*_N = 0.05, where $Ti^*_N = (Gd_N \times Dy_N)^{1/2}$) anomalies, thus suggesting a hydrothermal growth on previous Ca-plagioclase with loss of LREE during the albitisation process, as reported in [43]. The negative Ce anomaly could be related to preferential loss of Ce^{4+} during hydrothermal alteration of primary plagioclase. Distinct from cuttings, plagioclase from the dolerite sample (RN-19-2246 m) shows very low REE contents and the most marked positive Eu (Eu_N/Eu^*_N ~ 40) (figure 7) and Sr (Sr_N/Sr^*_N ~ 47) anomalies, thus suggesting a cumulus origin.

The results of EPMA and SIMS investigations of cpx are shown in tables 8 and 9, respectively. Cpx shows a great variability in terms of major constituents, such as FeO_T (range 5.98 - 20.27 wt%), MgO (11.86 - 18.05 wt%) and CaO (15.04 - 20.67 wt%), as well as in almost all trace elements analyzed (see table 9). In particular, the REE concentrations (figure 8) span from 2.3 to 10 (La_N), and from 8.4 to 34 (Yb_N), with the highest REE contents occurring in cuttings RN-17-750 and RN-17-1700. Common geochemical features are the following: i) LREE-depletion (La_N/Yb_N = 0.2 - 0.6) (figure 8); ii) negative Zr (Zr_N/Zr^*_N = 0.2 - 0.6), Sr (Sr_N/Sr^*_N = 0.1 - 0.4) and Eu (Eu_N/Eu^*_N = 0.7 - 1) anomalies. The clinopyroxene from the dolerite (RN-19-2246 m) shows very marked LREE/HREE fractionation (La_N/Yb_N ~ 0.1), no Eu anomaly (figure 8) and depletions in almost all other trace elements.

Epidote is an important index mineral in hydrothermal systems and its occurrence provides unequivocal evidence for temperature in excess of 240 °C at some stage during hydrothermal activity [44, 45, 46 and references therein). Epidote commonly forms in active geothermal systems as the product of reactions between hydrothermal solutions and Ca-bearing phases, as plagioclase, in the host rocks, and typically occurs as xenoblastic crystals and radiating crystal aggregates, and filling vesicles and veins as a secondary alteration mineral. In the Reykjanes geothermal field epidote has been found

Table 6. Electron probe data (wt%) for plagioclase.

sample	RN-17-GIOV-2									RN-17-400		RN-17-650		RN-17-750					
grain	plg 1A	plg 1B	plg 2B	plg 3A	plg 1D	plg 2B	plg 3E	plg 3B	plg 4C	plg 1l	plg 1N	plg 1S	plg 1D	plg 1B	plg 1C	plg 2D	plg 2B	plg 2'D	plg 3C
SiO_2	47.57	49.33	50.20	55.28	48.66	45.68	56.49	50.11	50.94	50.94	51.25	67.02	65.98	49.64	50.95	49.88	48.54	49.81	54.70
TiO_2	0.09	0.02	0.10	0.16	0.04	0.02	3.42	0.01	0.04	0.06	0.05	0.00	0.00	0.03	0.14	0.05	0.08	0.10	0.10
Al_2O_3	32.97	31.11	31.10	27.35	32.70	33.74	22.30	31.00	30.74	30.28	30.22	20.00	21.09	30.56	29.50	30.04	31.19	30.37	26.73
Cr_2O_3	0.01	0.02	0.02	0.03	0.00	0.08	0.01	0.02	0.04	0.00	0.00	0.00	0.03	0.00	0.04	0.00	0.07	0.01	0.02
FeO_T	0.55	0.48	0.73	1.12	0.63	0.35	3.73	0.75	0.90	0.90	0.87	0.03	0.11	0.56	0.75	0.79	0.65	0.85	0.95
MnO	0.00	0.00	0.00	0.02	0.07	0.00	0.03	0.00	0.02	0.00	0.02	0.01	0.00	0.00	0.04	0.00	0.02	0.05	0.07
MgO	0.07	0.20	0.21	0.13	0.24	0.24	0.18	0.28	0.33	0.28	0.28	0.00	0.01	0.19	0.20	0.22	0.22	0.40	0.15
CaO	17.18	16.32	15.78	10.96	15.40	18.75	9.48	15.57	15.14	14.38	14.16	1.60	2.43	15.42	14.53	14.80	15.64	15.04	11.17
Na_2O	1.76	2.59	2.58	5.25	2.43	1.08	4.49	2.65	3.11	3.22	3.26	10.44	9.94	2.86	3.14	3.19	2.61	2.75	5.06
K_2O	0.00	0.03	0.03	0.11	0.04	0.02	0.10	0.02	0.05	0.03	0.07	0.12	0.07	0.02	0.06	0.06	0.03	0.07	0.16
Total	100.20	100.11	100.76	100.41	100.22	99.96	100.23	100.41	101.30	100.10	100.16	99.22	99.67	99.29	99.37	99.04	99.05	99.45	99.12
An	84	78	77	53	78	90	53	76	73	71	70	8	12	75	72	72	77	75	54

sample		RN-17-1100								RN-17-1450							RN-17-1550	
grain	plg 2C	plg 7A	plg 7C	plg 8C	plg 5C	plg-1b	plg-x	plg-1ay	plg-1by	plg 3B	plg 4D	plg 4E	plg 6C	plg 6G	plg-1a-qz	plg-1b-op	plg 1B	plg 1C
SiO_2	55.24	54.01	55.78	49.99	53.51	51.56	51.16	51.08	51.43	54.40	56.18	52.49	52.78	55.99	48.30	51.36	53.19	53.04
TiO_2	0.03	0.03	0.07	0.00	0.14	0.07	0.03	0.07	0.01	0.04	0.02	0.08	0.01	0.22	0.00	0.02	0.09	0.12
Al_2O_3	28.10	29.54	28.40	31.92	29.03	30.67	30.84	30.42	30.79	28.12	27.24	30.12	29.09	25.48	32.68	30.03	28.62	29.04
Cr_2O_3	0.05	0.01	0.03	0.03	0.00	0.05	0.00	0.00	0.02	0.00	0.05	0.01	0.00	0.00	0.00	0.00	0.03	0.00
FeO_T	1.04	0.94	1.11	0.81	0.85	0.71	0.69	0.66	0.78	1.04	1.25	1.42	0.74	3.11	0.44	0.59	0.78	0.92
MnO	0.00	0.00	0.00	0.03	0.03	0.00	0.00	0.01	0.01	0.02	0.00	0.03	0.02	0.01	0.00	0.05	0.00	0.00
MgO	0.13	0.16	0.22	0.14	0.27	0.24	0.22	0.22	0.23	0.09	0.11	0.07	0.21	0.47	0.13	0.24	0.11	0.18
CaO	11.92	13.48	11.76	16.10	12.90	14.26	14.66	14.36	14.49	11.71	10.62	13.74	13.18	9.36	17.18	15.15	12.86	12.56
Na_2O	4.49	3.74	5.02	2.21	3.98	3.54	3.25	3.35	3.42	5.22	5.90	3.87	4.40	5.56	1.73	2.99	3.74	4.13
K_2O	0.11	0.08	0.08	0.04	0.09	0.06	0.05	0.06	0.06	0.09	0.11	0.03	0.08	0.14	0.04	0.04	0.08	0.10
Total	101.11	101.98	102.47	101.27	100.81	101.16	100.90	100.25	101.25	100.71	101.48	101.87	100.51	100.34	100.51	100.48	99.49	100.09
An	59	66	56	80	64	69	71	70	70	55	50	66	62	48	84	74	65	62

An = anorthite index of plg = 100 x [Ca/(Ca+Na+K)].

Table 6. continued.

sample						RN-17-1700								RN-17-2150				
grain	plg 1G	plg 1H	plg-2b-op	plg-2c	plg 2G	plg 2F	plg 3C	plg 3H	plg 4I	plg 4B	plg 5D	plg-2a-cc	plg-2b-cc	plg 6A	plg 6C	plg 6F	plg 4B	plg 3A
SiO_2	61.39	51.82	50.17	50.70	52.86	48.03	55.01	52.75	50.99	51.25	51.46	48.40	47.65	50.30	65.63	53.92	78.27	50.66
TiO_2	0.18	0.10	0.04	0.09	0.08	0.01	0.04	0.05	0.04	0.07	0.02	0.04	0.04	0.04	0.00	0.13	0.04	0.00
Al_2O_3	23.90	29.22	30.53	30.37	29.06	32.27	27.40	28.68	29.79	30.03	29.81	32.12	32.89	31.13	22.20	28.74	14.54	30.23
Cr_2O_3	0.00	0.00	0.00	0.01	0.00	0.05	0.00	0.00	0.00	0.00	0.00	0.00	0.00	0.00	0.02	0.00	0.01	0.00
FeO_T	1.19	0.98	0.78	0.91	0.83	0.59	0.77	0.94	1.15	0.88	1.02	0.54	0.52	0.90	0.34	0.89	0.33	1.19
MnO	0.00	0.07	0.02	0.02	0.06	0.02	0.00	0.02	0.00	0.03	0.02	0.02	0.01	0.00	0.01	0.02	0.02	0.03
MgO	0.06	0.16	0.13	0.21	0.13	0.20	0.11	0.18	0.13	0.16	0.14	0.19	0.17	0.11	0.01	0.35	0.01	0.11
CaO	6.11	13.71	14.68	14.66	12.97	16.54	11.17	13.05	14.28	14.06	13.45	17.11	17.06	15.20	3.33	12.97	1.57	15.05
Na_2O	7.22	3.70	3.08	3.06	3.96	1.85	4.93	3.92	3.46	3.19	3.63	1.71	1.77	3.22	10.29	4.33	6.89	3.22
K_2O	0.24	0.07	0.07	0.06	0.07	0.00	0.09	0.09	0.06	0.07	0.09	0.04	0.04	0.05	0.04	0.24	0.11	0.12
Total	100.28	99.83	99.50	100.08	100.02	99.56	99.51	99.68	99.91	99.73	99.65	100.16	100.14	100.95	101.87	101.58	101.79	100.62
An	31	67	72	72	64	83	55	64	69	71	67	85	84	72	15	62	11	72

sample					RN-17-2800											RN-17-3000				
grain	plg 3C	plg 5C	plg-1b	plg 2C	plg 2F	plg 2G	plg 2H	plg 2L	plg 3A	plg 3A	plg 3C	plg 3E	plg 3L	plg 3M	plg 3M	plg 1C	plg 1D	plg 1F	plg 2C	plg 2G	plg 2I
SiO_2	50.70	49.09	53.29	60.98	55.07	64.22	53.82	51.65	50.27	45.95	47.75	50.17	60.68	51.42	53.94	51.80	58.93	49.16	56.15	53.68	58.16
TiO_2	0.02	0.08	0.12	0.01	0.09	0.00	0.09	0.07	0.11	2.33	0.05	0.08	0.01	0.13	0.07	0.02	0.02	0.05	0.04	0.12	0.06
Al_2O_3	30.86	32.11	29.64	24.09	27.83	22.02	28.71	29.36	30.97	9.17	32.37	30.82	24.35	29.28	28.86	29.62	26.05	31.83	27.38	29.02	26.56
Cr_2O_3	0.00	0.01	0.01	0.00	0.00	0.00	0.12	0.00	0.06	0.06	0.02	0.00	0.00	0.00	0.01	0.00	0.00	0.00	0.01	0.00	0.01
FeO_T	1.29	1.18	0.95	0.83	1.02	0.96	1.20	0.86	1.00	9.85	0.85	0.82	0.45	1.18	0.92	0.82	0.83	0.66	0.75	1.04	0.83
MnO	0.04	0.09	0.02	0.01	0.00	0.03	0.03	0.07	0.11	0.17	0.00	0.00	0.02	0.00	0.03	0.01	0.03	0.00	0.03	0.00	0.00
MgO	0.08	0.11	0.12	0.03	0.06	0.24	0.12	0.22	0.12	17.17	0.06	0.03	0.05	0.13	0.08	0.28	0.14	0.40	0.05	0.06	0.06
CaO	15.20	16.13	12.79	6.68	11.82	3.82	12.47	14.12	15.13	11.22	16.97	14.95	6.83	13.39	11.85	13.99	8.64	16.49	10.26	12.36	8.90
Na_2O	2.93	2.67	4.31	7.65	4.82	8.73	4.50	3.56	3.06	3.46	2.17	3.32	7.79	3.98	5.15	3.52	6.57	2.16	5.93	4.45	6.50
K_2O	0.13	0.04	0.10	0.32	0.17	0.26	0.12	0.09	0.06	0.09	0.02	0.05	0.38	0.07	0.12	0.03	0.02	0.00	0.18	0.08	0.16
Total	101.24	101.51	101.34	100.59	100.89	100.28	101.18	100.01	100.89	99.48	100.25	100.24	100.57	99.57	101.03	100.09	101.23	100.75	100.79	100.81	101.25
An	74	77	62	32	57	19	60	68	73	64	81	71	32	65	55.6	69	42	81	48	60	43

Table 6. continued.

sample																	RN-19-2246		
grain	plg 3B	plg 3F	plg 3G	plg 3I	plg 3M	plg 3N	plg 4N	plg 5C	plg 5H	plg 5I	plg 5M	plg 5N	plg 5P	plg 5Q	plg 5R	plg-2c-ar	plg-3a-ar	plg 1P	plg 4A	plg 3A
SiO_2	50.50	63.14	61.03	50.66	64.77	54.02	68.25	47.40	50.43	48.06	47.81	47.52	48.34	48.03	51.01	53.58	49.14	55.54	48.76	52.64
TiO_2	0.00	0.00	0.08	0.10	0.00	0.10	0.00	0.05	0.06	0.05	0.07	0.04	0.05	0.04	0.07	0.08	0.05	0.02	0.03	0.06
Al_2O_3	29.71	22.32	23.40	29.59	22.15	27.05	19.75	32.56	30.28	32.19	32.31	33.16	32.31	31.95	30.22	27.89	31.15	27.64	31.32	29.41
Cr_2O_3	0.04	0.00	0.05	0.00	0.00	0.00	0.00	0.02	0.01	0.08	0.02	0.01	0.02	0.00	0.00	0.01	0.00	0.03	0.00	0.00
FeO_T	1.20	0.37	0.91	1.05	0.32	1.13	0.38	1.00	1.18	0.89	0.89	0.86	0.92	0.82	0.82	0.85	0.74	0.89	0.86	0.85
MnO	0.00	0.02	0.02	0.01	0.00	0.00	0.10	0.02	0.05	0.00	0.00	0.02	0.03	0.03	0.00	0.02	0.03	0.00	0.00	0.00
MgO	0.37	0.06	0.00	0.24	0.02	0.21	0.00	0.06	0.05	0.06	0.01	0.05	0.09	0.05	0.10	0.13	0.37	0.06	0.06	0.10
CaO	14.78	4.36	6.31	14.44	3.65	11.64	0.53	16.97	14.78	16.78	17.11	17.20	16.52	16.47	13.98	11.73	15.73	11.48	16.14	13.86
Na_2O	3.23	9.11	8.18	3.31	9.55	5.04	11.60	1.93	3.13	1.93	1.94	1.63	2.18	2.25	3.50	4.75	2.45	5.26	2.37	3.80
K_2O	0.03	0.05	0.26	0.01	0.09	0.13	0.06	0.06	0.09	0.06	0.01	0.05	0.01	0.05	0.03	0.12	0.01	0.06	0.01	0.04
Total	99.86	99.42	100.26	99.41	100.56	99.32	100.67	100.09	100.06	100.10	100.18	100.53	100.48	99.69	99.73	99.15	99.67	100.99	99.56	100.76
An	71.5	21	29	71	17	56	2	83	72	82	83	85	81	80	69	57	78	54	79	67

Table 7. Trace element concentrations (ppm) for plagioclase by SIMS.

Sample	RN-17-GIOV-2	RN-17-400				RN-17-650	RN-17-750				RN17-1100					
Grain	plg 1A	plg 1n	plg 1L	average (2)	1σ	plg 1D2	plg 2B	plg 3C	average (2)	1σ	plg 3A	plg 7C	plg 5D	plg 1b	plg x	plg 1ay
Sc	*2.1*	9.0	7.4	*8.2*	*1.1*	*3.2*	7.3	6.7	*6.7*	*0.42*	4.3	4.0	3.7	1.6	2.6	2.0
Ti	*215*	239	266	*253*	*19*	*4.3*	515	539	*446*	*17*	427	536	191	225	232	239
V	*4.6*	5.2	5.4	*5.3*	*0.14*	*4.9*	9.0	9.0	*8.0*		13	11	5.8	4.6	4.7	4.4
Cr	*0.60*	0.10	0.50	*0.30*	*0.28*	*0.40*		1.0	*0.67*					b.d.l.	0.10	0.80
Rb	*b.d.l.*	0.20	1.0	*0.60*	*0.57*	*0.40*	0.30	0.30	*0.20*					0.30	0.30	0.20
Sr	*178*	142	136	*139*	*4.2*	*372*	265	315	*270*	*35*	227	273	223	194	191	188
Y	*0.20*	0.48	0.39	*0.435*	*0.06*	*0.39*	0.62	0.67	*0.63*	*0.04*	0.30	0.40	0.20	0.10	0.20	0.20
Zr	*0.10*	0.11	0.08	*0.10*	*0.02*	*0.28*	0.70	0.20	*0.33*	*0.35*	0.30	0.40		0.10		
Nb							0.10		*0.10*					0.01	0.02	
Cs	*0.01*		0.05	*0.02*		*0.03*	0.03	0.04	*0.02*	*0.01*						0.01
Ba	*6.0*	12	10	*11*	*1.3*	*31*	13	27	*17*	*10*	15	38	8.3	8.9	9.3	10
La	*0.30*	0.33	0.31	*0.32*	*0.01*	*0.04*	0.39	0.47	*0.38*	*0.06*	0.22	0.55	0.25	0.30	0.30	0.26
Ce	*0.50*	0.53	0.50	*0.52*	*0.02*	*0.02*	0.69	1.06	*0.77*	*0.26*	0.70	0.74	0.50	0.53	0.48	0.42
Nd	*0.30*	0.23	0.13	*0.18*	*0.07*	*0.07*	0.46	0.51	*0.44*	*0.04*	0.39	0.39	0.31	0.20	0.22	0.17
Sm	*0.1*	0.08	0.10	*0.09*	*0.01*	*0.04*	0.19	0.20	*0.19*	*0.01*	0.11	0.12	0.08	0.07	0.04	0.04
Eu	*0.1*	0.19	0.15	*0.17*	*0.03*	*0.06*	0.22	0.47	*0.30*	*0.18*	0.42	0.53	0.50	0.20	0.2	0.23
Gd	*0.10*	0.06	0.05	*0.06*	*0.01*	*0.06*	0.20	0.18	*0.17*	*0.01*				0.10	0.05	0.05
Dy	*0.05*	0.07	0.08	*0.08*	*0.01*	*0.04*	0.11	0.10	*0.10*	*0.01*	0.06	0.06	0.05	0.04	0.03	0.03
Er	*0.02*	0.03	0.03	*0.03*		*0.03*	0.05	0.06	*0.05*	*0.01*	0.03	0.03	0.02	0.01	0.02	0.02
Yb	*0.01*	0.03	0.03	*0.03*		*0.03*	0.04	0.06	*0.05*	*0.01*	0.02	0.03	0.02	0.01	0.02	0.02
Hf	*0.01*	0.04	n.a.	*0.04*		n.a.	0.15	n.a.	*0.15*		n.a.	n.a.	n.a.	n.a.	n.a.	n.a.
Th	*0.07*	0.02	0.04	*0.03*	*0.01*	*0.04*	0.20	0.09	*0.33*	*0.08*	0.05	0.04		0.01	0.02	0.01
La_N/Yb_N	*21*	7.6	7.2	*7.4*	*0.33*	*0.92*	6.8	5.4	*5.3*	*0.9*	7.6	13	8.7	21	10	9.0
Eu_N/Eu_N*	*3.0*	8.3	6.4	*7.4*	*1.3*	*3.7*	3.4	7.5	*5.1*	*2.9*				7.3	14	16
$Ti_N/Ti*_N$	*1.5*	1.8	2.1	*2.0*	*0.18*	*0.04*	1.7	2.0	*1.8*	*0.19*				1.8	3.0	3.1
$Sr_N/Sr*_N$	*31*	27	36	*31*	*6.0*	*666*	32	29	*31*	*2.0*	29	34	38	40	39	47
$Zr_N/Zr*_N$	*0.04*	0.05	0.05	*0.05*	*0.01*	*0.35*	0.16	0.04	*0.08*	*0.08*	0.09	0.12		0.06		

$Eu_N/Eu*_N$ = Eu anomaly, with $Eu*_N = (Sm_N \times Gd_N)^{1/2}$; $Sr_N/Sr*_N$ = Sr anomaly, with $Sr*_N = (Ce_N \times Nd_N)^{1/2}$; $Zr_N/Zr*_N$ = Zr anomaly, with $Zr*_N = (Nd_N \times Sm_N)^{1/2}$; n.a. = not analyzed. Italic data represent both the average when more data are available and a single analysis when sample heterogeneity did not allow more investigations.

Table 7. continued.

sample				RN17-1450	RN-17-1550					RN-17-1700						
Grain	plg 1by	average (7)	1σ	plg-1a-qz	plg 1B	plg 1B	plg 2b	average (3)	1σ	plg 1D	plg 3D	plg 3C	plg 2a	plg 2b	average (5)	1σ
Sc	2.1	2.9	1.1	3.8	3.2	3.0	2.0	2.7	0.6	1.6	1.1	1.4	4.1	3.5	2.6	1.4
Ti	229	297	131	70	415	425	272	371	86	291	395	443	161	159	251	131
V	4.5	6.7	3.4	3.3	5.4	5.5	5.0	5.3	0.3	4.5	5.3	3.0	3.7	3.8	4.3	0.9
Cr	0.10	0.33	0.4	1.4	0.50	0.10	0.70	0.43	0.3	13			0.90	1.0	3.8	7.1
Rb	0.20	0.25	0.1	0.20	0.30	0.20	0.90	0.47	0.4	0.20	0.40	0.00	0.30	0.30	0.30	0.2
Sr	190	212	31	153	253	262	228	248	18	266	265	435	201	197	232	97
Y	0.20	0.23	0.1	0.20	0.19	0.24	0.30	0.24	0.1	0.25	0.17	0.36	0.20	0.20	0.21	0.1
Zr	0.10	0.13	0.2		0.09	0.02	0.05	0.05	0.03	0.10	0.06	0.08			0.08	0.02
Nb		0.015	0.01		0.02	0.01	0.01	0.01	0.01				0.01	0.02	0.015	0.01
Cs		0.01			0.02			0.02		0.03	0.03				0.03	0.003
Ba	8.6	14	11	4.6	18	22	28	22	5.4	16	18	46	7.5	7.3	12	16
La	0.28	0.31	0.1	0.12	0.48	0.57	0.40	0.48	0.1	0.32	0.34	1.33	0.23	0.30	0.30	0.5
Ce	0.52	0.56	0.1	0.30	0.86	0.99	0.73	0.86	0.1	0.55	0.80	2.4	0.49	0.50	0.59	0.8
Nd	0.19	0.27	0.1	0.18	0.44	0.38	0.22	0.35	0.1	0.32	0.32	0.88	0.29	0.26	0.30	0.3
Sm	0.05	0.07	0.03	0.04	0.09	0.08	0.06	0.08	0.02	0.07	0.06	0.11	0.07	0.07	0.07	0.02
Eu	0.20	0.33	0.2	0.18	0.32	0.27	0.41	0.33	0.1	0.13	0.32	1.64	0.10	0.12	0.17	0.7
Gd	0.06	0.07	0.02	0.04	0.05	0.07	0.09	0.07	0.02	0.05	0.06	0.10	0.06	0.06	0.06	0.02
Dy	0.03	0.04	0.01	0.03	0.04	0.04	0.05	0.04	0.01	0.05	0.04	0.08	0.05	0.04	0.05	0.02
Er	0.02	0.02	0.01	0.01	0.02	0.02	0.02	0.02		0.02	0.01	0.03	0.02	0.02	0.02	0.01
Yb	0.01	0.02	0.01	0.01	0.01	0.02	0.01	0.01	0.01	0.01	0.01	0.03	0.01	0.01	0.01	0.01
Hf	n.a.			n.a.	0.02	n.a.	n.a.	0.02		n.a.	n.a.	0.03	n.a.	n.a.	0.03	
Th	0.01	0.02	0.02	0.01	0.08	0.05	0.03	0.05	0.03	0.04	0.08	0.06	0.01	0.01	0.04	0.03
La_N/Yb_N	19	12	5.3	8.3	33	20	28	25	6.8	22	24	31	16	21	21	5.4
Eu_N/Eu_N^*	11	14	3.6	14	14	11	17	14	3.1	6.7	16	47	4.7	5.6	8.2	18
Ti_N/Ti^*_N	2.7	2.8	0.6	1.0	4.5	4.0	2.0	3.3	1.3	2.9	4.0	2.5	1.5	1.6	2.5	1.0
Sr_N/Sr^*_N	40	37	5.6	44	28	29	38	30	5.8	42	35	20	36	37	37	8.3
Zr_N/Zr^*_N	0.07	0.08	0.03		0.03	0.01	0.03	0.02	0.01	0.04	0.03	0.02			0.02	0.01

Table 7. continued.

Sample	RN-17-2150				RN-17-2800				RN-17-3000				RN-19-2246
Grain	plg 6A	plg 6F(2)	average (2)	1σ	plg 3A	plg 3M	average (2)	1σ	plg 5E	plg 5M	average 82)	1σ	plg 4A
Sc	2.3	5.0	3.6	1.9	2.4	2.3	2.4	0.07	2.4	1.9	2.1	0.33	2.7
Ti	485	166	326	226	352	553	453	142	358	239	299	84	140
V	4.3	4.2	4.3	0.05	4.0	5.1	4.6	0.78	3.9	3.5	3.7	0.28	5.2
Cr					0.10	0.40	0.25	0.21	0.20	0.30	0.25	0.07	0.20
Rb		0.58	0.58		0.10	0.10	0.10		0.20	0.10	0.15	0.07	b.d.l.
Sr	410	120	265	205	248	310	279	44	366	272	319	66	111
Y	0.40	0.49	0.45	0.06	0.28	0.30	0.29	0.01	0.23	0.50	0.37	0.19	0.12
Zr	0.20	0.23	0.22	0.02	0.05	0.04	0.05	0.007	0.10	0.10	0.10		0.03
Nb		0.01	0.01		0.01	0.02	0.015	0.007	0.01	0.02	0.015	0.007	
Cs									0.03		0.03		
Ba	45	27	36	12	11	23	17	9	17	13	15	3	2.8
La	1.16	1.22	1.19	0.0	0.38	0.48	0.43	0.07	0.51	0.74	0.63	0.16	0.07
Ce	2.15	2.41	2.28	0.2	0.67	0.85	0.76	0.13	0.88	1.68	1.28	0.57	0.19
Pr													
Nd	0.77	0.74	0.76	0.02	0.37	0.37	0.37		0.59	1.03	0.81	0.31	0.13
Sm	0.14	0.2	0.17	0.04	0.07	0.08	0.08	0.007	0.10	0.20	0.15	0.07	0.01
Eu	1.02	0.65	0.84	0.26	0.22	0.35	0.29	0.09		0.33	0.17		0.13
Gd	0.11	0.14	0.13	0.02	0.05	0.07	0.06	0.014	0.05	0.09	0.07	0.028	0.01
Tb													
Dy	0.08	0.09	0.09	0.007	0.04	0.05	0.05	0.007	0.04	0.07	0.06	0.021	0.01
Ho													
Er	0.03	0.04	0.04	0.007	0.02	0.03	0.03	0.007	0.02	0.04	0.03	0.014	0.01
Tm													
Yb	0.02	0.03	0.03	0.007	0.02	0.03	0.03	0.007	0.01	0.02	0.02	0.007	0.01
Lu													
Hf		0.08	0.08		0.01	0.00	0.01	0.007	0.17	0.06	0.12	0.08	0.04
Ta													
Pb													
Th	0.04	0.18	0.11	0.10	0.01	0.01	0.01		0.09	0.04	0.07	0.035	0.05
La_N/Yb_N	40	28	34	8	13	11	12	1.5	35	26	29	7	4.8
Eu_N/Eu_N^*	25	12	18	9	11	14	13	2.1		7.5	4.9		39
Ti_N/Ti^*_N	2.6	0.74	1.7	1.3	3.9	4.7	4.4	0.5	4.0	1.5	2.4	1.8	7.0
Sr_N/Sr^*_N	21	6.0	14	11	33	37	35	2.6	34	14	21	14	47
Zr_N/Zr^*_N	0.04	0.04	0.04	0.001	0.02	0.02	0.02	0.004	0.03	0.01	0.02	0.009	0.05

Table 8. Electron probe data (wt%) for clinopyroxene.

sample	RN-17-GIOV-2				RN-17-400			RN-17-750		RN-17-1100											
grain	cpx 3B	cpx 1C	cpx 2D	cpx 2F	cpx 1M	cpx 3M	cpx 3B	cpx 3M	cpx 3B	cpx 1A	cpx 1F	cpx 3E	cpx 2B	cpx 7B	cpx 7G	cpx 8B	cpx 5D	cpx 4A	cpx-1a	cpx-1c	cpx-3a
SiO_2	49.28	52.17	53.30	53.34	52.69	48.09	48.14	48.09	48.14	54.49	52.67	53.75	50.06	53.08	53.07	53.54	52.89	54.01	52.73	53.30	51.77
TiO_2	1.07	0.40	0.11	0.37	0.53	2.26	2.22	2.26	2.22	0.39	0.85	0.51	1.45	0.68	0.59	0.45	0.63	0.54	0.45	0.39	0.56
Al_2O_3	1.37	2.62	2.41	2.32	1.76	5.32	4.13	5.32	4.13	1.09	2.81	1.73	2.74	2.59	1.66	3.51	2.07	1.82	2.48	2.12	3.26
Cr_2O_3	0.0002	0.40	0.40	0.35	0.10	0.27	0.35	0.27	0.35	0.0002	0.02	0.11	0.0002	0.08	0.0002	0.15	0.11	0.02	0.11	0.10	0.12
FeO_T	24.50	7.98	5.23	6.94	8.63	9.27	12.36	9.27	12.36	14.13	10.99	8.94	16.36	10.21	10.06	9.37	8.80	8.94	9.65	7.61	8.25
MnO	0.60	0.26	0.13	0.15	0.26	0.26	0.32	0.26	0.32	0.39	0.21	0.21	0.42	0.25	0.24	0.38	0.30	0.31	0.33	0.24	0.24
MgO	8.80	18.48	18.78	17.89	16.25	13.65	14.04	13.65	14.04	18.68	14.87	16.20	11.99	15.58	15.12	14.08	15.97	16.10	16.49	17.28	16.33
CaO	14.65	17.30	20.34	19.44	19.01	20.07	17.21	20.07	17.21	11.88	18.89	19.16	17.12	18.14	18.86	18.68	18.83	19.95	18.21	19.45	19.77
Na_2O	0.06	0.27	0.12	0.26	0.30	0.28	0.31	0.28	0.31	0.15	0.18	0.26	0.21	0.27	0.14	0.48	0.08	0.08	0.20	0.24	0.23
K_2O	0.0001	0.0001	0.0001	0.01			0.02		0.02	0.0001	0.01	0.01	0.0001	0.004	0.0001	0.0001	0.0001	0.0001	0.03	0.01	
Total	100.98	99.88	100.81	100.66	99.52	99.48	99.10	99.48	99.10	100.72	101.50	100.87	100.34	100.90	99.73	100.63	99.68	101.77	100.70	100.74	100.54
Wo	45	35	40	45	39	43	37	43	37	46	39	39	37	38	39	47	39	40	37	39	40

sample	RN-17-1450															RN-17-1550			
grain	Cpx-3b	cpx 2F	cpx 5D	cpx 4A	cpx 4B	cpx 4F	cpx 4G	cpx 4H	cpx 6A	cpx 6D	cpx 6E	cpx 6F	cpx 5G	cpx-1a	cpx-2a	cpx-2b	cpx 1A	cpx 1I	cpx-1a	cpx-2a
SiO_2	53.02	50.06	53.99	53.85	54.22	52.99	53.00	53.75	50.11	52.34	53.23	52.08	52.38	52.44	52.48	53.41	50.39	50.35	51.30	51.60
TiO_2	0.41	0.03	0.32	0.18	0.20	0.38	0.29	0.28	1.15	0.64	0.32	0.32	0.33	1.03	0.71	0.61	1.17	1.27	0.90	0.82
Al_2O_3	2.50	4.31	1.10	1.98	1.32	3.08	2.33	1.86	3.52	3.73	1.66	3.96	3.19	3.38	3.52	2.04	2.19	2.64	2.78	2.77
Cr_2O_3	0.17	0.0002	0.0002	0.72	0.20	0.05	0.05	0.11	0.47	0.22	0.0002	1.12	0.22	0.40	0.75	0.06	0.01		0.07	0.12
FeO_T	7.47	12.67	7.25	4.13	6.02	7.82	9.53	9.43	10.39	7.01	6.10	5.33	6.35	7.59	6.76	7.96	15.15	12.63	9.50	8.68
MnO	0.23	0.29	0.26	0.11	0.38	0.40	0.21	0.22	0.21	0.24	0.25	0.08	0.19	0.19	0.18	0.20	0.50	0.34	0.25	0.26
MgO	16.98	12.73	17.37	17.69	14.94	16.30	16.78	17.41	16.18	16.77	15.69	16.40	17.13	15.40	16.08	16.30	14.46	13.86	15.36	15.88
CaO	20.14	18.99	20.63	21.53	22.83	19.51	18.06	17.92	19.51	19.45	22.63	20.65	20.76	20.31	20.44	20.21	15.33	17.75	19.12	19.36
Na_2O	0.25	0.81	0.31	0.16	0.25	0.24	0.35	0.12	0.34	0.12	0.28	0.22	0.11	0.22	0.25	0.24	0.21	0.28	0.29	0.27
K_2O		0.0001	0.0001	0.0001	0.0001	0.0001	0.01	0.0001	0.002	0.002	0.0001	0.01	0.0001	0.003	0.003		0.01	0.003	0.02	0.003
Total	101.18	99.89	101.23	100.36	100.35	100.77	100.63	101.09	101.89	100.53	100.16	100.17	100.67	100.97	101.17	101.03	99.42	99.10	99.59	99.76
Wo	40	41	41	44	47	40	37	36	39	40	46	43	42	42	42	41	32	38	40	40

Wo = wollastonite index, molar [(100 x Ca/(Ca+Mg+Fe^{2+}+Fe^{3+}+Mn)].

Table 8. continued.

sample	RN-17-1700									RN-17-2150									RN-17-2800	
grain	cpx 2D	cpx 2L	cpx 2B	cpx 2A	cpx 3F	cpx 3B	cpx 3A	cpx 4F	cpx 4L	cpx 6B	cpx 6E	cpx 4E	cpx 5A	cpx 5B	cpx 5G	cpx-1a	cpx-2a	cpx-2b	cpx 1a	cpx 1C
SiO_2	51.35	51.37	51.05	51.33	51.18	51.51	51.22	51.54	51.45	51.34	52.20	50.77	51.53	51.65	49.05	50.89	51.01	51.29	50.19	50.66
TiO_2	0.82	0.82	1.11	0.88	0.89	0.82	0.87	0.72	0.84	0.64	0.53	0.56	0.94	0.90	0.31	1.30	1.06	1.23	1.27	0.96
Al_2O_3	1.99	1.78	3.77	3.25	1.71	1.50	1.71	3.25	2.34	3.52	3.25	2.55	2.31	2.42	5.51	2.98	3.70	2.81	3.47	2.17
Cr_2O_3	0.05		0.23	0.18				0.25	0.02	0.11	0.12	0.0002	0.11	0.16	0.03	0.05	0.36	0.02	0.09	0.0002
FeO_T	11.28	12.49	8.10	7.52	15.43	15.42	14.28	7.55	10.19	8.28	8.30	10.40	8.23	7.83	14.27	9.75	7.91	10.99	11.20	12.21
MnO	0.32	0.30	0.15	0.26	0.59	0.53	0.47	0.16	0.30	0.25	0.35	0.45	0.14	0.26	0.30	0.27	0.22	0.32	0.36	0.35
MgO	15.12	15.12	15.57	15.62	14.85	14.43	13.98	16.47	15.46	15.62	15.98	14.24	16.11	16.09	9.29	15.22	15.48	15.41	14.45	13.21
CaO	18.17	16.99	19.41	19.85	14.95	15.21	17.13	19.29	18.29	20.35	20.32	20.70	20.33	20.36	19.56	19.73	20.23	18.13	18.66	19.33
Na_2O	0.21	0.26	0.30	0.20	0.22	0.22	0.22	0.23	0.21	0.30	0.12	0.28	0.34	0.19	1.19	0.33	0.27	0.29	0.39	0.30
K_2O		0.01	0.01		0.01	0.01	0.01			0.02	0.002	0.02	0.0001	0.0001	0.05		0.005	0.01	0.0001	0.0003
Total	99.30	99.13	99.70	99.09	99.84	99.66	99.90	99.44	99.10	100.42	101.16	99.98	100.04	99.85	99.55	100.52	100.25	100.50	100.10	99.20
Wo	38	35	41	42	31	32	36	40	38	42	41	42	41	42	45	40	42	37	39	41

sample													RN-17-3000								
grain	cpx 1D	cpx 1I	cpx 2E	cpx 2I	cpx 3F	cpx 3G	cpx 3H	cpx 3I	cpx-1a	cpx-2a	cpx-2b	cpx-1a	cpx 1A	cpx 2E	cpx 3D	cpx 5A	cpx 5B	cpx 5F	cpx 5L	cpx 5O	cpx-2b
SiO_2	50.84	53.11	48.81	53.16	50.60	52.16	52.12	51.42	51.75	50.71	50.43	51.82	50.78	50.79	53.14	50.26	50.12	51.84	50.75	50.94	51.78
TiO_2	1.12	0.17	1.44	0.11	1.35	0.97	1.04	1.14	0.76	1.18	1.20	0.69	1.11	1.01	0.12	1.15	1.21	0.73	1.14	1.40	0.90
Al_2O_3	2.31	0.65	3.55	0.88	2.48	1.78	2.21	2.27	2.39	3.16	2.89	2.43	2.57	3.53	0.92	3.45	3.48	2.69	3.52	2.90	2.22
Cr_2O_3	0.0002	0.0002	0.0002	0.03	0.0002	0.03	0.08	0.0002	0.30	0.08		0.38	0.0002	0.06	0.05	0.55	0.61	0.66	0.60	0.0002	0.04
FeO_T	13.44	11.78	12.77	11.21	10.85	10.27	9.96	9.76	7.33	8.98	9.94	7.27	12.25	10.04	12.14	7.46	7.45	6.72	7.56	9.43	8.87
MnO	0.41	0.56	0.24	0.48	0.26	0.20	0.28	0.30	0.26	0.26	0.28	0.24	0.38	0.42	0.98	0.20	0.14	0.11	0.09	0.26	0.23
MgO	13.38	11.81	13.79	12.17	14.59	16.29	15.59	15.51	16.11	15.32	15.10	16.51	13.53	14.66	12.33	15.45	15.89	15.94	15.72	15.42	16.09
CaO	19.20	22.19	18.59	22.52	19.57	19.19	19.67	19.72	20.57	20.35	19.60	20.13	19.38	19.34	20.17	20.37	20.73	20.91	20.63	19.57	19.00
Na_2O	0.22	0.08	0.26	0.18	0.36	0.25	0.30	0.34	0.26	0.34	0.35	0.24	0.38	0.34	0.23	0.19	0.39	0.28	0.27	0.36	0.28
K_2O	0.0001	0.05	0.004	0.04	0.0001	0.0001	0.01	0.0001	0.0007	0.01	0.0007	0.01	0.06	0.01	0.0001	0.0001	0.01	0.03	0.01	0.0001	
Total	100.93	100.39	99.45	100.77	100.05	101.13	101.27	100.46	99.74	100.39	99.79	99.73	100.43	100.20	100.09	99.08	100.02	99.90	100.28	100.29	99.41
Wo	39	46	39	46	40	38	40	40	42	42	40	41	40	40	42	43	43	43	43	40	39

Table 8. continued.

sample			**RN-19-2246**										
grain	cpx-1a	cpx-1b	cpx 1E	cpx 1F	cpx 2B	cpx 2C	cpx 3B	cpx 3C	cpx-2a	cpx-2b	cpx-1a	cpx-2a	cpx-2b
SiO_2	51.34	50.86	52.60	51.93	53.69	53.95	52.24	51.20	53.04	52.38	52.52	52.14	52.64
TiO_2	0.92	1.06	0.39	0.39	0.32	0.23	0.53	0.50	0.24	0.39	0.34	0.40	0.36
Al_2O_3	3.37	4.09	2.00	2.04	2.18	1.67	2.02	1.29	2.45	1.90	2.67	2.06	2.05
Cr_2O_3	0.46	0.65	0.02	0.06	0.12	0.08	0.02	0.01	0.12	0.02	0.23	0.00	0.01
FeO_T	7.27	7.19	11.70	11.94	7.49	6.58	13.89	20.27	7.44	11.32	5.98	9.39	10.98
MnO	0.21	0.18	0.39	0.30	0.21	0.10	0.28	0.43	0.29	0.37	0.22	0.33	0.39
MgO	15.48	15.31	17.12	17.10	18.01	18.05	15.13	11.86	17.98	16.25	16.93	16.37	16.42
CaO	20.41	20.43	16.77	16.54	19.24	20.27	17.63	15.04	18.68	17.51	20.67	18.81	17.82
Na_2O	0.29	0.28	0.21	0.18	0.22	0.23	0.28	0.17	0.21	0.23	0.18	0.22	0.19
K_2O	0.01	0.01		0.02					0.003			0.004	0.005
Total	99.76	100.05	101.20	100.50	101.47	100.74	102.01	100.78	100.46	100.37	99.74	99.73	100.86
Wo	43	43	34	33	38	45	35	32	38	36	42	38	36

Table 9. Trace element concentrations (ppm) for clinopyroxene by SIMS.

sample	RN-17-750	RN-17-1100								RN-17-1450				
grain	cpx 3M	cpx 7G	cpx 7B	cpx 1a	cpx 1c	cpx 3a	cpx 3b	average (6)	1σ	cpx 2a	cpx 2b	cpx 1a	average (3)	1σ
Sc	*123*	148	152	116	88	128	88	*120*	*28*	109	92	150	*117*	*30*
Ti	*12838*	3340	4462	3106	2378	3493	2430	*3201*	*772*	4310	3519	6804	*4878*	*1714*
V	*819*	712	865	411	284	427	310	*501*	*234*	328	279	930	*512*	*362*
Cr	*1424*	602	403	884	705	1379	1144	*853*	*361*	5573	1204	2735	*3171*	*2217*
Rb	*4.0*	2.8	3.4	3.5	3.4	3.7	3.5	*3.4*	*0.31*	3.8	4.1	4.2	*4.0*	*0.21*
Sr	*17*	8.7	9.1	10	8	10	9.0	*9.1*	*0.77*	12	10	13	*12*	*1.5*
Y	*37*	13	18	12	9.3	15	9.6	*13*	*3.4*	13	11	21	*15*	*5.3*
Zr	*71*	13	15	16	5.2	12	5.5	*11*	*4.7*	12	8.3	28	*16*	*11*
Nb	*0.67*	0.05	0.12	0.06	0.04	0.04	0.02	*0.06*	*0.03*	0.08	0.08	0.19	*0.12*	*0.06*
Cs			0.017			0.01		*0.014*	*0.005*	0.02			0.02	
Ba	*0.73*	1.00	1.74	1.80	0.10	0.10		*0.95*	*0.84*	0.70	0.20	0.10	*0.33*	*0.32*
La	*2.38*	0.52	0.59	0.61	0.3	0.48	0.25	*0.46*	*0.15*	0.88	0.33	0.79	*0.67*	*0.30*
Ce	*9.8*	1.90	2.21	1.90	1.20	2.12	1.21	*1.76*	*0.44*	2.92	1.67	3.6	*2.74*	*0.99*
Nd	*12.3*	2.78	3.7	2.37	2.29	3.5	1.64	*2.72*	*0.79*	3.9	2.65	5.2	*3.9*	*1.27*
Sm	*5.9*	1.12	1.77	1.23	1.01	1.69	0.79	*1.27*	*0.39*	1.71	1.16	1.91	*1.59*	*0.39*
Eu	*1.86*	0.34	0.58	0.47	0.34	0.67	0.33	*0.46*	*0.14*	0.67	0.43	0.74	*0.61*	*0.16*
Gd	*7.3*	1.71	2.56	1.54	1.50	2.40	1.31	*1.84*	*0.52*	2.35	1.71	3.05	*2.37*	*0.67*
Dy	*7.5*	2.28	3.4	2.01	1.62	2.50	1.48	*2.21*	*0.69*	2.09	1.75	3.5	*2.43*	*0.91*
Er	*4.3*	1.46	2.04	1.34	1.05	1.69	0.94	*1.42*	*0.41*	1.29	1.01	1.83	*1.38*	*0.42*
Yb	*3.7*	1.39	1.95	1.56	1.04	1.54	0.83	*1.39*	*0.40*	1.31	1.08	1.85	*1.41*	*0.40*
Hf	*3.7*	0.60	1.00	n.a.	n.a.	n.a.	n.a.	*0.80*	*0.28*	n.a.	n.a.	n.a.		
Th	*0.30*	0.03	0.08	0.03	0.02	0.03	0.02	*0.04*	*0.02*	0.04	0.05	0.04	*0.04*	*0.01*
La_N/Yb_N	*0.44*	0.26	0.21	0.27	0.20	0.22	0.21	*0.23*	*0.03*	0.47	0.21	0.30	*0.33*	*0.13*
Eu_N/Eu_N^*	*0.87*	0.75	0.83	1.04	0.84	1.01	0.99	*0.91*	*0.12*	1.01	0.93	0.93	*0.96*	*0.05*
Ti_N/Ti^*_N	*0.87*	0.85	0.76	0.88	0.76	0.71	0.87	*0.80*	*0.07*	0.97	1.02	1.05	*1.0*	*0.04*
Sr_N/Sr^*_N	*0.10*	0.25	0.21	0.32	0.32	0.24	0.43	*0.28*	*0.08*	0.24	0.32	0.20	*0.24*	*0.06*
Zr_N/Zr^*_N	*0.55*	0.48	0.39	0.61	0.22	0.32	0.32	*0.39*	*0.14*	0.31	0.31	0.59	*0.43*	*0.16*

Eu_N/Eu^*_N = Eu anomaly, with $Eu^*_N = (Sm_N \times Gd_N)^{1/2}$; Ti_N/Ti^*_N = Ti anomaly, with $Ti^*_N = (Gd_N \times Dy_N)^{1/2}$; Sr_N/Sr^*_N = Sr anomaly, with $Sr^*_N = (Ce_N \times Nd_N)^{1/2}$; Zr_N/Zr^*_N = Zr anomaly, with $Zr^*_N = (Nd_N \times Sm_N)^{1/2}$; n.a. = not analyzed. Italic data represent both the average when more data are available, and a single analysis when sample heterogeneity did not allow more investigations.

Table 9. continued.

sample	**RN-17-1550**							**RN-17-1700**					**RN-17-2150**	
grain	cpx 1F	cpx 1A	cpx 1a	cpx 1b	cpx 2a	average (5)	1σ	cpx 3B	cpx 3A	cpx 3F	average (3)	1σ	cpx 1a	cpx 2a
Sc	155	133	115	115	107	*125*	*19*	107	127	123	*119*	*10*	122	132
Ti	8745	6832	5819	6151	4852	*6480*	*1454*	5330	6086	5666	*5694*	*379*	8299	7129
V	540	399	453	474	399	*453*	*59*	214	242	197	*218*	*23*	434	360
Cr	29	26	223	166	766	*242*	*305*	16	16	17	*16*	*0.56*	190	4348
Rb	5.4	5.5	4.5	4.6	4.2	*4.8*	*0.58*	5.5	5.0	6.8	*5.8*	*0.93*	5.2	4.7
Sr	10	11	13	13	12	*12*	*1.3*	11	15	11	*12*	*2.1*	16	15
Y	36	35.2	20	21	17	*26*	*9.1*	48	56	62	*55*	*6.8*	22	18
Zr	35	31	17	18	13	*23*	*9.6*	30	38	39	*36*	*4.7*	19	20
Nb	0.26	0.15	0.06	0.09	0.09	*0.13*	*0.08*	0.13	0.13	0.21	*0.16*	*0.05*	0.09	0.14
Cs	0.018			0.01	0.01	*0.013*	*0.00*			0.031	*0.031*			0.02
Ba	0.09		0.20	0.10	0.20	*0.15*	*0.06*	0.03	0.20	0.13	*0.12*	*0.09*	0.10	0.10
La	1.80	1.29	0.72	1.01	1.28	*1.22*	*0.40*	1.18	2.13	1.63	*1.65*	*0.48*	0.73	0.70
Ce	6.7	5.7	3.4	4.4	3.7	*4.8*	*1.4*	5.8	8.9	7.9	*7.5*	*1.6*	3.8	3.2
Nd	8.8	8.4	4.6	5.4	5.0	*6.4*	*2.0*	9.1	13.9	13.2	*12.0*	*2.6*	5.8	5.1
Sm	3.6	3.4	1.97	1.90	1.71	*2.50*	*0.89*	3.9	5.6	5.7	*5.1*	*1.0*	2.99	2.14
Eu	1.36	1.31	0.80	0.88	0.71	*1.01*	*0.30*	1.49	1.86	1.84	*1.73*	*0.21*	1.04	0.94
Gd	4.5	5.1	2.75	2.73	2.43	*3.5*	*1.2*	7.1	9.2	9.8	*8.7*	*1.4*	3.9	2.89
Dy	6.2	6.2	3.3	3.5	2.87	*4.4*	*1.7*	8.5	10.6	11.0	*10.0*	*1.4*	4.1	3.2
Er	3.9	3.5	1.78	1.99	1.65	*2.55*	*1.0*	5.2	5.5	6.3	*5.6*	*0.57*	2.01	1.84
Yb	3.5	3.3	1.92	2.05	1.66	*2.5*	*0.86*	5.0	4.7	6.9	*5.6*	*1.2*	1.82	1.78
Hf	2.00	2.01	n.a.	n.a.	n.a.	*2.01*	*0.01*	2.45	2.66	2.91	*2.67*	*0.23*	n.a.	n.a.
Th	0.06	0.08	0.03	0.03	0.03	*0.05*	*0.02*	0.08	0.09	0.25	*0.14*	*0.10*	0.02	0.02
La_N/Yb_N	0.35	0.27	0.26	0.34	0.53	*0.35*	*0.11*	0.16	0.31	0.16	*0.21*	*0.09*	0.28	0.27
Eu_N/Eu_N^*	1.03	0.97	1.04	1.17	1.06	*1.05*	*0.07*	0.86	0.79	0.75	*0.79*	*0.06*	0.93	1.15
Ti_N/Ti^*_N	0.83	0.61	0.96	1.0	0.92	*0.86*	*0.16*	0.34	0.31	0.27	*0.31*	*0.04*	1.0	1.2
Sr_N/Sr^*_N	0.09	0.11	0.22	0.18	0.19	*0.16*	*0.06*	0.11	0.09	0.07	*0.09*	*0.02*	0.23	0.25
Zr_N/Zr^*_N	0.41	0.38	0.36	0.36	0.29	*0.36*	*0.04*	0.33	0.28	0.29	*0.30*	*0.03*	0.30	0.40

Table 9. continued.

sample				**RN-17-2800**						**RN-17-3000**					
grain	cpx 2b	average (3)	1σ	cpx 1a ar	cpx 1a	cpx 2a	cpx 2b	average (4)	1σ	cpx 5F	cpx 5B	cpx 1a	cpx 1b	average (4)	1σ
Sc	114	*123*	*9.1*	106	103	120	125	*113*	*11*	109	132	112	125	*120*	*11*
Ti	7815	*7748*	*588*	5096	4290	7576	8791	*6438*	*2101*	4850	6736	6218	7144	*6237*	*999*
V	392	*395*	*37*	294	260	418	434	*351*	*87*	287	394	369	388	*360*	*49*
Cr	272	*1604*	*2377*	2822	3054	573	100	*1637*	*1517*	3900	4654	2310	5111	*3994*	*1229*
Rb	6.2	*5.4*	*0.76*	3.7	3.8	4.3	6.2	*4.5*	*1.2*	2.8	2.5	4.5	4.4	*3.6*	*1.0*
Sr	14	*15*	*1.0*	15	13	16	16	*15*	*1.4*	14	19	16	17	*17*	*1.8*
Y	20	*20*	*1.8*	13	11	19	23	*16*	*5.4*	12	17	16	17	*15*	*2.6*
Zr	18	*19*	*0.90*	11	10	17	22	*15*	*5.6*	9.7	20	16	19	*16*	*4.7*
Nb	0.10	*0.11*	*0.03*	0.11	0.09	0.09	0.14	*0.11*	*0.02*	0.06	0.59	0.10	0.16	*0.23*	*0.25*
Cs	0.02	*0.02*									0.036	0.01		*0.03*	*0.02*
Ba	0.30	*0.17*	*0.12*	0.05		0.02	0.04	*0.04*	*0.02*	0.16	3.8	0.04	0.11	*1.02*	*1.8*
La	0.77	*0.73*	*0.04*	0.60	0.41	0.62	0.88	*0.63*	*0.19*	0.44	1.32	0.63	0.68	*0.77*	*0.38*
Ce	3.2	*3.4*	*0.32*	2.24	1.88	3.3	4.2	*2.92*	*1.1*	2.07	4.4	3.1	3.1	*3.2*	*1.0*
Nd	5.2	*5.3*	*0.39*	3.8	2.97	5.3	6.4	*4.6*	*1.5*	2.90	4.7	4.0	4.6	*4.0*	*0.81*
Sm	2.05	*2.39*	*0.52*	1.75	1.40	2.59	3.1	*2.21*	*0.77*	1.36	1.99	2.11	2.05	*1.88*	*0.35*
Eu	0.95	*0.98*	*0.06*	0.69	0.57	0.91	0.95	*0.78*	*0.18*	0.51	0.76	0.72	0.79	*0.70*	*0.13*
Gd	3.0	*3.3*	*0.54*	2.21	1.92	3.7	4.2	*3.0*	*1.1*	1.81	2.61	2.55	2.86	*2.46*	*0.45*
Dy	3.1	*3.5*	*0.52*	2.37	1.90	3.4	4.2	*2.97*	*1.0*	1.87	2.92	2.65	3.2	*2.65*	*0.56*
Er	1.78	*1.88*	*0.12*	1.37	1.15	1.81	2.30	*1.66*	*0.51*	1.20	1.86	1.51	1.37	*1.49*	*0.28*
Yb	1.97	*1.86*	*0.10*	1.36	0.96	1.92	2.26	*1.63*	*0.58*	1.20	1.66	1.43	1.10	*1.35*	*0.25*
Hf	n.a.			n.a.	n.a.	n.a.	n.a.			0.52	1.00	n.a.	n.a.	*0.76*	*0.34*
Th	0.02	*0.02*		0.03	0.02	0.02	0.03	*0.04*	*0.01*	0.07	0.13	0.03	0.02	*0.06*	*0.05*
La_N/Yb_N	0.27	*0.27*	*0.004*	0.31	0.30	0.22	0.27	*0.27*	*0.04*	0.25	0.55	0.31	0.43	*0.39*	*0.13*
Eu_N/Eu_N^*	1.16	*1.06*	*0.13*	1.07	1.06	0.89	0.80	*0.92*	*0.13*	0.99	1.01	0.94	0.99	*0.98*	*0.03*
Ti_N/Ti^*_N	1.3	*1.2*	*0.11*	1.1	1.1	1.1	1.1	*1.1*	*0.04*	1.3	1.2	1.2	1.2	*1.2*	*0.06*
Sr_N/Sr^*_N	0.23	*0.24*	*0.01*	0.34	0.37	0.26	0.21	*0.27*	*0.08*	0.39	0.27	0.30	0.30	*0.31*	*0.05*
Zr_N/Zr^*_N	0.37	*0.35*	*0.05*	0.27	0.31	0.31	0.32	*0.30*	*0.02*	0.32	0.43	0.36	0.41	*0.39*	*0.05*

Table 9. continued.

sample	**RN-19-2246**							
grain	cpx 2B	cpx 2a	cpx 2b	cpx 1a	cpx 2a	cpx 2b	average (6)	1σ
Sc	124	103	100	106	73	89	*111*	*17*
Ti	1960	2329	2323	1839	1322	1683	*2404*	*388*
V	410	401	399	341	267	324	*311*	*56*
Cr	384	21	20	1457	524	471	*362*	*527*
Rb	0.90	5.2	5.5	3.3	3.8	3.6	*6.0*	*1.6*
Sr	5.6	6.0	5.0	6.0	4.0	5.0	*6.2*	*0.77*
Y	11	14	14	9.9	7.7	9.9	*21*	*2.5*
Zr	3.5	4.0	3.4	4.4	2.1	3.3	*6.3*	*0.78*
Nb	0.02	0.03	0.02			0.03	*0.03*	*0.01*
Cs		0.02			0.02	0.02	*0.02*	
Ba		0.6	0.1		0.1		*0.27*	*0.29*
La	0.14	0.15	0.12	0.08	0.09	0.08	*0.46*	*0.03*
Ce	0.53	0.78	0.67	0.53	0.37	0.56	*1.58*	*0.14*
Nd	0.99	1.49	1.50	0.98	0.71	1.04	*2.42*	*0.31*
Sm	0.57	0.88	0.74	0.49	0.49	0.61	*1.27*	*0.15*
Eu	0.29	0.40	0.35	0.31	0.26	0.25	*0.57*	*0.06*
Gd	1.33	1.79	1.48	1.34	1.24	1.04	*2.22*	*0.25*
Dy	1.85	2.34	2.07	1.58	1.30	1.24	*3.1*	*0.44*
Er	1.16	1.29	1.46	0.89	0.80	1.13	*2.04*	*0.25*
Yb	1.07	1.77	1.62	0.91	0.87	1.13	*2.36*	*0.38*
Hf	0.31	n.a.	n.a.	n.a.	n.a.	n.a.	*0.31*	
Th	0.04	0.03	0.02	0.02	0.03	0.07	*0.04*	*0.02*
La_N/Yb_N	0.09	0.06	0.05	0.06	0.07	0.05	*0.13*	*0.02*
Eu_N/Eu_N*	1.01	0.97	1.02	1.16	1.01	0.95	*1.03*	*0.07*
$Ti_N/Ti*_N$	0.63	0.57	0.67	0.63	0.52	0.74	*0.46*	*0.08*
$Sr_N/Sr*_N$	0.52	0.37	0.33	0.56	0.52	0.44	*0.21*	*0.09*
$Zr_N/Zr*_N$	0.31	0.23	0.21	0.42	0.23	0.27	*0.24*	*0.08*

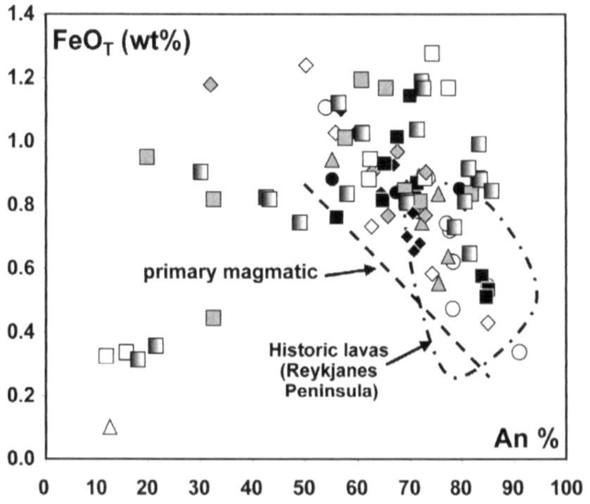

Figure 6. Plots of FeO_T (wt%) versus anorthite index (An %) of plagioclase relative to drill cuttings and core-sample from well RN-17 and RN-19, respectively (the symbols are the same as in figure 3). The field of igneous (primary magmatic) plagioclase from Hengill geothermal system, Reykjanes Peninsula, SW Iceland [60] and the field of plagioclase of historic lavas from Reykjanes Peninsula [61] are reported for comparison.

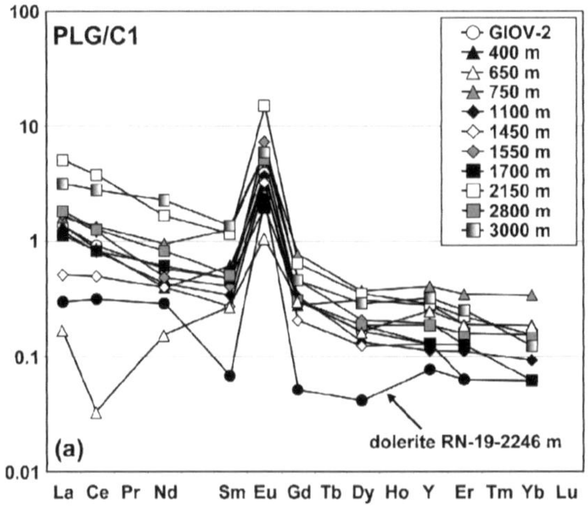

Figure 7. C1-chondrite-normalized REE patterns, obtained by SIMS, for plagioclase from wells RN-17 and RN-19 (Reykjanes geothermal system).

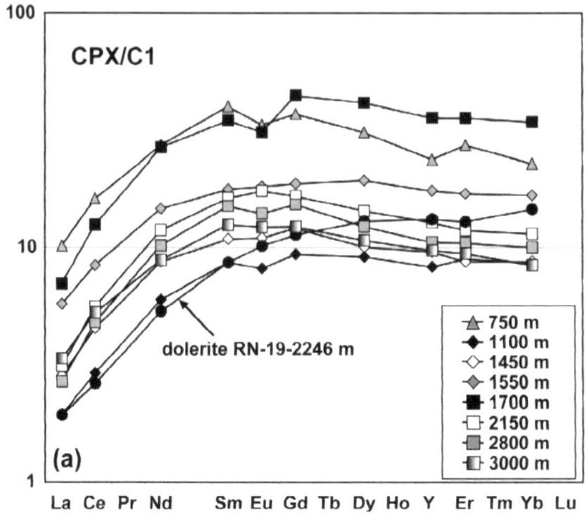

Figure 8. C1-chondrite-normalized REE patterns, obtained by SIMS, of clinopyroxene from wells RN-17 and RN-19 (Reykjanes geothermal system).

at depths of 312 m down to ~ 3000 m. Epidote shows noteworthy variations in Al_2O_3 (18.1 - 27.7 wt%) and FeO_T (7.97 - 18.4 wt%) concentrations (see table 10). Data by [17] in epidote crystals separated from drillhole-cuttings from three geothermal wells including RN-17, display complex chemical zoning, with Fe(III)-rich cores and Al-rich rims, and oscillatory zoning reflecting irregular fluctuations in Al-Fe(III) substitution. In hydrothermal environments the Fe(III)-Al composition of epidote is a sensitive function of temperature, pressure, bulk rock composition, fluid composition (especially redox state, CO_2 concentration and pH) [17]. The Al_2O_3 and FeO_T variability in the set of epidotes analyzed in the present work, is documented by SEM backscattered electron (BSE) image in figure 9 (see [16] for details), the highest Al and lowest Fe contents occurring in the deepest cutting RN-17-3000. The Al-poor (lighter) areas are more REE fractionated (La_N/Yb_N = 1.2 - 34) and are characterized by less variability in Eu anomalies (Eu_N/Eu^*_N = 4.9 - 26) than the Al-rich (dark areas (La_N/Yb_N = 0.48 - 7.8; Eu_N/Eu^*N = 4.2 - 45) (figures 10a and b) (see also table 11). The variations in the REE concentrations in our opinion are ascribed to changes in the fluid composition with time.

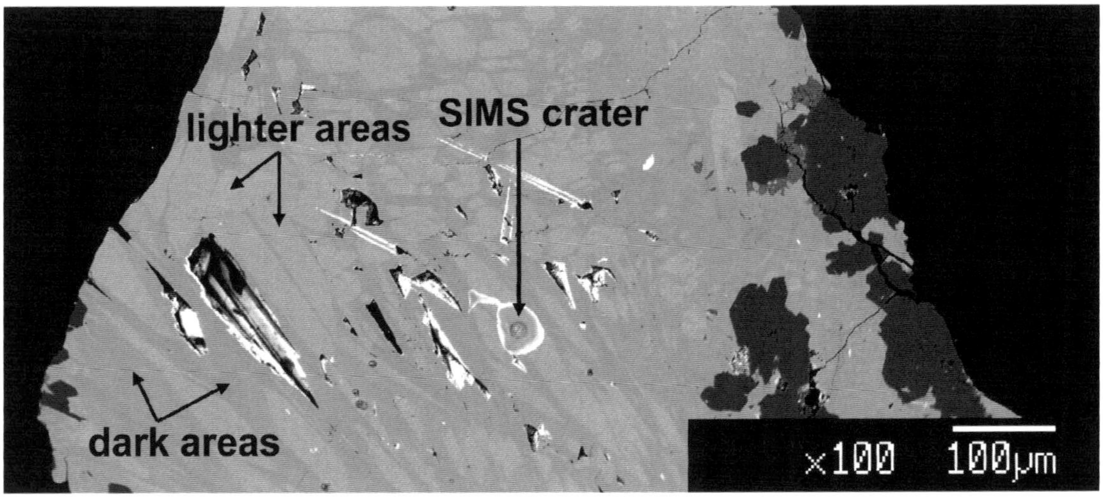

Figure 9. SEM backscattered electron images of epidote from cuttings RN-17-1100 (from [16]). The micrograph documents the occurrence of dark (Al-rich) and lighter (Fe_T-rich) areas in epidote. Note the SIMS crater after ion bombardment.

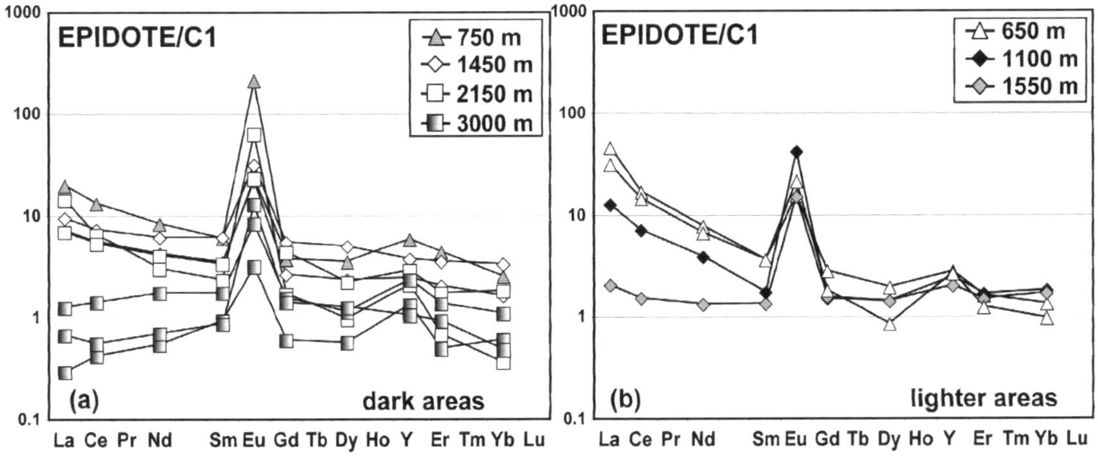

Figure 10. C1-chondrite-normalized REE (a,b) patterns of epidote from well RN-17 (Reykjanes geothermal system): diagrams (a) are relative to Al-rich (dark) areas; those (b) are relative to Fe_T-rich (lighter) areas. The data were obtained at the ion microprobe.

EMAS 2011: 12th European Workshop on Modern Developments in Microbeam Analysis IOP Publishing
IOP Conf. Series: Materials Science and Engineering **32** (2012) 012020 doi:10.1088/1757-899X/32/1/012020

Table 10. Electron probe data (wt%) for epidote.

sample	RN-17-650									RN-17-750		RN-17-1100			
spot	ep 1a	ep 1b	ep 1c	ep 2a	ep 2b	ep 3a	ep 3b	ep 3L	ep 3F	ep 2a	ep 2b	ep 1a	ep 1b	ep 2a	ep 2b
zone	lighter area	dark area	lighter area	lighter area	lighter area	lighter area	lighter area	dark area	dark area	dark area	dark area	lighter area	dark area	dark area	dark area
SiO_2	38.02	37.70	37.60	38.01	37.41	37.93	37.80	38.11	37.95	38.20	38.04	37.30	37.85	38.07	38.10
TiO_2	0.00	0.00	0.21	0.11	0.05	0.12	0.02	0.07	0.08	0.10	0.05	0.15	0.08	0.04	0.14
Al_2O_3	19.79	23.48	20.93	20.10	19.61	19.91	21.00	23.15	22.08	22.72	21.83	19.40	22.97	23.17	22.11
Cr_2O_3	0.05	0.02	0.02	0.00	0.00	0.00	0.06	0.00	0.00	0.01	0.04	0.00	0.00	0.01	0.00
FeO_T	16.56	13.11	15.20	16.66	16.83	17.25	15.78	12.48	13.98	14.25	15.15	17.46	13.34	13.41	14.26
MnO	0.09	0.30	0.20	0.16	0.12	0.13	0.16	0.49	0.25	0.21	0.09	0.04	0.21	0.12	0.15
MgO	0.07	0.05	0.06	0.06	0.06	0.05	0.05	0.14	0.08	0.04	0.05	0.02	0.06	0.08	0.23
CaO	23.12	23.07	23.17	23.16	22.86	23.22	23.17	23.37	23.11	23.33	23.38	23.08	22.98	23.37	23.03
Na_2O	0.001	0.01	0.004	0.02	0.02		0.01	0.003	0.01		0.002	0.005	0.01		0.01
K_2O							0.01			0.01	0.01			0.002	
Cl					0.01		0.003				0.001				
Total	97.70	97.74	97.40	98.28	96.97	98.60	98.04	97.81	97.54	98.86	98.64	97.45	97.51	98.26	98.02

sample			RN-17-1450						RN-17-1550						
spot	ep 3a	ep 3b	ep 3a	ep 3b	ep 3b(2)	ep 2a	ep 2b	ep 1a	ep 3A	ep 3B	ep 1a	ep 1b	ep 2a	ep 3a	ep 3b
zone	lighter area	lighter area	dark area	dark area	dark area	dark area	dark area	dark area	lighter area	lighter area	lighter area	dark area	dark area	dark area	dark area
SiO_2	37.76	37.07	37.37	37.54	37.74	38.93	38.59	40.78	37.34	37.47	37.87	38.36	39.37	38.64	38.44
TiO_2	0.00	0.27	0.05	0.02	0.01	0.05	0.10	0.06	0.02	0.01	0.01	0.00	0.08	0.02	0.01
Al_2O_3	21.28	18.88	22.18	23.58	22.36	22.92	21.95	22.07	18.08	19.65	21.86	22.99	25.01	24.16	23.61
Cr_2O_3	0.00	0.02	0.01	0.02	0.00			0.03	0.04	0.00	0.06	0.01		0.01	0.00
FeO_T	15.05	18.38	14.48	13.01	13.76	11.92	12.94	11.83	18.38	16.32	14.21	13.21	9.73	11.57	12.45
MnO	0.11	0.06	0.08	0.04	0.09	0.03	0.06	0.16	0.04	0.09	0.05	0.06	0.16	0.03	0.11
MgO	0.04	0.05	0.10	0.10	0.14	0.03	0.04	0.87	0.03	0.06	0.05	0.11	0.20	0.03	0.06
CaO	22.96	22.96	23.28	23.39	23.40	23.34	23.46	22.83	22.77	22.87	23.31	23.11	23.11	23.27	23.32
Na_2O	0.01	0.01	0.02	0.01	0.02	0.01	0.01	0.03	0.003	0.01	0.01	0.01	0.01	0.01	0.03
K_2O		0.005		0.002	0.003	0.01	0.01	0.01					0.004	0.003	0.002
Cl						0.004	0.005	0.01	0.004	0.01			0.01		
Total	97.21	97.71	97.56	97.71	97.52	97.23	97.16	98.65	96.70	96.49	97.44	97.86	97.68	97.75	98.03

Table 10. continued.

sample	**RN-17-1700**								**RN-17-2150**		
Spot	ep 1a	ep 1b	ep 2a	ep 2b	ep 3a	ep 3b	ep 4a	ep 4b	ep 3a	ep 3b	ep 1a
Zone	dark area	dark area	dark area	dark area	dark area	dark area	dark area	dark area	dark area	dark area	dark area
SiO_2	39.01	39.04	39.28	38.57	38.64	38.68	38.67	39.05	38.37	38.48	38.48
TiO_2	0.05	0.28	0.11	0.25	0.43	0.42	0.19	0.05	0.00	0.23	0.12
Al_2O_3	25.60	22.34	25.05	22.49	22.73	22.65	23.55	23.85	23.06	25.09	23.40
Cr_2O_3	0.01	0.03	0.00	0.03	0.02	0.04		0.00	0.02	0.00	0.01
FeO_T	9.73	12.09	9.91	12.37	12.52	12.40	11.77	11.37	13.82	11.32	13.24
MnO	0.34	0.07	0.23	0.07	0.04	0.06	0.07	0.07	0.11	0.29	0.07
MgO	0.04	0.08	0.02	0.07	0.05	0.04	0.04	0.02	0.02	0.04	0.08
CaO	23.17	23.53	23.21	23.29	23.65	23.37	23.53	23.43	23.46	23.14	23.59
Na_2O	0.003	0.002	0.01	0.002		0.01	0.01	0.002	0.02	0.03	0.03
K_2O		0.002		0.01	0.001			0.01	0.00	0.01	
Cl	0.001			0.01			0.001	0.005	0.01	0.003	0.001
Total	97.95	97.47	97.82	97.15	98.08	97.67	97.84	97.84	98.88	98.62	99.01
sample		**RN-17-2800**					**RN-17-3000**				
spot	ep 1b	ep 1a	ep 1b	ep 3a	ep 3b	ep 3c	ep 1a	ep 1b	ep 1c	ep 2a	ep 2b
zone	lighter area	dark area	dark area	dark area	dark area	lighter area	dark area	dark area	dark area	dark area	dark area
SiO_2	38.20	38.11	38.54	38.24	38.37	37.90	39.26	38.53	38.33	38.79	38.44
TiO_2	0.03	0.00	0.00	0.00	0.02	0.00	0.03	0.05	0.04	0.07	0.18
Al_2O_3	21.03	26.28	25.84	23.19	21.66	20.74	26.40	25.68	24.25	27.74	24.85
Cr_2O_3	0.01	0.03	0.00	0.00	0.04	0.04	0.00	0.00	0.01	0.00	0.05
FeO_T	15.17	10.29	10.69	12.94	14.70	15.79	8.99	9.99	11.83	7.97	10.75
MnO	0.14	0.40	0.25	0.23	0.08	0.03	0.12	0.05	0.33	0.19	0.09
MgO	0.12	0.06	0.07	0.09	0.12	0.03	0.16	0.03	0.19	0.03	0.05
CaO	23.10	23.56	23.91	22.99	23.00	23.28	23.32	23.76	23.10	23.77	23.67
Na_2O	0.02	0.005	0.002	0.01	0.01	0.01	0.004	0.01	0.01	0.01	0.004
K_2O	0.01	0.002	0.002	0.001		0.004	0.004	0.004			0.003
Cl		0.01									
Total	97.83	98.73	99.31	97.69	98.00	97.84	98.29	98.11	98.10	98.57	98.08

Table 11. Trace element concentrations (ppm) for epidote by SIMS.

Sample	RN-17-650	RN-17-650	RN-17-750	RN-17-1100	RN-17-1450		RN-17-1550	RN-17-2150		RN-17-3000		
Grain	ep 2a	ep 3b	ep 2a	ep 1a	ep 2a	ep 2b	ep 2b	ep 3a	ep 3b	ep 1a	ep 1b	ep 2a
Zone	lighter area	lighter area	dark area	lighter area	dark area	dark area	lighter area	dark area	dark area	dark area	dark area	dark area
Sc	13	2.7	5.9	7.0	5.6	5.2	5.5	5.3	6.1	9.3	7.8	2.9
Ti	526	237	590	785	267	558	623	180	957	529	652	230
V	163	105	274	514	257	256	337	300	393	276	223	134
Cr	1.7	9.5	1.1	2.1	0.8	0.9	3.2	1.4	27	9.2	9.5	0.80
Rb	8.4	5.5	6.9	7.3	4.5	3.8	4.6	5.3	5.2	3.7	4.4	3.1
Sr	604	752	655	746	720	796	778	1313	795	694	594	587
Y	4.2	4.4	9.3	3.9	3.9	5.9	3.2	3.3	4.6	3.7	2.1	1.67
Zr	0.5	1.1	0.3	4.2	4.0	3.8	19	0.30	5.0	3.4	4.4	0.7
Nb	0.44	0.21	0.34	0.97	0.12	0.32	0.48	0.11	0.14	0.04	0.23	0.02
Cs				0.01			0.02			0.00	0.01	
Ba	17	27	6.9	28	6.2	12	2.8	19	16	6.2	3.9	2.8
La	4.2	11	4.8	3.0	1.7	2.3	0.5	3.4	1.7	0.30	0.07	0.16
Ce	4.9	10	8.1	4.4	3.4	4.5	0.93	4.0	3.3	0.86	0.26	0.34
Nd	1.71	3.6	3.9	1.8	1.9	2.8	0.61	1.4	1.9	0.80	0.25	0.32
Sm	0.43	0.54	0.90	0.26	0.52	0.92	0.20	0.35	0.50	0.26	0.14	0.13
Eu	1.22	0.88	12	2.4	1.3	1.8	0.86	3.6	1.3	0.48	0.18	0.74
Gd	0.31	0.56	0.75	0.30	0.53	1.1	0.31	0.34	0.87	0.31	0.12	0.28
Dy	0.21	0.48	0.87	0.35	0.58	1.2	0.35	0.24	0.55	0.28	0.14	0.31
Er	0.25	0.26	0.71	0.27	0.33	0.57	0.24	0.11	0.28	0.22	0.08	0.15
Yb	0.20	0.22	0.42	0.30	0.27	0.55	0.28	0.06	0.3	0.18	0.10	0.08
Th	0.04	0.03	0.04	0.03	0.04	0.04	0.11	0.02	0.02	0.06	0.02	0.02
La_N/Yb_N	14.6	34	7.8	6.9	4.4	2.9	1.2	39	3.8	1.2	0.48	1.4
Eu_N/Eu_N^*	10.1	4.9	45	26	7.3	5.4	10	32	6.1	5.1	4.2	12

Eu_N/Eu^*_N = Eu anomaly, with $Eu^*_N = (Sm_N \times Gd_N)^{1/2}$.

On the basis of a petrographic study [13] amphibole was found from 1022 m down to the bottom of the well RN-17 as amygdale and vein mineral, and as clinopyroxene replacement. However, we did not find amphibole in the RN-17-1550 and RN-17-1700 cuttings. For this reason, our data on amphibole are limited to samples RN-17-1450, -2150, -2800 and -3000, along with the dolerite dyke RN-19-2246. EPMA data collected in our set of amphiboles are shown in table 12. On the basis of the diagram Si (atoms per formula units, apfu) versus X_{Mg} [X_{Mg} = Mg/(Mg+Fe^{2+})] by [47], amphibole is characterized by a large compositional variability, with the widest Si and XMg ranges occurring in the deepest cutting RN-17-3000 (Si = 7.1 - 7.9 apfu, X_{Mg} = 0.34 - 0.66). This compositional variability is that of amphibole of a sheeted dike complex from Hess Deep (Equatorial Pacific; [48]) and from Hole 504B (Costa Rica Rift; [49]) (figure 11). [48] interpreted it as the product of early interaction with magmatic fluids exsolved locally during the final stages of crystallisation. La_N/Sm_N, Dy_N/Yb_N and Eu anomaly (Eu_N/Eu^*_N) range from 0.09 to 1.3, 0.40 to 1.9 and 0.6 to 3.2, respectively. The results of SIMS analyses are collected in table 13. Based on SIMS REE abundance and shape of REE patterns, two main compositional groups are distinguished (figure 12). One group, consisting of actinolite from samples RN-17-2150, -2800, -3000 and RN-19-2246, is characterized by low REE contents and LREE depletion (La_N/Yb_N = 0.08 - 0.39). The second group, including Mg-hornblende from the deepest cutting RN-17-3000, shows the highest REE enrichments and positive LREE/HREE fractionation (La_N/Yb_N = 1.6). The occurrence of positive and negative Eu anomalies may be related to the alteration of clinopyroxene, on one hand, and to the buffering effect of plagioclase, on the other hand.

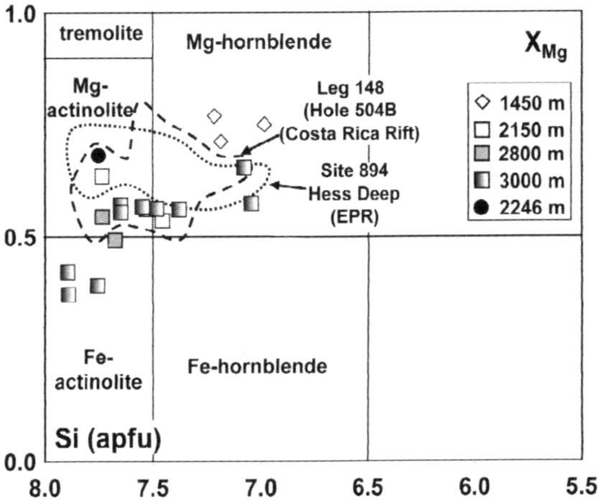

Figure 11. Amphibole classification [after [47]] of cuttings and core-sample from wells RN-17 and RN-19, respectively. The fields of amphiboles from Leg 148 (Hole 504B, Costa Rica Rift; [49]) and Site 894 Hess Deep (Cocos-Nazca rift valley, East Pacific Rise; [62]) geothermal systems are reported for comparison.

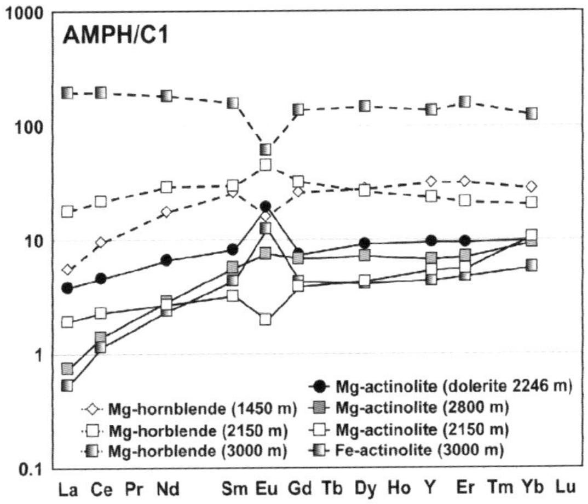

Figure 12. C1-chondrite-normalized REE patterns, obtained by SIMS, for amphibole from wells RN-17 and RN-19 (Reykjanes geothermal system).

5. Discussion

The investigation of crushed lithology of drillhole cuttings in the well RN-17 of the Reykjanes peninsula has precluded detailed analysis of paragenetic relationships among minerals and mineralized vein systems. Nevertheless the results of our multi-analytical approach, represented by EPMA, SIMS

Table 12. Electron probe data (wt%) for amphibole.

sample	**RN-17-1450**			**RN-17-2150**		**RN-17-2800**		**RN17-3000**		
spot	Mg-bl 1A	Mg-hbl 1A2	Mg-hbl 1D	Mg-hbl 1a-ar	Mg-act ar-2a	Mg-act 2b-ar	Mg-act 2c-ar	Mg-hbl 1B	Mg-hbl 1E	Mg-act 4B
SiO_2	50.05	50.7	48.88	50.78	53.48	52.17	51.79	48.94	47.78	51.11
TiO_2	0.39	0.50	0.42	0.29	0.21	0.31	0.17	0.39	0.68	0.29
Al_2O_3	6.42	7.14	9.14	4.47	2.26	2.21	2.36	7.87	7.33	4.08
Cr_2O_3	0.10	0.00	0.05	0.01	0.00	0.00	0.00	0.10	0.00	0.00
FeO_T	11.50	9.07	9.69	17.88	14.69	17.97	21.20	13.24	16.24	16.34
MnO	0.33	0.31	0.33	0.42	0.29	0.47	0.50	0.52	0.58	0.62
MgO	16.37	17.45	16.76	11.87	14.72	12.39	11.75	14.35	12.60	12.05
CaO	11.82	10.90	10.68	11.68	11.86	10.94	10.03	11.51	11.50	12.70
Na_2O	0.86	1.07	1.27	1.02	0.30	0.60	0.58	1.06	1.06	0.29
K_2O	0.10	0.08	0.11	0.06	0.04	0.07	0.06	0.08	0.03	0.03
Cl						0.03	0.03			
Total	97.93	97.22	97.32	98.48	97.86	97.14	98.44	98.05	97.79	97.51
Mg#	0.72	0.77	0.75	0.54	0.64	0.55	0.50	0.66	0.58	0.57

sample									**RN-19-2246**
spot	Mg-act 4C	Mg-hbl 4D	Mg-hbl 4F	Mg-act 4I	Mg-act 4O	Fe-act 1a-ar	Fe-act 2a-ar	Fe-act 2a2-ar	Mg-act 1C
SiO_2	51.47	49.91	51.08	52.30	50.88	51.55	52.72	52.56	54.26
TiO_2	0.35	0.52	0.16	0.38	0.20	0.00	0.05	0.05	0.20
Al_2O_3	3.22	4.99	4.57	3.55	3.72	1.87	0.77	0.64	2.42
Cr_2O_3	0.06	0.11	0.02	0.00	0.04	0.00	0.00	0.01	0.00
FeO_T	16.24	16.73	16.70	16.87	16.78	24.29	23.66	25.99	13.67
MnO	0.79	0.75	0.81	0.84	0.81	0.74	0.42	0.52	0.23
MgO	12.44	12.28	12.32	12.05	12.69	8.96	9.91	8.82	16.79
CaO	11.37	11.18	11.83	11.46	11.40	10.82	10.58	10.35	10.01
Na_2O	0.34	0.91	0.59	0.57	0.28	0.32	0.24	0.23	0.27
K_2O	0.06	0.02	0.03	0.03	0.05	0.03	0.02	0.03	0.10
Cl									
Total	96.35	97.40	98.10	98.05	96.84	98.59	98.38	99.20	97.94
Mg#	0.58	0.57	0.57	0.56	0.57	0.40	0.43	0.38	0.69

Mg# = $Mg/(Mg+Fe^{2+})$; Mg-hbl = Mg-horblende; Mg-act = Mg-actinolite; Fe-act = Fe-actinolite.

IOP Conf. Series: Materials Science and Engineering **32** (2012) 012020 doi:10.1088/1757-899X/32/1/012020

Table 13. Trace element concentrations (ppm) for amphibole by SIMS.

sample	**RN-17-1450**				**RN-17-2800**						
grain	amph 1A Mg-hornblende	amph 1D Mg-hornblende	average (2)	1σ	amph-2b-actinolite	amph-2c-actinolite	average	1σ	amph 1D actinolite	average	1σ
Sc	62	51	*57*	*7.8*	10	7.5	*8.8*	*1.8*	66	*68*	*2.5*
Ti	1643	2335	*1989*	*489*	752	526	*639*	*159*	1593	*1415*	*252*
V	257	337	*297*	*57*	107	128	*118*	*15*	244	*239*	*5.9*
Cr	170	198	*184*	*20*	b.d.l.	30.1	*30*		5.5	*5.5*	
Rb	1.0	1.2	*1.1*	*0.14*	9.7	15	*12*	*3.6*	2.6	*3.0*	*0.57*
Sr	3.5	5.9	*4.7*	*1.7*	13	6.0	*10*	*4.9*	5.8	*5.9*	*0.14*
Y	46	51	*49*	*3.5*	11	10	*10*	*0.42*	14	*15*	*1.27*
Zr	39.1	34.2	*37*	*3.5*	105	48	*76*	*40*	1.4	*1.4*	*0.07*
Nb	1.19	1.64	*1.4*	*0.32*	2.0	1.7	*1.9*	*0.18*	0.04	*0.04*	*0.00*
Cs	0.035	0.052	*0.04*	*0.01*	0.0	0.02	*0.01*	*0.01*	0.08	*0.07*	*0.02*
Ba	11.7	19.3	*16*	*5.4*	5.9	3.6	*4.8*	*1.63*	11	*10*	*1.41*
La	0.92	1.68	*1.3*	*0.54*	0.23	0.12	*0.18*	*0.08*	0.92	*0.89*	*0.04*
Ce	4.31	7.16	*5.7*	*2.0*	0.94	0.75	*0.85*	*0.13*	2.72	*2.75*	*0.04*
Nd	7.14	8.65	*7.9*	*1.1*	1.36	1.24	*1.30*	*0.08*	3.0	*3.0*	*0.02*
Sm	3.67	3.91	*3.8*	*0.17*	0.81	0.86	*0.84*	*0.04*	1.13	*1.20*	*0.10*
Eu	0.83	0.95	*0.89*	*0.08*	0.46	0.38	*0.42*	*0.06*	1.1	*1.08*	*0.04*
Gd	4.56	5.63	*5.1*	*0.76*	1.28	1.39	*1.34*	*0.08*	1.39	*1.45*	*0.08*
Dy	6.38	7.01	*6.7*	*0.45*	1.74	1.70	*1.72*	*0.03*	2.02	*2.20*	*0.25*
Er	4.48	5.44	*5.0*	*0.68*	1.16	1.07	*1.12*	*0.06*	1.49	*1.49*	
Yb	3.83	5.22	*4.5*	*1.0*	1.49	1.53	*1.51*	*0.03*	1.51	*1.61*	*0.13*
Hf	2.54	2.45	*2.5*	*0.06*	n.a.	n.a.			0.29	*0.29*	
Th	0.25	0.08	*0.17*	*0.12*	0.08	0.05	*0.07*	*0.02*	0.03	*0.03*	
La_N/Yb_N	0.17	0.22	*0.20*	*0.04*	0.11	0.05	*0.08*	*0.04*	0.42	*0.39*	*0.05*
Eu_N/Eu_N*	0.62	0.61	*0.62*	*0.001*	1.37	1.06	*1.21*	*0.22*	2.67	*2.48*	*0.26*
$Ti_N/Ti*_N$	0.15	0.19	*0.17*	*0.02*	0.25	0.17	*0.21*	*0.06*	0.48	*0.40*	*0.11*
$Sr_N/Sr*_N$	0.04	0.05	*0.05*	*0.01*	0.77	0.42	*0.61*	*0.25*	0.14	*0.14*	*0.003*
$Zr_N/Zr*_N$	0.50	0.39	*0.44*	*0.08*	6.53	3.0	*4.8*	*2.5*	0.05	*0.05*	*0.004*

Table 13. continued.

sample	**RN-17-3000**					**RN-19-2246**			
grain	amph 4D Mg-hornblende	amph-1a-ar Fe-actinolite	amph-1a-ar (spot 2) Fe-actinolite	average (2) Fe-actinolite	1σ	amph 1C actinolite	amph 1D actinolite	average (2)	1σ
Sc	*43*	3.1	2.2	*2.7*	*0.64*	70	66	*68*	*2.5*
Ti	*564*	362	360	*361*	*1.1*	1238	1593	*1415*	*252*
V	*388*	136	133	*134*	*1.8*	235	244	*239*	*5.9*
Cr	*0.30*	21	18	*20*	*1.9*	b.d.l.	5.5	*5.5*	
Rb	*0.1*	9.1	8.2	*8.7*	*0.64*	3.4	2.6	*3.0*	*0.57*
Sr	*762*	16	14	*15*	*1.4*	6.0	5.8	*5.9*	*0.14*
Y	*209*	8.0	5.6	*6.8*	*1.7*	16	14	*15*	*1.27*
Zr	*79*	20	22	*21*	*1.1*	1.3	1.4	*1.4*	*0.07*
Nb	*0.20*	0.74	0.70	*0.72*	*0.03*	0.04	0.04	*0.04*	*0.00*
Cs	*0.03*	0.02		*0.02*		0.05	0.08	*0.07*	*0.02*
Ba	*11*	5.2	3.5	*4.4*	*1.2*	9.0	11	*10*	*1.41*
La	*45.9*	0.13	0.12	*0.13*	*0.01*	0.86	0.92	*0.89*	*0.04*
Ce	*118*	0.78	0.60	*0.69*	*0.13*	2.78	2.72	*2.75*	*0.04*
Nd	*82.4*	1.09	1.05	*1.07*	*0.03*	2.99	3.0	*3.0*	*0.02*
Sm	*23*	0.73	0.55	*0.64*	*0.13*	1.27	1.13	*1.20*	*0.10*
Eu	*3.40*	0.81	0.58	*0.70*	*0.16*	1.05	1.1	*1.08*	*0.04*
Gd	*26.7*	0.93	0.75	*0.84*	*0.13*	1.51	1.39	*1.45*	*0.08*
Dy	*35*	1.15	0.84	*1.00*	*0.22*	2.37	2.02	*2.20*	*0.25*
Er	*24.7*	0.90	0.61	*0.76*	*0.21*	1.49	1.49	*1.49*	
Yb	*19.8*	1.13	0.72	*0.93*	*0.29*	1.70	1.51	*1.61*	*0.13*
Hf	*5.4*	n.a.	n.a.			0.29	0.29	*0.29*	
Th	*2.3*	0.06	0.09	*0.08*	*0.02*	0.03	0.03	*0.03*	
La_N/Yb_N	*1.61*	0.08	0.12	*0.09*	*0.03*	0.35	0.42	*0.39*	*0.05*
Eu_N/Eu_N^*	*0.42*	3.0	2.74	*2.88*	*0.17*	2.30	2.67	*2.48*	*0.26*
Ti_N/Ti^*_N	*0.01*	0.18	0.23	*0.20*	*0.04*	0.33	0.48	*0.40*	*0.11*
Sr_N/Sr^*_N	*0.52*	1.2	1.2	*1.2*	*0.01*	0.14	0.14	*0.14*	*0.003*
Zr_N/Zr^*_N	*0.12*	1.5	1.9	*1.6*	*0.28*	0.04	0.05	*0.05*	*0.004*

Mg-hbl = Mg-hornblende; Mg-act = Mg-actinolite; Fe-act = Fe-actinolite; Eu_N/Eu^*_N = Eu anomaly, with $Eu^*_N = (Sm_N \times Gd_N)^{1/2}$; Ti_N/Ti^*_N = Ti anomaly, with $Ti^*_N = (Gd_N \times Dy_N)^{1/2}$; Sr_N/Sr^*_N = Sr anomaly, with $Sr^*_N = (Ce_N \times Nd_N)^{1/2}$; Zr_N/Zr^*_N = Zr anomaly, with $Zr^*_N = (Nd_N \times Sm_N)^{1/2}$; n.a. = not analyzed. Italic data represent both the average when more data are available, and a single analysis when sample heterogeneity did not allow more investigations.

(as in situ techniques) and XRF, ICP-MS and TIMS (as bulk techniques) allowed us to get insights into the geochemical characteristics of the well, listed in the following:

1) significant variations of REE, K_2O, Rb, Ba and B contents, LREE/HREE fractionations, and Sr isotope ratios in the cuttings, whereas Li contents are almost constant and Nd isotope ratios show the same variations as observed in Reykjanes basalts;
2) variable plagioclase and clinopyroxene compositions in terms of major (such as: anorthite, K, Mg and Fe contents for plagioclase; Mg#, Ti and Cr for clinopyroxene) and trace elements (such as REE, Ba and Sr) as well as of Sr isotope ratios;
3) the lowest Sr isotope ratios in some of the most REE-enriched cuttings (RN-17-1550 and -2800);
4) the presence of epidote from 312 m down to 3000 m depth, and of amphibole from 1450 m down to 3000 m depth;
5) variable U concentrations and U/Th fractionations.

REE patterns of Ca-plagioclase and, more significantly, of clinopyroxene are those of magmatic minerals even if REE contents and LREE/HREE fractionations can be affected by various degrees of hydrothermal alteration. The REE concentrations of hydrothermal fluids are generally low when compared to rocks: thus significant REE depletions of altered basalts are expected only at very high fluid rock ratios ([50] and reference therein).

According to [51] and [52], Sr isotope composition of hydrothermal waters of the Reykjanes Peninsula and, in general, in Iceland falls along a mixing line between fluids of meteoric origin and seawater systems with a decrease in fluid salinity from west to east. Sr isotopic data of the Reykjanes geothermal system have $^{87}Sr/^{86}Sr = \sim 0.7040$, lower than seawater ($^{87}Sr/^{86}Sr = 0.7092$) and higher than those of fresh Reykjanes Peninsula basalts ($^{87}Sr/^{86}Sr \sim 0.7032$). [51] and [52] argued that during water/rock interaction, Sr is leached from basalts by hot seawater but ~ 75 % of the leached Sr is re-deposited in secondary minerals of which epidote and chlorite are likely the most important. The Sr and Nd isotope decoupling of the altered basalts, along with that of the elemental B and Li (figure 2a), and the variable contents of K, Ba, Rb and B would be the result of the interaction with a fluid enriched in these elements. The down-hole variation in K_2O-concentration (figure 13a), Rb_N/La_N (figure 13b), Ba_N/La_N (figure 13c) of the cuttings would seem to indicate that the alteration has been more significant in the shallower parts (e.g., RN-17-400 and -650 samples) of the well RN-17. The water/rock ratio during the interaction between the geothermal fluids and the Reykjanes basalts has been estimated using the modelling of [53] and [54] (see Appendix I, for details). Our results indicate that the water/rock ratio (N) varies between 0 and 3.6 (figure 14) decreasing, though not systematically, with depth. The results of the modelling of the Sr isotope variations of cuttings are in agreement with a series of observations: i) the decrease of Sr isotopic disequilibria between minerals and whole-rocks with depth (figures 15a and b); ii) the decrease of $^{87}Sr/^{86}Sr$ values of epidote from 0.70478 (RN-17-1100) to 0.70373 (RN-17-2150) with a corresponding increase of Sr content from 750 to 1300 ppm; and iii) the systematic decrease of the U/Th ratio with increasing depth (figure 16a), with decreasing water/rock ratio (figure 16b), and with decreasing Sr isotope ratio (figure 16c). The fluid responsible for the hydrothermal alteration interacts and exchanges Sr and U with the host basalts, decreasing down-hole its amount relative to the rocks.

Fluid composition at Reykjanes has evolved through time as a result of changing proportions of meteoric water influx as well as differing pressure and temperature conditions [14, 55, 56]. Stable isotope analysis of fluids and alteration minerals reveals significant hydrothermal interaction with basalts as well as inputs of both shallow and deep meteoric fluids to the primarily high salinity fluids throughout the history of the geothermal system [12, 14, 55-57]. The lithology of the Reykjanes drill field is characterized by phreatic submarine tuffs, hyaloclastite breccias and shallow marine sediments in the uppermost 1 km, followed by deeper marine pillow basalt formations at lower levels. Intrusive rock intensity varies considerably with increasing depth, increases in general but not gradually, nor have apparent magma chambers been detected at Reykjanes by any geophysical means so far [56, 58]. The effects of temperature, increasing with depth, would explain the occurrence of alteration minerals, such as epidote and amphibole, and strongly altered minerals, such as Na-rich plagioclases, also in the

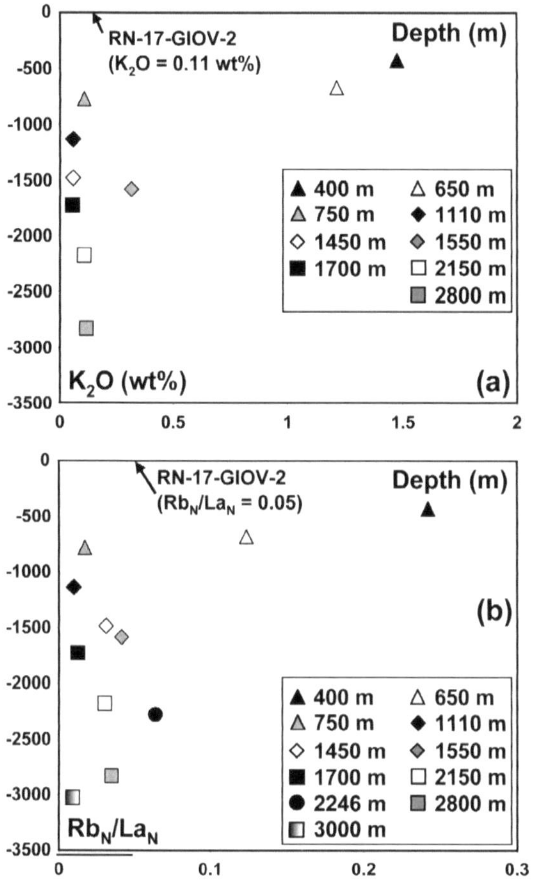

Figure 13. Plots of (a) K_2O (wt%), (b) Rb_N/La_N and (c) Ba_N/La_N versus depth to document the alteration degree of our samples. The K, Rb and Ba alteration process seems to be more consistent in the shallower samples (400, 650 m). The arrows indicate the position of surface lava RN-17-GIOV-2.

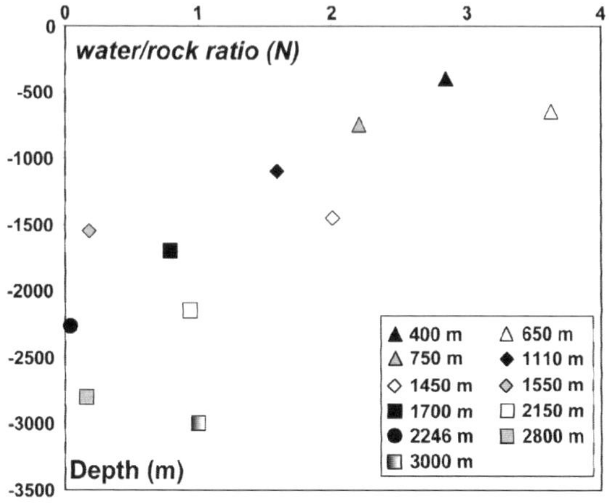

Figure 14. Water/rock ratio (N) versus depth for the studied samples. See the text and Appendix I for more details.

deep cuttings. Moreover, the occurrence of epidote in the deeper cuttings, such as RN-17-2150 and -3000, can account for the higher bulk rock $^{87}Sr/^{86}Sr$ isotope ratios of these samples with respect to clinopyroxene from the same depths (figure 15b). The occurrence of strongly altered primary minerals down to the well bottom is an evidence of the persistent fluid/rock interaction at depth.

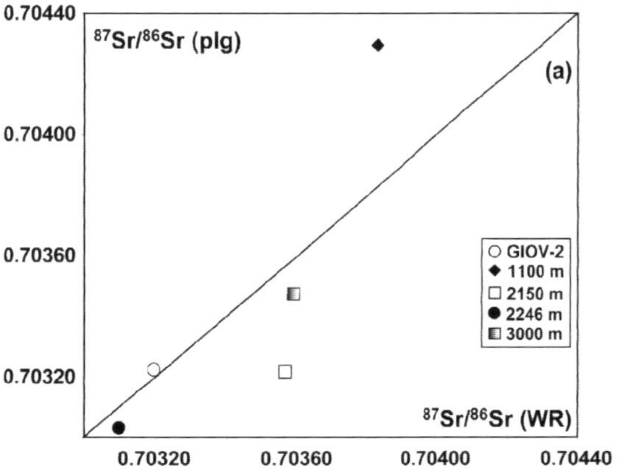

Figure 15. $^{87}Sr/^{86}Sr$ in plaioclaseg (a), and $^{87}Sr/^{86}Sr$ in clinopyroxene (b) versus whole-rock (WR) $^{87}Sr/^{86}Sr$ for selected cuttings from well RN-17 and RN-19. The higher Sr isotope compositions in the whole rock cuttings suggest the presence of secondary phase(s) enriched in ^{87}Sr.

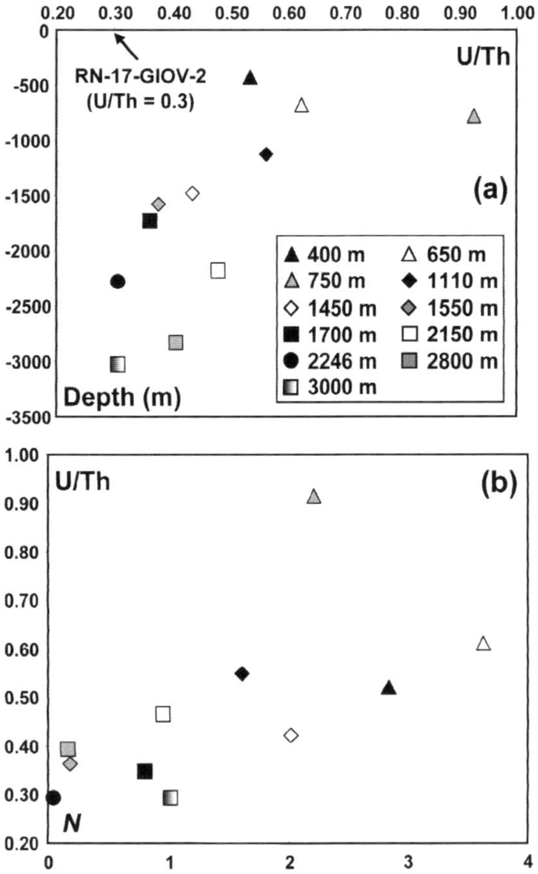

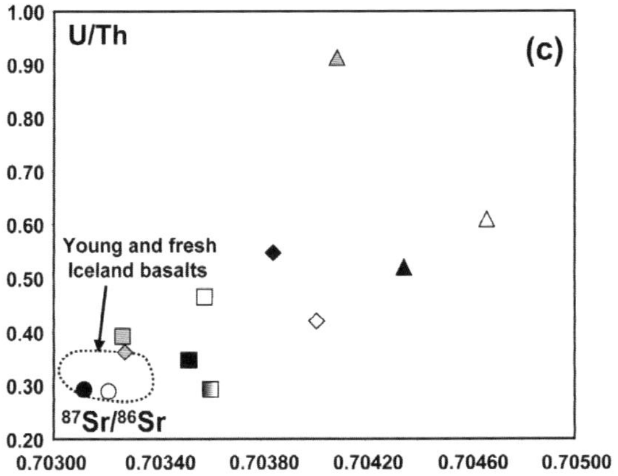

Figure 16. Whole-rock U/Th versus depth (a), water/rock ratio N (b), and $^{87}Sr/^{86}Sr$ (**c**), for the studied samples. The arrow in (a) indicates the position of the surface lava RN-17-GIOV-2. The field of young and fresh Iceland basalts [63], [28] and [64] is reported for comparison in (c).

6. Conclusions

In the present study, bulk-rock analyses (XRF, ICP-MS and TIMS) and in-situ (EMPA, SIMS) investigations have been carried out on 10 drill cutting samples (depth 400 - 3000 m) and mineral phases, i.e., plagioclase, clinopyroxene, epidote and amphibole, from well RN-17 drilled in the Reykjanes geothermal system. In addition, one unaltered lava flow (RN-17-GIOV-2) and one core dolerite dyke (well RN-19) from 2246 m depth have been considered for comparison.

The data on trace elements derived by the ion microprobe analysis emphasize the important role played by SIMS to complement the more traditional bulk rock analysis in the investigation of complex geological areas. Whole rock and mineral phases (plagioclase, clinopyroxene, epidote and amphibole) indicate that the cuttings were affected by hydrothermal alteration that has been able to modify the Sr isotope ratios and the concentration of K, Rb, Cs, Ba, U, B. Seawater/rock ratio (N), evaluated according to [53, 54] for Sr isotope, achieved its maximum for RN-17-650, decreasing although not systematically with depth, with the minimum value for cutting RN-17-2800. The occurrence of Na-rich plagioclase, epidote and amphibole down to the well root, on the other hand, is an evidence of the persistence of fluid/rock interaction at depth.

Acknowledgements

This research has been partly funded to L.P. Ottolini by CNR grants (year 2008): RSTL: Study of basaltic crust of Iceland (Iceland Deep Drilling Project, IDDP). Additional fundings derived from SIMS-CNR IGG (Pv) research projects (year 2010).

Many thanks to G.H. Gudfinnsson (ISOR, Iceland) for his accurate revision of the draft, and to two anonymous reviewers for useful comments on the final version of the paper.

References

[1] Gillis K M, Thompson G and Kelley D S 1993 *J. Geophys. Res. B* **98** 19597.
[2] Vanko D A and Laverne C 1998 *Earth Planet. Sci. Lett.* **162** 27
[3] Laverne C, Agrinier P, Hermitte D and Bohn M 2001 *Chem. Geol.* **181** 73
[4] Jakobsson S 1972 *Lithos* **5** 365
[5] Pálmason G and Saemundsson K 1974 *Ann. Rev. Earth Planet. Sci.* **2** 25
[6] Jakobsson S, Jónsson J and Shido F 1978 *J. Petrol.* **19** 669
[7] Clifton A E and Kattenhorn S A 2006 *Tectonophysics* **419** 27
[8] Franzson H, Thordarson S, Björnsson G, Gudlaugsson S Th, Richter B, Fridleifsson G Ó and Thorhallsson S 2002 *Proc. 27th Workshop on Geothermal Reservoir-Engineering, SGP-TR-171. (Stanford, CA: Stanford University)* (http://pangea.stanford.edu/ERE/pdf/IGAstandard/SGW/2002/Franzson.pdf)
[9] Marks N, Schiffman P, Zierenberg R A, Franzson H and Fridleifsson G Ó 2010 *J. Volcanol. Geotherm. Res.* **189** 172
[10] Marks N, Schiffman P, Zierenberg R A, Elders W A, Fridleifsson G Ó and Franzson H 2010 *Proc. World Geothermal Congress 2010 (Bali, Indonesia, 25-29 April 2010)*
[11] Björnsson S, Arnórsson S and Tómasson J 1970 *Geothermics* **56** 2380
[12] Sveinbjörnsdóttir Á E, Coleman M L and Yardley B W D 1986 *Contrib. Mineral. Petrol.* **94** 99
[13] Fridleifsson G Ó, Blischke A, Kristjansson B R, Einarsson G M, Jónasson H, Franzson H, Sigurdsson O, Danielssen P E, Jónsson S J, Thordarson S, Thorhallsson S, Hardardottir V and Egilsson Th 2005 *Reykjanes, well report RN-17 & RN-17ST. Drilling operations of well RN-17 and sidetrack from surface down to 3082.4 m; and IDDP_ICDP spot coring in well RN-19 at 2248 m depth.* ISOR-2005/007 (Iceland GeoSurvey), 198 p. (including appendices)
[14] Pope E C, Bird D K, Arnórsson S, Fridriksson Th, Elders W A and Fridleifsson G Ó 2009 *Geochim. Cosmochim. Acta* **73** 4468

[15] Raffone N, Ottolini L, Tonarini S, Gianelli G and Fridleifsson G Ó 2008 *Microchim. Acta* **161** 307
[16] Raffone N, Ottolini L P, Tonarini S, Gianelli G and Fridleifsson G Ó 2010 *IOP Conf. Ser.: Mater. Sci. and Engng.* **7** 012026
[17] Freedman A J E, Bird D K, Arnórsson S, Fridriksson T, Elders W A and Fridleifsson G Ó 2009 *Amer. J. Sci.* **309** 788
[18] Bottazzi P, Ottolini L, Vannucci R and Zanetti A 1994 *Secondary ion mass spectrometry, SIMS IX Proceedings.* Benninghoven A, Nihei Y, Shimizu N and Werner H W, eds. (Chichester: Wiley) 413
[19] Trua T, Clocchiatti R, Schiano P, Ottolini L and Marani M 2010 *Lithos* **118** 1
[20] Rautenschlein M, Jenner G A, Hertogen J, Hofmann A W, Kerrich R, Schmincke H-U and White W M 1985 *Earth Planet. Sci. Lett.* **75** 369
[21] Tanaka T, Togashi S, Kamioka H, Amakawa H, Kagami H, Hamamoto T, Yuhara M et al. 2000 *Chem. Geol.* **168** 279
[22] Wood D A 1977 *Ph.D.-thesis.* (London: University of London) ; unpublished
[23] Wood D A 1979 *Geochim. Cosmochim. Acta* **43** 1031
[24] Zindler A, Hart S R, Frey F A and Jakobsson S P 1979 *Earth Planet. Sci. Lett.* **45** 249
[25] Lindstrom M M and Haskin L A 1981 *Geochim. Cosmochim. Acta* **45** 15
[26] Metrich N, Sigurdsson H, Meyer P S and Devine J D 1991 *Contrib. Mineral. Petrol.* **107** 435
[27] MacDonald R, McGarvie D W, Pinkerton H, Smith R L and Palacz Z A 1990 *J. Petrol.* **31** 429
[28] Hémond Ch, Arndt N T, Lichtenstein U, Hofmann A W, Óskarsson N and Steinthorsson S 1993 *J. Geophys. Res.* **98** 15833
[29] Gurenko A A and Chaussidon M 1997 *Chem. Geol.* **135** 21
[30] Gurenko A A and Chaussidon M 2002 *Earth Planet. Sci. Lett.* **205** 63
[31] Hilton D R, Gronvold K, Macpherson C G and Castillo P R 1999 *Earth Planet. Sci. Lett.* **173** 53
[32] Ryan J G and Langmuir C H 1987 *Geochim. Cosmochim. Acta* **51** 1727
[33] Ryan J G, Leeman W P, Morris J D and Langmuir C H 1996 *Geochim. Cosmochim. Acta* **60** 415
[34] Langmuir C H, Bender J F and Batiza R 1977 *Earth Planet. Sci. Lett.* **36** 133
[35] Bender J F, Langmuir C H and Hanson G N 1984 *J. Petrol.* **25** 213
[36] Ryan J G and Langmuir C H 1993 *Geochim. Cosmochim. Acta* **57** 1489
[37] Chaussidon M and Jambon A 1994 *Earth Planet. Sci. Lett.* **121** 277
[38] Hofmann A W, Feigenson M D and Raczek I 1984 *Contrib. Mineral. Petrol.* **88** 24
[39] Thirlwall M F, Gee M A M, Taylor R N and Murton B J 2004 *Geochim. Cosmochim. Acta* **68** 361
[40] Gee M A M, Thirlwall M F, Taylor R N, Lowry D and Murton B J 1998 *J. Petrol.* **39** 819
[41] Murton B J, Taylor R N and Thirlwall M F 2002 *J. Petrol.* **43** 1987
[42] Furman T, Frey F A and Park K-H 1991 *Contrib. Mineral. Petrol.* **109** 19
[43] Baker J H 1985 *Mineral. Mag.* **49** 107
[44] Kristmannsdóttir H 1979 *International clay conference 1978*. Mortland and Farmer, eds. (Amsterdam: Elsevier Sci. Publ. Company), pp. 359-367
[45] Absar A 1991 *J. Geol. Soc. India* **38** 625
[46] Martinez-Serrano R G 2002 *Geothermics* **31** 579
[47] Leake B E, Maresch W V, Nickel E H, Rock N M S, Schumacher J C, Smith D C, Stephenson N C N, Ungaretti L, Whittaker E J W and Youzhi G 1997 *Canad. Mineral.* **35** 219
[48] Gillis K M, Muehlenbachs K, Stewart M and Karson J 2001 *J. Geophys. Res.* **106** 26311
[49] Alt J C, Honnorez J, Laverne C and Emmermann R 1986 *J. Geophys. Res. B* **91** 10309
[50] Bach W and Irber W 1998 *Chem. Geol.* **151** 309
[51] Ederfield H and Greaves M J 1981 *Geochim. Cosmochim. Acta* **45** 2201

[52] Cohen R S and O'Nions R K 1982 *J. Petrol.* **23** 299
[53] Taylor H P 1977 *J. Geol. Soc. Lond.* **133** 509
[54] McCulloch M T, Gregory R T, Wasserburg G J and Taylor M P J 1980 *Earth Planet. Sci. Lett.* **46** 201
[55] Arnórsson S 1995 *Geothermics* **24** 561
[56] Fridleifsson G Ó, Ármannsson H, Árnason K, Bjarnason I Þ and Gíslason G 2003 *Iceland deep drilling project, report-part I-III, IDDP feasibility report, geosciences - Site selection.* pp. 1-103
[57] Ólafsson J and Riley J 1978 *Chem. Geol.* **2** 219
[58] Fridleifsson G Ó and Elders W A 2005 *Geothermics* **34** 269
[59] Anders E and Grevesse N 1989 *Geochim. Cosmochim. Acta* **53** 197
[60] Larsson D, Grönvold K, Oskarsson N and Gunnlaugsson E 2002 *J. Volcanol. Geotherm. Res.* **114** 275
[61] Peate D W, Baker J A, Jakobsson S P, Waight T E, Ken A J R, Grassineau N V and Skovgaard A C 2009 *Contrib. Mineral. Petrol.* **157** 359
[62] Gillis K M 1996 in: *Proc. ODP Science Results 147.* Gillis M C, Allan K M and Meyer P S, eds. pp. 59-75
[63] Hémond Ch, Codomines M, Foucade S, Allègre C J, Oskarsson N and Javoy M 1988 *Earth Planet. Sci. Lett.* **87** 273
[64] Chauvel C and Hémond C 2000 *Geochem. Geophys. Geosyst.* **1**

IOP Conf. Series: Materials Science and Engineering **32** (2012) 012020 doi:10.1088/1757-899X/32/1/012020

Appendix I

Evaluation of seawater/rock ratio (N) using [53] and [54] equation:

$$N = [(\delta^f_r - \delta^i_r)/(\delta^i_w - \delta^f_w)]* C^i_r/C^i w$$

where:
δ^f_r = Sr isotope composition for each altered rock
δ^i_r = Sr isotope composition for unaltered rock ($^{87}Sr/^{86}Sr$ = 0.703198 for RN-17-GIOV-2)
δ^i_w = Sr isotope composition of unmodified seawater ($^{87}Sr/^{86}Sr$ = 0.70916)
δ^f_w = Sr isotope composition of modified seawater ($^{87}Sr/^{86}Sr$ = 0.7040 for Reykjanes system; [51, 52])
C^i_r = Sr concentration in unaltered rock in µg/g (Sr = 100 ppm for RN-17-GIOV-2)
C^i_w = Sr concentration in seawater (Sr = 7.8 ppm)

and assuming that:

$$\delta^f_r = \delta^f_w - \Delta\text{water-rock}$$

with isotopic fractionation factors ($\Delta_{water\text{-}rock}$) = 0 for $^{87}Sr/^{86}Sr$ ratio, results:

Sample	δ^f_r	N
RN-17-400	0.704333	2.82
RN-17-650	0.704650	3.61
RN-17-750	0.704075	2.18
RN-17-1100	0.703830	1.57
RN-17-1450	0.703997	1.99
RN-17-1550	0.703261	0.16
RN-17-1700	0.703507	0.77
RN-17-2150	0.703567	0.92
RN-17-2800	0.703253	0.14
RN-17-3000	0.703590	0.97
RN-19-2246	0.703101	0

IOP Conf. Series: Materials Science and Engineering **32** (2012) 012021 doi:10.1088/1757-899X/32/1/012021

Combining trace elements micro-analysis in deposited dredged sediments: EPMA and μ-XRF analysis

A Poitevin[1,4], C Lerouge[1], G Wille[1], P Bataillard[1], P. Quinn[2] and L Hennet[3]
[1] BRGM, 3 avenue Claude Guillemin, P.O. Box 36009, FR-45060 Orleans cedex 2, France
[2] Diamond Light Source Ltd., Didcot OX11 0DE, Great Britain
[3] CEMHTI, CNRS, FR-45071 Orleans cedex 2, France

E-mail: a.poitevin@brgm.fr

Abstract. Since deposited dredged sediments are rich in metallic contaminants, they present a risk for environment. This work aims to study dredged sediments chemical composition, identify metal-carrier minerals and understand their mobility. Combining chemical and spectroscopic techniques at multi-scale for an integrative approach of trace elements (zinc, lead, iron) behaviour is therefore necessary. The global mineralogy and the chemistry of the sediment were determined by X-ray diffraction and fluorescence (XRF), respectively. Zn and Pb enriched fractions were separated using a sequential chemical extraction procedure and measured by inductively coupled plasma atomic emission and mass spectroscopy. Microanalyses using scanning electron microscopy (SEM), electron microprobe microanalysis (EPMA), combined with synchrotron radiation X-ray fluorescence (μ-XRF) were carried out to characterize mineralogical phases and identify Zn and Pb carrier minerals. Iron oxyhydroxides and iron sulphides were consistently identify as Zn and Pb carriers. The assumption that carbonate fraction was the major Zn carried phase, as demonstrated by chemical extraction results, was not verified by EPMA or μ-XRF.

1. Introduction

Sediments originating from periodic dredging of waterways were traditionally deposited on soil without specific precaution or planed treatments. It may be of environmental concern especially when they came from areas historically contaminated with trace elements (e.g., Zn, Pb) like in the north French coal basin and considering the volume of sediments dredged annually (~ 200000 m^3). The chemical risk is the breakdown of trace metals-carrier phases (electronic structure: oxidation state and nature of chemical bonds; crystal chemistry) liberating toxic elements which can migrate to the underlying soil and groundwater. In this framework, dredged sediments are currently studied to determine trace element contents, their distribution among phases, and the stability of these phases.

The study site is a ten years old dredged sediment deposit from the Canal de Lens (northern France) located in the district of Courrières (62). The whole amount of contaminated sediments deposited on soil reaches 3 metres. The main purpose of this work is to better understand and constrain mobility and bioavailability of trace elements along an annual cycle, more particularly to identify and characterize the Pb- and Zn-carrier minerals by combining chemical and spectroscopic techniques. Sequential chemical methods [1] have been traditionally applied to various contaminated

[4] To whom any correspondence should be addressed.

soils [2, 3]. It is now widely recognized that this indirect and operational approach should be combined with other physical techniques, which allow direct identification of metal forms, such as scanning electron microscopy (SEM), electron probe microanalysis (EPMA), and synchrotron radiation techniques such as X-ray absorption spectroscopy (XAS), diffraction (XRD) and fluorescence (XRF) [4-7].

In this framework, the study presented here integrates three approaches: (i) traditional chemical sequential extraction method; (ii) reconstruction of the mineralogy of the sediment combining X-ray diffraction (XRD) and chemical analyses of major elements by X-ray fluorescence (XRF) of bulk sediment; (iii) identification of Zn and Pb-carrier minerals using optical and SEM observations, followed by quantification of Zn and Pb contents using EPMA and μ-XRF.

2. Materials and methods

2.1. Sample collection and preparation

Samples were collected at the deposit surface (0 - 20 cm). They were conditioned in closed plastic bag with limited air at 4 °C on the field then in a fridge at the laboratory. They were homogenized and divided into two parts. The first one was directly used for chemical sequential extractions. The other one was dried at 105 °C during 48 h. This dried sediment was for one part used for chemical bulk analyses (X-ray, ICP-AES, XRD), and for a second part embedded in a polyacrylamide resin (LR White®) and mounted into polished thin sections for microanalysis (SEM, EPMA, μ-XRF).

2.2. Scanning electron microscopy

Observations, analyses and elemental mapping were carried out with a Jeol JSM 6100 scanning electron microscope coupled with an energy-dispersive X-ray spectrometer (Kevex Quantum) using an acceleration voltage of 25 kV. Prior to the analysis, samples were coated with a 20 nm carbon conductive layer.

2.3. Electron probe microanalysis

EPMA analyses and elemental mapping were carried out on carbon-coated polished thin sections using a Cameca SX-50 electron microprobe. Elemental mapping of major elements (Si, Al, Ca, Fe) were performed with a 15 kV acceleration voltage and a beam current of 15 nA. EPMA analyses of Ca, Zn and Pb on carbonate grains were carried out under analytical conditions of 25 kV and 50 nA. Ca, Zn and Pb peak intensities were measured with counting times of 240 s for punctual analyses and about 600 s for analyses along transects to improve detection limits. Elemental analyses were performed by employing 4 spectrometers. Ca Kα and Pb Mα were analyzed using a PET (pentaerythritol) crystal, and Zn Kα was analyzed using a LiF (lithium fluoride) crystal. Standards used included both well characterized natural minerals and synthetic ones: apatite ($CaPO_4$) for Ca, sphalerite (ZnS) for Zn, and galena (PbS) for Pb. In these conditions, detection limits, calculated from [9], are in carbonate about 175 mg/kg for Ca and 125 mg/kg for Zn. Pb was not detected. The PAP matrix correction model was used for quantitative microanalysis [8]. Background subtractions were determined for each analysis by taking into account potential lines interferences.

2.4. X-ray diffraction

The nature of crystalline phases present in the samples was determined by XRD using a Siemens D5000 diffractometer working in a Bragg-Bentano geometry and equipped with a scintillation counter. Diffraction patterns were recorded using Co $K\alpha_1$ radiation (40 kV, 30 mA) over a 2θ angular range of 4 to 84° for powders and 2 to 36° for clayey fractions. In both cases, 2θ step of 0.02° and counting time of 1 s were applied. X-ray diagrams were analyzed using Diffrac-Plus software in an unquantitative way.

2.5. X-ray and μ-X-ray fluorescence

The global chemical composition was determined by XRF using a wavelength-dispersive X-ray spectrometer Panalytical PW2400. Prior to the measurements, powders were heated at 1000 °C during 2 hours to measure the loss of ignition, then mixed with lithium bromide and lithium tetraborate and then heated at 1200 °C to vitrify samples. Concentrations were calculated from a calibration line of synthesized or natural standard concentrations. Detection limits varied in the range of 200 mg/kg (manganese oxide MnO) to 2000 mg/kg (for alumina Al_2O_3 and silica SiO_2).

μ-XRF experiments were carried out at the I18 microfocus beamline at Diamond Light Source (Didcot, UK). The beam size was about 3 μm x 3 μm and the incident energy was 16.5 keV. A 9-elements germanium solid state detector placed perpendicularly to the beam was used to measure the fluorescence signal from the sample. Elemental mappings on polished thin sections were performed in the 4 - 16.5 keV energy range. The data treatment was carried out using the PyMCA software.

2.6. Sequential chemical extractions protocol

Chemical analysis was performed onto three replicates of each dried bulk and wet surface sediments. A 5-step extraction procedure was applied on 0.5 g of wet deposited sediment according to the method described by Piou et al. [8] and summarized in table 1.

Table 1. Sequential chemical extraction procedure adapted from Piou et al. [10].

Fraction	Procedure	Mineralogical extracted compartment
F1 : Leachable	Calcium nitrate, ambient temperature, 24 h	Leachable
F2 : Acid-soluble	Ammonium acetate, pH 5, ambient temperature, 5 h	Acid-soluble fraction, assumed carbonates
F3 : Reducible	Hydroxylamine hydrochloride (3/4) in acetic acid (1/4), 96 °C, 6 h	Mn and Fe oxyhydroxides
F4 : Oxidisable	Hydrogen peroxide and nitric acid, 96 °C, 2 h; then, hydrogen peroxide, 85 °C, ½ h.; and ammonium acetate in nitric acid addition at ambient temperature	Organic matter and sulphides
F5 : Residual	Aqua regia	Silicate-like quartz

At each step, concentrations are measured in the liquid extracted fraction by inductively coupled plasma atomic emission spectroscopy (ICP-AES) and mass spectroscopy, which provides the lowest detection limits. ICP-AES measurements are also applied on dried bulk sediment after chemical attack using aqua-regia.

3. Results

3.1. Chemical composition and mineralogical characterisation

The sediment essentially consists of a silty (~ 40 %), clay (~ 20 - 25 %), carbonate (~ 20 %) and organic (~ 10 - 15 %) fractions. The XRD pattern presented in figure 1 indicates that sediments consist of major quartz and calcite with minor feldspar (albite, microcline), dolomite, and clay fraction dominated by illite, illite-smectite mixed layer, chlorite, and kaolinite, and iron sulphides like pyrite.

IOP Conf. Series: Materials Science and Engineering **32** (2012) 012021 doi:10.1088/1757-899X/32/1/012021

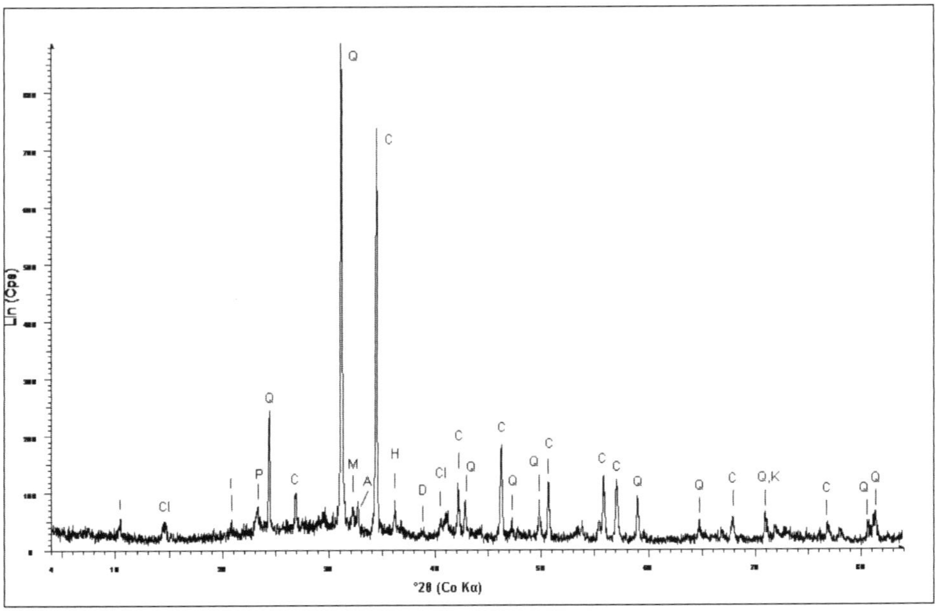

Figure 1. XRD pattern for the bulk sediment (surface sample). Mineralogical phases identified are quartz (Q), calcite (C), kaolinite (K), illite (I), chlorite (Ch), dolomite (D), hematite (H), pyrite (P), albite (A) and microcline (M).

3.2. Chemical sequential extractions

Results are presented in figure 2 and table 2. Elemental concentrations of Zn, Pb and Fe measured in the bulk sediment are lower than in the sum of extracted fractions (figure 2). This point will be discussed in the following. Concentrations in the leachable fraction (table 1) are null (Pb) or too low (Fe, Zn) with regard to the others to be represented in the figure. Fe is mostly concentrated (~ 60 %) in the reducible fraction assumed to be Fe- and Mn-oxyhydroxide fraction. This fraction exhibits ~ 80 % of Pb and ~ 50 % of Zn. The acid-soluble (assumed carbonates) fraction concentrates ~ 10 % of Pb and 35 % of Zn.

Table 2. Elemental concentrations (mg/kg) measured by ICP-AES in extracted fractions and bulk sediment and uncertainties (mg/kg). Detection Limits (DL) are reported for each element.

Extractants	Zn (mg/kg)	Pb (mg/kg)	Fe (mg/kg)
DL	*1*	*5*	*2*
Calcium nitrate	12 ± 1	< DL	< DL
Ammonium acetate	747 ± 20	62 ± 2	31 ± 4
Hydroxylamine hydrochloride	1331 ± 88	473 ± 11	17437 ± 544
Hydrogen peroxide, nitric acid	128 ± 17	58 ± 14	5760 ± 939
Aqua regia (residual)	36 ± 4	< DL	6394 ± 2154
Aqua regia (bulk sediment)	1685 ± 10	446 ± 14	17113 ± 251

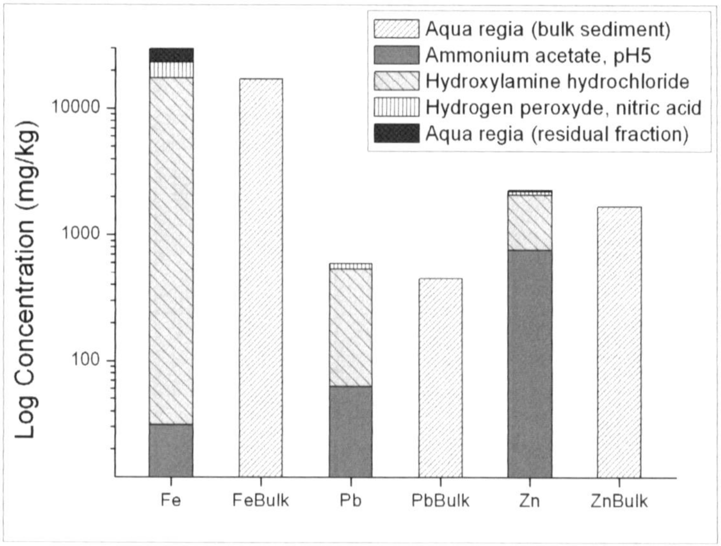

Figure 2. Fe, Pb and Zn concentrations (mg/kg) measured by ICP-AES in extracted fractions and bulk sediment. Uncertainties are reported in table 1.

3.3. Micro-analyses by SEM-EDS, EPMA and µ-XRF

Several zones of interest in embedded sample were observed by SEM. Backscattered electron images and EDS spectra are used for mineralogical phase's identification. An example is presented in figure 3, resuming various phases encountered in the sediment.

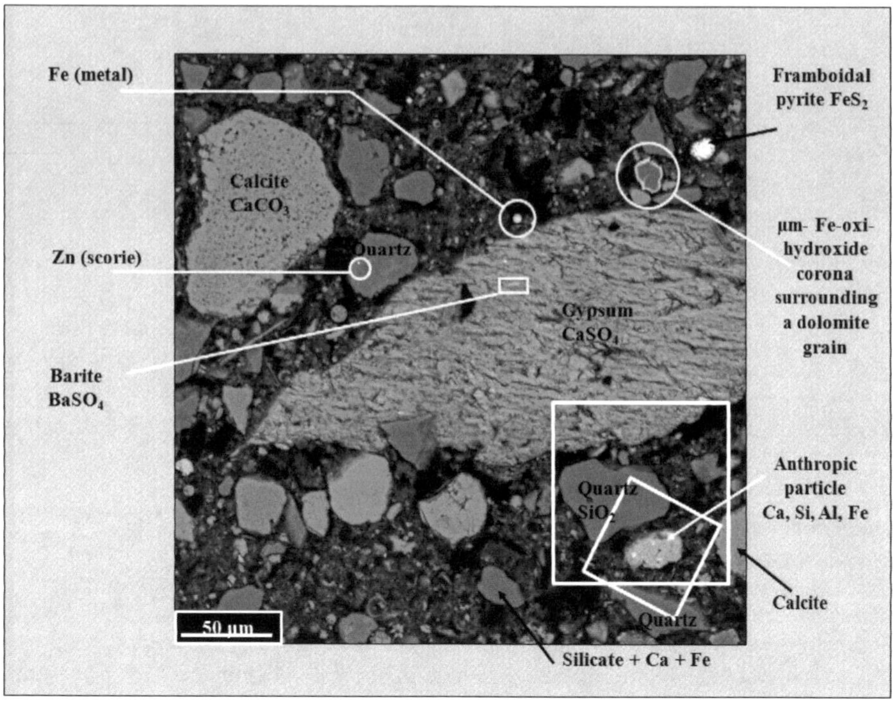

Figure 3. BSE image and zooms (white squares) in the embedded sample.

Microanalyses and elemental mappings among zones of interest were carried out using EPMA (figure 4). Maps presented in the figure (HV = 15 kV, beam current = 15 nA, dwell time 100 ms/pixel, 256 x 256 pixels) give the distribution of Si, Al, Fe and Ca. The presence of micro-carbonates (calcite-like) distributed among the matrix, silicates (quartz), and an undetermined Al-, Ca-, Si-, Fe-rich particle are highlighted. No Zn or Pb signal was detected under these analytical conditions. The same region was analyzed by μ-XRF (see figure 5) with Zn and Pb detection limits (DL) below the EPMA ones. From these results, Zn and Pb seem to be diffusive among the matrix. No Ca-carbonate (calcite) correlations with Pb and Zn were underlined here.

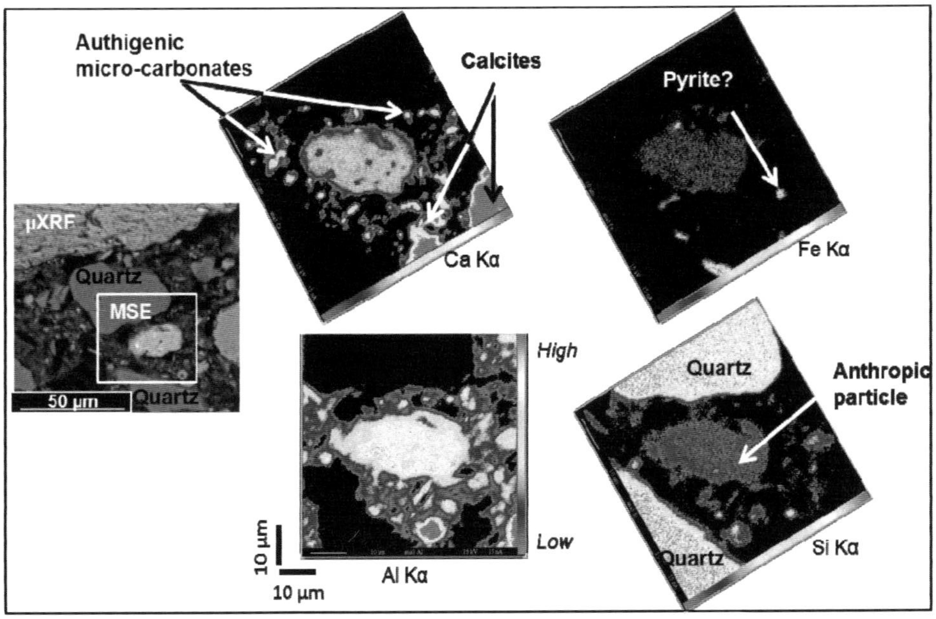

Figure 4. Elemental distribution of Ca, Fe, Al and Si by EPMA (50 μm x 50 μm).

Quantitative EPMA analyses carried out on several calcite grains (45 points of analyses performed in 3 areas in a single thin section) showed Zn content in the range of 0 to 1478 mg/kg with a mean value of ~ 160 mg/kg. Furthermore, well crystallized calcites (inherited round grains, 20 - 100 μm size) are not correlated with detected Zn or Pb. It suggests that different carbonate generations (inherited well crystallized grains, authigenic micro-carbonates) coexist and only some type of calcium carbonate may carry some Zn or Pb in trace concentrations. Thus, in figure 3, a dolomite grain surrounded by a μ-Fe-oxyhydroxide corona can be observed. Dolomite does not present any Zn or Pb content, whereas the Fe-rich corona exhibit significant Zn and Pb amounts. Others Fe-rich phases carrying Zn and Pb have also been noticed: reduced phases such as Fe-sulphides (pyrite FeS2, figure 5).

4. Discussion

The correlation between carbonates and trace elements (Zn and Pb), highlighted by chemical extraction, is not consistent with microanalyses on defined areas. Assuming 20 % of carbonates in the sample and a Zn concentration of 750 mg/kg measured in the acid-soluble extraction fraction (assumed to be the carbonate-rich fraction), the Zn amount attended in carbonates would be 3750 mg/kg. Yet, the detection of Zn in carbonates by EPMA is about 20 times lower (considering a

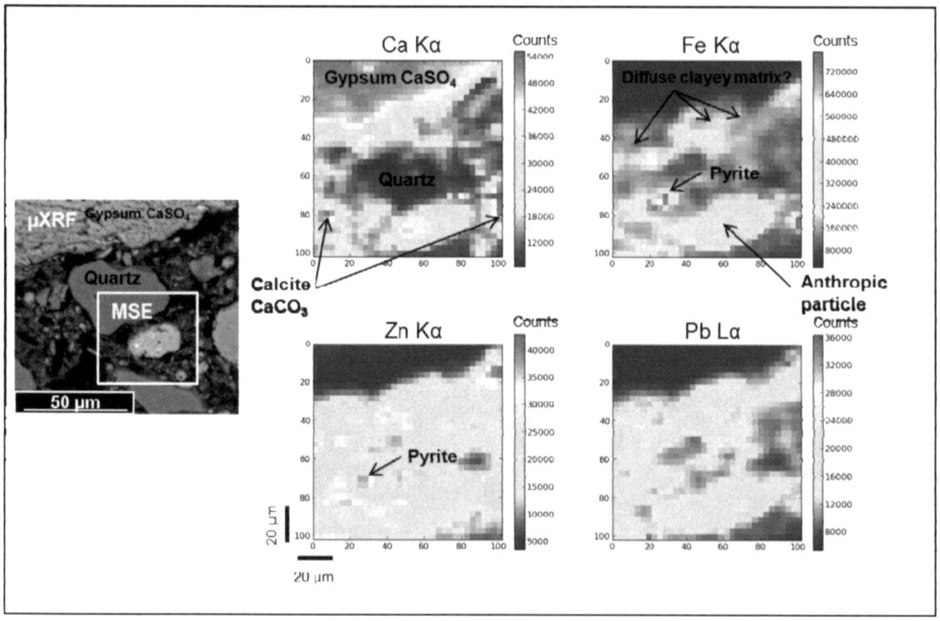

Figure 5. Ca, Fe, Zn, and Pb mappings (100 µm x 100 µm) by µ-XRF using synchrotron radiation.

Zn mean value in carbonates of 160 mg/kg). Some hypotheses can be done to explain this inconsistency:

- Sequential chemical extraction may not be selective extractions: some phases, sensitive to ammonium acetate attack, can enrich the acid-soluble fraction by dissolution (such as phosphate phases since 5 % of P is extracted in this fraction) and lead to Zn and/or Pb releasing.
- An unreasonable difference (up to 30 %) between bulk amounts of Fe, Zn and Pb and sums of concentrations in the 5 extraction fractions is observed. It can be explained by the use of aqua-regia to determine bulk concentration in a pseudo-total way, leading to an under-estimation of global concentrations in samples. Moreover, samples heterogeneity (in chemical compositions, spatial distribution on the field, crystalline structures, grain sizes, nugget effects for example) can induce different types of contacts between reactants and reagents involving various dissolution pathways.

Another chemical extraction procedure is in progress, in order to determine compositions by ICP-AES and µ-XRF and to compare them respectively in liquid extracted fractions and solid residues of each extraction step. It will allow us determining whether the carbonate fraction is totally attacked and solubilized or whether the recalcitrant forms of carbonates remain in other fraction(s) (ankerite $Ca(Fe^{2+},Mg,Mn^{2+})(CO_3)^2$, siderite $FeCO_3$, …). Another scope will be to estimate the potential selectivity of extractants. For instance, we'll study if some phases are attacked concomitantly with carbonates during the acid-soluble attack. In the same way, the use of aqua-regia and alkaline fusion protocols to extract total concentrations and validate the pseudo-total behaviour of aqua-regia extractant will be tested.

The correlation of Fe with Zn and/or Pb has been demonstrated among mineralogical phases under different oxidation states. Fe speciation may be a key parameter on the dynamic of trace elements mobility. As demonstrated in this study, Fe-oxyhydroxides and Fe-sulphides are Zn and Pb carriers.

The fine grained matrix, assumed to be silicates and clays but also amorphous phases like organic matter, can be trace element carriers by adsorption or interfoliar substitutions on clays or creation of organo-minerals complexes.

5. Conclusion

The characterisation of the global mineralogy of the sediment and identification of trace metal carriers has been carried out by combining different approaches: chemical extraction analysis and spectroscopic techniques such as SEM-EDS, EPMA, XRD, XRF and synchrotron radiation μ-XRF.

Chemical sequential extractions carried out on the bulk sediment show a high Zn content in the reducible extracted fraction and a minor but significant Zn content in the acid-soluble extracted fraction (~ 3750 mg/kg) assumed to be carbonates. Pb is more concentrated in the reducible fraction than in the acid-soluble fraction. Microanalysis using EPMA and synchrotron radiation μ-XRF show a lower Zn content in carbonates (about 20 times lower), and a well correlated Zn and Pb association with Fe and a fine grained silicate matrix enriched in Zn and Pb.

On the one hand, the trapping capacity of the carbonate fraction, expected by the chemical extraction, is not validated by microanalysis carried out on inherited and well crystallized carbonates, nor on neo-formed and micro-sized ones. This discrepancy could be explained by dissolution of unexpected mineralogical phases during the acid-soluble extraction step leading to reassign Zn and Pb initial distribution to phases soluble at ammonium acetate, like phosphates. Nevertheless, the study of solid residues, compared with results on liquid extracted fractions, is in progress to establish a geochemical distribution balance. On the other hand, Zn-Fe and Pb-Fe associations had been revealed among phases like Fe-oxyhydroxides and Fe-sulphides. Finally, data processing of Fe speciation will be pointed out by XAS analysis carried out on Fe and Zn-rich phases, especially by studying XANES spectra at Fe Kα edge.

Acknowledgments

This work was financially supported by the BRGM and the Regional Council of the Région Centre. The authors are also grateful to the MMA (BRGM) and I18 beamline (Diamond Light Source) staffs for technical help and to E. Veron (CEMHTI) for the SEM observations.

References

[1] Tessier A, Campbell P G C and Bisson M 1979 *Anal. Chem.* **51** 844-51
[2] Gomez-Ariza J L, Giràldez I, Sànchez-Rodas D and Morales E 1999 *Int. J. Environ. Anal. Chem.* **414** 151-64
[3] Leleyter L and Baraud F 2005 *C. R. Geoscience* **337** 571-79
[4] Isaure M P, Laboudigue A, Manceau A, Sarret G, Tiffreau C, Trocellier P, Lamble G, Hazemann J L and Chateigner D 2002 *Geochim. Cosmochim. Acta* **66** 1549-67
[5] Kirpichtchikova T A, Manceau A, Spadini L, Panfili F, Marcus M A and Jacquet T 2006 *Geochim. Cosmochim. Acta* **70** 2163-90
[6] Sarret G, Balesdent J, Bouziri L, Garnier J M, Marcus M A, Geoffroy N, Panfili F and Manceau A 2004 *Environ. Sci. Technol.* **38** 2792-801
[7] Manceau A, Lanson B, Schlegel M L, Harge J C, Musso M, Eybert-Bérard L, Hazemann J L, Chateigner D and Lamble G M 2000 *Am. J. Sci.* **300** 289-343
[8] Pouchou J L and Pichoir F 1991 Quantitative analysis of homogeneous or stratified microvolumes applying the model "PAP". in: *Electron probe quantitation.* Heinrich K F J and Newbury D E, eds. (New York: Plenum Press) 31-75
[9] Ancey M, Bastenaire F and Tixier R 1978 *Application des méthodes statistiques en microanalyse. Ecole d'été de Microanalyse et de Microscopie Electronique à Balayage.* (St. Martin d'Heres: Editions de Physique)
[10] Piou S, Bataillard P, Laboudigue A, Férard J F and Masfaraud J F 2009 *Environ. Res.* **109** 712-20

Shielded field emission EPMA for microanalysis of radioactive materials

R Restani[1] **and A Wälchli**
Paul Scherrer Institut, Hotlaboratory Division, CH-5232 Villigen PSI

E-mail: renato.restani@psi.ch

Abstract. A standard Jeol 8500F electron probe microanalyzer (EPMA) with field emission gun (FEG) has been installed in a lead shielded cabin of the laboratory and customized for remote control. The wavelength-dispersive X-ray spectrometers (WDS) have been additionally shielded against β,γ-radiation with tungsten alloy modules. These measures have proven to be very effective for the examination of specimens with an activity of less than 20 GBq ^{137}Cs. The background intensity is dependent on crystal type, spectrometer and sample position. The efficiency of the spectrometer with Xe-sealed counter is higher but also its sensitivity to the specimen's γ-radiation than the Ar/CH_4 flow counter. The possibility of applying high beam currents increases the peak-to-background ratio for the analysis of low concentrations in the active sample. The high electron beam density of the FE-gun allows for the recording of high resolution elemental X-ray maps in the range of a few tenths of a micrometre at even medium acceleration voltage and thus short acquisition time. It is thus a valuable instrument in post irradiation examinations of spent fuel rods and materials irradiated by other sources.

1. Introduction

Electron probe microanalysis (EPMA) is a versatile method in post-irradiation examinations for materials characterisation. The necessity to shield the wavelength-dispersive X-ray spectrometers (WDS) against strong β,γ-radiation has already been successfully handled in the late seventies with the Cameca MS 46 and was already used for dissolved fission gas analysis in irradiated fuel [1-3]. The examination with the Jeol 733 required a sophisticated sample preparation to very small samples [4]. In the eighties and early nineties followed the shielded Camebax R [5, 6] and SX-50R [7, 8] respectively. The last one is equipped with a massive stage shielding and spectrometers separated from the specimen chamber by narrow tubes with window sleeves. The large Rowland circle of 180 mm offers space for additional spectrometer shielding and a big detector-crystal-to-specimen distance. The electron gun has been equipped with a tungsten filament. The successor instrument Cameca SX-100R, just a few years on the market, can also be equipped with a LaB_6 cathode. A field emission gun (FEG) with even smaller probe diameter and higher current at low accelerating voltage is not yet provided due to constructive changes necessary related to the vacuum system.

It has been our incentive to select and modify a FEG EPMA for the examination of nuclear materials in cooperation with the necessary manufacturers and partners to accomplish this task which were found in Jeol Germany and remX GmbH.

[1] To whom any correspondence should be addressed.

2. Design of shielded FEG EPMA

The selected EPMA is a standard Jeol 8500F with Schottky field emission gun. The instrument had to be installed in a lead shielded cabin at the end of a sample preparation cell line (figure 1). Jeol Germany divided the electronics cabinets and installed long distance cables with the appropriate signalling to the control desk outside the cabin. Afterwards modular internal shields made of tungsten alloy on the stage, at spectrometers ports around the specimen chamber and around the pole shoe were constructed by the project leader, remX GmbH (table 1, figures 2 and 3). The relatively small WDS-spectrometers with 140 mm Rowland circle could not be additionally shielded. The contractor organized also the optical window, the motorized aperture from Deben UK Ltd., and was also engaged in infrastructural equipments. The shielded sample holder is loaded with the polished sample in the adjacent cell and transferred to the cabin via a rotary gate and put in place by cell manipulators in the airlock chamber of the EPMA. These manipulators also operate the rod manipulator of the EPMA to transfer the sample into the recipient.

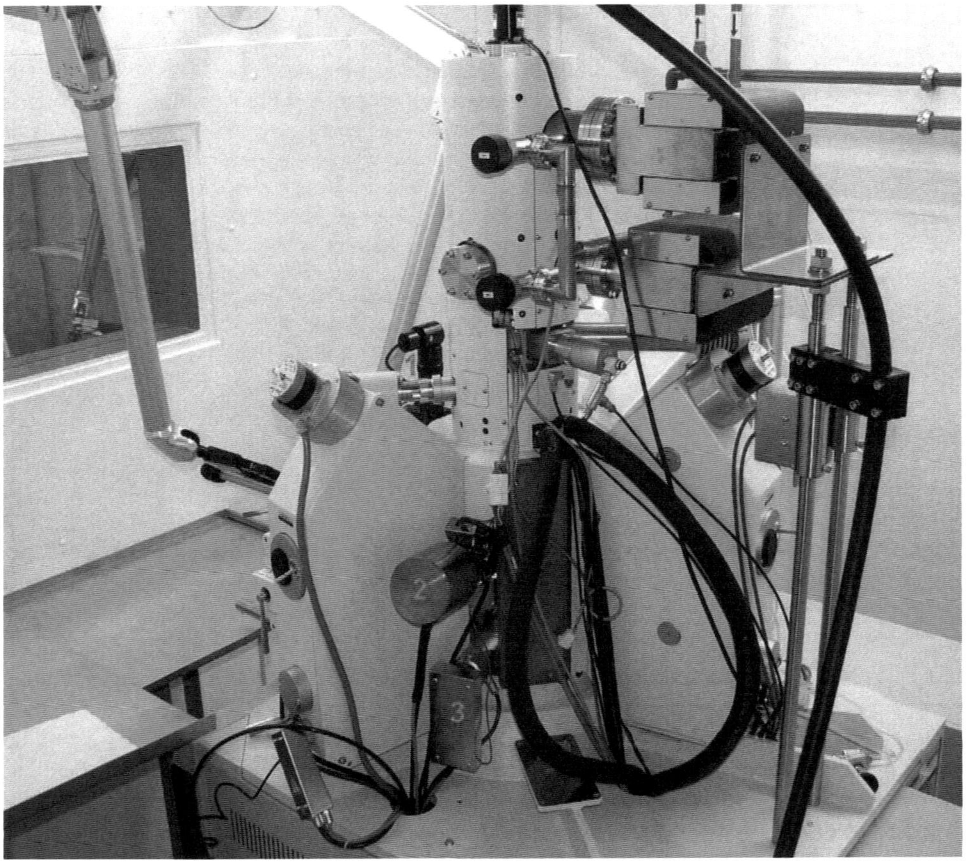

Figure 1. Modified and shielded field emission Jeol 8500F EPMA with 4 WDS-spectrometers (2 of them hidden on photo) in a lead shielded cabin (back view). 1: optical microscope; 2: SE-detector; 3: BSE-detector

A compromise between the driveway of the stage and spectrometers and the extent and thus effectiveness of the shielding had to be found. The windowless space between the spectrometers and chamber for the crystal drive was fit to the dimensions of the crystal frame and is the only opening.

Table 1. Internal EPMA shielding of JEOL 8500F with tungsten alloy (W content > 90 wt%, density 18 g/cm^3).

Shielding place	Port between spectrometer and chamber	Plates around pole shoe	Stage + specimen holder
Number of shielding units	5 (vertical)	2 (horizontal, including a 2.5 mm β-shield of Al)	2
Height (cm)	15		5
Width (cm)	8		8
Depth (cm)	5		
Length (cm)			13
Thickness (cm)		1.5 + 0.5	

Figure 2. Specimen chamber with vertical tungsten alloy shield blocks at spectrometer ports and a tungsten-aluminium plate around the pole piece.

Figure 3. Shielded stage (horseshoe-shaped) with the shielded specimen holder inside containing drillings for 1-inch and 17 mm specimens. X-Y-Z axes drives underneath.

For the spectrometers the lower limit is at about 75 mm between sample and crystal (eq. Bragg angle sin θ ≈ 0.27). There are two spectrometers with LIF/PET diffracting crystals and sealed Xe-counters. For light elements two spectrometers with TAP/LDE1 and TAP/LDE2 crystals with Ar/CH_4 flow counters are attached.

3. Effectiveness of shielding

The effect of the tungsten module shields were tested with an encapsulated highly active fuel specimen (γ-source). The radiation background is mainly dependent on the shielding environment and the distance between detector and the active specimen. This means that the positions of the spectrometers, crystal-detector and sample are decisive but also the crystal and counter type have an influence. It proved that the rear spectrometers (pos. 3 and 4, spectrometers on the right side of figure 1) are better shielded than the ones in front at pos. 1 and 2 (figure 4). The dependency on the sample position is given in figure 5. It also shows that at certain border positions the background level can change in the opposite direction than is expected from the crystal-detector position.

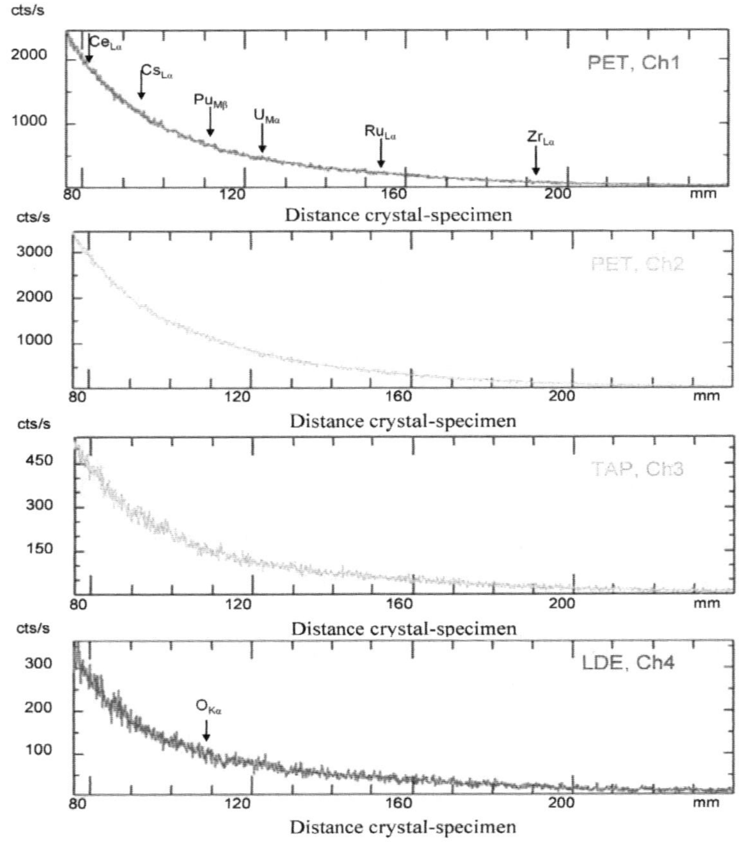

Figure 4. Radiation background measured on 4 spectrometers with a γ-source of 200 mGy/h in 0.1 m. Channels 1 and 2 with Xe-counters, channels 3 and 4 with Ar/CH_4. See theoretical positions of some X-ray lines, conversion: sin θ = distance (mm)/280.

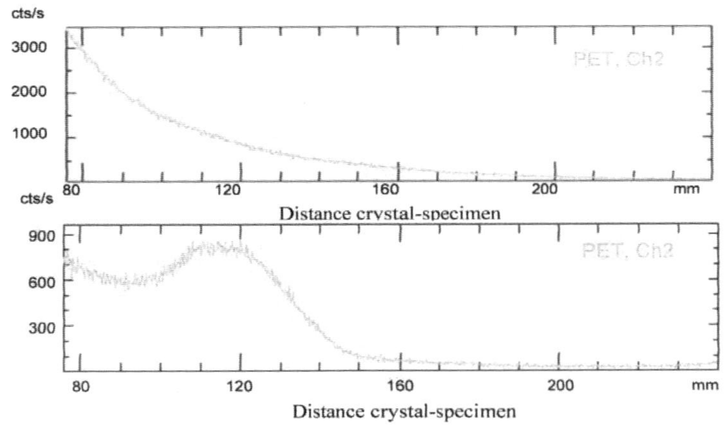

Figure 5. Background characteristics at 2 different sample positions measured with a γ-source of 200 mGy/h in 0.1 m.

The benefit of the tungsten plate around the pole piece is shown on figure 6. A thin Al-plate screwed on the shielding (figure 2) to capture the electrons from β^--decay and reduce the subsequent bremsstrahlung had no additional positive influence.

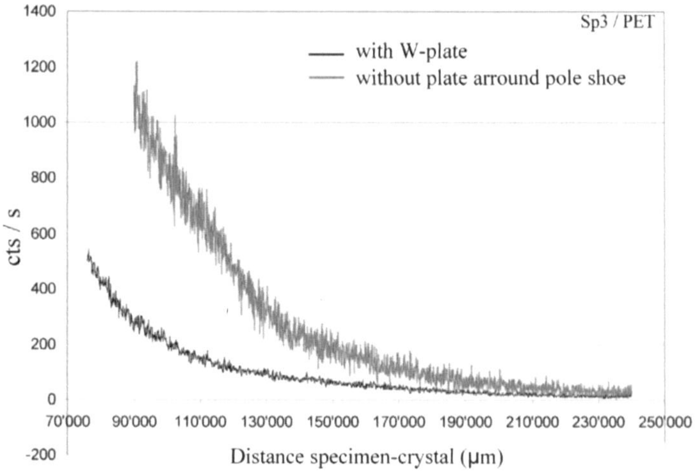

Figure 6. Effect of a 1.5 cm thick tungsten shield plate around the pole piece on the radiation background level of a rear spectrometer, mounted in addition to other shields (γ-radiation source: 200 mGy/h in 0.1 m).

The comparison of the background level for the different counters is given in figure 7. The Ar/CH_4-counter is less susceptible than the Xe-counter. Nonetheless it was decided to maintain the Xe-counters for spectrometers with LIF/PET crystals because of its higher efficiency and often longer lifetime as no impurities could accidentally be brought in.

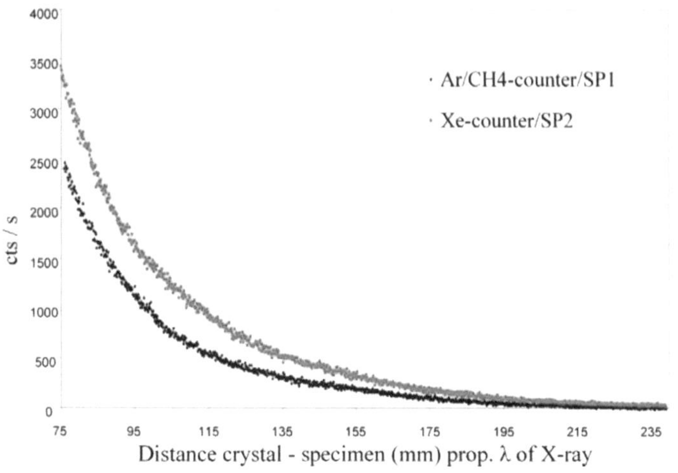

Figure 7. Sensitivity of different detectors to a γ-radiation source of 200 mGy/h in 0.1 m.

The detector characteristics require to run the pulse height analyzer in the differential mode to further discriminate the radiation impact. An energy window of about 60 % of the voltage range is typically applied. This means that the pulse height distribution must always be checked for each element and analysis setup to avoid pulse height depression and broadening. The result of all these measures is shown on a spectrum taken from a small irradiated UO_2-fuel sample ($\approx$ 80 mm_3) which had an equivalent activity of 5 GBq ^{137}Cs (figure 8). The acquisition was recorded on a spectrometer with PET crystal applying a typically higher voltage and current and a medium bias. The rise in background due to the radiation which has about doubled for the higher energy characteristic X-rays is not too high to impede the quantitative measurement of the low content fission products. For the Nd Lα-line the background is at the limit as the detector is only partly shielded at this position (top end) but Nd can also be measured with the LIF crystal. The spectrometer efficiencies for typical element concentrations in an irradiated fuel are given in table 2. The spectral resolution with respect to the small Rowland circle is good but is of lower importance than the spectrometer efficiency achievable under radiation conditions. Nonetheless the number of fission products causes interferences from subordinate lines as well as from uranium satellites which must be accounted for in the microprobe analysis (figure 8). Also a careful selection of positions for background measurements has to be made which has to be adapted to the standardisation procedure as well. Often the background can only be measured on one side of the peak with linear or curved extrapolation depending on the background gradient (e.g., for Pu, Cs, Nd, Ce, Ru).

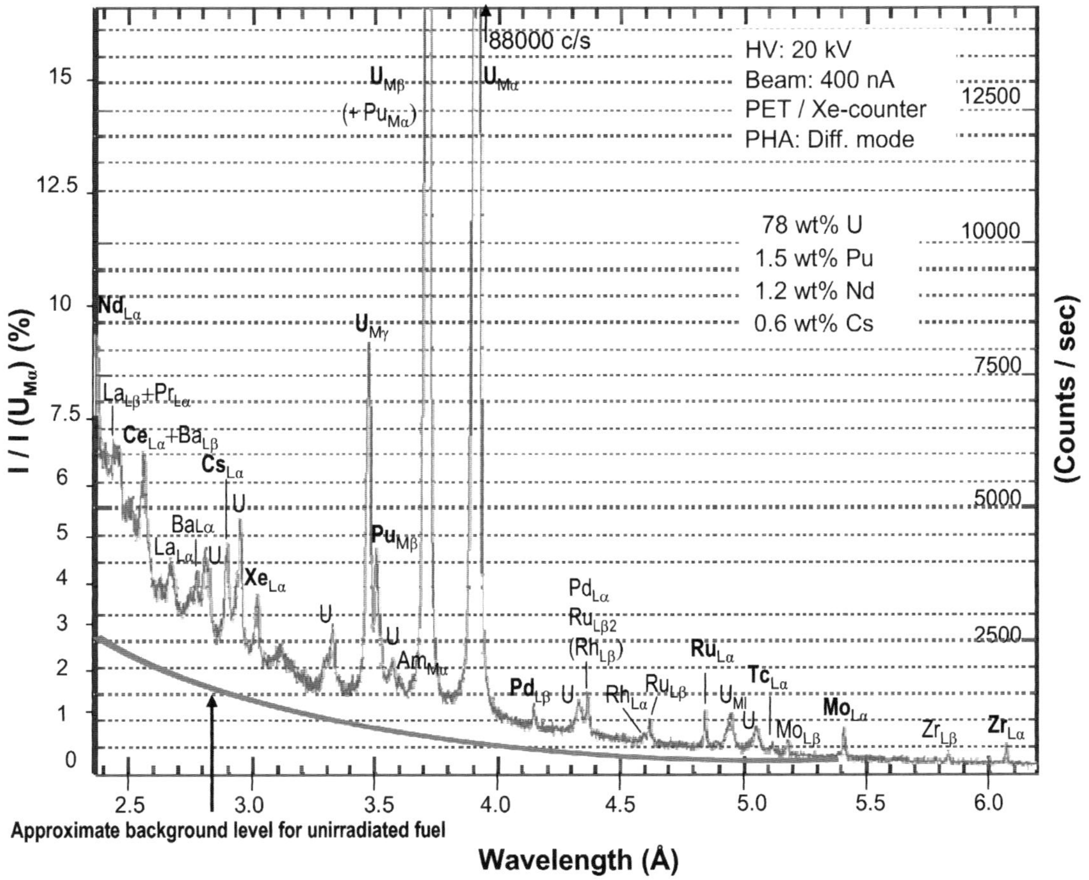

Figure 8. WDS-spectrum of a small high burn-up UO_2-fuel piece (equiv. 5 GBq ^{137}Cs) with peaks from U and its satellites, Pu and Am from the nuclear breeding reaction and the fission products. Left Y-axis in percent of the U Mα peak count rate. Background level increased due to radiation impact.

4. Spatial resolution

The advantage of the FE gun is the small probe diameter which is the limiting factor for scanning electron microscopy of the surface (SEM). The high beam density provides high X-ray signal yields important for elemental dot mapping and analysis at low accelerating voltage to further increase the resolution. For standard analysis it is however more important to limit the dwell time and the exposure time of the specimen to the electron beam for the sake of the delicate instruments components and particularly avoiding contamination with the volatile ^{137}Cs. For a once-through

Table 2. Spectrometer efficiency (PET) for an irradiated UO_2-fuel specimen (0.85 g) with a burn-up of 64 MWd/kgU, equiv. 5 GBq ^{137}Cs; 50 mGy/h in 0.1 m. Condition: 20 kV, 350 nA.

Jeol 8500F shielded / PET crystal (Xe counter)			
Element (wt%), line	Peak/BG-ratio	Yield at 20kV c/s/nA)	Resolution $\Delta\lambda/\lambda$ (FWHM)
Nd (0.8 wt%), Lα	1.2	3.9 (4.4)[1]	
Cs (0.5 wt%), Lα	1.45	2.7 (3.9)[1]	
Pu (1 wt%), Mβ	1.95	2.9 (5.0)[1]	
U (80 wt%), Mα	79	255	≈0.0037 (12eV)
Ru (0.35 wt%), Lα	≈2.2	≈1.5	
Zr in Zircaloy-2, Lα (98 wt% Zr)	205 / -[2]	135 / 170[2]	

[1] Yield after interference correction, in brackets gross yield.
[2] Yields on different spectrometers opposite to each other.

analysis which has to measure also very low concentrations the accelerating voltage is therefore typically set to 15 - 20 kV to achieve acceptable counting and exposure times.

4.1. SEM resolution
High resolution surface topography of fuel must usually be made with a SEM [9-14]. Fine grain structure in high burn-up fuel zone can now also be displayed with this FEG EPMA with a resolution of 10 - 20 nm for pore observation (figure 9). In this context tests are underway with a tilt (maximum 200) and rotary sample holder for fractography. As the focus is on surface imaging it has only a small shielding ring around the sample holder. Only samples with a contamination free surface can be accepted.

4.2. X-ray mapping
Elemental mapping is widely used to record phase distribution and elemental agglomerations in the examination of irradiated materials (e.g., [3, 7, 8, 11, 14]). For precipitates and interfaces sub micron resolution is often requested.

4.2.1. Fuel surface. Figures 10a and 10b show the start of grain re-crystallisation (SEM images) in high burn-up zones of a MOX fuel with the intermetallic precipitation of 4d elements in the range of 0.1 to 0.2 μm at grain faces. The Ru maps were acquired in only half an hour for 256 x 256 pixels with a beam of 20 kV and 350 nA resulting in a spatial X-ray resolution in the order of 0.3 μm even under these stringent beam conditions.

Figure 10c acquired with the same conditions as before shows the precipitation of fission products in the form of oxides (Ba, Cs, Mo) and especially the accumulation of Ba(Cs) at fuel grain boundaries with an apparent concentration of 1 - 3 wt%. The semi-quantitative mappings are the result from further evaluation by quantitative point analysis taking into account the interference of U on Ba when present (figure 8, [15]) but not the local height variation and the related shadow effects. In this case they are of minor importance as they do not determine the basic element pattern.

Figure 11 shows the precipitation of Am and the fission product Ba at grain boundaries of an irradiated inert matrix fuel revealed also by the phase contrast of the backscattered electron image (BSE).

4.2.2. Zircaloy-2 rod. Secondary phase particles (SPPs) in the zircaloy base metal, which are in the size range of 0.1 - 0.3 μm, are improving the corrosion resistance. Density, size and elemental ratios also have an influence on the undesirable hydrogen uptake during the corrosion process. Transmission

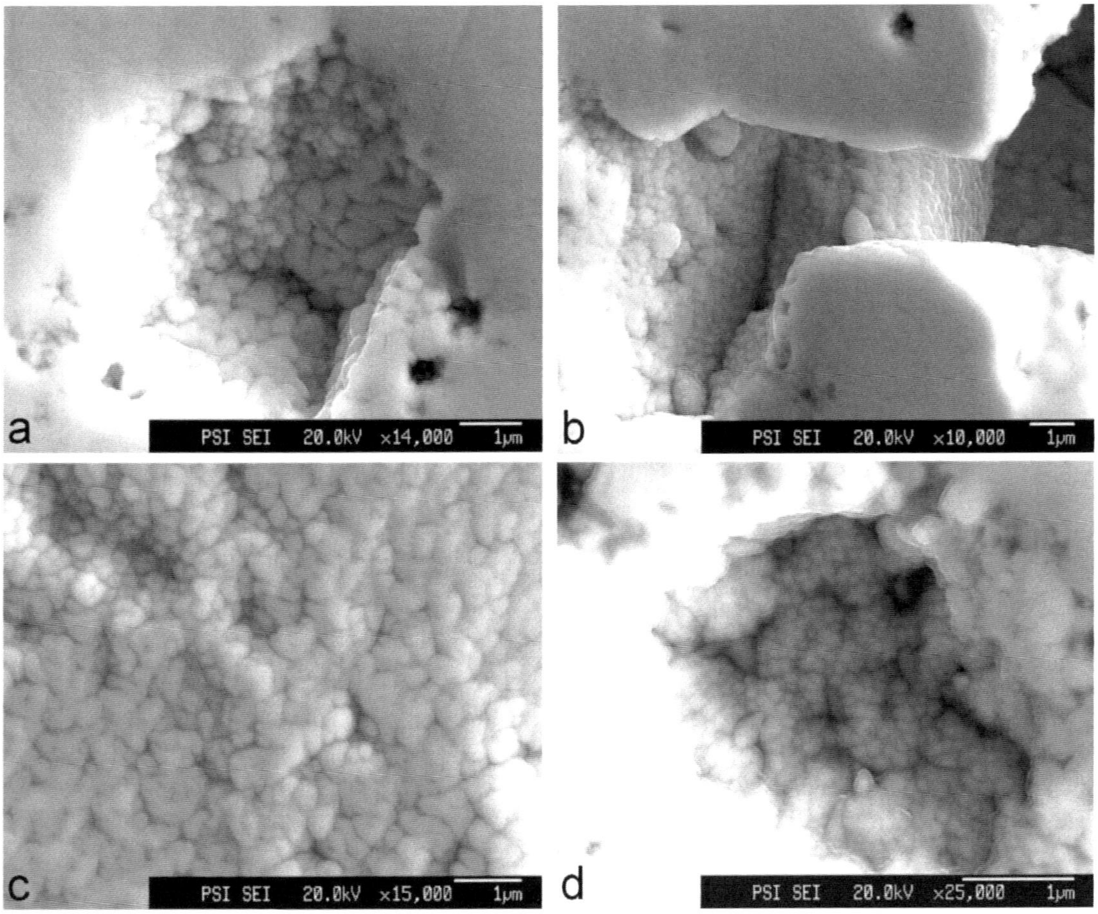

Figure 9. SEM from shielded FEG EPMA. Development of grain subdivision in high burn-up UO_2-fuel zone (a-c) and fully subdivided grains of 0.1 - 0.2 μm (d).

electron microscopy (TEM), SEM-EDS [16], and Auger microprobe [17] have been used for examination requiring often extensive sample preparation. Wide local examinations by FEG EPMA X-ray mappings give information about the state and local distribution of the SPPs during irradiation and its dissolution. Figure 12 shows local elemental mappings on a rod cross-section by an ad hoc measurement. Lower voltage with much longer acquisition time would be needed in a diligent survey. Oxide deposits on fuel rods (CRUD). In boiling water reactors deposits from suspended particles and solutes are formed on the fuel rods up to some micrometres [18]. It can have a detrimental influence on the fuel rod behaviour through decrease in thermal conductivity and if flaked off can contaminate the coolant system. It is desirable to know its composition for possible change in water chemistry. The BSE image in figure 13 shows there are at least two phases. Spot quantitative analysis revealed they are made up of pure hematite and mixed ferritic spinel phases. The applied μ-XRD method [19] confirmed that the spinels are composed of Zn-ferrite on the one hand and where other 3d metal ions are incorporated a franklinite type spinel is formed. The quality of the semi-quantitative mappings (figure 13), measured with 15 kV and 220 nA in half an hour for each element, allows assessing the fraction of the different phases.

To complete the information the corresponding signal efficiencies for CRUD measurements on the spectrometer with LIF crystal are given on table 3.

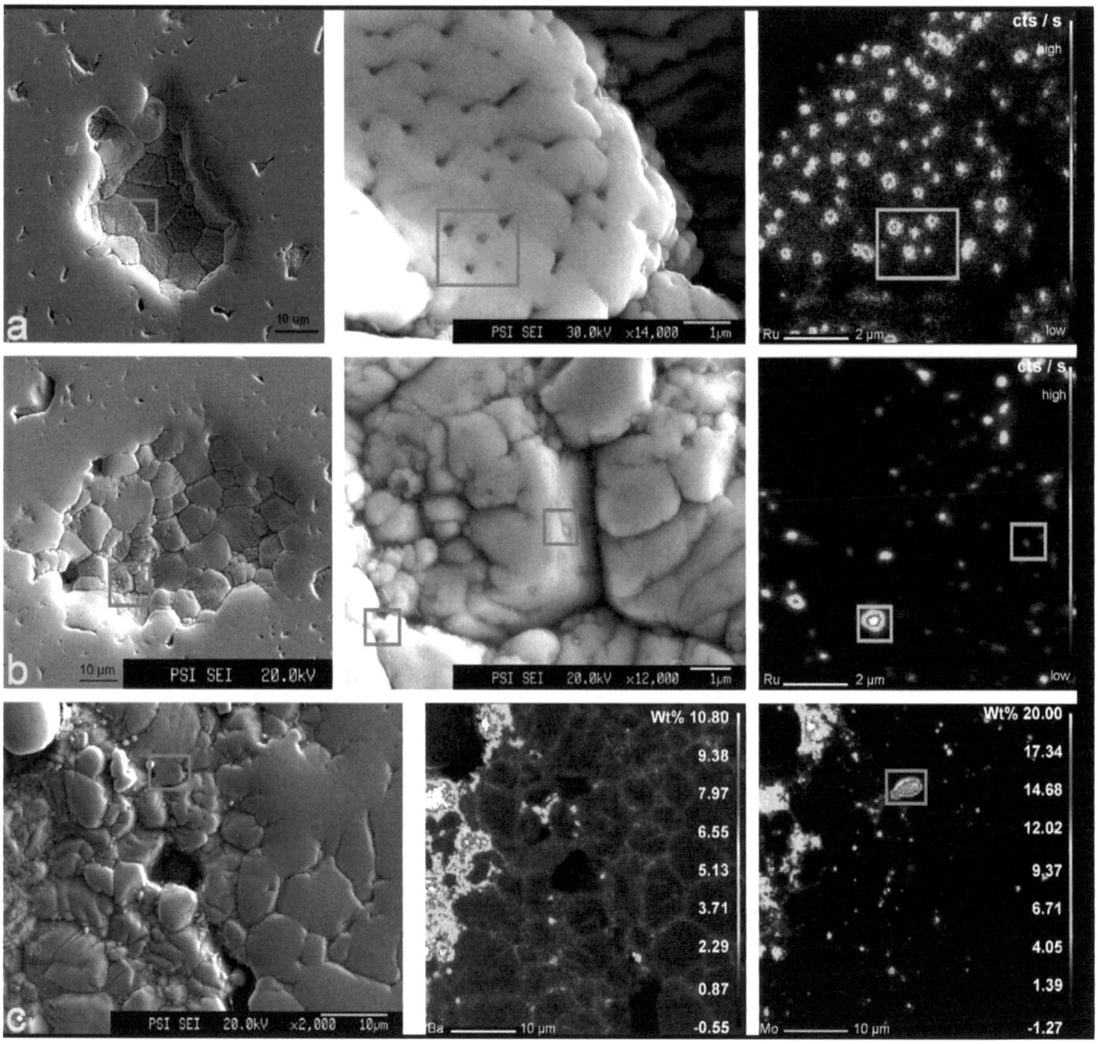

Figure 10. SEM and X-ray mappings of MOX fuel. Series a-b of high burn-up zone with start of grain subdivision and precipitation of intermetallic fission products (Ru, Rh, Mo, Pd, Tc). Serie c: Precipitation of fission product oxides (Ba, Mo, a.o.) in the fuel and at grain boundaries (semi-quantitative mappings; to note that height variations and shadow effects are not corrected).

Table 3. Spectrometer efficiency (LIF) for oxide deposits on irradiated Zircaloy-2 rod (CRUD) with fuel, equiv. 5 GBq ^{137}Cs; 50 mGy/h in 0.1 m. Condition: 15 kV, 350 nA.

Jeol 8500F shielded / LIF crystal (Xe counter)		
Element (wt%)	Peak/BG-ratio	Yield at 15kV (c/s/nA)
Fe in Fe_2O_3, 70 wt% Fe	82	145
Zn in $ZnFe_2O_4$, 27 wt% Zn	8.8	40
Fe in $ZnFe_2O_4$, 46 wt% Fe	56	92

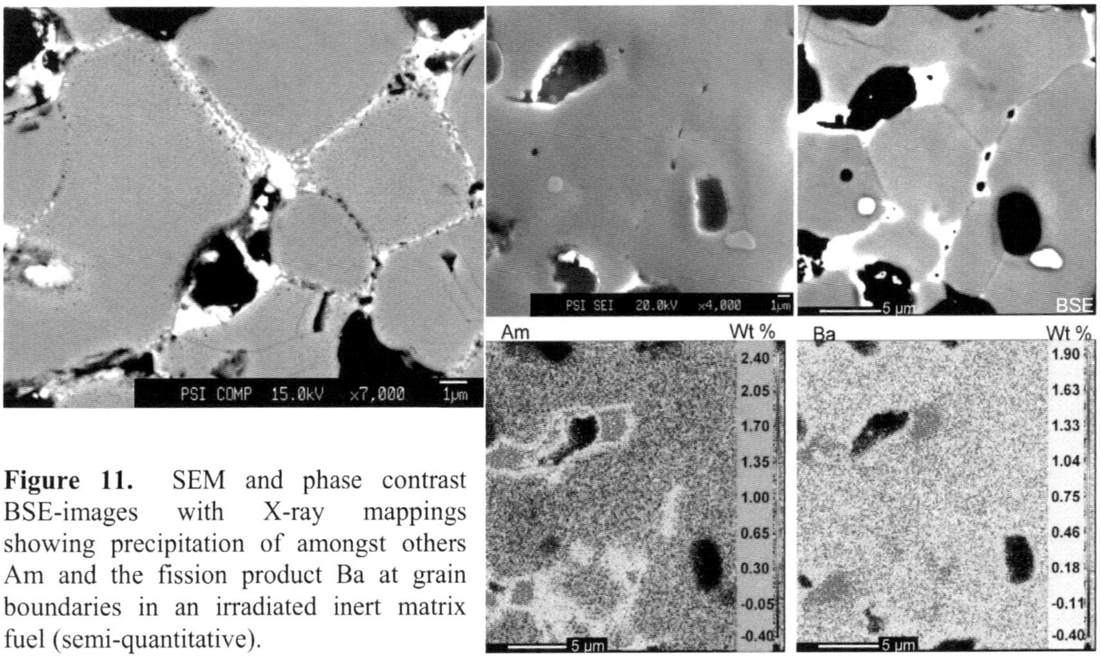

Figure 11. SEM and phase contrast BSE-images with X-ray mappings showing precipitation of amongst others Am and the fission product Ba at grain boundaries in an irradiated inert matrix fuel (semi-quantitative).

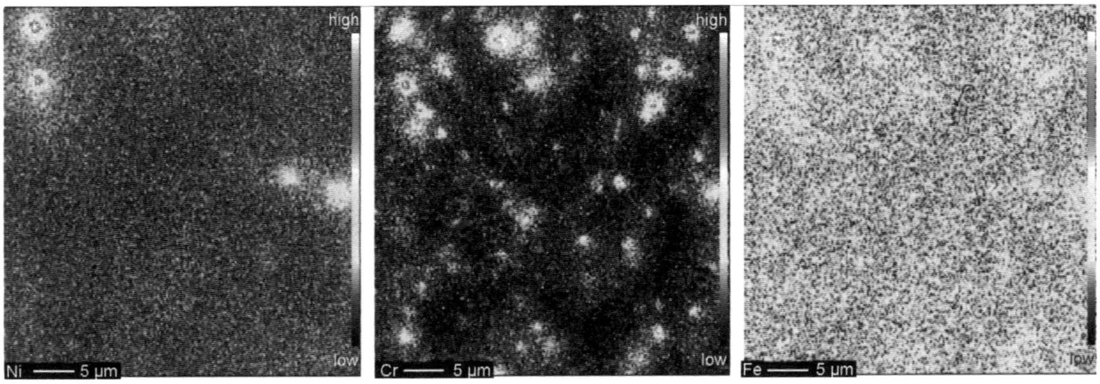

Figure 12. X-ray mappings of secondary phase particles in irradiated Zircaloy-2 rod known to be made up of $Zr_2(Fe,Ni)$ and $Zr(Fe,Cr)_2$ are partly distinguishable. Particle sizes on mappings comparable with actual size of 0.1 to 0.3 μm. Particles just below surface also detected. Fe partly dissolved in the base metal. Beam condition: 20 kV, 450 nA, 200 x 200 pixels, 85 ms/pixel.

5. Conclusion

The performance goal by modifying and shielding a FEG EPMA for the microanalysis of high active materials has been realized. The radiation background could be reduced to a level that the spectrometer efficiencies allow to quantitatively measure the composition of nuclear materials down to very low element concentrations in an even smaller micro-area.

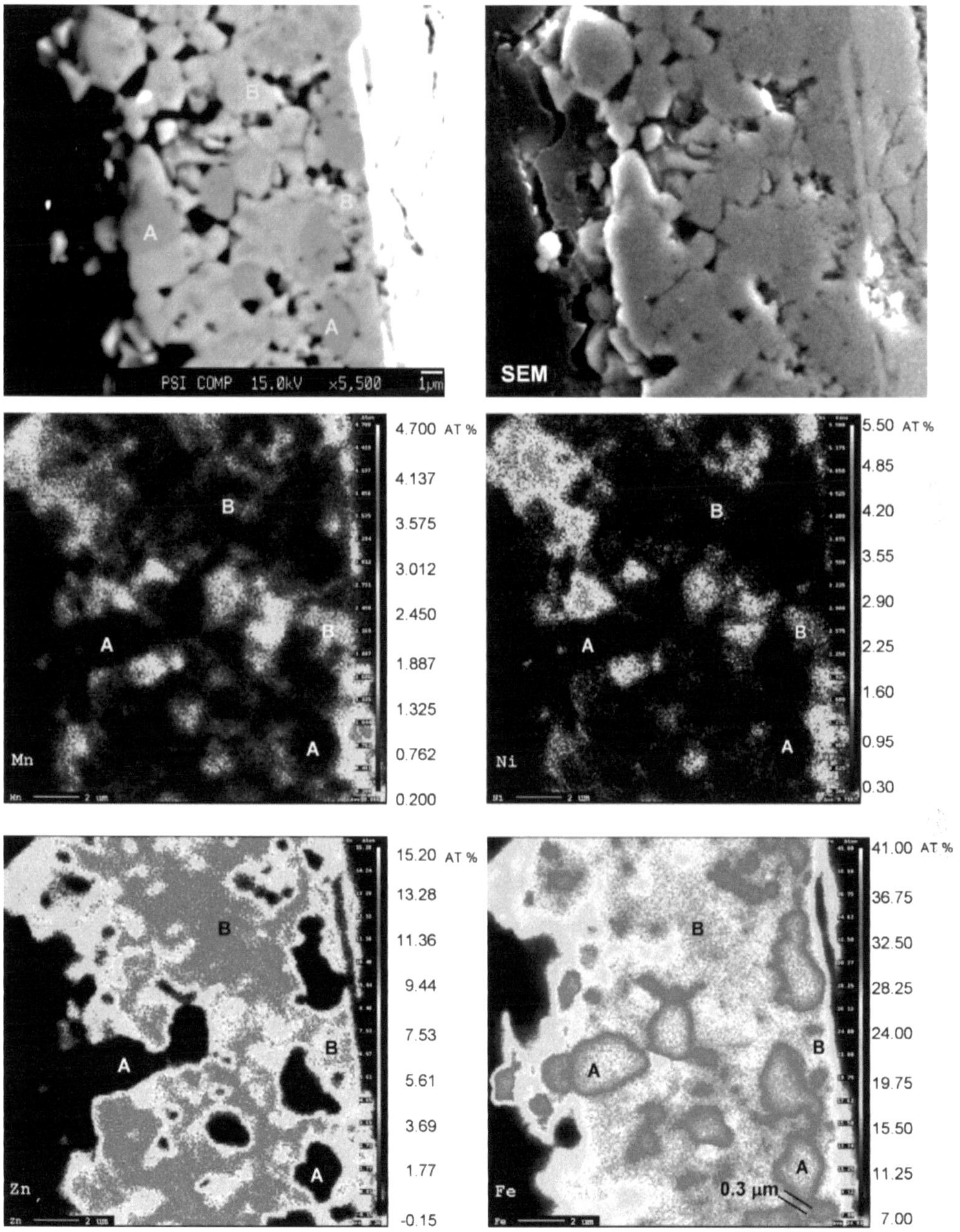

Figure 13. BSE-, SEM- and semi-quantitative X-ray images of oxide deposits on fuel rods (CRUD). A: hematite; B: spinels. Note: Mn, Zn, Fe with LIF crystal (Kα-lines), Ni with TAP crystal (Lα-line).

The quality of SEM and BSE imaging at high magnification is considerably improved and can only be reached with a SEM so far. Fractography is possible although SEM will stay the first choice. As there is no containment for the instrument only specimens with 'clean' surfaces can be accepted for examination to avoid contamination to spread.

The X-ray resolution is increased down to sub micrometre range thanks to small probe diameter and particularly high beam density. An examination of surface details such as precipitates and interfaces with elemental X-ray mapping is straightforward and reduces the need to apply other methods and/or extensive sample preparation. Alternatively faster acquisition is possible.

In operation one must be aware of the compact design with higher radiation impact on some of the delicate components such as crystals and detectors. This requires amongst others faster acquisition also to reduce possible spreading of volatile active nuclides which again often affords a higher beam voltage to increase the detection limits.

Acknowledgments
We wish to thank for the good collaboration with the manufacturers of Jeol Germany and remX GmbH and our Hotlab operating group to make this project a success and especially M. Mostert, remX for getting all well done and J. Börder, A. Neumann, Th. Königs, Jeol to get it all run.

References

[1] Kleykamp H 1979 *J. Nucl. Mater.* **80** 13
[2] Walker C T 1979 *J. Nucl. Mater.* **80** 190
[3] Brémier S, Chareau J M and Walker C T 2006 *EMAS 2006 - 7th Regional Workshop on Electron Microanalysis Today (Karlsruhe, Germany; 13-16 May 2006)*
[4] Pearce J H, Summerling R and R. Hargraves R 1983 *J. Nucl. Mater.* **116** 1
[5] Lamontagne J, Desgranges L, Valot C, Noirot J, Blay Th, I. Roure I and B. Pasquet B 2006 *Microchim. Acta* **155** 183
[6] Verwerft M 2000 *J. Nucl. Mater.* **282** 97
[7] Perrot M 1995 Thèse - Diplôme Universitaire Supérieur de Recherche. (Paris: Université Paris Sud) UFR scientifique d'Orsay, préparé au CEA/CEN de Saclay, 29. sept. 1995
[8] Restani R, Martin M, Kivel N and Gavillet D 2009 *J. Nucl. Mater.* **385** 435
[9] Une K, Nogita K, Kashibe S and Immamura M 1992 *J. Nucl. Mater.* **188** 65
[10] Walker C T, Kameyama T, Kitajima S and Kinoshita M 1992 *J. Nucl. Mater.* **188** 73
[11] Fisher S B, White R J, Cook P M A, Brémier S, Corcoran S R C, Stratton R, Walker C T, Ivision P K and Palmer I D 2002 *J. Nucl. Mater.* **306** 153
[12] Manzel R and Walker C T 2002 *J. Nucl. Mater.* **301** 170
[13] Spino J, Stalios A D, Santa Cruz H and Baron D 2006 *J. Nucl. Mater.* **354** 66
[14] Noirot J, Desgranges L and Lamontagne J 2008 *J. Nucl. Mater.* **372** 318
[15] Kleykamp H 1981 *Z. Naturforsch.* **36a** 1388
[16] Proff C, Abolhassani S and Lemaignan C 2011 *J. Nucl. Mater.* **416** 125
[17] Hatano Y and Sugisaki M 1996 *J. Nucl. Sci. Technol.* **33** 829
[18] Uchida S, Asakura Y, Ohsumi K, Miki M, Aizawa M, Matsushima Y and Yonezawa K 1987 *J. Nucl. Sci. Technol.* **24** 385
[19] Orlov A, Restani R, Kuri G, Degueldre C and Valizadeh S 2010 *Nucl. Instrum. Methods B* **268** 297

IOP Conf. Series: Materials Science and Engineering **32** (2012) 012023 doi:10.1088/1757-899X/32/1/012023

Measurement of palladium crust thickness on catalyst by EPMA

L Sorbier[1,3], A-S Gay[1], A Fécant[1], M Moreaud[1] and N Brodusch[2]
[1] IFP Energies nouvelles, Rond-point de l'échangeur de Solaize, P.O. Box 3, FR-69360 Solaize, France
[2] McGill University, M.H. Wong Bldg., 3610 University Street, CA-H3A 2B2 Montréal, Quebec, Canada

E-mail: loic.sorbier@ifpen.fr

Abstract. Selective hydrogenation is a key process in petrochemistry to obtain good feedstock for polymers synthesis. Common catalysts for this process consist in metallic palladium deposited with an eggshell distribution on porous alumina. For this system, the catalytic activity is known to be in strong relation with the thickness of the palladium crust. Typical catalyst consists of 2 - 4 mm diameter spherical beads having a 200 - 400 μm thick palladium crust and a total palladium amount of about 0.3 to 0.5 wt%. The palladium distribution in the catalyst bead can be easily characterized by electron probe microanalysis (EPMA) using polished cross-sections of the beads trough their diameter. By measuring the local concentration of palladium on several points along the bead diameter we obtain the distribution profile of palladium in the bead. Two strategies are proposed to measure this palladium crust thickness by EPMA. First the crust thickness is defined by the distance to the catalyst bead surface containing a fixed amount of total palladium (for example 95 % or 98 %). Second, the palladium profile is modelled by a parameterized analytical function from which a crust thickness can be extracted. Catalytic tests on four samples having different palladium crust thicknesses confirm the strong relation between activity and crust thickness. However the crust thickness containing 98 % of the palladium content shows the best correlation with activity.

1. Introduction

Selective hydrogenation is a key process in petrochemistry. For example, pyrolysis gasoline, a naphta cracking by-product contains highly reactive species such as conjugated dialkenes, styrene and indene that needs to be hydrogenated before being blended to the gasoline pool. This hydrogenation should not be completed to maintain a good octane number. Other important applications occur in polymer synthesis. Polymerisation catalysts for polystyrene production are very sensitive to the presence of phenylacetylene in the feedstock. Therefore, a selective hydrogenation of phenylacetylene in styrene is mandatory in the feedstock before polymerisation of styrene. For polyethylene or polypropylene synthesis, the feedstock needs to be rid of alkynes compound without transforming the desired products, ethylene or propylene. To obtain a good selectivity and a reasonable activity, heterogeneous catalysts are used. Active phases for selective hydrogenation of unsaturated products are noble metals (Pd, Pt, Au, Ag) and transition metals like nickel and cobalt. The industrial processes are commonly

[3] To whom any correspondence should be addressed.

using Pd heterogeneous catalysts in fixed bed reactors. These catalysts are typically 2 - 4 mm diameter alumina beads impregnated with 0.1 - 0.5 wt% palladium.

It has been known for a long time that the distribution of active phases in catalyst beads have an influence on the selectivity for successive reactions and on the deactivation by a self-fouling reaction [1-3]. Experimentally, it has been shown that selective hydrogenation palladium eggshell catalysts have a better activity [4], selectivity [4] and a lower deactivation rate by cocking [5] than homogeneously distributed catalysts with the same palladium loading. It is therefore of primary importance to have a robust method to measure the palladium crust thickness in order to design a new generation of catalysts with increased performances. This method should be able to measure typical crust thicknesses in the 100 - 500 µm range. Electron probe microanalysis (EPMA) has a lateral resolution on the micrometre scale and a good sensitivity yielding detection limit in the tenth of ppm range. It is then well suited to study the distribution of palladium in catalyst beads. It has been effectively employed by several authors in the profile mode [5-8]. EPMA measurements on membranes, a system close to eggshell catalyst, are also reported [9, 10]. Other methods such as optical microscopy have also been used [7, 8]. It is nevertheless surprising that no attempt has been made to extract a quantitative measurement of the crust thickness. The exception is a patent where inventors are using crust thickness measured by EPMA (with data processed as exposed in section 3.2) and transmission electron microscopy to define an optimal particles density [11]. We propose to define and compare some methodologies to measure the palladium thickness crust by EPMA. We will show how these thicknesses correlate with catalytic activity. At first, we describe the experimental methods for the preparation of samples with varying crust thicknesses, the EPMA measurements and the catalytic tests. A section is devoted to the description of the EPMA data processing to extract a crust thickness. The results are then presented and discussed, followed by a conclusion.

2. Experimental

2.1. Sample preparation

The samples preparation method used an aqueous solution of palladium nitrate (Engelhard). Incipient wetness impregnation was realized on alumina beads (2 - 4 mm diameter, total porous volume 0.63 cm^3/g, SBET = 70 m^2/g) in a rotating beaker. Four catalysts were prepared following different ways: sequential impregnation and/or addition of soda in order to pre-neutralize the acidic palladium nitrate solution. After each impregnation, the catalysts were dried at 120 °C overnight and calcined under airflow at 450 °C during 2 h. The palladium loading for all samples was 0.2 wt%. In table 1, catalysts preparation protocols are presented.

Table 1. Catalysts preparation.

Catalyst	Pd impregnation solution (pre-neutralisation by soda)		
	Number of impregnation	pH	T in °C
a	1	1.18	21.4
b	2	1.21	21.2
c	1	2.12	20.0
d	3	2.16	20.3

2.2. EPMA measurements

EPMA on porous samples [12, 13] and particularly on catalysts [14] is known to yield a total of concentration lower than 100 %. The main reasons are carbon contamination during the embedding of catalyst beads in resin and surface roughness due to the internal porosity of the catalyst [15]. Embedding in resin can lead to 10 to 30 wt% of residual carbon coming from the physical or chemical adsorption of resin components on the high specific surface of the catalyst. Moreover, even with a perfect polishing, the internal porosity of the sample generates a rough surface. These two effects affect in a different way the lines of the analyzed elements. By a careful preparation it is possible to reduce the carbon contamination and to obtain a quantitative analysis on well polished mesoporous samples. Nevertheless, macroporous carriers such as the one employed in this study always keep a rough surface yielding a total of concentrations of about 80 to 90 %. The concentrations of palladium are then expected to be 10 - 20 % lower than expected. Nevertheless, as there is no obvious reasons that there exists a gradient of roughness along the catalyst beads diameter, concentration profiles along the diameter should remain not distorted. For the measurement of bulk palladium content, X-ray fluorescence gives an accurate value.

To prepare the samples for EPMA, epoxy resin (Struers Caldofix) was pre-polymerized in oven at 70 °C for about 20 minutes. Ten catalyst beads per sample were placed in 2.5 cm diameter cylindrical plastic moulds and the samples were embedded when the resin becomes viscous. The polymerisation was continued overnight in oven at 70 °C. The moulds were polished with SiC paper under water solvent up to the beads diameters and finished with a 1 μm diamond suspension. The preparations were rinsed with pure ethanol and coated by an about 20 nm thick carbon layer.

EPMA analysis was performed on a Jeol JXA 8800R fitted with four wavelength-dispersive X-ray spectrometers. High voltage was set to 20 kV and a 200 nA probe current was used in order to have a good sensitivity on palladium in a reasonable counting time. Pd was measured using the Lα-line on a PETH crystal and Al using the second order Kα-line on a TAP crystal not to saturate the pulse height analyzer. Counting time on samples was 20 s on peak and 10 s on backgrounds (measured on each side of the peak and linearly interpolated) yielding a detection limit in the order of 300 ppm for Pd, an order of magnitude below the bulk concentration. Pure Pd (Goodfellow, wire 99.95 % purity) and pure alumina (Goodfellow, sapphire pellet Ø 3 mm, 99.9 % purity) were used as standards. Standards and samples were coated by carbon at the same time. Standards were acquired with at least 500000 counts on net peak.

Analysis points were selected as follows. Five profiles were selected along the diameter of five catalyst pellets. For each profile, the crust thickness was evaluated by measuring only Pd Lα- and Al Kα-peaks intensities with counting time of 1 s and a step size of 1 μm. Semi-quantitative analysis points were chosen along the diameter to have 15 points on the crust equally spaced with a narrow step size (5 to 20 μm), 10 points equally spaced with a wide step size in the core (50 to 100 μm) and 15 points with a narrow step in the opposite crust. The first and last points of the profile are acquired in the surrounding resin to be sure than the total diameter of the pellet has been measured. Data acquisition was performed overnight for a duration of about 8 hours per sample.

2.3. Catalytic test

To relate the crust thickness with the catalytic performance of the prepared catalyst, their activities were evaluated in selective hydrogenation of a 50/50 wt% mixture of isoprene and styrene diluted in n-heptane. The reaction scheme is illustrated in figures 1 and 2.

Catalytic tests are done in a perfectly agitated Grignard type reactor presented in figure 3. The catalyst is contained in a cylindrical basket communicating with the reactor by numerous holes with diameters of 1.2 mm. The bottom part of the reactor is filled in gloves bag under argon with a weighted mass (corresponding to about 2 cm^3) of freshly reduced catalyst, and covered by about 200 ml of previously degassed n-heptane. The bottom part is then removed from the gloves bag and the reactor is sealed. The surrounding atmosphere is purged 3 times by hydrogen at 35 bars. The reactor is then put back at atmospheric pressure for feedstock injection of 34 g of a 50/50 wt% styrene-isoprene mixture. The reactor is agitated and a sampling of the liquid phase (sample 0) is done. The temperature is increased to 45 °C, the agitation is stopped and hydrogen at a pressure of 35 bars is admitted in the reactor. The test begins when the agitation is switch on again.

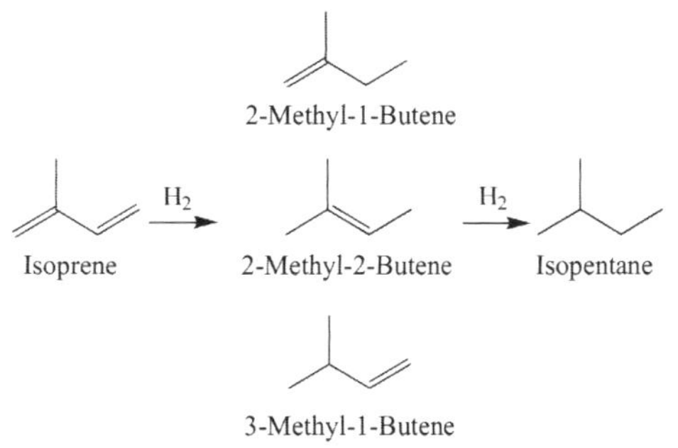

Figure 1. Reactions during hydrogenation of isoprene on catalyst containing palladium.

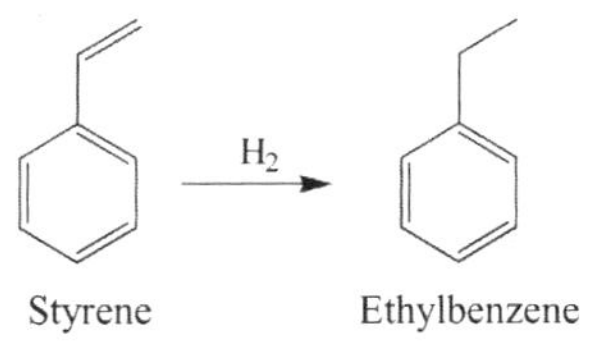

Figure 2. Reactions during hydrogenation of styrene on catalyst containing palladium.

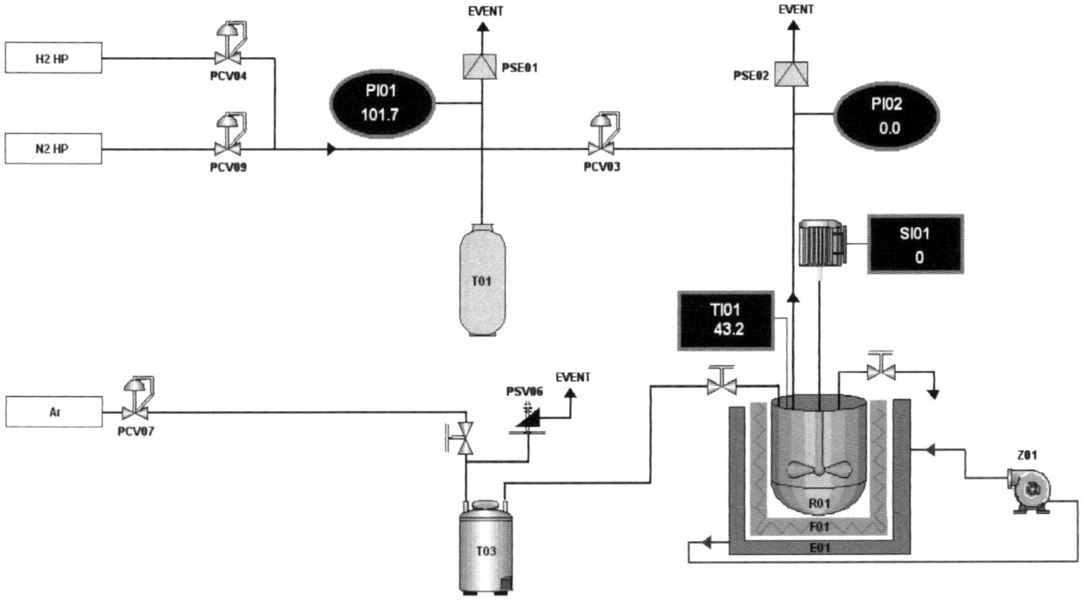

Figure 3. Schematic diagram of the catalytic test unit.

Activity is followed both by surveying hydrogen consumption and by the analysis of samples of the liquid phase by gas chromatography. The activity of the catalyst A may be expressed by the speed of hydrogen pressure drop in the reactor:

$$A = \frac{V_{H_2} P_{H_2}}{24 V_{Cata}} \quad (1)$$

where A is in mol.min^{-1}.cm^{-3}, V_{H_2} is the hydrogen volume in the reactor (in cm^3), P_{H_2} is the pressure drop by unit time (bar.min^{-1}) in the reactor and V_{Cata} the volume of catalyst (in cm^3).

3. Data processing of EPMA measurements

3.1. Profiles

We note $c_i(x)$ with $x \in [0, 2R_i]$ the concentration profile of palladium inside the catalyst bead i along its diameter, R_i being the radius of the bead i. c_i is measured in a discrete and finite set of positions $_j$ along the diameter. The positions are defined as xi,j and the associated concentrations ci,j=c(xi,j) for j∈{1, ..., Ni}, where Ni is the number of points for the concentration profile of bead i. The mean palladium profile C(x) is computed by linear interpolation of the five beads profiles assuming an homothetic distribution of palladium. To be more precise, the algorithm can be described as follows:

- Look for $i$max such that $N = N_{imax} = \max\{Ni, i = 1, \ldots, 5\}$
- Compute $R = \frac{1}{5}\sum_{i=1}^{5} R_i$
- Define a grid $x_j = x_{i\max,j}$ and $\overline{x}_j = x_{i\max,j}/2R_{i\max}$ for $j \in \{1, ..., N\}$
- Define a grid $\overline{x}_{i,j} = x_{i,j}/2R_i$ for $i,j \in \{1, ..., 5\} \times \{1, ..., N_i\}$
- For each $j \in \{1, ..., N\}$
  - Set $C_j = 0$
  - For each $i \in \{1, ..., 5\}$
    - Find $k \in \{1, \ldots, N_i - 1\}$ such as $\overline{x}_{i,k} \leq \overline{x}_j \leq \overline{x}_{i,k+1}$
    - Set $C_j = C_j + \frac{c_{i,k+1} - c_{i,k}}{\overline{x}_{i,k+1} - \overline{x}_{i,k}}\left(\overline{x}_j - \overline{x}_{i,k}\right)$
  - Set $C_j = \frac{C_j}{5}$

A simple binary search algorithm [16] is efficient enough to perform the search for k. As an illustration, figure 4 shows a mean profile obtained from the measurement of five profiles on five different beads.

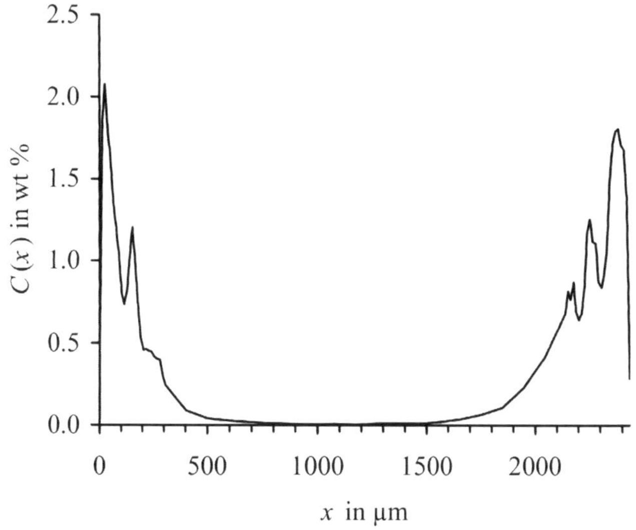

Figure 4. Example of mean Pd profile.

3.2. Fraction of palladium content

Figure 5 shows the geometry of an idealized perfectly spherical catalyst pellet, polished at its diameter. We define $Q(x)$, $x\in[0, R]$ as the amount of palladium in the corona having an inner radius x and an outer radius R. Assuming a spherical symmetry for the palladium distribution we have:

$$Q(x)=2\pi\left[\int_0^x c(R-y)(R-y)^2\,dy+\int_{2R-x}^{2R} c(y-R)(y-R)^2\,dy\right] \tag{2}$$

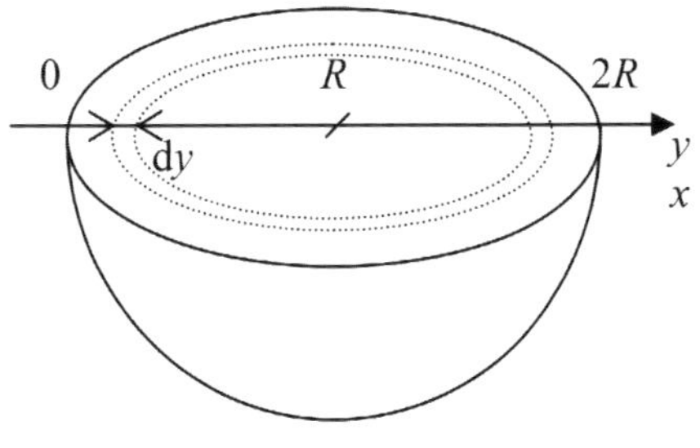

Figure 5. Elementary volume for palladium quantity integration.

We evaluate Q by the trapezoidal rule and linear interpolation of c using the following algorithm:

- Define a grid $\bar{y}_k=(k-1)/999$ for $k\in\{1, ..., 1000\}$
- Set $i=1,\ j=N-1,\ S=0$
- For each $k\in\{1, ..., 999\}$
  - Compute $r_1=R\bar{y}_k$, if $r_1>x_i$ then $i=i+1$, compute $C_1=c_i+\dfrac{c_{i+1}-c_i}{x_{i+1}-x_i}(r_1-x_i)$
  - Compute $r_2=R\bar{y}_{k+1}$, if $r_2>x_j$ then $i=i+1$, compute $C_2=c_i+\dfrac{c_{i+1}-c_i}{x_{i+1}-x_i}(r_2-x_i)$
  - Compute $r_3=R(2-\bar{y}_k)$, if $r_3<x_j$ then $j=j-1$, compute $C_3=c_j+\dfrac{c_{j+1}-c_j}{x_{j+1}-x_j}(r_3-x_j)$
  - Compute $r_4=R(2-\bar{y}_{k+1})$, if $r_4<x_j$ then $j=j-1$, compute $C_4=c_j+\dfrac{c_{j+1}-c_j}{x_{j+1}-x_j}(r_4-x_j)$
  - Compute $S=S+\dfrac{\pi}{1000}\left(C_1(R-r_1)^2+C_2(R-r_2)^2+C_3(R-r_3)^2+C_4(R-r_4)^2\right)$
  - Set $Q_k=Q(R\bar{y}_k)=S$

Once Q is known for each $\bar{y}_k$, it is easy to find the thickness of the corona containing a fixed proportion of total palladium. For example, to find t_{80}, the thickness of the corona containing 80 % of total palladium we use the following algorithm:

- Find $k\in\{1, ..., 1000\}$ such as $Q_k\le 0.8Q_N\le Q_{k+1}$
- $t_{80}=R\left(\bar{y}_k+\dfrac{\bar{y}_{k+1}-\bar{y}_k}{Q_{k+1}-Q_k}(0.8Q_N-Q_k)\right)$

As an illustration, figure 6 shows the relative quantity of palladium versus the corona thickness for the profile presented in figure 4.

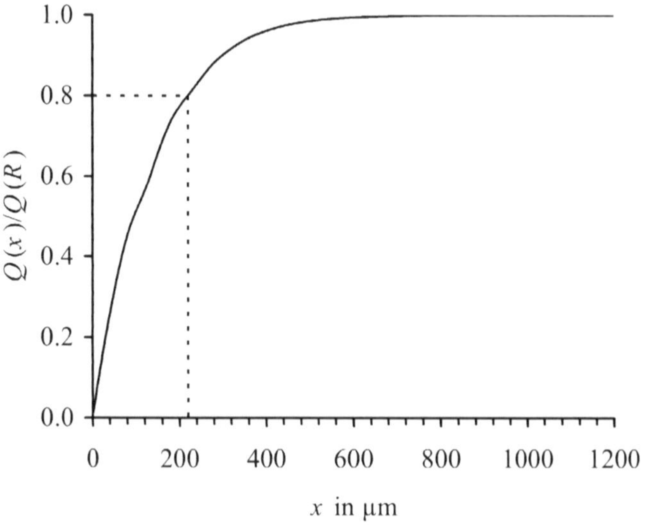

Figure 6. Example of fraction of palladium contained in a corona versus its thickness. Dotted lines define t_{80}.

3.3. Fit by an analytical function

Another strategy to define a crust thickness is to fit the measured profile by a parameterized analytical function. If well chosen, one parameter can give a measure of the thickness. To have a simpler form for the analytical function, we symmetrize the profile by its centre (assuming a spherical symmetry for the palladium distribution):

$$\bar{c}(x) = \frac{1}{2}\left[c(x) + c(R - x)\right] \tag{3}$$

We evaluate $\bar{c}(x)$ by linear interpolation of the profile using the following algorithm:

- Set $n = N/2$
- For each $i \in \{1, \ldots, n\}$
  - Find k such as $\bar{x}_k \leq 1 - \bar{x}_i \leq \bar{x}_{k+1}$
  - Compute $C_1 = c_k + \dfrac{c_{k+1} - c_k}{\bar{x}_{k+1} - \bar{x}_k}\left(1 - \bar{x}_i - \bar{x}_k\right)$
  - Set $\bar{c}_i = \dfrac{1}{2}\left(c_i + C_1\right)$

One proposal for a parameterized analytical function (monotonous decreasing function) is given by equation 4:

$$f(x) = a_1 \exp(-a_2 x) + a_3 \tag{4}$$

The term a_3 is present to take into account large crust thicknesses (nearly homogeneous palladium distribution). We define the crust thickness t_{fit} as the point where the tangent of f in $x = 0$ intercepts the abscissa axis:

$$t_{\text{fit}} = \frac{a_1 + a_3}{a_1 a_2} \tag{5}$$

The fit of the experimental profile is done in two steps. A first approximation $\widetilde{a}_i$ of the parameters is performed by the following algorithm (see Appendix A for the derivation of equations 6 and 7):

- Assume $\widetilde{a}_3 = \frac{1}{3}\left(\widetilde{c}_n + \widetilde{c}_{n-1} + \widetilde{c}_{n-2}\right)$
- Solve numerically equation 6 for a_2

$$\frac{\sum_{i=1}^{n}\left(\widetilde{c}_i - \widetilde{a}_3\right)\exp\left(-a_2 x_i\right)}{\sum_{i=1}^{n}\exp\left(-2a_2 x_i\right)} - \frac{\sum_{i=1}^{n} x_i\left(\widetilde{c}_i - \widetilde{a}_3\right)\exp\left(-a_2 x_i\right)}{\sum_{i=1}^{n} x_i \exp\left(-2a_2 x_i\right)} = 0 \tag{6}$$

- Evaluate $\widetilde{a}_1$ from equation 7

$$\widetilde{a}_1 = \frac{\sum_{i=1}^{n}\left(\widetilde{c}_i - \widetilde{a}_3\right)\exp\left(-\widetilde{a}_2 x_i\right)}{\sum_{i=1}^{n}\exp\left(-2\widetilde{a}_2 x_i\right)} \tag{7}$$

In a second step, these values are used as input parameters in a Levenberg-Marquardt algorithm [17]. Uncertainties of the measured data, σ_c, are estimated neglecting the uncertainties of the standards and using peak and background counts on the sample by the procedure explained at the end of Appendix B. Once the parameters are optimized, t_{fit} is easily calculated using equation 5. As an example, figure 7 shows such a fit.

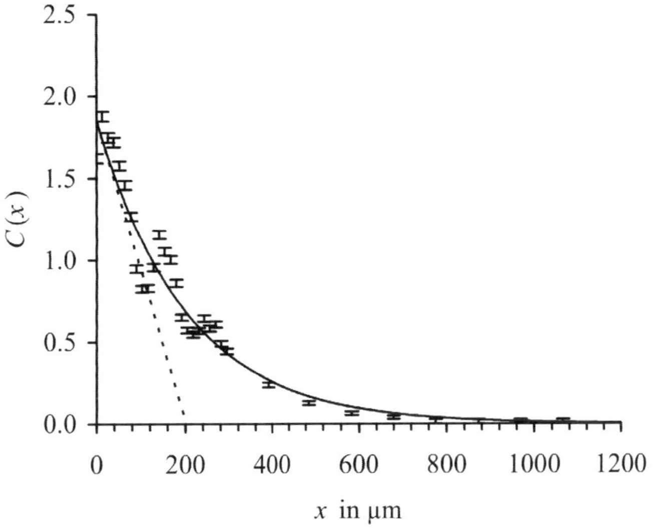

Figure 7. Example of an analytical fit of a palladium profile. The intercept of the tangent of the curve at $x = 0$ (in dotted line) and the abscissa axis defines the crust thickness t_{fit}.

4. Results and discussion

For the four prepared catalysts described in section 2.1, we have measured the palladium profile by EPMA as explained in section 2.2. EPMA profiles were processed as described in section 3. The measurement of apparent activity has been performed with catalytic test presented in section 2.3. The results obtained are reported in table 2. As expected, there is a clear effect of the palladium crust thickness on the measured apparent activity. The thinner the crust, the larger the apparent activity.

Table 2. Catalytic activities and Pd crust thicknesses determined by EPMA.

Catalyst	Activity 10^{-3} mol.min^{-1}.cm^{-3}	Pd crust thickness - µm				
	A	t_{95}	t_{97}	t_{98}	t_{99}	t_{fit}
a	5.2	251	275	294	326	208
b	6.2	137	154	171	196	126
c	9.6	68	79	89	133	29
d	9.8	41	67	90	128	9

In the internal diffusion limited regime (which is the case for the conditions of the test), there is a relation between the apparent measured activity A and the characteristics of the catalyst [18]:

$$A = \frac{\alpha \times A_i}{d} \sqrt{S_{\mathrm{BET}} V_{\mathrm{p}}} \tag{8}$$

where α is a constant, d is the bead diameter, A_i is the intrinsic activity, S_{BET} is the specific surface, and V_p the total porous volume. Equation (8) is valid for homogeneously distributed palladium in catalyst beads. For an eggshell distribution of palladium, the bead diameter should be replaced by the crust thickness because this is the working zone for the chemical reaction. This is true if the density of active sites is constant in the crust. Hence, we expect an affine relationship between the apparent activity and the inverse crust thickness t:

$$A = \frac{\beta}{t} + \delta \tag{9}$$

where β and δ are constants.

Figure 8 shows the activity plotted versus the inverse palladium crust thickness. In figure 8, least square linear fits are superimposed on the experimental points. Linear least square fit for the analytical fit is rather of poor quality. One explanation is a wrong form for the analytical function. It is worth noticing that choosing a multiple of t_{fit} would give the same poor regression coefficient. On the contrary there is a much stronger linear relation between apparent activity and thicknesses determined by cumulative palladium content. Figure 9 shows that the threshold choice of 98 % is nearly optimal. Choosing a higher threshold leads to large uncertainty since the $Q(x)$ curve is almost flat and sensitive to the large uncertainties of measurement of low palladium concentration in this area. Lower threshold gives a worse correlation since a part of the catalytically active palladium is not taken into account. However we can notice that the extrapolation of the linear fits for infinite crust thickness ($1/t \rightarrow 0$) shows a residual activity, δ, which should correspond to the apparent catalytic activity of catalyst with homogeneous repartition of the metal in the whole volume of the beads.

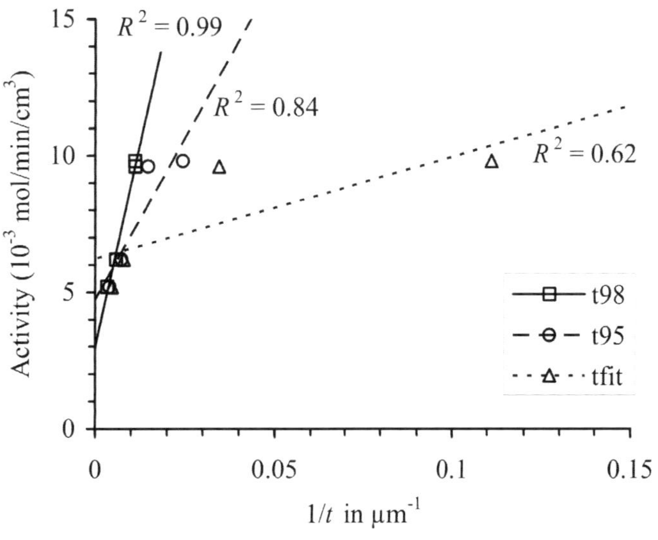

Figure 8. Apparent activity plotted versus inverse palladium crust thickness for the four prepared catalysts. Lines corresponds to least square linear fits with linear regression coefficient R indicated close to the lines.

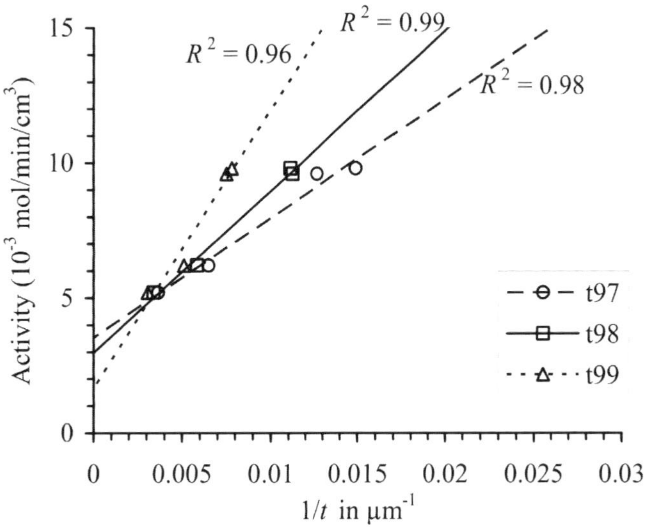

Figure 9. Apparent activity plotted versus inverse palladium crust thickness for the four prepared catalysts. Lines corresponds to least square linear fits with linear regression coefficient R indicated close to the lines.

5. Conclusion.

Four palladium catalysts were prepared with varying palladium crust thicknesses. Their catalytic activities were measured in selective hydrogenation of a styrene-isoprene mixture diluted in n-heptane. The catalyst palladium concentration profiles along the catalyst beads diameter have been measured by EPMA. We have set up two strategies to extract a quantitative crust thickness from these concentration profiles. The first strategy is based on the computation of the cumulative palladium content versus the distance to the bead surface. The second strategy uses fit of experimental profile to an exponential decaying function. As expected, the thinner is the crust, the better is the activity. We have shown that the analytical fit approach does not yield to a satisfactory correlation between activity and crust thickness. On the contrary, cumulative palladium curves using a relative 98 % threshold give a very good linear fit between activity and palladium crust thickness. This results in an efficient way to predict activities of new generation of catalysts without testing all of them. Yet, EPMA requires rather long preparation and acquisition times. It is then difficult to use this technique on heterogeneous batches where the sampling (number of beads, number of profiles along different directions in one bead) must be large to be representative. Since palladium is coloured (brown) and alumina carrier is white it should be possible to measure the crust thickness by optical microscopy and correlate with the results of EPMA and activities. This will be the motivation for a future work.

References

[1] Shadman-Yazdi F and Petersen E 1972 *Chem. Eng. Sci.* **27** 227-37
[2] Corbett W and Luss D 1974 *Chem. Eng. Sci.* **29** 1473-83
[3] Villadsen J 1976 *Chem. Eng. Sci.* **31** 1212-13
[4] Berenblyum A, Mund S, Karelskii V, Goranskaya T, Zolotukhin V and Lakhman L 1986 *Kinet. Catal.* **27** 184-87
[5] Lin T B and Chou T C 1994 *Appl. Catal. A* **108** 7-19
[6] Lin T B and Chou T C 1995 *Ind. Eng. Chem. Res.* **34** 128-34
[7] Chang J C and Chou T C 1997 *Appl. Catal. A* **156** 193-205
[8] Shao Z, Li C, Chen X, Pang M, Wang X and Liang C 2010 *ChemCatChem* **2** 1555-58
[9] Kim D, Sauk J, Byun J, Lee K S and Kim H 2007 *Solid State Ionics* **178** 865-70
[10 Seshimo M, Ozawa M, Sone M, Sakurai M and Kameyama H 2008 *J. Membr. Sci.* **324** 181-87
[11] Fischer L, Petit-Clair C, Thomazeau C, Sorbier L and Verdon C 2007 Catalyst comprising palladium and selective hydrogenation application thereof. *French Patent FR 2.922.784*
[12] Ichinokawa T, Kobayashi H and Nakajima M 1969 *Jap. JJAP* **8** 1563-68
[13] Lakis R, Lyman C and Goldstein J 1992 *Proc. 50th Annual Meeting of the Electron Microscopy Society of America (Boston)* (San Francisco: San Francisco Press), pp. 1660-61
[14] Lakis R, Vicenzi E and Allen F 1996 *Proc. 54th Annual Meeting of the Electron Microscopy Society of America (Minneapolis)* (San Francisco: San Francisco Press), pp. 512-13
[15] Sorbier L, Rosenberg E and Merlet C 2004 *Microsc. Microanal.* **10** 745-52
[16] Lewis G, Boynton N and Burton F 1981 *Inform. Proc. Lett.* **13** 4-7
[17] Press W, Teukolsky S, Vetterling W and Flannery B 1992 *Numerical recipes in FORTRAN: the art of scientific computing.* (Cambridge: Cambridge University Press)
[18] Le Page J F 1978 *Catalyse de contact.* (Paris: Technip)
[19] Ancey M, Bastenaire F and Tixier R 1977 *J. Phys. D: Appl. Phys.* **10** 817-30

Appendix A. Derivation of eqs. (6) and (7).

We want to minimize χ^2, the quadratic sum of distance between the experimental points $(x_i, \bar{c}_i)$ and the analytical model $(x_i, f_i = f(x_i))$ with:

$$\chi^2 = \sum_{i=1}^{n} (\bar{c}_i - f_i)^2 \tag{A.1}$$

The best parameters are found when the partial derivative of χ^2 are equal to 0, that is:

$$\frac{\partial \chi^2}{\partial a_j} = 0 \text{ for } j \in \{1,2,3\} \tag{A.2}$$

Substituting by the expression for χ^2 of equation (A.1) leads to:

$$\sum_{i=1}^{n} f_i \left.\frac{\partial f}{\partial a_j}\right|_{x_i} = \sum_{i=1}^{n} \bar{c}_i \left.\frac{\partial f}{\partial a_j}\right|_{x_i} \text{ for } j \in \{1,2,3\} \tag{A.4}$$

Equation (A.4) is a set of non-linear equations. Even if a solution can be found numerically, it is hard to bracket the roots to ensure that it is the global and not a local minimum of χ^2. We will use an approximate solution and refine it using a Levenberg-Marquardt algorithm. Suppose that the palladium crust is thin enough compared to the bead's radius. A good estimator of a_3 is:

$$\tilde{a}_3 = \sum_{i=n-2}^{n} \bar{c}_i \tag{A.5}$$

Then eq. (A.4) reads:

$$\begin{cases} a_1 \sum_{i=1}^{n} \exp(-2a_2 x_i) = \sum_{i=1}^{n} (\bar{c}_i - \tilde{a}_3)\exp(-a_2 x_i) \\ a_1 \sum_{i=1}^{n} x_i \exp(-2a_2 x_i) = \sum_{i=1}^{n} x_i (\bar{c}_i - \tilde{a}_3)\exp(-a_2 x_i) \end{cases} \tag{A.6}$$

There is no analytical expression for a_2, nevertheless it is possible to find numerically its value by equalling the two values for a1 in the system of eqs. (A.6):

$$\frac{\sum_{i=1}^{n} (\bar{c}_i - \tilde{a}_3)\exp(-a_2 x_i)}{\sum_{i=1}^{n} \exp(-2a_2 x_i)} - \frac{\sum_{i=1}^{n} x_i (\bar{c}_i - \tilde{a}_3)\exp(-a_2 x_i)}{\sum_{i=1}^{n} x_i \exp(-2a_2 x_i)} = 0 \tag{A.7}$$

Equation (A.7), equivalent to eq. (6), can be numerically solved for a_2, for example with Brent's method 0. The bracketing of the root of this equation is done as follows. The largest crust thickness should be in the order of two bead's radius, so the minimal value for $a_2 = 0.5/(x_n - x_1)$. The crust thickness should not be smaller than half the spacing of the two first points. Thus the maximal value for a_2 is $2/(x_2 - x_1)$. Once $\tilde{a}_2$ the optimal value of a_2 is known, $\tilde{a}_1$ can be calculated using the first equation of (A.6).

$$\tilde{a}_1 = \frac{\sum_{i=1}^{n} (\bar{c}_i - \tilde{a}_3)\exp(-\tilde{a}_2 x_i)}{\sum_{i=1}^{n} \exp(-2\tilde{a}_2 x_i)} \tag{A.8}$$

If no root is found for eq. (A.7), the values in table 3 are used as feeds for the Levenberg-Marquardt algorithm. In practice, there is always a solution in eq. (A.7) for well marked crusts comprising at least 3 points.

Table 3. Estimation of parameters when no root is found in eq. (A.7).

Parameter	Estimator
$\tilde{a}_1$	$\bar{c}_1 - \bar{c}_n$
$\tilde{a}_2$	$2.0/(x_2 - x_1)$
$\tilde{a}_3$	$\bar{c}_n$

Appendix B. Uncertainties and detection limit.

The problem of the confidence interval of the concentration and the detection limit in microbeam analysis has been addressed by Ancey et al. [19]. The method has one drawback: counting on standards has to be known which is not straightforward if automatic data analysis of numerous data sets is targeted. When the measured counts on standards are high, uncertainty is only governed by measurements on the sample. Thus the expression for the detection limit can be simplified.

For one element, let's call y_c the measure of the concentration on the sample (close to the detection limit) and y_b the measure of concentration on the same matrix which is not containing the element (blank). We will assume that $\Delta y = y_c - y_b$ is close to a normal law of mean $m_c - m_b$ and variance $\sigma^2 = \sigma_c^2 + \sigma_b^2$. If the element is not present $m_c - m_b = 0$ and $\sigma^2 = 2\sigma_b$. Let's call:

$$z = \frac{\Delta y}{\sqrt{2}\sigma_b} \tag{B.1}$$

The variable z follows a reduced normal law. We define the decision limit L_c, the concentration for which, if $y > L_c$, the probability to affirm that the element is present whereas it is absent is lower than a predefined probability α. We have:

$$P(z > u_{1-\alpha}) = \alpha \tag{B.2}$$

where $u_{1-\alpha}$ are tabulated values obtained from the cumulative normal law. By replacing z by its definition, eq. (B.2) reads:

$$P\left(y_c > \sqrt{2}\sigma_b u_{1-\alpha}\right) = \alpha \tag{B.3}$$

which means that:

$$L_c = \sqrt{2}\sigma_b u_{1-\alpha} \tag{B.4}$$

At the decision limit, there is a low risk (lower than α) to declare an element present whereas it is absent. Yet, the risk to declare that the element is absent whereas it is present is large ($\beta = 0.5$). One way to diminish the risk is to notice that z' follows a reduced normal law with:

$$z' = \frac{\Delta y - L_c}{\sqrt{2}\sigma_b} \tag{B.5}$$

Then we have:

$$P(z' < u_{1-\beta}) = \beta \tag{B.6}$$

which is equivalent to:

$$P\left(y_c < L_c + \sqrt{2}\sigma_b u_{1-\beta}\right) = \beta \tag{B.7}$$

This gives the definition of Ld, the detection limit:

$$L_d = \sqrt{2}\sigma_b\left(u_{1-\alpha} + u_{1-\beta}\right) \tag{B.8}$$

Figures 10 and 11 shows graphically the definition of L_c and L_d.

A measure in X-ray microanalysis results from a measure of the peak subtracted by a measure of the background. The distribution of measure on a blank is not normal but close to a normal of null mean and $2N_B$ variance where N_B is the total number of counts on the background.

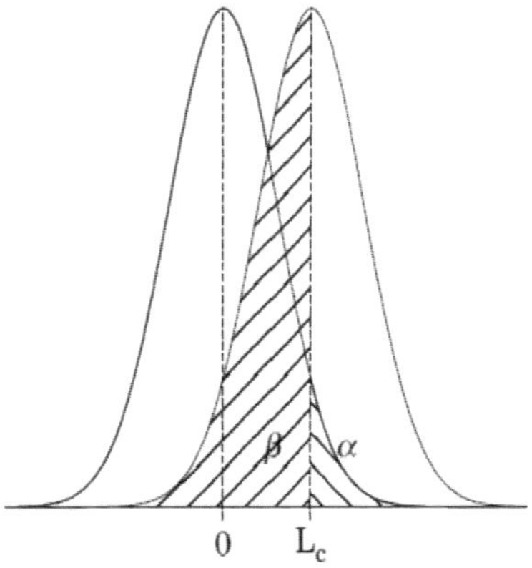

Figure 10. Decision limit L_c.

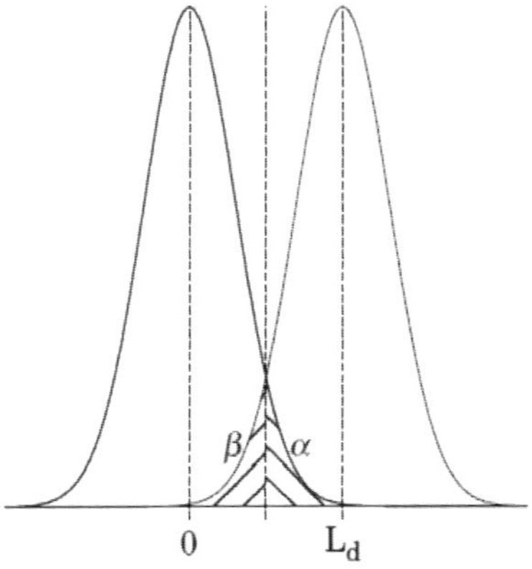

Figure 11. Detection limit L_d.

We suppose that the concentration of the element is low so that there is a linear relation between the concentration and the intensity of X-ray. Choosing $\alpha = \beta = 5$ %, we have $u_{1-\alpha} = u_{1-\beta} \approx 1.6459$ and then:

$$L_d = 4.65 \frac{c}{N_P - N_B} \sqrt{2N_B} \tag{B.9}$$

where c is the measured concentration, N_P is the total number of counts on the peak, and N_B the total number of counts on the background. The eq. (B.9) is meaningful only if c is sufficiently low to have a linear relation between concentration and X-ray intensity and $c > L_d$. Therefore, we calculate the detection limit for all the points of the profile. We look for the 20 lowest concentrations larger than their associated detection limit and take the mean to have the overall detection limit.

To evaluate an uncertainty σ_c on each of the points of the concentration profile c for the Levenberg-Marquardt algorithm, we proceed as follows:

- we calculate the overall detection limit L_d as explained above
- if $c \leq L_d$ then we set $c = L_d/2$ and $\sigma_c = L_d/2$
- otherwise we set $\sigma_c = 2\frac{c\sqrt{N_P + N_B}}{N_P - N_B}$ assuming a local linear relation between concentration and intensity.

IOP Conf. Series: Materials Science and Engineering **32** (2012) 012024 doi:10.1088/1757-899X/32/1/012024

Systematic discrepancies in Monte Carlo predictions of *k*-ratios emitted from thin films on substrates

P Statham[1,4], X Llovet[2] and P Duncumb[3]
[1] Oxford Instruments Nanoanalysis Ltd., Halifax Road, High Wycombe HP12 3SE, Great Britain
[2] University of Barcelona, Centres Científics i Tecnològics (CCiT), C/ Lluís Solé i Sabarís 1-3, ES-08028 Barcelona, Spain
[3] Willow House, Great Shelford CB22 5LZ, Great Britain

E-mail: peter.statham@oxinst.com

Abstract. We have assessed the reliability of different Monte Carlo simulation programmes using the two available Bastin-Heijligers databases of thin-film measurements by EPMA. The MC simulation programmes tested include Curgenven-Duncumb MSMC, NISTMonte, Casino and PENELOPE. Plots of the ratio of calculated to measured k-ratios ("k_{calc}/k_{meas}") against various parameters reveal error trends that are not apparent in simple error histograms. The results indicate that the MC programmes perform quite differently on the same dataset. However, they appear to show a similar pronounced trend with a "hockey stick" shape in the "k_{calc}/k_{meas} versus k_{meas}" plots. The most sophisticated programme PENELOPE gives the closest correspondence with experiment but still shows a tendency to underestimate experimental *k*-ratios by 10 % for films that are thin compared to the electron range. We have investigated potential causes for this systematic behaviour and extended the study to data not collected by Bastin and Heijligers.

1. Introduction

The thickness and composition of thin layers on a substrate can be determined by electron probe microanalysis (EPMA) by measuring the "*k*-ratio" for the X-ray characteristic lines emitted when the same electron beam strikes the specimen and pure material. For each line, the *k*-ratio is defined as I_s/I_p, where I_s is intensity from specimen and I_p is the intensity from bulk pure material.

A theoretical simulation model is commonly used to predict *k*-ratios that would be obtained from a specific structure of thin layers on a substrate. By an iterative process, a structure is found that gives *k*-ratios consistent with those measured. The accuracy of the deduced structure is therefore directly dependent on the accuracy of the theoretical model. Analytical $\varphi(\rho z)$-based models, which have been successfully deployed in many applications, are well known to have problems when there are any large atomic number differences between the various layers or the substrate [1]. In principle, Monte Carlo (MC) simulation should be better for calculating *k*-ratios because it deals explicitly with the change in properties at boundaries between materials and indeed good results have been demonstrated on some structures [2]. Although MC simulation is widely used and assumed to be accurate, there are very few reports that show under what conditions inaccuracies could occur for thin film applications.

[4] To whom any correspondence should be addressed.

There is a wide range of sophistication in Monte Carlo programmes and many offer several options for the equations to be used. It is not clear what sophistication is necessary, what accuracy can be achieved or what options are important for modelling X-ray emission from thin films on substrates. This prompted the current investigation into the accuracy of simulation models.

We started compiling measurement results from literature searches but the examples were diverse and it was not easy to detect any trends in behaviour between different programmes. However, we found the comprehensive set of measurements on Pd and Al films by Bastin and Heijligers [3] ("B&H") particularly valuable and we compared the predicted k-ratios from various programmes with those measured by B&H.

In this study, we have assessed the reliability of different MC simulation programmes using the two available B&H databases. The MC simulation programmes tested include Curgenven-Duncumb MSMC, NISTMonte, Casino and PENELOPE. Plots of the ratio of calculated to measured k-ratios ("k_{calc}/k_{meas}") against various parameters reveal error trends that are not apparent in simple error histograms. The results indicate that different MC programmes perform quite differently on the same dataset. However, they appear to show a similar pronounced trend with a "hockey stick" shape in the "k_{calc}/k_{meas} versus k_{meas}" plots. The most sophisticated programme, PENELOPE, gives the closest correspondence with experiment but still shows a tendency to underestimate experimental k-ratios by 10 % for films that are thin compared to the electron range. We have investigated potential causes for this systematic behaviour and extended the study to data not collected by Bastin and Heijligers.

2. Methods and simulation models

2.1. Dataset used for visualising error trends

The mass thickness of each film is needed in order to simulate X-ray intensities and B&H had difficulty in determining what mass thicknesses to put in their tables. Since their choice was influenced by the results of MC simulations, the tabulated mass thicknesses cannot be used in an unbiased test of MC simulation accuracy. B&H also found that density of films varied and was not the same as bulk density so linear thickness (e.g., from TEM or XRD) was not a good indicator of mass thickness. The only independent measure of mass thickness was Rutherford backscattering spectrometry, RBS, so here we have used RBS mass thickness results as the basis for the MC simulations.

The B&H databases for Al and Pd contain measurements of 6 different mass thicknesses of film on a variety of substrates Z = 4 to Z = 83 at beam currents from 3 – 30 kV. When all the results are accumulated in a single error histogram or plot, it is impossible to see whether there are any combinations that cause particular bias. Therefore we chose a subset of data that showed all extremes of combinations, as indicated in table 1.

Table 1. Subset data set for visualisation.

Film element	Substrate element			Film thickness ($\propto$g/cm^2)		
	Low	Medium	High	Low	Medium	High
Pd	C	Mo	Au	11.04	44.9	180.5
Al	C	Si	Au	3.27	21.9	82.0

2.2. Monte Carlo simulation programmes

The MC simulation programmes used in this investigation include the single scattering model (SS) as described in David Joy's book [4], Curgenven-Duncumb MSMC (see ref.[4], chapter 4), NISTMonte [6], CASINO [5] and PENELOPE [9].

3. Results

3.1. Comparison of MC codes

Figure 1 shows of the ratio of calculated to measured *k*-ratios ("k_{calc}/k_{meas}") against measured k-ratios ("k_{meas}") for different MC simulation programmes. The Pd Lα intensity has been calculated for both a thin film of appropriate thickness on substrate and bulk pure Pd. Circles of different size are used to represent results from low, medium and high thickness and different colours are used to represent each substrate.

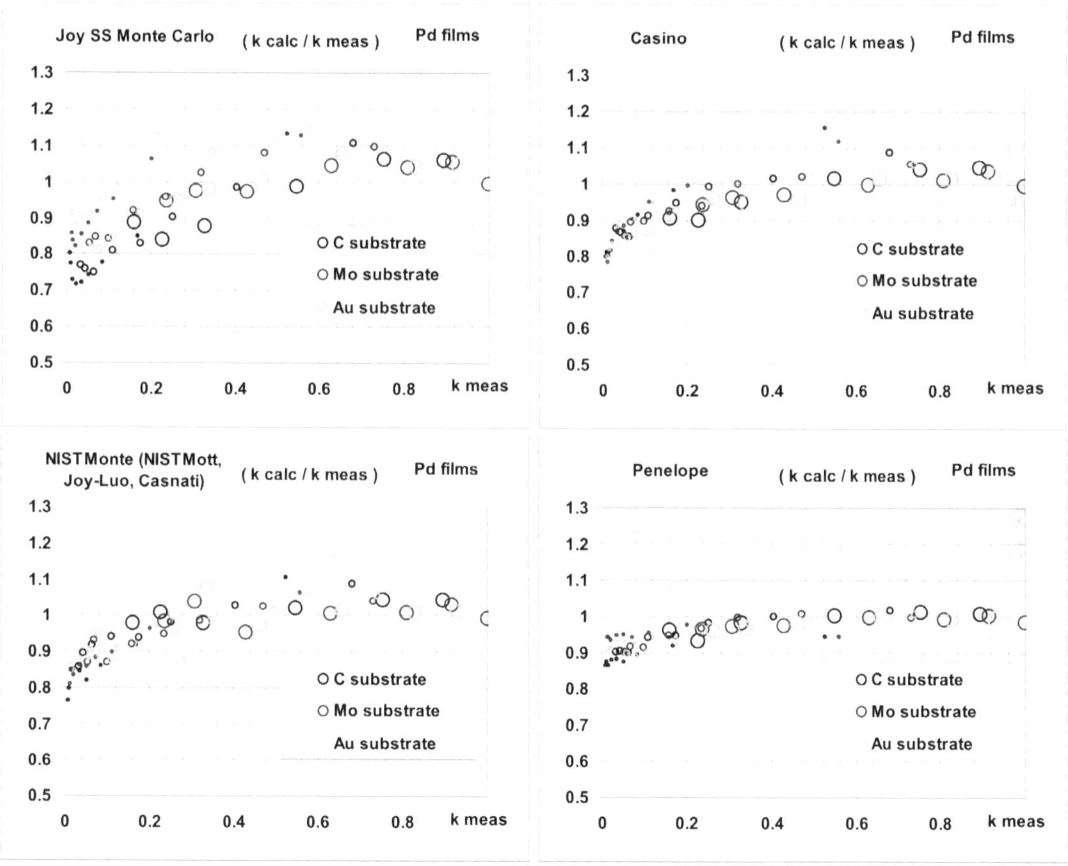

Figure 1. Ratio of calculated to measured *k*-ratios versus measured *k*-ratios obtained by using different MC programmes, as indicated in the figures, for Pd films on C, Mo and Au. Circles of different size represent results from low, medium and high thickness. Experimental data taken from ref. [3].

The ratio of predicted and measured *k*-ratios should be 1 but in fact shows a large scatter and distinct trend when plotted against the measured *k*-ratio. As expected, CASINO with default settings does a better job than the simple SS Monte Carlo. Other than a few points associated with the thinnest film at low kV, most of the results cluster quite closely together but "k_{calc}/k_{meas}" still shows a pronounced downward trend towards low *k*-ratios.

Another popular programme is NISTMonte and the results are shown here using the default NISTMott scattering cross-section and the same Casnati ionisation cross-section that is used in the

Casino simulation. Whereas there are some notable differences, "k_{calc}/k_{meas}" shows the same downward trend towards low *k*-ratios.

Joy-SS, MSMC, Casino and NISTMonte all use the same continuous slowing down approximation (CSDA) for energy loss because of its advantages in speeding computation. The most sophisticated Monte Carlo programme available in the public domain is PENELOPE where the philosophy has been to use the best physics available and avoid approximations wherever practical. The results with PENELOPE are demonstrably better than the other programmes. Nevertheless, there is still a downward trend or "droop" at low *k*-ratios.

If the "droop" was due to something fundamental in MC simulation, it should occur for other element layers. When the B&H data for Al layers was investigated using the RBS thicknesses (figure 2), NISTMonte produced a similar systematic trend to that seen with the Pd data. With PENELOPE, there is still evidence of the "droop" for the Au substrate and a tendency to underestimate *k*-ratios for the thinner films but several of the results for C and Si substrates were good so we really need more data to determine if there really is something fundamental to be addressed in the MC simulation.

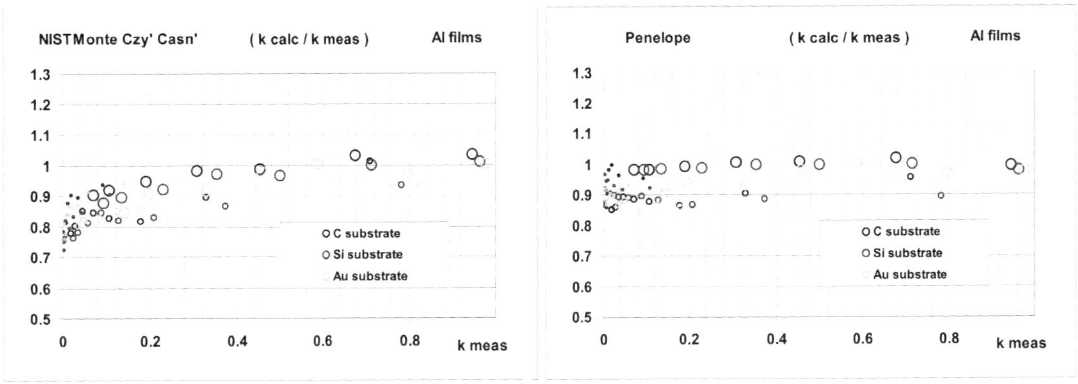

Figure 2. Ratio of calculated to measured *k*-ratios versus measured *k*-ratios obtained by using NISTMonte (left) and PENELOPE (right), for Al films on C, Si and Au. Circles of different size represent results from low, medium and high thickness. Experimental data taken from ref. [3].

We considered other mechanisms that might explain the "droop" such as fluorescence or fast secondaries. For Pd Lα, the only substantial fluorescence effect would be from bremsstrahlung and that would be more than an order of magnitude larger for the Au substrate compared to the C substrate. If this was the cause, there would be a distinct separation of results on C and Au substrates. For incident beam energies < 30 keV in SEM, fast secondary excitation can be significant for light elements but we would expect any effect on Pd Lα to be less than 1 % relative. Whereas there are a number of parameters that affect the overall shape of the error curve and the positioning of results for some combinations of kV and substrate, we have not found a reason for the systematic "droop" at low *k*-ratios. The improved treatment of electron scattering and ionisation cross-section by PENELOPE may go some way to reduce this effect and these are obvious areas for further exploration. Unfortunately, we have no opportunity to check the equipment used by B&H to see if some count rate effect was depressing the bulk measurements compared to the thin film but that seems unlikely. Also, we could not check the method use to correct for background on the B&H measurements. However, if this was a source of error, we would have expected a big difference between results on C and Au substrates for the same film yet this was not observed.

3.2. Sensitivity to different parameterisations

The simplified "multiple scattering" approach based on Curgenven and Duncumb's method (ref. [4] chap. 4), MSMC, has been useful for investigating the sensitivity of simulated results to different parameterisations. Whereas changing the formulae for inner-shell ionisation cross-section had a small effect, stopping power was much more important. Figure 3 shows that the MSMC with Joy-Luo modification to Bethe stopping power and Berger-Seltzer mean ionisation energy (J) gave both a large scatter on results and a pronounced downward trend at low k-ratio. By empirical adjustment of the stopping power formulation at low energies (Strim) it was possible to reduce the scatter substantially, although the systematic trend remained.

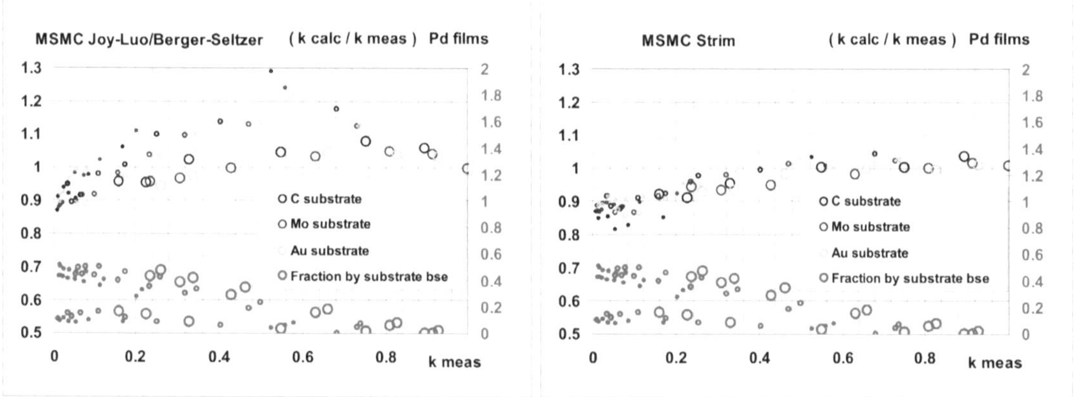

Figure 3. Ratio of calculated to measured *k*-ratios versus measured *k*-ratios obtained by using the MC code MSMC with Joy-Luo modification of Bethe stopping power and Berger-Selzer mean ionisation energy (left) and an empirical adjustment of Bethe stopping power (Strim) at low energies (right), for Pd films on C, Mo and Au. Experimental data taken from ref. [3].

It is interesting that the trend and scatter obtained by optimising the MSMC simulation are similar to those obtained by using PENELOPE, suggesting that perhaps there is something fundamental about the "droop" at low *k*-ratio. This fundamental trend could be related to a poor description of energy straggling i.e., the fluctuation in energy-loss after a given path length, which is disregarded in the MC codes employing the CSDA and accounted for approximately in PENELOPE [9].

Another possibility for the trend was that excitation of the surface layer by exiting backscattered electrons, bse, was not accurately predicted. A rough estimate of the emission fraction due to this cause, right axis, has been plotted using red circles, simply to demonstrate where the contribution is high or low. It is clear that even though there is a wide range of bse excitation, there is no obvious connection between kcalc/kmeas and the bse emission fraction.

3.3. Möller et al.'s experimental measurements

The only publication we have found that includes an RBS measurement of the film was some work by Möller et al [7]. Although the original paper only gives the results from thin film analysis, Möller's original thesis contained some *k*-ratios measured for one of the Cu films at different kV [10]. Simulations using NISTMonte showed very different results depending on whether Cu Lα or Cu Kα *k*-ratios were used. PENELOPE gave much more consistent results for Cu Lα and Cu Kα. The downward trend at low *k*-ratio is possibly evident, but calculated *k*-ratios are systematically high (figure 4).

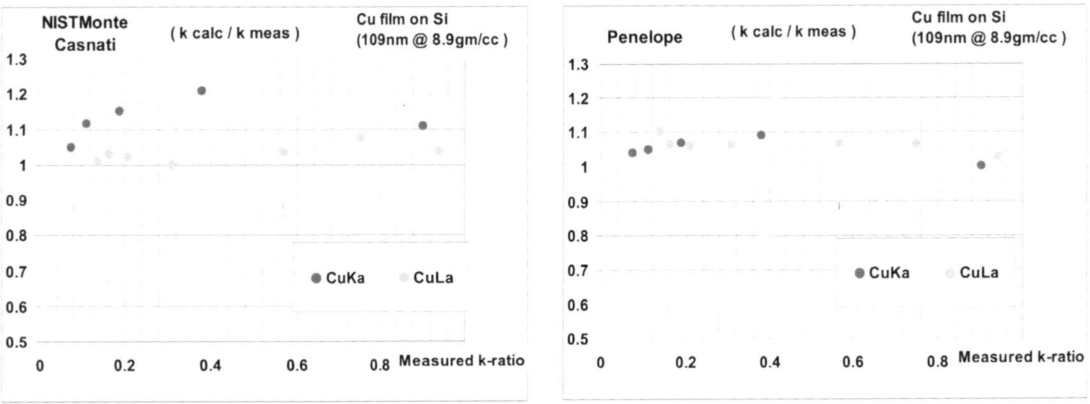

Figure 4. Ratio of calculated to measured *k*-ratios versus measured *k*-ratios obtained by using NISTMonte (left) and PENELOPE (right) for Cu films on Si. Experimental data taken from ref. [7].

3.4. φ(0) measurements

Measurements of φ(0) provide a method to check the accuracy of simulation of bse enhancement of emission from a surface layer and figure 5 shows the PENELOPE prediction to be well within 10 % of the measured value in general [8]. There seem to be some systematic trends for Ge/Fe(L), Au/Ta(M) and Au/W(M). If such a trend was responsible for the "droop" in *k*-ratios we would expect to see some connection between the level of bse enhancement and the error in "k_{calc}". The red circles in the MSMC plots above (figure 3) show that for the Pd data there is no obvious correlation between bse enhancement and error in "k_{calc}". Note that "*k*-ratio" compares the intensity to that from bulk pure element whereas φ(0) ratios the intensity to that from a free standing film. Therefore, φ(0) measurements would not show if there was any problem in predicting the relative intensities from unsupported film and bulk.

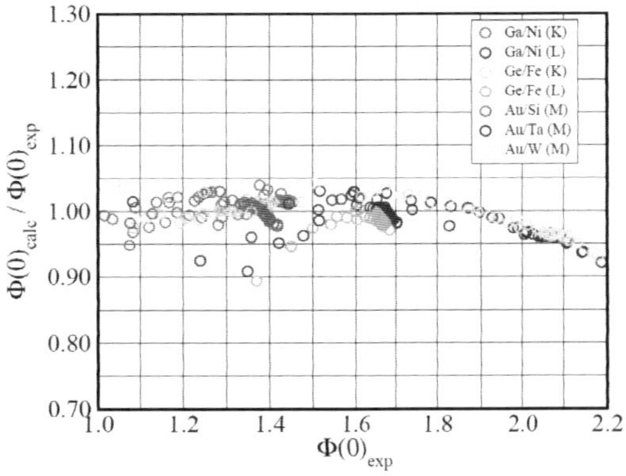

Figure 5. Ratio of calculated to measured φ(0) versus measured φ(0) obtained by PENELOPE for different tracer/substrate combinations. Experimental data taken from ref. [8].

4. Conclusions

Popular Monte Carlo programmes (CASINO and NISTMonte) are fast and very useful for general application but are not accurate enough for thin film measurement if errors < 20 % are required. PENELOPE is much slower but performs consistently better than any other Monte Carlo programme on the tests we performed. Systematic errors found with PENELOPE using B&H data suggest there may be inaccuracies as much as 10 %, particularly when simulating the intensity of films relative to bulk pure material when the thickness is much less than the electron range. Although this trend has only been seen in one set of measurements, we could find no other comprehensive sets to corroborate this hypothesis. Therefore, for testing MC programmes, it would be useful to have some more systematic measurement data.

The problem with making measurements is that it is difficult to prepare specimens with known mass thicknesses of pure elements. Crystal oscillators need calibrating and different materials have different sticking coefficients. Film density can differ between bulk and thin films so linear thickness measurements like TEM or XRD are not reliable indicators of mass thickness. RBS is the only technique used so far that also measures mass thickness and any *k*-ratio measurements that include RBS would be useful to extend this study.

Acknowledgments

Financial support from the Spanish Ministerio de Ciencia y Innovación (project no. FPA2009-14091-C02-01) is gratefully acknowledged.

References

[1] Pouchou J L 2002 *Microchim. Acta* **138** 133-52
[2] Ammann N and Karduck P 1990 *Microbeam analysis.* (San Francisco: San Francisco Press) pp. 150-4
[3] Bastin G F and Heijligers H J M 2000 *X-ray Spectrom.* **29** 212-38 and 373-97
[4] Joy D C 1995 *Monte Carlo modeling for electron microscopy and microanalysis.* (New York: Oxford University Press)
[5] Casino programme 2007 see: http://www.gel.usherbrooke.ca/casino/What.html version and Drouin D et al. 2007 *Scanning* **29** 92-101
[6] NISTMonte programme 2007 see: http://www.cstl.nist.gov/div837/837.02/epq/index.html
[7] Möller et al. 1995 *Mikrochim. Acta* **119** 41-7
[8] Merlet C and Llovet X 2001 *X-ray Spectrom.* **40** 47-54
[9] Salvat F, Fernández-Varea J M and Sempau J 2009 *PENELOPE-2008: A code system for Monte Carlo simulation of electron and photon transport.* OECD/NEA Data Bank (Issy-les-Moulineaux: OECD)
[10] Weinbruch S 2010 Personal communication

EBSD study of the microstructure evolution in a commercially pure aluminium severely deformed by ECAP

B Tolaminejad[1,3], **F Brisset**[2] **and T Baudin**[2]
[1] Iran University of Science and Technology, School of Metallurgy & Materials Engineering, Tehran, Iran
[2] CNRS - Université Paris-Sud, ICMMO - Institut de Chimie Moléculaire et des Matériaux d'Orsay, 15, rue Georges Clémenceau, bât. 410, FR-91405 Orsay Cedex, France

E-mail: btolaminejad@iust.ac.ir

Abstract. Equal channel angular pressing (ECAP) is a promising technique for production of ultrafine-grain materials of a few hundred nanometres size. In the present paper commercially pure aluminium billets were processed by ECAP up to 8 passes using the route BC in which the samples were rotated about the longitudinal axis by 90° in the same direction after each consecutive pressing. Electron backscattered diffraction (EBSD) was used to evaluate the microstructure of the aluminium. EBSD scans revealed a fairly homogeneous ultrafine grained microstructure after 8 passes. This analysis also showed that the fraction of high angle grain boundaries and average grain size were about 70 % and 1500 nm, respectively. Then, tensile properties of the aluminium billets were evaluated. The results represent that the yield stress of aluminium is increased significantly by about four times after application of four consecutive passes of ECAP and then it remains approximately constant.

1. Introduction

Several techniques for producing ultrafine-grained materials for structural applications have been introduced, especially in the last decades [1, 2]. The advantages of fabricating materials with sub-micrometre size grained microstructure as structural components lie in their increased mechanical properties such as strength, ductility, and low-temperature superplasticity [3-5]. Moreover, severe plastic deformation (SPD) has recently been developed and passed from the laboratory stage to industrial condition to produce ultrafine grain and nano grain size microstructures.

Various SPD processes such as accumulative roll bonding (ARB) [6, 7] or high pressure torsion (HPT) [8, 9] have been developed. But, the most versatile, and potentially the most popular among materials scientists, appears to be equal channel angular pressing (ECAP) [10, 11].

The ECAP die initially is charged with a billet inserted in the vertical channel and then pushed into the horizontal channel with the aid of a punch (figure 1). Under this condition a large shear strain is induced in the billet, while the cross-section of the billet does not experience any change. Equation (1) presents the equivalent strain imposed on the billet after N passes of ECAP process versus a die angle (φ) and an outer curved corner angle (ψ) [12]:

[3] To whom any correspondence should be addressed.

$$\bar{\varepsilon}_N = \frac{N}{\sqrt{3}}\left(2\cot\left(\frac{\varphi+\psi}{2}\right)+\psi\,\mathrm{cosec}\left(\frac{\varphi+\psi}{2}\right)\right) \tag{1}$$

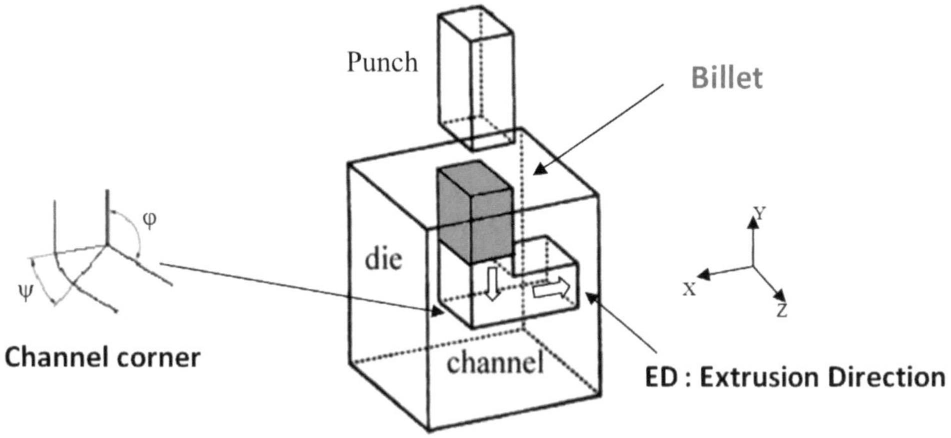

Figure 1. Schematic of equal channel angular pressing (ED // X).

2. Material and experimental procedures

The material used in this research consisted of commercial pure aluminium (1175) billet. The dimensions of the billet were 70×18×18 mm^3. All billets were annealed at 450 °C for 30 min before being subjected to the ECAP process. In order to perform the ECAP tests, a split die with an inner angle of φ =90°, outer corner angle of ψ = 20°, and a square cross section of 18×18 mm^2 was made. The split design not only eased the ECAP but also facilitated the removal of the entangled billet. ECAP tests via route BC (90° clockwise rotation of the specimen after each pass) [13] were carried out at room temperature under a constant press velocity of 3 mm/s. Tests were executed up to 8 passes using MoS_2 spray as lubricant. To study the microstructure of the aluminium billets after subjecting to different number of passes, specimens with dimensions of 16×4×2 mm^3 were cut at the middle of the deformed billet. For EBSD analysis the X-Y surface of samples was polished by electropolishing in an $HClO_4$ + 900 ml CH_3OH solution. The EBSD measurements were carried out using orientation imaging microscopy OIM™ system mounted on a FEG-SEM Zeiss Supra 55 VP. The step size and scan area for mapping were chosen as 100 nm and 100×100 μm^2, respectively. Tensile test specimens were machined from the deformed billet according to ASTM E-8 standard. These specimens were subjected to tensile test at a strain rate of 10-3 s^{-1} using the Instron tensile testing machine model 5582.

3. Results and discussion

Figure 2 exhibits the EBSD maps obtained from the samples subjected to 1, 2, 4 and 8 passes. The numbers shown on the upper left part of each maps row represent the number of the ECAP passes. The inverse pole figure (IPF) maps in figure 2a indicate the distribution of crystallographic directions perpendicular to the observed area. Figures 2b and 2c show the corresponding grain colour maps with tolerance of 15° and 2°, respectively.

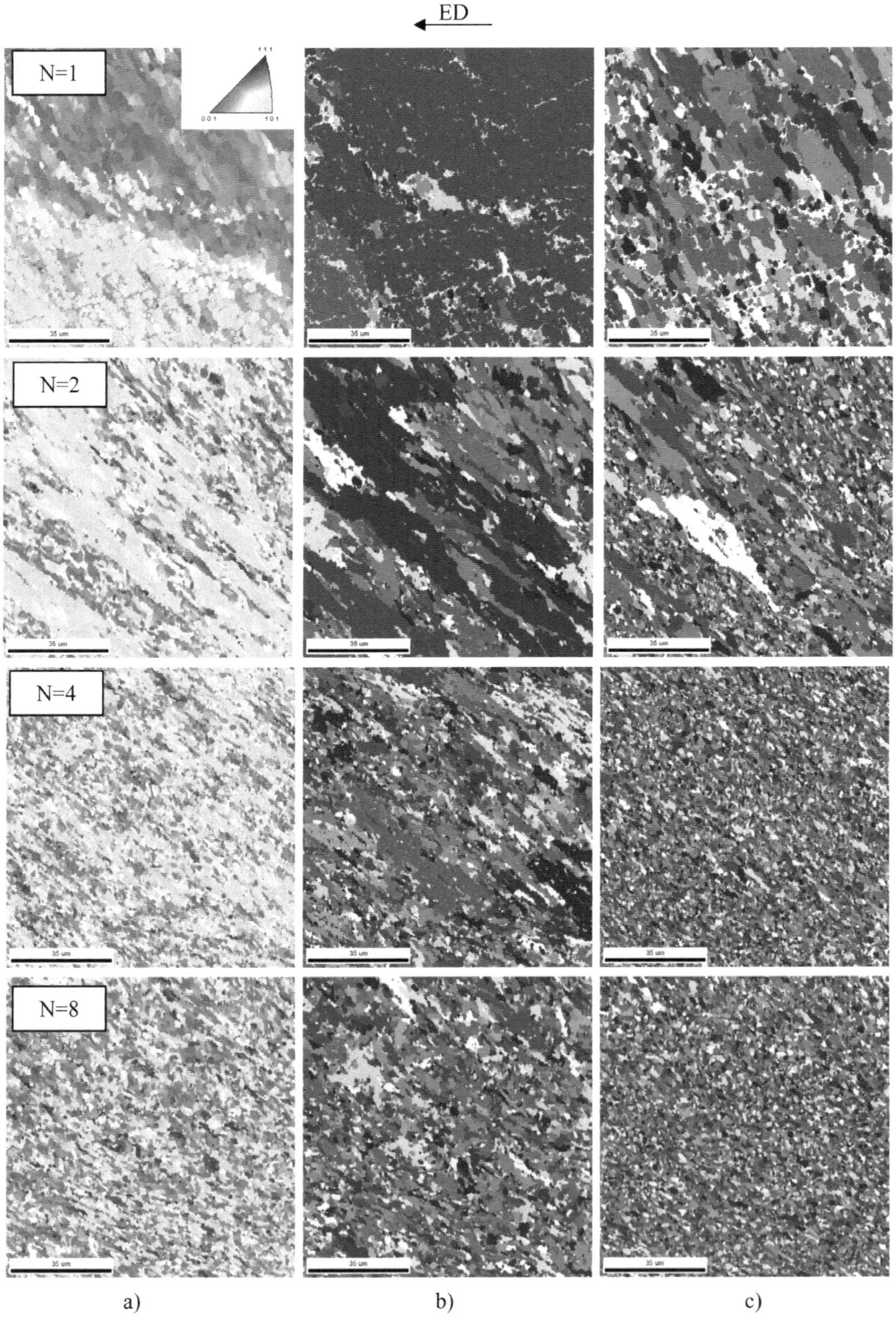

Figure 2. EBSD maps obtained in the centre region of the ECAP samples subjected to 1, 2, 4 and 8 passes. a) Inverse pole figure maps; b) grain colours maps (15° tol.); and c) grain colours maps (2° tol.).

Figure 2a demonstrates that after the first pass, the microstructure consists of elongated sub-grains of mainly {001}<100> and {110}<001> texture components with <uvw>//ED (see the IPF maps for N > 1). Also, referring to Figures 2b and 2c, it is obvious that a sub-grain substructure is generated by means of one ECAP pass. After 2 passes, the colour maps show that the large elongated regions which are visible in the microstructure of the first pass sample are broken into equi-axed or small elongated grains. Figure 2a shows that when pressing is continued up to 4 passes, the microstructure mainly contains smaller equi-axed grains. This phenomenon is justified as the rotation of the samples according to route BC by +90°, causes the shearing of the deformed grain on two orthogonal planes in reverse direction. Indeed, the strain required for restoration of equi-axed grains is the summation of shear strain in each pass. The shear strains generating at the first four passes are ε_{xz} at the first pass; $\varepsilon_{xz} + \varepsilon_{yz}$ at the second pass; ε_{yz} at the third pass and (0) for the fourth pass [14]. However, it is apparent that, even after 4 passes, the microstructure is heterogeneous in terms of grain morphology and grain alignments (figures 2b and 2c). In addition, several microstructural features are worth noting from the inspection of colour maps of eight-pass ECAP sample. First, the microstructure becomes fairly homogeneous. This was also stated by Sun et al. [15] through a TEM investigation of 1050 aluminium processed to 8 passes via route BC. Mishra et al. [16] reported similar microstructure for ECAP deformed copper up to 8 passes. Moreover, comparable result was explained by Dalla Torre et al. [17] in which they reported a cell size remained approximately constant during the first ECAP passes in copper while the misorientation angle increased. Second, various colours are seen on the IPF map, indicating a weak texture after 8 passes. However, the orientation distribution function calculation shows that the {001}<100> texture component becomes majority in this analyzed area. Moreover, it can be seen that the microstructure of 4 and 8 passes are very similar using 2° map (figure 2c). However, a few larger grains are still visible, demonstrating that the grain refinement is not fully developed in some portions of same orientations (15° map) [18].

Figures 3 and 4 show the statistical variations of grain size and misorientation angles obtained from the EBSD analysis after consecutive passes, respectively. The average grain size is about 3.5 μm, 2.0 μm, 1.5 μm and ~ 1.5 μm after 1, 2, 4 and 8 passes, consistently. In the same time, the percentage of high angle grain boundaries, i.e., with a misorientation angle greater than 15°, increases from 46.5 %, 50.8 %, 65.9 % to 71.3 %. This means that after 8 passes, the microstructure consists of low fraction of low angle grain boundaries (~ 30 %) as was already shown in figure 2, qualitatively.

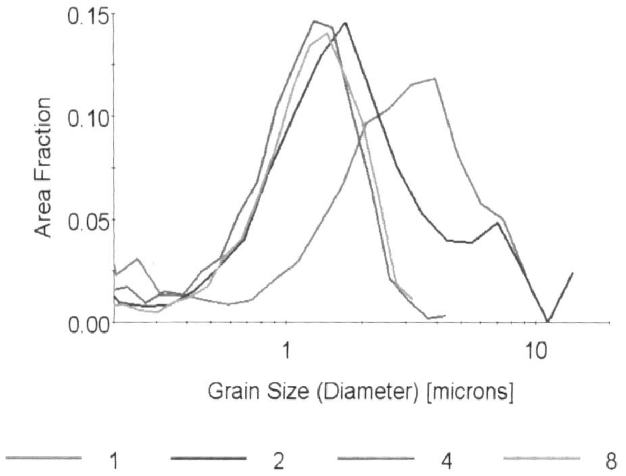

Figure 3. Statistical variation of grain size of the samples subjected to 1 - 8 ECAP passes.

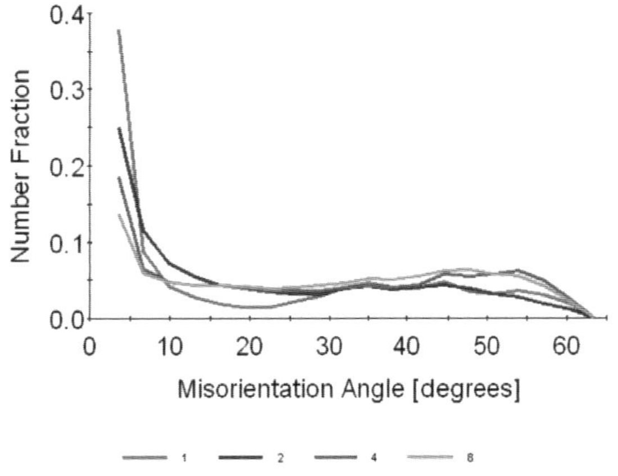

Figure 4. Statistical variation of misorientation angle of the samples subjected to 1-8 ECAP passes.

Referring to figure 5, the yield strength is significantly increased due to a possible severe increase of the dislocation density and the initiation of production of smaller grains with high angle misorientation during pass one. However, after that the yield strength is stabilized to some extent. In fact, aluminium is a high stacking fault-energy metal. This results in the ease of dislocation cross-slip, which in turn facilitates dislocation annihilation and dynamic recovery (DRV) [19, 20]. However, continuous dynamic re-crystallisation (CDRX) as the conversion of low-angle dislocation cell boundaries into high-angle boundaries forming sub-grains is widely accepted in aluminium [21].

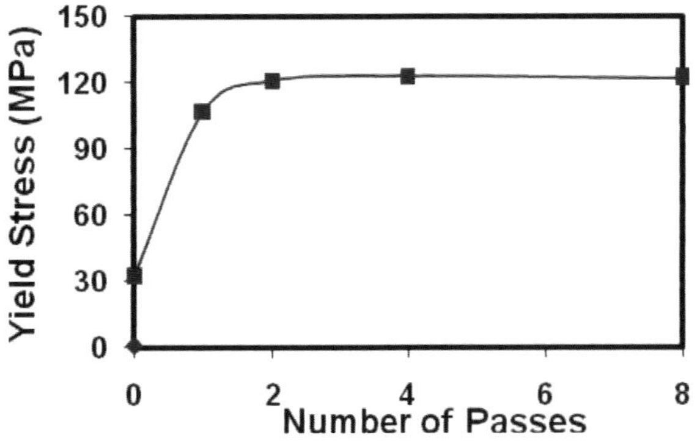

Figure 5. Variation of the yield stress as a function of ECAP pass number.

4. Conclusion

1. The ECAP process increases the aluminium yield strength up to four times and the rate of increase in strength significantly decreased after the first pass and become insignificant after fourth pass.
2. The results show that the processing of commercial pure aluminium (1175) should be conducted through eight passes to attain a fairly homogeneous equi-axed configuration of grains.
3. The vast drop in grain size saturates after the 4th pass, where the (sub-)grain size is 1.5 μm, however, the angle of misorientation kept on developing.

References

[1] Zhu Y T and Lowe T C 2000 *Mater. Sci. Engng. A* 291 46
[2] Segal V M, Reznikov V I, Drobyshevskiy A E and Kopylov V I 1981 *Russian Metallurgy* **1** 99
[3] Berbon P B, Tsenev N K, Valiev R Z, Furukawa M, Horita Z, Nemoto M and Langdon T G 1998 *Metall. Mater. Trans. A* **29** 2237
[4] Chang J Y, Yoon J S and Kim G H 2001 *Scripta Mater.* **45** 347
[5] Shan A, Moon I G and Park J W 2002 *J. Mater. Proc. Technol.* **122** 255
[6] Kuśnierz J, Mathon M H, Bogucka J, Faryna M, Jasieński Z, Penelle R and Baudin T 2006 *Archives of Metall. Mater.* **51** 237
[7] Kuśnierz J, Bogucka J, Mathon M H and Baudin T 2008 *Archives of Metall. Mater.* **53** 1
[8] Loucif A, Figueiredo R B, Baudin T, Brisset F and Langdon T G 2010 *Mater. Sci. Engng. A* **527** 4864
[9] Loucif A, Figueiredo R B, Baudin T, Brisset F and Langdon T G 2011 Mater. Sci. Forum **667-669** 223
[10] Segal V M 1995 *Mater. Sci. Engng. A* **197** 157
[11] Guo Z, Solas D, Etter A L, Baudin T and Penelle R 2003 *Mater. Sci. Forum* **426-432** 2723
[12] Iwahashi Y, Wang J, Horita Z, Nemoto M and Langdon T G 1996 *Scrip. Mater.* **35** 143
[13] Valiev R Z and Langdon T G 2006 *Prog. Mater. Sci.* **51** 881
[14] Gholinia A, Prangnell P B and Markushev M V 2000 *Acta Mater.* **48** 1115
[15] Sun P, Kao P and Chang C 2004 *Met. Trans. A* **35** 1359
[16] Mishra A, Kad B K, Gregori F and Meyers M A 2007 *Acta Mater.* **55** 13
[17] Dalla Torre F, Lapovok R, Sandlin J, Thompson P F, Davis C H J and Pereloma E V 2004 *Acta Mater.* **52** 4819
[18] Etter A L, Baudin T, Rey C and Penelle R 2006 *Mater. Charact.* **56** 19
[19] Prangnell P B, Bowen J R and Gholinia A 2001 in: *Proc. 22nd Symp. Science of nanocrystalline and metastable alloys, structure properties and modelling.* Dinesen et al., eds. (Roskilde: Riso Nat. Lab.) pp. 105
[20] Wei W, Wei K X and Fan G J 2008 *Acta Mater.* **56** 4771
[21] Skrotzki W, Scheerbaum N, Oertel C G, Brokmeier H G, Suwas S and Toth L S 2007 *Acta Mater.* **55** 2211

Geometry correction procedure for quantitative X-ray microanalysis of low-Z number matrices

G Tylko[1]
Jagiellonian University, Institute of Zoology, Department of Cell Biology and Imaging, ul. Ingardena 6, PL-30060 Krakow, Poland

E-mail: grzegorz.tylko@uj.edu.pl

Abstract. Low-Z element matrices (Z<10) are typical for biological materials. H, C, N and O are the main contributors enriched with elements from Na to Ca (biological elements). In biological EPMA, protein-based standards in a "bulk" form are used to assess elemental concentrations; however, the size of an unknown specimen does not always meet the criterion of "bulk specimen" in one of its dimension. Then, specimen geometry affects accurate and precise calculation of elements. Here, Monte Carlo simulation was applied to assess the effects of specimen geometry on the efficiency of X-ray emission from cellulose of increasing width (from 0.5 to 25 µm) containing "biological elements" and exposed to 10 keV electrons. The normalized intensities of emitted X-rays were plotted against specimen width and the double decay exponential function fitted. 150 simulations were performed using randomly defined concentrations of "biological elements" to estimate relative errors for uncorrected and geometrically corrected results. The contents of "biological elements" calculated from simulated intensities revealed ± 30 % of relative error if no correction was applied that decreased to ± 5 % after correction for specimen geometry. It seems that the procedure might be suitable for many biologically derived materials and polymers.

1. Introduction

Quantification of elements in biological X-ray microanalysis relays on standards composed of protein matrix with elements of interests homogeneously distributed within. Their concentrations should be comparable to those usually found in tissues or cells [1]. Then the standard resembles an unknown specimen of interest, especially in the value of the mean atomic number, and the correction for Z and A is not necessary. Further, to diminish influence of specimen topography and/or analytical conditions (beam current or analytical time) on element quantification, the peak-to-background method (P/B) is used. However, P/B method fails when any extraneous radiation contributes to the background [2], what is the case when semi-thick specimens are exposed to the electron beam.

It is pointed out that quantitative analysis of semi-thick biological specimens is the most challenging part of biological EPMA and there is not remedy to perform calculation of elements in such samples on the basis of the bulk standards [2, 3]. The semi-thick specimen characterizes inefficient absorption of incoming electrons due to the thickness of the material. There, a certain proportion of electrons escapes from the specimen and does not participate in X-ray generation. On the second hand, radiation characteristic for low-Z number elements emitted from semi-thick materials

[1] To whom any correspondence should be addressed.

is less absorbed by the matrix on their way to the X-ray detector increasing the value of minimum detection limit for those elements.

The Monte Carlo approach was found to be useful in quantitative X-ray microanalysis of many materials [4, 5] as well as to assess an influence of topography [6] or specimen conductivity [7] on X-ray intensities generated and emitted from rough specimens and insulators, respectively. Unfortunately, little is known about the application of the Monte Carlo approach to determine experimental conditions for EPMA of soft tissues and cells. Scott and Ritchie [8] demonstrated effects of accelerating voltage on determination of structures in single-cell organisms prepared to FIB-EDS examination. Ro et al. [9] has shown the applicability of reverse Monte Carlo quantification of elements in particles composed of low-Z elements (C, N, O), typical for biological materials. Tylko et al. [10] has found that protein matrix is a reliable choice to compose standards for estimation of elemental concentration in cellulose (carbohydrate) materials.

The aim of the study was to search for geometry correction factors that might increase accuracy of quantitative X-ray microanalysis of semi-thick low-Z number materials. Here, the Monte Carlo simulation approach was used to calculate an amount of characteristic X-rays emitted from "biological elements" (Na and K) homogeneously distributed in laterally semi-thick cellulose (from 0.5 μm to 25 μm) but infinite in depth dimension, when exposed to 10 keV electron beam. Furthermore, the usefulness of a correction approach obtained was verified on the basis of relative error propagation. The test was performed after numerous simulations of cellulose materials composed of different concentrations of "biological elements" and different in their lateral dimensions.

2. Material and methods

The cellulose matrix used for Monte Carlo simulation contained H (5.4 %), C (37.0 %), O (43.7 %), Na (1.9 %), Mg (0.8 %), P (1.1 %), S (2.6 %), Cl (3.4 %), K (3.6 %), and Ca (0.5 %) and was treated as the standard for further experiments. The calculated density of the material was 0.88 g/cm_3 with G-factor value at the level of 3.98 [3]. The Monte Carlo simulation was performed using Casino software v2.42 [11] to collect emitted intensities of X-rays generated in a cellulose matrix of increasing lateral size starting from 0.5 μm to 25 μm. The composition of the material chosen in the study reflects concentrations of elements, typically used in standards for biological X-ray microanalysis [5, 10].

The accelerating voltage applied in the tests was 10 keV, since it is the most commonly used electron energy in biological studies. The simulation conditions were the following: 100 nm of the electron beam diameter, the Si(Li)-detector positioned at 25° of take-off angle, 90° of azimuth angle, and tilt angle 0° in relation to the electron impact. Mott cross-sections [12] were chosen for the calculation of elastic scattering of electrons. The electron energy loss was based on continuous slowing down [13], whereas effective ionisation cross-section for characteristic photons was calculated according to Casnati et al. [14].

The intensity of characteristic X-rays emitted from the cellulose material of the increasing size was normalized to the intensity emitted from the bulk cellulose specimen. Thus, the constant value of the intensities was determined reflecting the minimal lateral dimensions of the specimen that can be treated as a bulk. A distribution of normalized characteristic X-ray intensities in relation to the specimen size was fitted to double exponential decay function using QtiPlot 0.9.8 (Ion Vasilief, released 2010/06/29). $\chi 2$ test evaluated fitting quality and values of the function parameters.

The applicability of correction function obtained for the elements of interest was verified on the basis of 150 simulations of biological materials composed of cellulose matrix with various values of the weighted sum of Z^2/A for all biological elements present in the simulated volume (G-factor) [1, 3]:

$$\overline{Z^2 / A} = \sum \left(\left(Z_s^2 / A_s \right) \cdot C_s \right) \tag{1}$$

where Z_s is atomic number of element present in simulated material, A_s its atomic weight, and C_s reflects its concentration (a fraction in a material). On the basis of the eq. (1), the values of G-factors used in simulations ranged from 3.60 to 4.94. Sizes of simulated materials were assigned between 0.5 and 25 μm using random sampling method. Next, the concentration of elements were quantified from the intensities obtained by means of the calibration curves (figure 1) and the results compared to the expected values (the standard mentioned above with G-factor = 3.98). The relative errors were calculated for all data. Finally, the results were corrected using the expression determined from the fitting process and the fractions of relative errors determined again.

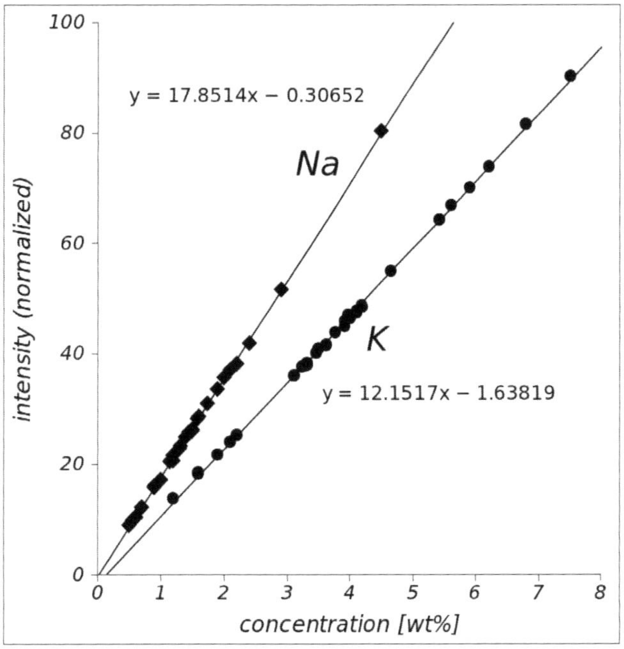

Figure 1. The calibration curves obtained from Monte Carlo simulation of increasing concentration of sodium (♦) and potassium (●) distributed homogeneously in bulk cellulose matrix to calculate the content of both elements in the experimental specimens of different composition and lateral size.

3. Results and discussion

The Monte Carlo simulations were performed on the model specimens composed of light element (H, C, N and O) matrix with homogeneously distributed elements from Na to Ca. The Z-dimension of the models exceeded significantly the penetration depth of 10 keV electrons, thus the specimens might be treated as "bulk". X and Y dimensions, however, increased from 0.5 μm to 25 μm. The lowest lateral size (0.5 μm in width) of the model was far below the range of electron penetration and X-ray production as calculated for Na on the basis of Anderson and Hasler equation [15], whereas the highest width (25 μm) chosen in the study was almost 10-fold of the range of X-ray production in model material. Thus, the highest width of the specimen used met the criterion of "bulk" specimen in both X- and Y-dimensions [16].

The distribution of X-ray intensities emitted from elements of interest in relation to increasing width of the model specimen fit well to double exponential decay function (2):

$$y = a_1 e^{\left(\frac{-x}{b_1}\right)} - a_2 e^{\left(\frac{-x}{b_2}\right)} + y_0 \qquad (2)$$

where a_1, a_2, b_1, b_2 and y_0 are parameters calculated on the basis of the fitting process. It was found the values of a_1, a_2, b_1, and b_2 strongly depend on Z-numbers of elements. Since a_1 and a_2 describe peak maximum for both exponential functions (figure 2), the highest value of these parameters was related to sodium (Na), whereas the lowest value for potassium (K). Similar relationship concerns b_1- and b_2-parameters that describe the rates of function decay. Their values decrease together with an increase of Z-number (table 1).

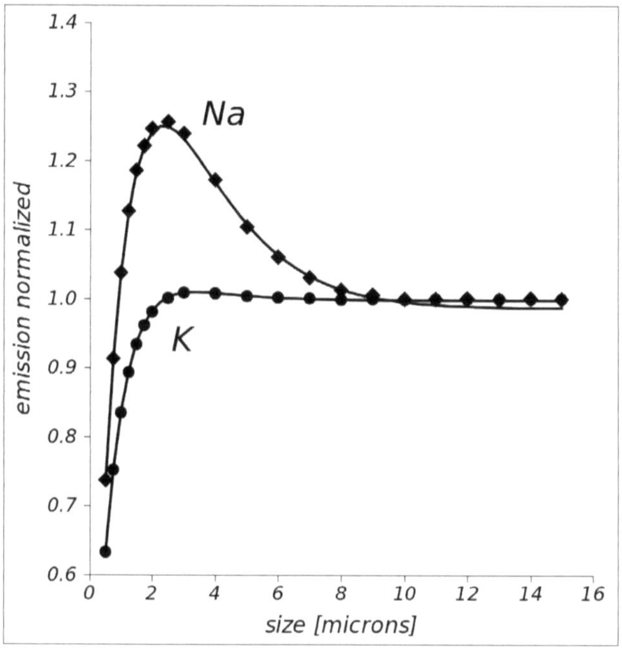

Figure 2. Distribution of normalized X-ray intensities emitted from increasing widths of cellulose material calculated for sodium, Na Kα (♦) and potassium, K Kα (●). Solid lines represent double decay exponential functions fitted to data obtained from simulations.

Table 1. a_1, a_2, b_1, b_2 and y_0 parameters as well as χ^2 values obtained after fitting simulation data to double exponential decay function.

Parameters	sodium (Na)	potassium (K)
a_1	213.5	24.1
a_2	214.3	24.9
B_1	1.47	1.01
B_2	1.46	1.00
χ^2	$1.1 \cdot 10^{-5}$	$8.2 \cdot 10^{-7}$

It has been shown that Na Kα intensity dramatically increases with the width of the model specimen due to mass increase of irradiated volume. More electrons participate in ionisation process and in consequence, more X-rays are generated. The same effect was observed for potassium (K Kα). However, the intensities of Na Kα emitted from cellulose model were significantly higher in relation to that noted for K Kα. It is probably the result of larger excitation volume for Na Kα since critical ionisation energy for that element is significantly lower in comparison to K critical ionisation energy [15]. It results in Na Kα emission from deeper parts of a specimen. Nevertheless, there is effective attenuation of low energy X-rays (sodium - 1.04 keV) by the matrix of the material tested, the structure of the model confines absorption effect significantly. It is due to relatively short distance for X-rays to travel through the matrix from their generation site to the EDS detector. Lower X- and Y-dimensions of the model in comparison to those of the bulk form, lead to more efficient detection of Na characteristic X-rays.

The peak of Na Kα intensity emitted from cellulose matrix used in this study has been observed at the thickness range of 2 - 4 μm. It is also the case for K emission, however, the intensity peak is not so pronounced because of negligible contribution of absorption effect on rather high energy X-rays of K Kα (3.313 keV) (figure 2). Similar phenomenon is observed at detector-faced parts of specimens with a highly rough surface or polished specimens but tilted towards the detector [6, 16].

The double decay exponential functions determined for X-ray intensities of Na and K present in cellulose matrix were found to be reliable tools to correct the results obtained after simulations of different organic materials with different size. 150 model specimens were randomly determined due to their composition and X-Y dimensions. G-factor of the cellulose random models ranged between 3.60 and 4.94. Dimensions of the cellulose models in X and Y axis were chosen from the range of 0.5 to 25 μm. The relative error calculated from uncorrected data obtained after simulations performed on the random models varied between ± 30 % for sodium. Calculation of potassium data however, showed strong underestimation of the results that reached -35 % (figure 3, left panel). When double decay exponential function presented above was applied and the results recalculated, the distribution of relative error decrease to ± 4 % for Na and to ± 5 % for K (figure 3, right panel). It is satisfactory for quantitative calculations of elements in biological materials and other low-Z number bio-derived composites or polymers.

It seems that the geometric effects for small organic structures or semi-thick biological specimens might be corrected efficiently. Many years ago, Armstrong and Buseck [17] developed the method that calculates shape factors for some basic specimen shapes that are partially transparent for electrons. Their correction procedure reduced relative errors to 10 % in comparison to 60 % relative error when commonly used ZAF correction procedure was applied [18]. Wróblewski et al. [19] showed that peak-to-background (P/B) method is the most suitable for quantitative analysis of semi-thick biological specimens. Recently, the Monte Carlo simulation tools were applied to determine influences of specimen geometry on quantitative EPMA as well as to search for correction procedures that might minimise analytical errors [9, 20, 21]. It was also the case in this study, however, specimens of interest were semi-thick in X-Y dimensions and "bulk" in Z-direction. Moreover, the specimen was composed with cellulose, low-Z number element matrix characteristic e.g., for plant cell walls. However, low-Z number matrices are typical for other biologically derived components, organic polymers or some biomaterials (poly(L-lactide-co-glycolide; PLGA) [1, 10, 22].

In conclusion, the double decay exponential function (eq. 2) was found to properly describe changes in characteristic X-ray intensities emitted from different widths of low-Z number matrices. Application of that function to correct data obtained from laterally thin and semi-thick biological specimens decreases relative error of results from ± 30 % to the level of ± 5 %. Moreover, 25 % difference in a value of G-factor (weighted Z^2/A) between the standard and unknown biological specimens has no effect on accuracy of elemental calculations and other corrections for differences in Z-number or X-ray absorption are not necessary.

Acknowledgments
The author wishing to acknowledge financial support from Polish Ministry of Science and Higher Education grant no. K/ZDS/001725.

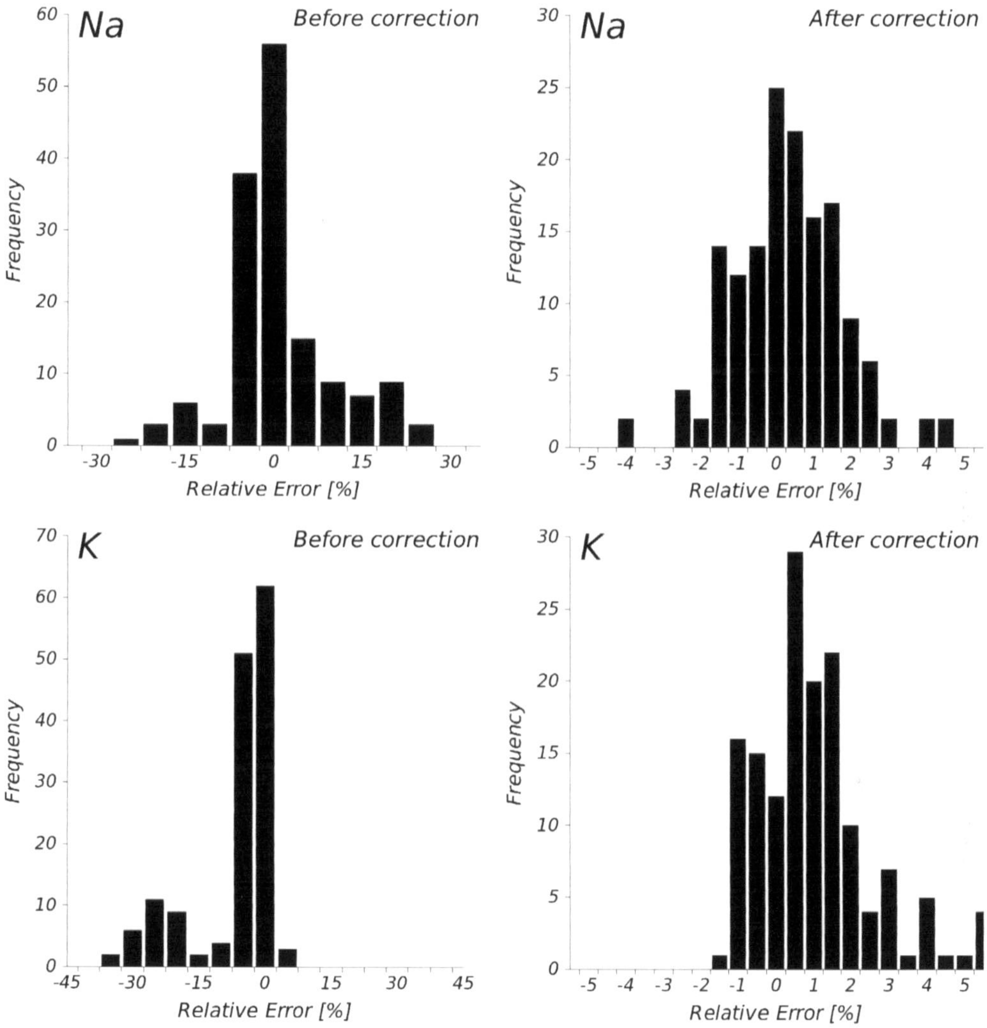

Figure 3. Distributions of relative errors from non-corrected (left) and corrected (right) data obtained after simulations of Na and K X-rays emitted from increasing widths of cellulose matrix.

References

[1] Warley A 1997 *X-ray microanalysis for biologists.* (London: Portland Press Ltd.)
[2] Roomans G M 1990 *X-ray microanalysis.* (London: Biophys. Electr. Microsc. Academic Press)
[3] Roomans G M 1988 *J. Elect. Microsc. Tech.* **9** 19
[4] Llovet X, Riveros J A and Salvat F 1996 *Mikrochim. Acta* **13** 409
[5] Janik P, Tylko G and Turnau K 2010 *Microsc. Res. Tech.* **73** 1134
[6] Gauvin R and Lifshin E 2000 *Mikrochim. Acta* **132** 201
[7] Demers H and Gauvin R 2004 *Microsc. Microanal.* **10** 776
[8] Scott K and Ritchie N W M 2009 *J. Microsc.* **233** 331
[9] Ro Ch-U, Kim H K and Van Grieken R 2004 *Anal. Chem.* **76** 1322

[10] Tylko G, Dunchak S, Banach Z and Turnau K 2010 *IOP Conf. Ser.: Mater. Sci. Engng.* **7** 012028
[11] Hovington 1997
[12] Czyzewski Z, MacCallum D O, Romig A and Joy D C 1990 *J. Appl. Phys.* **68** 3066
[13] Joy D C and Luo S 1989 *Scanning* **11** 176
[14] Casnati E, Tartari A and Baraldi C 1982 *J. Phys. B* **15** 155
[15] Anderson C A and Hasler M F 1966 *Proc. 4th Int. Conf. on X-ray Optics and Microanalysis.* (Paris: Herman) pp. 310
[16] Goldstein J I, Lyman C E, Newbury D E, Lifshin E, Echlin P, Sawyer L, Joy D C and Michael J R 2003 *Scanning electron microscopy and X-ray microanalysis.* (New York: Kluwer Academic/Plenum Publishers)
[17] Armstrong J T and Buseck P R 1975 *Anal. Chem.* **47** 2178
[18] Armstrong J T 1991 *Electron probe quantification.* (New York: Plenum Press)
[19] Wróblewski J, Müller R M, Wróblewski R and Roomans G M 1983 *Histochemistry* **77** 447
[20] Ro Ch-U, Osán J, Szalóki I, de Hoog J, Worobiec A and Van Grieken R 2003 *Anal. Chem.* **75** 851
[21] Ritchie N W M 2010 *Microsc. Microanal.* **16** 248
[22] Pamula E, Kokoszka J, Cholewa-Kowalska K, Laczka M, Kantor L, Niedzwiedzki L, Reilly G C, Filipowska J, Madej W, Kolodziejczyk M, Tylko G, Osyczka A M 2011 *Ann. Biomed. Engng.* **39** 2114

Mineral surface - organic matter interactions: basics and applications

G Valdrè[1], D Moro and G Ulian
Università di Bologna, Dipartimento di Scienze della Terra e Geologico-Ambientali, Piazza di Porta San Donato 1, IT-40126 Bologna, Italy

E-mail: giovanni.valdre@unibo.it

Abstract. The ability to control the binding of biological and organic molecules to a crystal surface is central in several fields; for example, in biotechnology, catalysis, molecular microarrays, biosensors preparation and environmental sciences. The nano-morphology and nanostructure at the surface may have physico-chemical properties that are very different from those of the underlying mineral substrate. Recent developments in scanning probe microscopy (SPM) have widened the spectrum of possible investigations that can be performed at the nanometric level on the surface of minerals. They range from the study of physical properties such as surface potential, electric field topological determination, Brønsted–Lowry site distributions, to chemical and spectroscopic analysis in air, in liquid or in gaseous environments. After an introduction to SPM modes of operation and new SPM-based technological developments, we will present recent examples of applications in the study of interactions between organic matter and mineral surface and report on the advances in knowledge that have been made by the use of scanning probe microscopy.

1. Introduction

The surface properties of minerals may affect important interaction processes, for example organic matter interactions, coagulation, aggregation, sedimentation, filtration, catalysis, and ionic transport in porous media. The nano-morphology and nanostructure at the surface may have physico-chemical properties that are very different from those of the underlying mineral substrates. The overall effective reactivity can be greatly affected, regulating the rate of particle solubility as well as the surface sorption of inorganic and organic components. For instance, cell adhesion and subsequent biofilm formation may be controlled by the nanostructure properties rather than by those of the whole bulk particle [1].

In addition, the properties of water adsorbed onto layered silicates and clay surfaces are significantly influenced by the mineral surface where the net charge, cation occupancy and distribution affect the dynamics and surface distribution of water molecules. Ulcinas et al. [2] have shown that cations adsorbed on the surface of a muscovite act as pinning centres restricting the mobility of sub-nanometre confined water molecules.

Studies of clay mineral surfaces showed that particle size, shape, composition, charge and surface area actively drive mineral-biomolecule interactions. For instance these potentially catalytic surfaces can induce RNA polymerisation and vesicles formation [3, 4].

---

[1] To whom any correspondence should be addressed.

Characterisation techniques with a high spatial resolution are needed to analyse in detail these physico-chemical properties and the local interactions at the mineral surface. One of the most promising and powerful technique is scanning probe microscopy (SPM) combined with high resolution microanalysis.

The latest advances in SPM give us the ability to measure at the nanometric level some physical properties of the mineral surface, such as electric field topology and surface potential as well as carry out chemical and spectroscopic analysis in air, in liquid and in gaseous environments. Furthermore, single-biomolecule interactions on the mineral phase can be characterized with previously unattainable force sensitivity (< pN).

In this paper, new technological developments which stand behind SPM will be introduced. Then, we will present recent achievements in the field of the characterisation and application of nano-morphology, surface potential and cleavage patterns of minerals, especially the behaviour of some layered silicates, oxides, hydroxides, apatites and synthetic compounds. Analysis of the interaction of these mineral surfaces with organic molecules will be reviewed respect to single-molecule imaging and the correlation between the mineral nano-morphology, surface properties and molecule absorption, self-assembling and nano-patterning. A last section will be focused on some computational studies of the mineral surfaces and the analysis of their interactions with organic molecules, by means of *ab initio* and molecular dynamics (MD) modelling.

2. Mineral nano-topography and cleavage patterns

Atomic force microscopy (AFM), a SPM methodology which can attain atomic resolution, is the most suitable analytical method for characterising the surface of materials. The information obtained from the surface can complement those of the bulk usually obtained by the well-established electron microscopy and microprobe techniques [5-7]. Figure 1 shows a schematic diagram of an atomic force microscope. A tip integrated near the free end of a cantilever probes the surface of the sample. A piezoelectric scanner controls the sample-to-tip distance (z) while scanning in the x,y-plane. A dither piezo vibrates the cantilever and an optical system (laser and position sensitive photodetector) measures its static and dynamic parameters (deflection, amplitude, frequency, phase, etc.). Feedback electronics and data processors are used to image the surface.

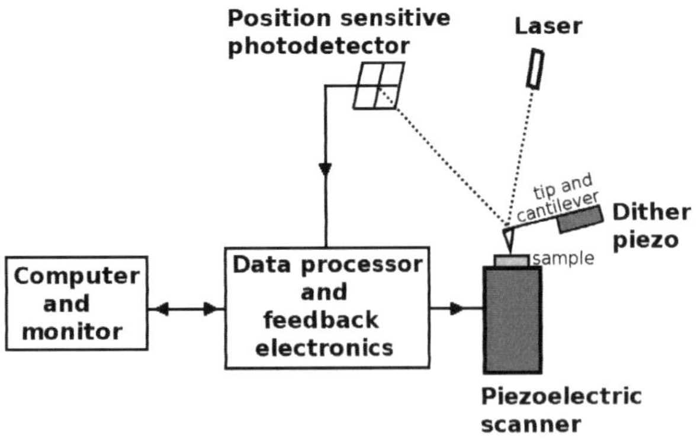

Figure 1. Schematic diagram of the main components of an atomic force microscope.

In AFM there are basically two main modes of operation: the amplitude modulation (AM-AFM) and the frequency modulation (FM-AFM).

In AM-AFM the dither piezo mechanically excites the cantilever at its fundamental frequency and the oscillatory motion is detected by the optical system [8]. AM-AFM adopts large amplitude of oscillation (tens of nm) to probe the sample surface. The oscillation amplitude is set as feedback parameter to image the sample topography, by means of a feedback loop using a lock-in amplifier.

Furthermore, material-based properties (e.g., composition, adhesion, friction and visco-elasticity) may be qualitatively imaged by using the phase signal, ϕ, that is the phase shift between the excitation signal and the response of the cantilever during the scan.

In the frequency modulation mode (FM-AFM) an increase in sensitivity is achieved with respect to AM-AFM [9]. The resonant frequency of the cantilever, which is modified by the tip-surface interaction, is measured to produce an image of the sample surface. The cantilever is kept oscillating at its acting resonant frequency, achieved by means of a phase-locked loop. The spatial dependence of the frequency shift induced in the cantilever motion by the tip-sample interaction is the source of topographical contrast. This mode of operation has achieved sub-atomic resolution in ultra-high vacuum [10]. It is a promising technique in the imaging of organic systems; for example, single- and double-stranded DNA. Recently, high resolution was also achieved in liquid both on minerals (e.g., the surface of muscovite was atomically resolved in water) and biomolecules (e.g., the direct observation of hydration layers on a lipid bilayer and β-strands constituting an amyloid fibril) [11].

Sugimoto et al. [12] using FM-AFM attained chemical identification of individual surface atoms in ultrahigh vacuum at room temperature by detecting the short-range chemical forces between the tip and surface atoms.

As an example of layered silicates nano-morphology we present the surface cleavage pattern of a clinochlore. Clinochlore is a layered silicate where a brucite-like positively charged octahedral sheet about 0.5 nm thick is sandwiched between two negatively charged, 1 nm thick tetrahedral-octahedral-tetrahedral (TOT) layers, by means of weak electrostatic forces. Figure 2 shows an example of an AFM observation of the surface of a freshly-cleaved clinochlore surface where the brucite-like sheets, B, and siloxane surface, S, of the TOT layers are visible. The surface was imaged in air by AM-AFM (figure 2a). The section profile, figure 2b, reveals the typical surface steps related to the mineral structure. The map of the phase shift signal, ϕ is shown in figure 2c and clearly shows the presence of the two different materials (a brucite-like sheet, B, and a siloxane surface, S).

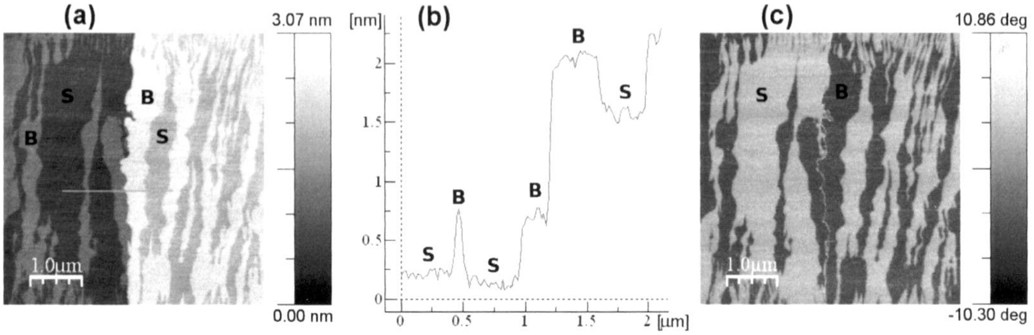

Figure 2. a) AM-AFM image of the surface nanomorphology of as-cleaved clinochlore. Brucite-like sheets (B), 0.5 nm thick, extending in width from a few nanometres to micrometres are resolved above the siloxane surface (S) of the TOT layer. b) Profile of the section along the horizontal line in the image (a). Both brucite-like and TOT layers are identified by thickness measurements. c) Phase image of the same area revealing the presence of two different materials.

3. Mineral surface potential and charge distributions

The surface potential can be measured by using three main SPM methods: static and dynamic electric force microscopy (EFM) and Kelvin probe force microscopy (KPFM).

Static EFM measures the cantilever deflection caused by the electrostatic force acting between the probe and the specimen surface. The effective electrostatic force depends on the voltage difference between the probe and the specimen and also on both tip and cantilever geometry (shape and size).

Consequently, both tip and cantilever contributions on the total force gradient must be taken into account for a complete understanding of the experimental data.

The electrostatic force interaction F between a conductive tip and sample is given by

$$F = -\frac{1}{2}\frac{dC(z)}{dz}V^2$$

where $C(z)$ is the system capacitance, z is the tip–sample distance, and V the tip–sample electric potential difference. At equilibrium, the reacting elastic force of the cantilever balances the electrostatic force. Thus, the measured deflection is directly related to the electrostatic force acting on the cantilever system.

In static EFM a bi-dimensional matrix of force curves can be acquired by means of a force volume experiment. The cantilever deflection, ΔL, during the approach-withdrawal cycle is related to the electrostatic force, F_E, by means of the elastic constant of the cantilever k_L, so that $F_E = k_L \Delta L$.

The relation between the probe's deflection and the mineral surface potential can be found by numerical simulation of the experiment [13, 14]. The electro-mechanical interaction between the probe and the mineral surface can be studied by 3D finite element analysis.

In dynamic mode EFM, a piezo-mechanically driven conductive cantilever is vibrated above the sample surface. In this case, the detection of frequency, amplitude and phase is used to study the tip-surface interactions.

In Kelvin probe force microscopy (KPFM) it is possible to measure the surface potential by three methods: dual pass mode [15], single pass mode [16] and single pass in multifrequency [17].

The surface potential distribution of the cleaved surface of a clinochlore, measured by two-pass KPFM, is shown in figures 3a and b. T he potential measurement was performed with the probe at 30 nm above the mineral surface. Figure 3c presents single-pass KPFM image of clinochlore revealing highly resolved surface potential features < 25 nm in size.

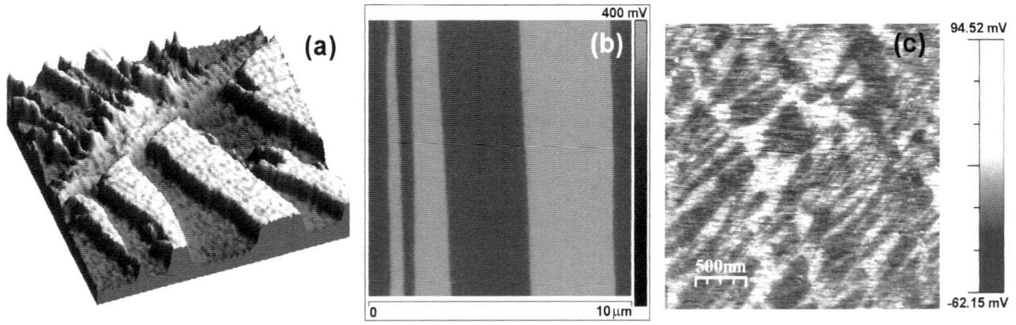

Figure 3. a) 3D, and b) 2D 10 μm x 10 μm dual-pass KPFM images of the surface potential of micrometres long brucite-like and TOT stripes on a clinochlore surface. c) 2D single pass Kelvin probe image showing the surface potential of clinochlore with details less than 25 nm in size.

Very recently, KPFM supported by EPMA and quantum mechanical modelling has revealed the existence of 2D arrays of zeolitic-type Brønsted–Lowry sites on atomically flat Al-rich chlorites [18]. This approach was used to correlate at the nanoscale the contrast due to the surface potential (related to the Brønsted-Lowry proton) with the surface morphology and crystal-chemistry.

Another important challenge is the investigation at the nanoscale of the surface properties of a mineral in a liquid environment because most organic-mineral interactions may occur in liquid. The AFM signal phase shift ϕ, that is the phase lag between the excitation signal and the response of the cantilever during a scan, may be related to various material-based properties such as, composition, adhesion, friction, visco-elasticity and also the surface potential [19]. For instance, by monitoring point by point the phase shift ϕ on a mineral surface in a liquid environment, we can attain some indication on the surface-liquid interface at various pH. In AFM, the pH of a solution in contact with a mineral surface can be easily controlled by means of a fluid cell. We have preliminarily investigated the surface of a clinochlore using the phase signal, ϕ at room temperature (25 °C) in a water solution at three different values of pH. Figure 4 shows the phase image of the same surface of a clinochlore obtained at pH 4.2, 5.5 and 9.0. The difference between the phase measured on the TOT siloxane and that on the brucite-like surface ($\Delta\phi$) is reported for each image. Because clinochlore is made of two opposite charged layers, the combined effect of adhesion, visco-elasticity and surface potential may lead to a not-linear $\Delta\phi$ (pH) function.

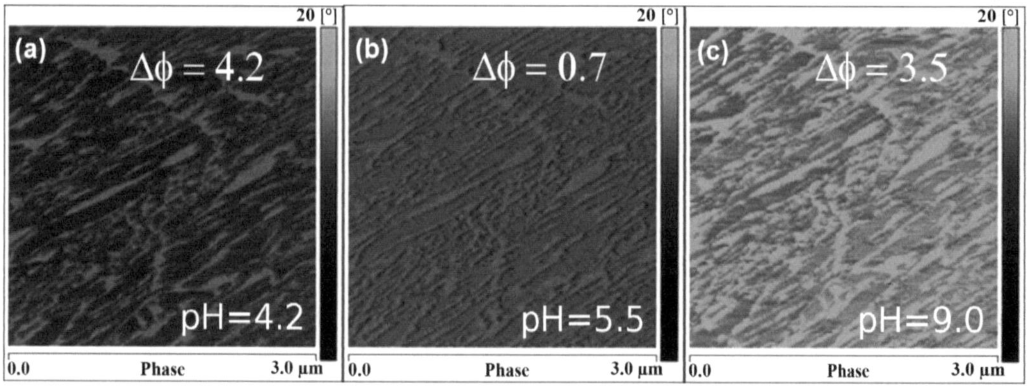

Figure 4. AFM phase image of the surface of a clinochlore in liquid at three different pH. $\Delta\phi = \phi$(TOT siloxane) - ϕ(Brucite-like sheet).

4. Nano-mechanical and visco-elastic properties of a mineral surface

New SPM-based techniques have been developed to measure the mineral topography whilst achieving a simultaneous acquisition of the sample properties such as surface adhesion, surface charge distribution and visco-elastic material behaviour.

These methods apply a modulation of the z-offset (i.e., sample to SPM tip distance) by means of a sinusoidal signal of frequency from 0.1 to 5 kHz and amplitude in the range 10 - 500 nm. The maximum tip-sample force is the basis of feedback control. Therefore, the user can control the maximum tip-sample repulsive force, which can range down to the piconewton. To analyze the sample characteristics it is necessary to transform the time function curves in separation function ones (figure 5a). A wide variety of properties can be obtained from the curve analysis, such as elastic modulus, tip-sample adhesion, energy dissipation and the maximum deformation of the sample. As shown by Pittenger et al. [20], the elastic modulus can be determined in a range from 700 kPa to 70 GPa by fitting the retract curve (dashed in figure 5a) using the Derjaguin-Muller-Toporov model described in detail in specific literature [21].

For the measurement of visco-elastic properties of liquids at the surface of minerals transverse dynamic force microscopy (TDFM), also known as shear force microscopy, can be employed. It is an SPM-based method where a probe is mounted vertically and perpendicularly to the sample surface and

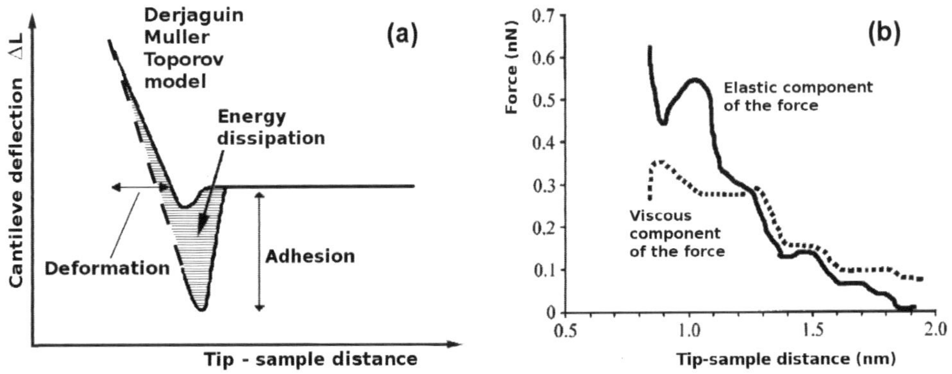

Figure 5. a) Example of a SPM force curve F = k_L.ΔL, versus tip-sample distance. Tip-sample adhesion, energy dissipation and maximum deformation of the sample can be directly derived from the force curve. The elastic modulus of the sample can be determined by fitting the dashed part of the curve with the Derjaguin-Muller-Toporov model [20, 21]. b) Viscous and elastic forces trends calculated from the approach curves on a freshly cleaved muscovite surface in pure water. The dissipative viscous force dominates over the elastic component at the start of the interaction, but becomes less significant when the tip-sample separation is less than 1.5 nm.

set in horizontal oscillation using a dither piezo [22-24]. This technique is of great importance for the study of the shear properties of liquids confined to their molecular dimensions. Nano-confined liquids can reveal completely different properties from those of the bulk. In many phenomena, such as the swelling of clays, viscosity and lubrication, this behaviour has fundamental consequences. Ion adsorption/desorption from the mineral surface can also be investigated.

Figure 5b shows the elastic and viscous forces calculated from the approaching curve to a freshly cleaved muscovite sample in pure water. It is to be noted that the dissipative viscous component dominates for tip-sample separations greater than 1.6 nm and that the elastic force increases rapidly with the decreasing separation. A different visco-elastic behaviour of the confined liquid pertains inside each molecular layer and it is also dependent on the number of water layers. TDFM can be used also to investigate the role of surface adsorbed cations on the visco-elastic response of nano-confined water on muscovite and other layered minerals [2].

5. Organic molecules interaction with layered silicates

Since the advent of high resolution techniques, the interactions that arise between a mineral phase (adsorbent) and biomolecules (amino acids, proteins, lipids and nucleic acids) have attracted the interest of many researchers. Silicates are among the most investigated minerals and attract a large variety of different analytical methods.

Muscovite, biotite, brucite, chlorite and vermiculite were studied by SPM and compared in terms of surface affinity, self-assembly and nano-patterning of adsorbed DNA molecules by Valdrè et al. [25]. A remarkable variety of DNA adsorption mechanisms were observed and related to differences in the surface structure and chemistry of these layered silicate minerals. For example, the investigations on di- and trioctahedral micas revealed different deposition patterns in terms of the amount of adsorbed molecules and their configurations (e.g., low charge trioctahedral layered silicates present about 80 % DNA coverage in comparison to 30 % for dioctahedral ones). Later work focused on the clinochlore surface as a substrate for nucleotide and DNA adsorption [4, 26]. The nucleotide deposition on clinochlore was systematically analyzed by AFM and the presence of agglomerates, ordered structures and alignments of nucleotides were found (figure 6a). In particular, the edges and ledges of the brucite-like $Mg(OH)_2$ sheet provide preferential adhesion and filament-forming sites for nucleotides.

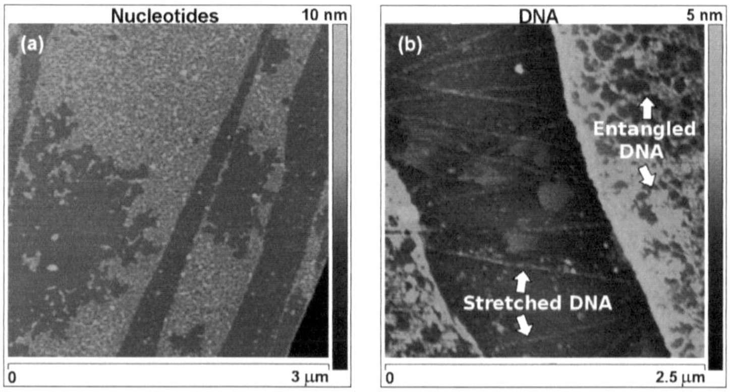

Figure 6. a) AFM image of the surface of a clinochlore with both brucite-like and TOT areas. Nucleotides (bright contrast) are visibly adsorbed only on the brucite-like sheets. b) AFM image showing the results of DNA interaction with the surface of a clinochlore crystal. No DNA is adsorbed on the TOT areas. Only single filaments of DNA bridge in a stretched conformation between two brucite-like areas over a TOT surface. Dense DNA networks are clearly visible on the brucite-like regions. The average DNA length is about 15 μm.

The DNA adsorption on the clinochlore surface showed (i) the positively charged brucite-like $Mg(OH)_2$ sheets with about 50 % coverage (high DNA-affinity), and (ii) the negatively charged siloxane zones where the DNA molecules were almost completely absent. As shown in figure 6b, DNA molecules bridge between two adjacent $Mg(OH)_2$ brucite-like parallel stripes. The DNA molecules appear stretched (linear conformation) by comparison to the DNA which is not bridged, but relaxed in entangled and bundled structures. As in the case of nucleotides, DNA tends to agglomerate and order at the edges of the $Mg(OH)_2$ sheets.

The "binary" behaviour of clinochlore surface could be an ideal model platform to understand the mechanisms regulating the adsorption of molecules at surfaces. For instance, surfactants exhibit a hydrophobic chain and a positively charged hydrophilic head group. The competition between the hydrophobic and electrostatic forces can be investigated thanks to the simultaneous presence of positively charged hydrophobic areas (brucite-like) and negatively charged hydrophilic areas (TOT). This approach improves the understanding of how exactly these two forces balance each other and determine the structure and morphology of the adsorbed layer. These investigations are fundamental for the characterisation of the effective hydrophobic or hydrophilic behaviour of surfactants.

Biotite, chlorite and vermiculite were also considered as support for bioadsorption of red blood cells [27]. Their fixing behaviour was compared to that of commercial mica (muscovite) without the use of any fixative and binding chemical agent. Red blood cells interactions with biotite and vermiculite surfaces are similar, but the latter exhibits a higher surface adsorption density. On the contrary, the sample cells directly deposited on muscovite under the same experimental conditions are loosely bound. On biotite and vermiculite the cell adhesion is strong enough to allow the observation of membrane channels at the nanoscale.

More recently, Alessandrini et al. [28], and Jurak and Chibowski [29], used AFM to investigate muscovite as a substrate for mono and bilayer lipid systems. From the thermodynamic analysis of the excess area and excess free energy of the lipid mixture they found that lateral phase separation takes place for the components, this was confirmed by SPM.

6. Molecular interaction with oxides, hydroxides and synthetic compounds

Attenuated total reflectance Fourier transform infrared (ATR-FTIR) spectroscopy has been used to probe the interaction of phospholipids comprising phosphodiester or terminal phosphate functionalities with goethite (α-FeOOH) and hematite (α-Fe_2O_3) surfaces, and to assess the impacts of zwitterionic, basic and acidic polar head groups [30]. Interaction with iron (oxyhydr)oxides involves electrostatic attraction, phosphodiester H-bridging interactions, and P-OFe covalent bond formation. Results suggest that these domains may promote initial bacterial cell interaction with iron (oxyhydr)oxides through electrostatic attraction, followed by the formation of inner-sphere phosphate complexes on oxide surfaces. Very recently, the adsorption of lipids on the surface of alumina, titania and iron oxide films have been studied to create models of a cell membrane (a lipid bilayer) [31]. The researchers discovered that titania is a robust and convenient support to achieve the objective.

The adsorption of glycine, lysine, and other small biomolecules on synthetic silica surfaces were investigated by Stievano et al. [32]. The adsorption process from the view-point of the surface has been considered, using previously gained knowledge on molecular identification of surface functional groups. From the different spectroscopic signatures of the amino acid/adsorption site complexes, different adsorption mechanisms were postulated. On silica surfaces, the authors found cooperative hydrogen bonds similar to molecular recognition processes. The presence of molecular recognition is also revealed by experiments on the co-adsorption of different aminoacids, where strong selectivity effects may be observed. Such phenomena can be of practical significance in the fields of biofilm formation and prebiotic chemistry.

7. Interaction of organic molecules with apatites

The term 'apatites' refers to a mineral family whose building block is constituted by calcium orthophosphates. The basic unit formula is $Ca_5(PO_4)_3$[OH,F,Cl]. Despite the low presence in the earth's crust, calcium orthophosphates are of special significance because they represent the inorganic part of calcified tissues of mammals. Owing to this characteristic, since 1920 Albee [33] has studied them as possible biomaterials for bone prosthetic and repair. These minerals have also been adopted as substrate for protein and DNA adsorption/chromatography, because of their capability to bind these kind of biomolecules. From this perspective, the mineral hydroxyapatite (HA) is the most studied and used one, due to its structural similarity to the mineral phase of bone and tooth.

Hydroxyapatite (HA) has been widely used for protein and nucleic acid chromatography since its first introduction by Tiselius et al. in 1956 [34]. It is well established that HA is a multifunctional material containing both negatively charged binding sites (the so-called P-sites) and positively charged binding sites (the so-called C-sites). This combination results in unique selectivity and affords the ability to separate complex mixtures. Accordingly, basic proteins in their cationic form are adsorbed mainly on the P-sites primarily as a result of electrostatic interactions. A limitation of hydroxyapatite is its modest stability in acidic conditions. Only very recently were experimental and phenomenological models of the HA chromatography proposed [35].

Thanks to the studies in the chromatography field, researchers have proposed new ways to synthesize hydroxyapatite. Simple or complex organic molecules, also amino acids, can limit the growth of a crystal along certain directions. For example, the effect of L-serine in supersaturated solutions of calcium phosphate was investigated under plethostatic conditions [36]. The inhibitory effect of L-serine was found to be due to blocking of a portion of the active growth sites by adsorption. Kinetics measurements in the presence of L-serine as well as adsorption isotherm analysis suggested Langmuir-type adsorption of L-serine on the surface of hydroxyapatite with a relatively low affinity for the substrate. Adsorption experiments showed that at pH 7.4 considerable adsorption of

L-serine onto HA takes place, whereas at pH 10.0 the adsorption was negligible, suggesting that electrostatic interactions are dominant. On the basis of the experimental results, a model was proposed according to which L-serine absorbs on the surface of HA through electrostatic attractions exerted between one negative site of the HA surface, e.g., phosphate or hydroxyl ion, and the positively charged protonated amino group of one L-serine molecule, forming a surface ion pair.

Recent studies by Dangaria et al. [37] have demonstrated that the nano-morphology of hydroxyapatite induces stem cell differentiation. Dimension and structure of extracellular matrix surfaces revealed powerful influences on cell shape, adhesion, and gene expression. They showed that natural tooth root topographies induce integrin-mediated extracellular matrix signalling cascades in tandem with cell elongation and polarisation to generate physiological periodontium-like tissues. Together, these studies illustrated the capacity of natural extracellular surface topographies to instruct progenitor cell populations to fully regenerate complex cellular and structural morphologies of tissues once lost to disease. The researchers suggested the use of this strategy for the re-implantation of teeth lost due to trauma or as a novel approach for tooth replacement using tooth-shaped replicas.

8. *Ab initio* static and molecular dynamics simulations

In this section we provide some general notions on *ab initio* and molecular dynamics simulations and examples of current studies concerning the interaction of organic molecules with mineral layer substrates.

Computational simulations can work side by side with SPM-based methods to achieve an atomic scale understanding of the interactions between adsorbate and adsorbent. The most adopted computational models are *ab initio* (or static), which are very accurate but have high computational costs, and molecular dynamics (MD) methods, less expensive and with temporal information. From the simulations it is possible to obtain geometric features of the system and many thermodynamics (energy, enthalpy, etc.) and physics (band gap, vibrational frequencies, phononic and plasmonic spectres, etc.) correlated properties. A comparison between known experimental data and the calculated ones is needed to evaluate the simulation model and the additional information that it provides. We present here some results related to the minerals previously reported.

With the aid of the software GAUSSIAN03, quantum chemistry simulations of organic molecules onto single (silica, alumina, titania), binary (silica/alumina, silica/titania), and ternary (alumina/silica/titania) fumed oxides were conducted [38]. The authors adopted the cluster approach to simulate the adsorption reaction and the effects of morphology and surface composition of the materials. The results provided a better understanding of the adsorbate effects on the surface of the material, the type of hydrogen bonds and the energy of the adsorption process, with good agreement with experimental results.

The polymerisation of amino acids on mineral surface, from both the experimental and theoretical points of view, was recently reviewed [39]. The comparison of the different results was made to obtain a general understanding of the effect of the surface and the mechanism of polymerisation. The researchers studied basically amino acids in interaction with silicates, oxides and sulphides. The amino acid polymerisation process is favoured by adsorption, at least when is followed by drying at moderately high temperatures. It is unclear if adsorption without drying could result in significant polymerisation, which unfortunately would require quite a lot of ageing experiments. The proposed mechanisms of amino acid adsorption on oxide surfaces at the molecular level are being verified. However, the situation is rather complicated because many different mechanisms may come into play depending on the specific surface, the specific amino acid, and the experimental conditions (pH, ion strength, etc.) under consideration.

The same experimental interest in the apatite mineral family also involved theoretical studies. The hexagonal form of hydroxyapatite (HA) was exhaustively analyzed (bulk and surface) with *ab initio* simulations [40]. Afterwards, the surfaces were explored with water and amino acids: the first to mimic the aqueous environment in which the biomaterial is employed, the latter was chosen as the more suitable and simplest building block for the more complex collagen fibrils [41]. Simulations were conducted with density functional theory; glycine and lysine amino acids and glutamic acid were adsorpted from the gas-phase. From theoretical observations, the authors discovered that the water molecule is adsorbed and dissociated on Ca^{2+}-ions and, after that process, the surface is immediately passivated. In recent years the HA surface was explored with formic acid and alendronic acid to better understand the effect of these molecules on the structure [42]. The first one is involved in tooth decay, the second is a pharmacological molecule used to cure osteoporosis. It was found that formic acid binds strongly at Ca^{2+}-sites, triggering the dissolution of HA and alendronic acid binds and dissociates at the surface, creating a protective layer against dissolution effects.

A work by Roark and Feller [43] concerned the MD simulations of a phospholipid bilayer interacting with the solid surface of hydroxylated nano-porous amorphous silica, carried out over a range of lipid-solid substrate distances. The porous solid surface allowed the water layer to maintain equal pressures above and below the membrane bilayer, dynamically adjusting the H_2O layer thickness. Qualitative estimates of the force between the surfaces leads to an estimated lipid-silicon distance in very good agreement with the results of neutron scattering experiments.

One further MD case study is that of a complex hydroxyapatite/tropocollagen simulating bone tissue [44]. It was stated that the interfacial interactions between mineral and organic phases and the structural effects due to staggered arrangement significantly affect the strength of biomaterials. The authors simulated the complex HA/TC and applied transversal and longitudinal stretching to the system to test the strength of the biomaterial resistance. The simulations were conducted in a non-hydrated, hydrated and ionized water environments. The latter is obtained by addition of Ca^{2+} in the hydrated environment without any counter-anion. The theoretical MD results confirmed the experimental data on the stabilising effect of water and calcium ions on the complex [45].

9. Conclusions

SPM-based methods are useful tools to complement electron microscopy and microbeam analysis for the surface characterisation of materials. They are suitable to investigate the surface nano-topography and physical-chemical properties of minerals with nanometric spatial resolution. Because of their low environment constraints and almost absent induced material damage, they allow the interaction between organic matter and inorganic phases to be analyzed, providing information about the specific mineral reactivity. For example, a remarkable variety of nucleotides, DNA and RNA adsorption mechanisms were observed and related to differences in the surface structure and chemistry of layered silicate minerals.

To support the SPM experimental results, the organic-mineral interaction can be modelled using theoretical methods. The modelled systems can be studied with static (*ab initio*) or molecular dynamics techniques. However, at present, there are only a few direct comparisons between SPM and computational results. Good agreement with quantum mechanics simulations have been reported for silicon, clinochlore and pentacene [10, 18, 46].

References

[1] Kendall T A, Na C, Jun Y S and Martin S T 2008 *Langmuir* **24** 2519
[2] Ulcinas A, Valdre G, Snitka V, Miles M J, Claesson P M and Antognozzi M 2011 *Langmuir* **27** 10351
[3] Hanczyc M M, Mansy S S and Szostak J W 2006 *Origins Life Evol. B* **37** 67
[4] Valdrè G 2007 *Eur. J. Mineral.* **19** 309
[5] Bocchi G and Valdrè G 1993 *J. Inorg. Biochem.* **49** 209
[6] Borgia G C, Brown R J S, Fantazzini P, Mesini E and Valdrè G 1992 *Nuovo Cimento* **14** 745
[7] Gatti A M, Valdrè G and Tombesi A 1996 *J. Biomed. Mater. Res.* **31** 475
[8] Lee M and Jhe W 2006 *Phys. Rev. Lett.* **97** 036104
[9] Giessibl F J 2003 *Rev. Mod. Phys.* **75** 494
[10] Giessibl F J, Hembacher S, Bielefeldt H and Mannhart J 2000 *Science* **289** 422
[11] Fukuma T 2009 *Jpn. J. Appl. Phys.* **48** 08JA01

[12] Sugimoto Y, Pou P, Abe M, Jelinek P, Perez R, Morita S and Custance O 2007 *Nature* **446** 64
[13] Valdrè G and Moro D 2008 *Nanotechnology* **19** 405501
[14] Valdrè G and Moro D 2008 *Nanotechnology* **19** 405502
[15] Elings V B and Gurley J A 1994 *U.S. Patent US5308974*
[16] Ziegler D, Rychen J, Naujoks N and Stemmer A 2007 *Nanotechnology* **18** 225505
[17] Stark R W, Naujoks N and Stemmer A 2007 *Nanotechnology* **17** 065502
[18] Valdrè G, Tosoni S and Moro D 2011 *Am. Mineral.* **96** 1461
[19] Baumann M and Stark R W 2010 *Ultramicroscopy* **110** 578
[20] Pittenger B, Erina N and Su C 2010 Veeco Application Note AN128 Rev. A0
[21] Maugis D 2000 *Contact, adhesion and rupture of elastic solids.* (Berlin: Springer-Verlag)
[22] Antognozzi M, Humphris A D L and Miles M J 2001 *Appl. Phys. Lett.* **78** 300
[23] Antognozzi M, Protti A, Miles M J and Valdrè G 2003 *GeoActa* **2** 101
[24] Antognozzi M, Ulcinas A, Picco L, Simpson S H, Heard P J, Szczelkun M D, Brenner B and Miles M J 2008 *Nanotechnology* **19** 384002
[25] Valdrè G, Antognozzi M, Wotherspoon A and Miles M J 2004 *Philos. Mag. Lett.* **84** 539
[26] Antognozzi M, Wotherspoon A, Hayes J M, Miles M J, Szczelkun M D and Valdrè G 2006 *Nanotechnology* **17** 3897
[27] Valdrè G and Fabbrizioli S 2006 *Scanning* **28** (2) 72
[28] Alessandrini A, Valdrè G, Valdrè U and Muscatello M 2007 *Chem. Phys. Lipids* **146** (2) 111
[29] Jurak M and Chibowski E 2010 *Colloids Surf. B* **75** 165
[30] Cagnasso M, Boero V, Franchini M A and Chorover J 2010 *Colloids Surf. B* **76** 456
[31] Nellis B A, Satcher J H and Risbud S H 2011 *Acta Biomater.* **7** 380
[32] Stievano L, Piao L Y, Lopes I, Meng M, Costa D and Lambert J F 2007 *Eur. J. Miner.* **19** 321
[33] Albee F H 1920 *Ann. Surg.* **71** 32
[34] Tiselius A, Hjerten S and Levin O 1956 *Arch. Biochem. Biophys.* **65** 132
[35] Bankston T E, Dattolo L and Carta G 2010 *J. Chromatogr. A* **1217** 2123
[36] Spanos N, Klepetsanis P G and Koutsoukos P G 2001 *J. Colloid Interf. Sci.* **236** 260
[37] Dangaria S J, Ito Y, Yin L, Valdrè G, X L and Diekwisch T G H 2011 *Tissue Eng. Part A* **17** 279
[38] Gun'ko V M, Yurchenko G R, Turov V V, Goncharuk E V, Zarko V I, Zabuga A G, Matkovsky A K, Oranska O I, Leboda R, Skubiszewska-Zie J, Janusz W, Phillips G J and Mikhalovsky S V 2010 *J. Colloid Interf. Sci.* **348** 546
[39] Lambert J F 2008 *Origins Life Evol. B* **38** 211
[40] Corno M, Busco C, Civalleri B and Ugliengo P 2006 *Phys. Chem. Chem. Phys.* **8** 2464
[41] Corno M, Rimola A, Bolis V and Ugliengo P 2010 *Phys. Chem. Chem. Phys.* **12** 6309
[42] Canepa P, Chiatti F, Corno M, Sakhno Y, Martra G and Ugliengo P 2011 *Phys. Chem. Chem. Phys.* **13** 1099
[43] Roark M and Feller S E 2008 *Langmuir* **24** 12469
[44] Dubey D K and Tomar V 2009 *J. Mech. Phys. Solids* **57** 1702
[45] Fantner G E, Hassenkam T, Kindt J H, Weaver J C, Birkedal H, Pechenik L, Cutroni J A, Cidade G A G, Stucky G D, Morse D E and Hansma P K 2005 *Nat. Mater.* **4** 612
[46] Gross L, Mohn F, Moll N, Liljeroth P and Meyer G 2009 *Science* **325** 1110

IOP Conf. Series: Materials Science and Engineering **32** (2012) 012028 doi:10.1088/1757-899X/32/1/012028

Microbeam analysis of irradiated nuclear fuel

C T Walker[1], S Brémier, P Pöml, D Papaioannou and P W D Bottomley
European Commission, JRC, Institute for Transuranium Elements (ITU), Hermann-von-Helmholtz Platz 1, P.O. Box 2340, DE-76125 Karlsruhe, Germany

E-mail: clive.walker@ec.europa.eu

Abstract. Microbeam analysis is widely used in the nuclear power industry. It is used to characterise the as-fabricated fuel, for routine post-irradiated examination and for research into the mechanisms of phenomena that limit the energy production of the fuel. The techniques most commonly used are wavelength-dispersive electron probe microanalysis, scanning electron microscopy and secondary ion mass spectrometry. Other microbeam analysis techniques that have been successfully applied to irradiated nuclear fuel are transmission and replica electron microscopy, X-ray fluorescence and micro X.-ray diffraction. Specific examples illustrating the past and present use of microbeam analysis in nuclear research establishments are presented with emphasis on the unique results they provide. As an aid to understanding, some basic facts about nuclear fuel rods and their irradiation are first given. This is followed by a description of features that set apart the microbeam analysis of high radioactive materials from standard practice.

1. Introduction

Microbeam analysis is widely used in the nuclear power industry. It is used to characterise as-fabricated fuel, for routine post-irradiation examination and for research into the mechanisms of phenomena that limit the energy production of the fuel. The techniques most commonly used are wavelength-dispersive X-ray electron probe microanalysis (WDS-EPMA), scanning electron microscopy (SEM), and secondary ion mass spectrometry (SIMS). Other microbeam analysis techniques that have been successfully applied to irradiated nuclear fuel are transmission and replica electron microscopy (TEM and REM), X-ray fluorescence (XRF), and micro X-ray diffraction (μ-XRD). SEM, TEM and REM have been mainly used to study the evolution of fission gas bubbles that cause the fuel to swell during irradiation [1-3], and XRD has been used to investigate the change in lattice parameter caused by irradiation damage and the build-up of fission products during irradiation and to assess their influence on the transformation of the fuel microstructure at high burn-up (i.e., after prolonged irradiation) [4].

WDS-EPMA is the microbeam analysis technique most widely available in nuclear research centres around the world. Nevertheless, from the perspective of the investigation of irradiated nuclear fuel the technique has a number of drawbacks. The main one is that it does not measure isotopes. A further shortcoming is the detection limit, which for a radioactive sample is at best 200 ppm [5]. In addition, the fission gas Kr cannot be measured because the second order $M\alpha_1$ U X-ray line coincides with the Kr $L\alpha_1$ line and the Kr K X-ray lines have a high critical excitation energy of 14.3 keV. A major failing is that, owing to the shallow electron penetration in nuclear fuel, the fission gas

[1] To whom any correspondence should be addressed.

trapped in pores and bubbles larger than about 0.1 μm can not be detected [6, 7]. These deficiencies have been overcome by combining EPMA with XRF [8] and SIMS [9].

This paper presents specific examples illustrating the past and present use of microbeam analysis in nuclear research establishments with emphasis on the unique results it has provided. Since this is a specialist area of application, each example includes a preamble giving a brief description of the subject area. As an aid to understanding, some basic facts about nuclear fuel rods and their irradiation are first given. This is followed by a description of features that set apart the microbeam analysis of highly radioactive materials from standard practise.

2. Light water fuel rod design and irradiation

A large amount of microbeam analysis has been carried out on light water reactor (LWR) fuel either from full length fuel rods from power reactors or from short rods (≈ 50 cm in length), containing fresh fuel or made by re-fabricating spent fuel rods, that have been irradiated in material test reactors.

Enriched uranium dioxide is the principal fuel for light water reactors, although most of the reactors in Europe contain some MOX fuel (UO_2 mixed with a few percent of PuO_2). For light water reactors the fuel is usually fabricated in the form of pellets, approximately 10 mm dia. x 10 mm high. For irradiation these are introduced into a Zircaloy tube.

A standard full length fuel rod is about 1 cm in diameter and is more than 3 m in length and contains about 300 fuel pellets. At one end or both ends of the fuel stack there is free space called a plenum which contains a steel spring. The spring holds the fuel pellets in place, whilst the space in which the spring sits provides a free volume in which the fission gas released from the fuel during irradiation can accumulate. The fuel rod is generally filled with helium gas up to a pressure of 25 bar. This provides a path for the conduction of heat across the small engineering gap that exists between the fuel pellets and the Zircaloy tube.

LWR fuel rods are usually loaded in the reactor in batches referred to as assemblies. PWR (pressurised water reactor) assemblies contain between 180 and 265 fuel rods, and in the reactor core there is space for between 120 and 195 fuel assemblies. As a rule fresh fuel assemblies are loaded close to the core periphery and as the irradiation progresses they are moved closer to the centre of the reactor core as assemblies that have reached their end of life are discharged from the reactor. Each assembly remains at a given location in the reactor core for about one year, which is referred to as an irradiation cycle. During each reactor cycle the rod power decreases since the concentration of fissile material in the fuel decreases. For example, in the course of an irradiation lasting five years the rod power can fall by more than half, from a maximum of about 40 kW/m to less than 15 kW/m.

The power or heat generated in a fuel rod is the result of the nuclear chain reaction called fission. The fission reaction written below is typical of a UO_2 power reactor fuel.

$$^{235}_{92}U + ^{1}_{0}n \rightarrow ^{236}_{92}U^{*} \rightarrow ^{147}_{57}La + ^{87}_{35}Br + 2^{1}_{0}n + 200MeV$$

The fissile isotope is ^{235}U. This absorbs a neutron to form ^{236}U in an excited state which then splits forming two fission products; in the example given, lanthanum with a mass of 147 and bromine with a mass of 87. At the same time two neutrons are emitted that sustain the chain reaction, and approximately 200 MeV of energy is released. The fraction of the heavy elements uranium and plutonium consumed by fission during irradiation is referred to as the burn-up (see subsection 5.1).

Few engineering materials are exposed to such harsh operating conditions as nuclear fuel. During irradiation, steep temperature and stress gradients exist in the fuel. For example, along the radius of the fuel, over a distance of less than 5 mm, the temperature changes from about 350 °C at the fuel surface to around 1000 °C at the fuel centre. This corresponds to a temperature gradient of about 1300 °C/cm.

Figure 1 shows the appearance of a sample taken from an UO_2 fuel rod for EPMA. The fuel rod had been subjected to a slow temperature ramp in which the reactor power had been raised in a matter of hours from 24 to 42 kW/m. It can be seen that the fuel is highly porous and contains a number of large cracks.

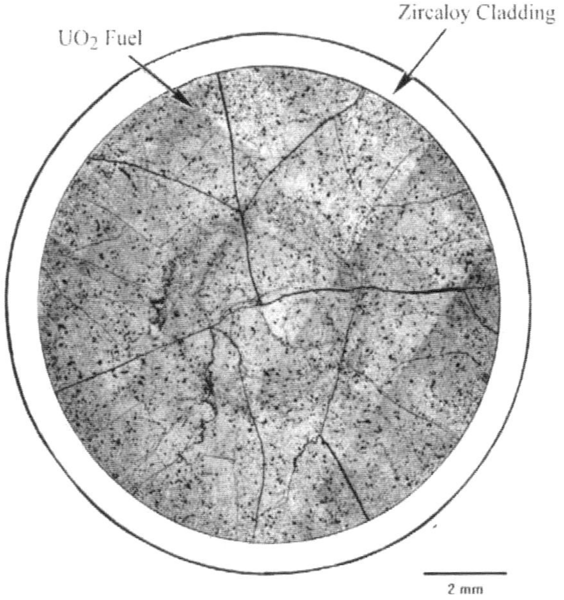

Figure 1. A sample of UO_2 nuclear fuel prepared for EPMA (after Walker, ref. [5]). The sample is from a fuel rod that had seen a reactor power excursion to 42 kW/m. The fuel is highly porous and contains a number of large cracks.

3. Handling requirements for highly radioactive materials

Irradiated nuclear fuels emit α-particles (^{4}He nuclei) with a penetrating power of a few cm in air, β-rays (electrons or positrons) with a wide range of energies, which can penetrate thin metal foil, and γ-rays that can travel several metres in air. As a result sample preparation (cutting, grinding and polishing) is carried out in a hot cell using manipulators and the instruments are shielded with heavy metal. In addition, an extra step, decontamination, is introduced in the sample preparation procedure at the Institute for Transuranium Elements (ITU) to avoid carry-over of radioactive particles to clean areas of the laboratory and to the specimen chamber or sample stage of the instrument. Decontamination involves repeated washing of the sample in ultrasonically agitated ethanol absolute to remove fission product caesium smeared over the surface during polishing, and any stray fuel particles.

The heavy metal shielding of the instruments falls into two categories; biological shielding to protect the operator from radiation exposure and detector shielding to reduce the background signal generated by β- and γ-rays. Figure 2 shows the Cameca SX100R electron microprobe used for the analysis of irradiated materials at ITU. The instrument is housed in a lead cell measuring 2 m wide by 3 m deep with walls 10 cm thick. This enables the analysis of samples of nuclear fuel with activities of up to 75 GBq to be carried out with full protection. The irradiation dose rate received by the operator is less than 5 $\mu Sv h^{-1}$.

A particular feature of the SX100R microprobe is the specimen carriage which in essence is a solid block of INERMET (W 92.5 wt%, Cu, Ni 7.5 wt%, density 17.6 g cm^{-3}) of 20 cm diameter. To change the filament of the electron gun during the analysis of a radioactive sample, the specimen carriage is transferred by remote control from the instrument to an INERMET storage flask located in front of the machine. This allows the operator to safely enter the lead cell in the presence of the sample.

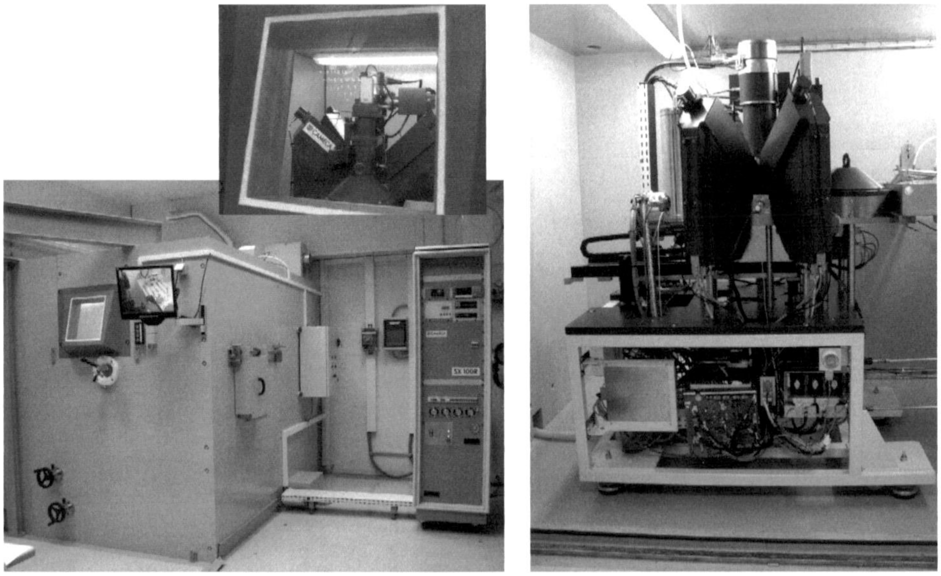

Figure 2. The Cameca SX100R electron microprobe for the analysis of highly irradiated materials at the Institute for Transuranium Elements.

Other special modifications have been made to the Cameca IMS 6F SIMS and the Jeol 6400 SEM used for the investigation of irradiated nuclear materials at ITU. Since material is sputtered from the sample surface in SIMS, contamination of some internal parts of the instrument cannot be avoided. Thus, the sample introduction port and specimen chamber are enclosed in a stainless steel glove box and the immersion lens has been modified to enable its removal for regular inspection and cleaning [10]. In the case of the SEM instrument the electron detector is fitted with a shielded electron spectrometer that filters out the low energy secondary electrons from the β-rays emitted by the fuel sample [11, 12]. This significantly improves the image quality.

4. EPMA procedures for nuclear fuel

4.1. Fission products dissolved in the fuel matrix

The procedure adopted for the analysis of fission products in solution in the fuel matrix follows the rules for trace element analysis which prescribe that the net peak count for the sample and the counting time for the sample background should be as large as practically possible to ensure a low detection limit. A high peak-to-background ratio is obtained by the use of a high electron acceleration of 25 kV and a high beam current of 250 nA. As to the counting time, fission product concentrations are calculated from *k*-ratios that are the average of six peak and eight background determinations of 50s duration (i.e., total peak count duration 5 min, total background count duration nearly 7 min). Under these conditions, the confidence limit on the measured concentration at a significance level of 99 % is about 5 % relative at 0.5 wt% and 10 - 20 % relative at 0.05 wt%. The limit of detection lies in the range 250 - 750 ppm.

4.2. Fission gas xenon

The analysis of xenon is complicated by the fact that part of the gas retained in the fuel matrix is contained in small bubbles, the size and distribution of which affect the emitted X-ray intensity [6, 7]. It has been calculated that the emitted X-ray intensity begins to fall when the bubble size exceeds about 10 nm and that the X-ray intensity emitted from bubbles with a radius of 100 nm is 50 % of that obtained from an atomic dispersion of the gas [13]. Thus, the xenon concentration measured by EPMA is mainly that dissolved in the fuel matrix as an atomic dispersion – only part of the gas present in bubbles is detected.

A further complication is that a suitable standard for xenon has so far not been found. At ITU this problem has been circumvented by using antimony as a standard for xenon. A correction factor is applied to the measured Sb $L\alpha_1$ X-ray intensity to obtain the corresponding reference Xe $L\alpha_1$ intensity under the specific analysis conditions employed. The correction factor is derived by interpolating the intensity for pure solid xenon (xenon has been shown to be solid in nanometre size bubbles) from the intensities of elements on each side of xenon in the periodic table [14].

4.3. Plutonium and the minor actinides americium and curium

X-ray line interference frequently occurs amongst the M-lines of the actinides and can be troublesome when the sample contains two or more actinide elements. Kleykamp has published the M-spectra for U, Np, Pu, Am [15] and Cm [16]. He found that the U $M\beta$ line interferes with the Pu $M\alpha_1$ line, the $UM\gamma_2$ interferes with the Pu $M\beta$ line, the Np $M\beta$ line interferes with the Am $M\alpha_1$ line and the U $M\gamma_2$ and the Pu $M\beta$ line interfere with the Cm $M\alpha_1$ line.

Although the U $M\gamma_2$ is a relatively minor line (its intensity is 2 % of that of the $M\alpha_1$ line) its interference with the Pu $M\beta$ line has to be considered in situations where the concentration of uranium is high and the concentration of Pu is low (e.g., in the determination of Pu in irradiated UO_2 nuclear fuel). The correction procedure adopted is described in ref. [17] and can be employed for other interferences.

For accurate quantitative EPMA standards are required. For reasons of safety and cost, use of standards of the minor actinides is not feasible. (The specific activity of Am (At. No. 95) is about fifty times higher than that of Pu and the specific activity of Cm (At. No. 96) is about 1000x higher. Moreover, Cm also emits neutrons. Both actinides currently cost around 150,000 US$ per gram.) PuO_2, therefore, is used as a standard for the analysis of Am and Cm and a correction coefficient is applied to the measured M-line intensity [5]. The correction coefficient is derived simply by extrapolating the X-ray intensities obtained from UO_2, PuO_2 and NpO_2.

5. Examples of current and past applications

In the case of irradiated nuclear fuel, the information provided by EPMA is frequently used to validate the predictions of fuel performance codes (e.g., TRANURANUS [18], ENIGMA [19] and CARO [20]). Such codes are used by fuel vendors and the operators of nuclear power plants to obtain licences from their national regulatory authority.

5.1. Determination of the local burn-up in irradiated nuclear fuel

The fraction of uranium and plutonium consumed by fission in an irradiation is termed the burn-up. Accordingly, burn-up increases with the duration of the irradiation. For power reactors burn-up is expressed as the cumulative heat output from a nuclear fuel, i.e., in megawatt days per kilogram heavy metal (U + Pu). The rare-earth fission product neodymium can be used as an indicator of the local burn-up because it is immobile in the fuel [21]. (The neodymium nuclide remains in the lattice site where it comes to rest following fission.) Consequently, as a rule of thumb, the local concentration of neodymium in the fuel increases linearly with the local burn-up.

The radial distribution of the burn-up in a UO_2 fuel derived from the local concentration of neodymium measured by EPMA is shown in figure 3. The fuel has a cross-section average burn-up of 102 MWd/kgHM. It can be seen from the plot that in the interior of the fuel the burn-up is slightly less than the average value, whereas at the fuel pellet surface it increases dramatically and is at least 2.5x higher than the average value. Before EPMA was used for local burn-up analysis these facts were not fully appreciated.

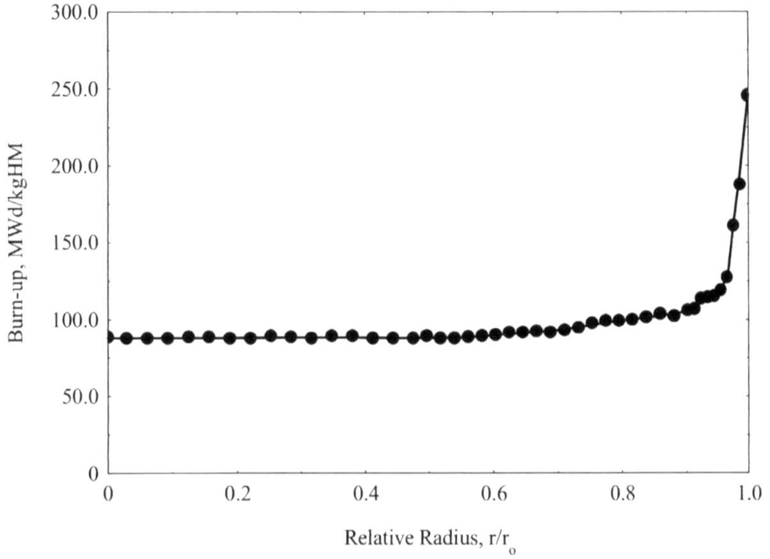

Figure 3. The radial burn-up distribution in a UO_2 nuclear fuel pellet with an average burn-up of about 100 MWd/kgHM (reprinted from ref. [22] with permission of Elsevier). The local burn-up increases sharply at the fuel surface due to the fission of plutonium created from uranium by neutron capture.

5.2. Fission gas release

Together the inert gases xenon and krypton constitute about 30 % of the fission product inventory in a nuclear fuel. At low and moderate burn-up, the gas atoms are uniformly dispersed in the uranium dioxide lattice at temperatures below about 1000 °C and therefore cause few problems. Above 1000 °C part of the gas precipitates as bubbles and part is released to the rod free volume. The formation of gas bubbles causes the fuel to expand or swell. At high burn-up swelling can result in increased mechanical interaction between the fuel and its metal containment tube, which can eventually lead to rod failure. The presence of gas bubbles also decreases the thermal conductivity of the fuel leading to an increase in its operating temperature. Alternatively, high levels of fission gas release can cause over-pressurisation of the fuel rod and eventual failure. Moreover, any accumulation of xenon in the narrow gap between the fuel pellet and the containment tube decreases heat transfer to the coolant and increases the fuel temperature aggravating fission gas release.

Microbeam analysis is used to investigate the mechanisms controlling fission gas release. TEM and SEM are used to study the nucleation and growth of gas bubbles and the formation of gas release channels (see e.g., refs. [1, 2]). EPMA is used to measure the concentration of xenon retained in the fuel matrix. Since the electron penetration depth in UO_2 is small (approx. 0.5 μm at E_0 = 25 kV), EPMA detects only a small fraction of the gas present in bubbles larger than 0.1 μm in size. The fraction of gas contained in bubbles and pores, however, has been obtained by combining EPMA with XRF. The XRF measurements were performed at the Risø National Laboratory in Denmark [23] using a rotating anode, high intensity X-ray generator (Rigaku®) to produce a 50 kV primary X-ray beam. X-rays of this energy excite the Xe Kα characteristic X-ray line to a depth of 18 μm in the fuel. As a result of the high X-ray penetration depth, all the xenon present in the fuel sample can be detected by XRF.

The fission gas krypton which is not detected by EPMA can be measured by SIMS. The first measurement of the radial distribution of krypton in an irradiated fuel has recently been carried out by SIMS and has confirmed the assumption that krypton behaves similarly to xenon [9].

Figure 4 shows vermicular gas pores on a UO_2 grain face as revealed by SEM. The structure was formed by the interlinkage of individual spherical bubbles. When the channels contact a grain edge an escape tunnel forms and the grain surface is drained of gas. The radial distribution of xenon in an irradiated nuclear fuel as measured by EPMA and XRF is shown in figure 5. XRF measures all the gas present in the fuel, whereas EPMA is considered to measure only the gas retained in the UO_2 matrix and intragranular bubbles smaller than 0.1 μm in size. Thus, the difference between the two radial profiles approximates to the amount of gas contained in bubbles and pores, which are situated mainly on the grain boundaries. In the case shown, about 85 % of the retained gas is present in the fuel matrix; the remainder is located on grain boundaries. The percentage of gas released to the rod free volume is obtained by subtracting the average integral xenon concentration given by the XRF profile from the created amount and is equal to about 22 %.

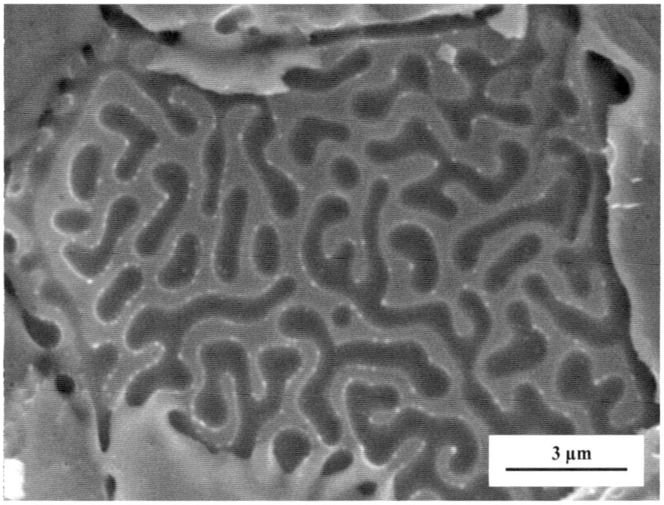

Figure 4. SEM micrograph of vermicular gas pores on a grain face in a UO_2 nuclear fuel formed by the interconnection of individual spherical bubbles (reprinted from ref. [24] with permission of Elsevier).

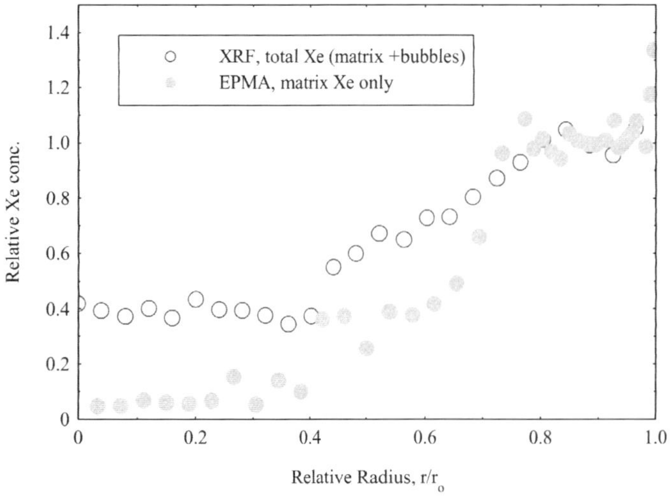

Figure 5. The radial distribution of xenon retained in a UO_2 fuel that had been subjected to a reactor power excursion to 40 Wm^{-1} at a burn-up of 50 MWd/kgHM (redrawn from ref. [23] with permission of Elsevier). The XRF profile gives the percentage of xenon released from the fuel and EPMA gives the percentage of xenon retained in the fuel matrix. The difference between the two profiles corresponds to the amount of gas in bubbles and pores larger than 0.1 μm in size.

5.3. Re-crystallisation of nuclear fuel at high burn-up

When the pellet burn-up of UO_2 fuel exceeds about 45 MWd/kgHM the UO_2 grains at the fuel surface begin to re-crystallize. The transformed microstructure (figure 6), which is termed the high burn-up structure, exhibits a pronounced decrease in grain size and high volume density of small facetted pores [25]. The driving force for re-crystallisation is generally accepted to be the build-up of stored energy resulting from the accumulation of radiation damage at low temperature [26, 27]. The discovery of the high burn-up structure initially caused concern amongst nuclear fuel technologists who predicted that its presence would lead to a general deterioration in fuel performance and to the fragmentation of the fuel and enhanced fission gas release during a reactor accident. These fears, however, have proven to be unfounded. In fact, the high burn-up structure has significant positive attributes, such as a higher thermal conductivity and increased plasticity.

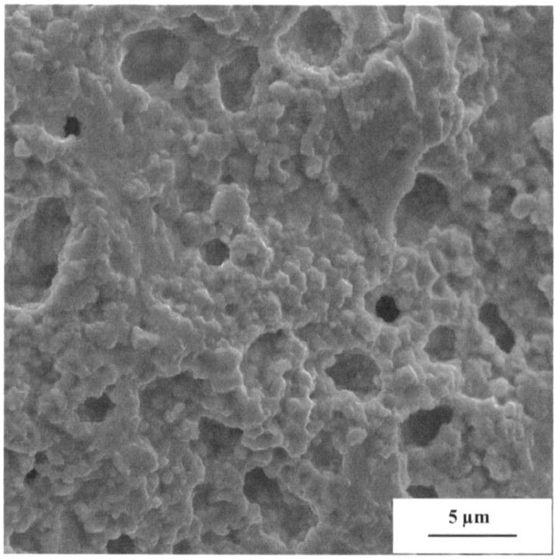

Figure 6. The high burn-up structure in the outer region of a UO_2 fuel pellet as revealed by scanning electron microscopy (reprinted from ref. [22] with permission from Elsevier).

Almost all that is known about the characteristics of the high burn-up structure, the mechanism of its formation and the processes contributing to its evolution have been obtained from studies employing microbeam analysis techniques. Lattice parameter measurements using micro-XRD have shown that the lattice strain is highest at the radial position where nucleation of the high burn-up structure begins as would be expected if the transformation of the microstructure is initiated by radiation damage (see figure 4 of ref. [28]). EPMA has revealed that when re-crystallisation occurs nearly all the fission gas is swept out of the fuel matrix [21]. Consequently, the radial extent of re-crystallisation, and the burn-up and temperature thresholds for the process can be established from the distance over which xenon depletion occurs (figure 7) [29]. Using XRF and more recently SIMS [9] it has been confirmed that the fission gas lost from the matrix is contained in the pores of the high burn-up structure (figure 8) and that little gas, if any, is released to the rod free volume either when re-crystallisation occurs, or with increase in burn-up above the threshold value once the high burn-up structure is established (figure 9).

An important question with implications for the behaviour of the high burn-up structure during a reactor accident is the pressure of the gas in the pores. The Xe intensity spikes in the SIMS depth profile in figure 8 can be converted to concentration using the matrix intensity level which is known from EPMA to correspond to about 0.25 wt%. Knowing the size distribution of the pores in the high burn-up structure, the gas pressure can then be estimated from the Xe concentration using a hard sphere equation of state [9]. With this method the average pressure of the gas in the pores was found to be 45 MPa. Thus, the pores of the high burn-up structure are over over-pressurized, but generally

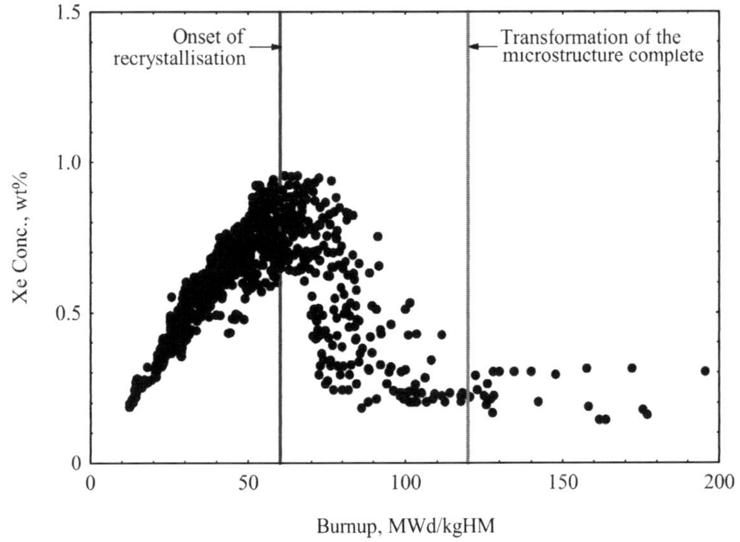

Figure 7. EPMA results for the local concentration of xenon in the outer region of UO_2 fuel related to the local burn-up (reprinted from ref. [29] with permission of Elsevier). The peak concentration at 60 to 75 MWd/kgHM marks the burn-up threshold for the formation of the high burn-up structure

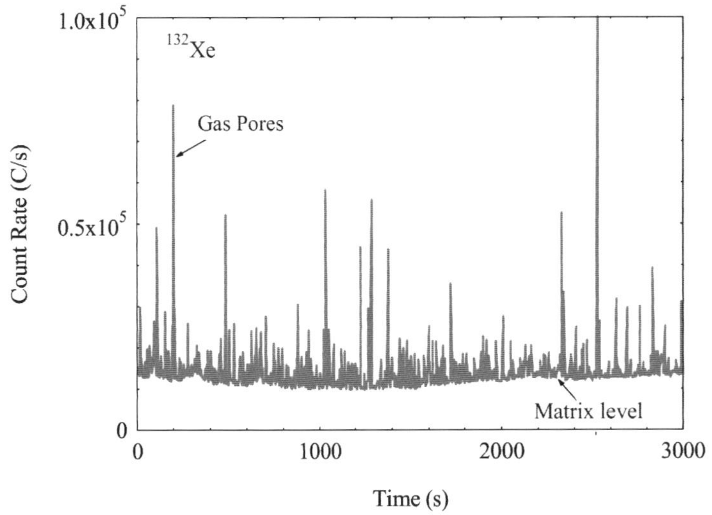

Figure 8. SIMS depth profile for ^{132}Xe in the high burn-up structure produced with a $\left(^{16}O\right)_2^+$ primary ion beam at a high current density of 154 mA/cm^2 (after Walker et al. ref. [9]). The intensity spikes are caused by the expulsion of gas from the pores in the microstructure. Time is equivalent to erosion depth.

not excessively so. Consequently, it would appear that few of the pores present in the high burn-up structure are sufficiently over-pressurized to initiate fragmentation of the fuel during a reactor accident.

5.4. Hydrogen pick-up by the Zircaloy cladding of LWR fuel rods

Standard Zircaloy is Zr with 1 - 2 wt% Sn and trace amounts of Fe and Cr as alloying elements. The material has a propensity to absorb hydrogen which can result in the formation of zirconium hydride. The occurrence of such hydrides can lead to embrittlement of the cladding during irradiation and eventual premature failure of the fuel rod. The hydrogen present in irradiated Zircaloy cladding is mainly a by-product of its corrosion by water, which is the coolant medium in the reactor and is in direct contact with the fuel rods. The chemical reaction can be written as

$$Zr + 2H_2O \rightarrow ZrO_2 + 2H_2 \quad (1)$$

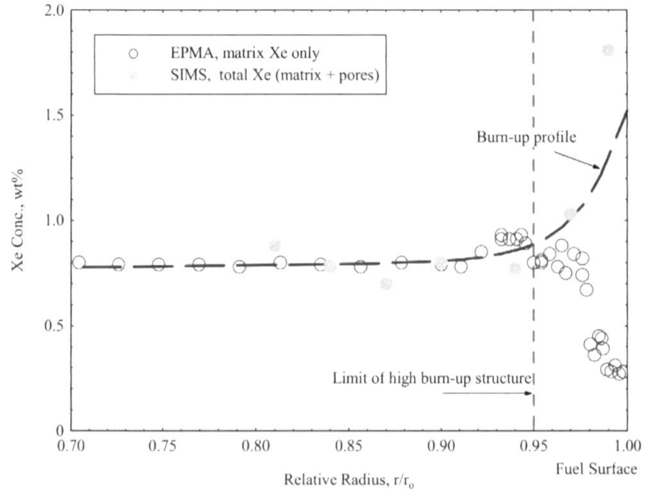

Figure 9. The local concentration of retained xenon in the outer region of UO_2 fuel as measured by SIMS and EPMA (reprinted from ref. [9] with permission of Elsevier). The steep increase in the SIMS profile at the fuel surface indicates that almost all the gas missing from the UO_2 matrix is contained in the pores of the high burn-up structure. Note that in the interval between $r/r_o = 0.80$ and 0.95, the gas is present as an atomic dispersion.

Only part of the hydrogen formed actually diffuses into the cladding. This so-called pick-up fraction depends on the alloying elements in the Zircaloy and can vary from 5 to 25 % of the amount created by the dissociation of water. During the oxidation of Zircaloy cladding, δ-phase hydride stringers ($ZrH_{1.9}$)form when the solid solubility limit for hydrogen, which is about 150 ppm at 350 °C, the highest temperature of the coolant in a PWR, is exceeded. The length and orientation of these platelets or stringers are important in determining the mechanical properties of the cladding during irradiation. Moreover, the steep radial temperature gradient in the cladding (~ 625 °C/cm) often produces a dense accumulation of hydrides at the outer surface of the cladding that can accelerate the corrosion process.

Backscattered electron imaging has been used to investigate the amount, location and morphology of zirconium hydrides in irradiated Zircaloy cladding and laboratory-hydrided samples [30, 31]. Since the mean atomic number of zirconium hydride is much lower than that of zirconium there is a visible contrast difference between the hydrides and the matrix. Hence, the platelets can clearly be distinguished in the backscattered electron image of the cladding tube cross-section (see figure 10). By applying quantitative image analysis techniques the local area fraction of hydrides in the backscattered electron image can be estimated and the local concentration of hydrogen in the cladding calculated assuming that all the hydrogen present is contained in the hydrides. Schrire and Pearce [30] have shown that using backscatter electron imaging in combination with quantitative image analysis the local hydrogen concentration in irradiated cladding can be determined with a spatial resolution of <100 μm, enabling the hydrogen concentration gradient in the cladding to be measured.

The orientation of a hydride platelet in Zircaloy mainly determines whether it acts as a nucleation point for a crack. Tensile stress results in radially-orientated hydrides, whereas under stress free conditions the hydrides are circumferentially-orientated. The lattice orientation of zirconium hydrides in Zircaloy cladding has been investigated using electron backscattering diffraction (EBSD) [32]. This technique allows determination of the lattice orientation of polycrystalline materials with a spatial resolution of 50 nm and an angle resolution of <1°. EBSD has shown that the most common crystallographic relationship for both radially and circumferentially orientated hydrides is $(001)_{matrix}//\{111\}_{hydride}$. It has also revealed that the change in the orientation of the hydrides caused by tensile stress is due to a change in the preferred sites for hydride precipitation and not to a change in the crystallographic relationship between the hydride and the matrix. The effect of tensile stress can be explained by a decrease in the compressive elastic strain energy which results in a decrease in the solid solubility of hydrogen at a given temperature.

Figure 10. Backscattered electron image showing zirconium hydride platelets (dark phase) in Zircaloy cladding hydrided in the laboratory. The accumulation of hydrides at the sample surface indicates the presence of a steep hydrogen concentration gradient in this region.

5.5. Effect of water chemistry on the corrosion rate of Zircaloy cladding

In a light water reactor the water acts not only as the coolant, but also as the moderator, and reflector. Boron in the form of boric acid is added to the water to control reactivity of the reactor by absorbing neutrons, and Li in the form of lithium hydroxide is added to counter the increase in acidity that this causes. For reasons that are not fully understood enhanced corrosion of Zircaloy cladding may occur if the concentration of Li in coolant exceeds a certain level. Both lithium and boron are taken up by the oxide as it grows, and their levels of incorporation can be used to gauge the Li and B content of the water. The incorporation of Li and B in the external oxide layer of corroded Zircaloy cladding has been investigated by secondary ion mass spectrometry.

Figure 11 shows results of SIMS depth profiling for ^{7}Li and ^{11}B in the external oxide layer of the Zircaloy cladding of a PWR fuel rod obtained by Gebhardt [33]. It can be seen that Li and B are not homogeneously distributed in the oxide. At the outer surface of the oxide a steep increase in the concentration of Li is accompanied by a sharp fall in the concentration of B. In contrast, at the oxide/metal interface the concentrations of both Li and B decrease dramatically. At this location a thin layer of dense oxide exists which is known to control the corrosion rate. The concentrations of Li and B in the oxide ranged from 8 to 23 ppm and 60 to 190 ppm, respectively. The measurements were made on an Atomika 4000 quadrupole SIMS using a liquid gallium primary ion source to obtain high spatial resolution. The energy of the primary ions was 25 keV and the beam current was 100 to 500 ρA.

5.6. Characterisation of corium

Corium is a lava-like molten material that forms when the reactor core melts during a severe nuclear accident such as that which occurred at Chernobyl in Russia in 1986, and recently at the Fukushima power plant in Japan. It is essentially a binary solid-solution of UO_2 and ZrO_2 containing small amounts of iron and chromium from steel components, silver from control rods and fission products. The degradation of irradiated UO_2 fuel rods during a core melt-down is being studied in the International Phebus-FP (Fission Product) Programme at the CEA, France [34]. In this programme, encapsulated test assemblies of up to 18, one metre long, pre-irradiated fuel rods of known burn-up are heated in the driver core of the Phebus Reactor until melting occurs. A steam rich atmosphere

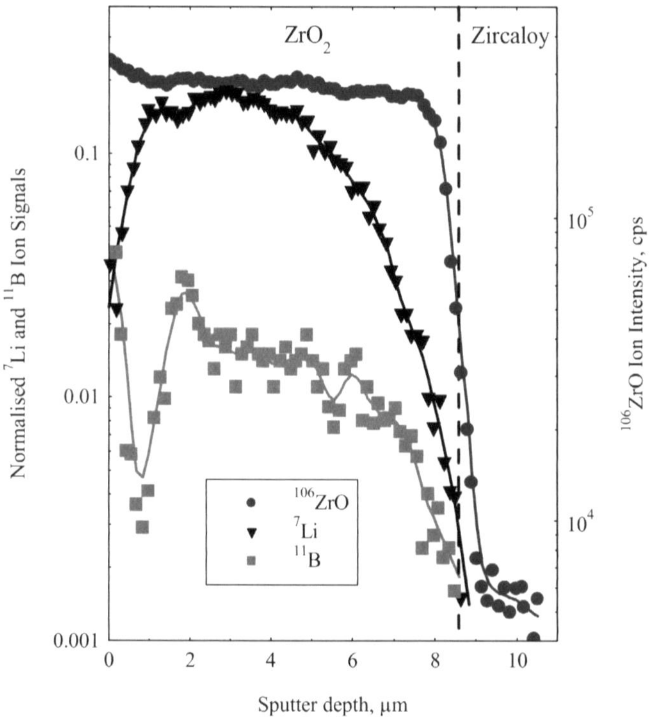

Figure 11. SIMS depth profiles for ^{7}Li and ^{11}B in the external oxide layer of the Zircaloy cladding of a PWR fuel rod (redrawn from ref. [32] with kind permission from Springer Science + Business Media).

(simulating conditions prevailing during a loss of coolant accident) is provided by injecting water vapour into the assembly.

Figure 12 shows a false colour X-ray radiograph of the Phebus FPT2 fuel rod assembly [35] at the end of the test. The material density increases in the order black-blue (void), green-yellow (porous fuel or structural materials) red (UO_2 fuel or corium). It is evident from this X-ray radiograph that at mid-height where the temperature in the test was highest the fuel rods have melted completely creating a large cavity (black-blue). Below this is a solidified pool of corium (red) that was created by the fusing of the fuel rods and the downward flow of the molten mass. At the top and bottom of the test assembly, segments of fuel rods that have not melted but are oxidized and degraded remain in place (green-yellow).

EPMA and SIMS are being used in the post-test examinations to determine the chemical state of the reacting materials, the composition of the corium, and the influence of trace elements (fission products, steel constituents, and control rod components) on its solidification temperature (see e.g., ref. [36]). Large area, false colour EPMA X-ray maps for U, Zr and Ag in the zone containing degraded fuel rods below the corium pool in the FPT2 test assembly are shown in figure 13. The area examined contained two fuel rods revealed by the two large, yellow discs in the U map. The Zircaloy cladding (green and yellow in the Zr map) has oxidized and partially peeled away from the fuel pellets. Corium (light-blue in the U map and dark-blue in the Zr map) has flowed down from the pool above and has penetrated the open space between the two fuel rods. Silver inclusions of varying size (yellow in the Ag map) are present amongst the Zr containing phases in the bottom left of the area mapped. The source of this element is the Ag-15In-5Cd control rod that was located at the centre of the test assembly.

EPMA point analysis has shown that the solidified corium melt is a homogeneous solid-solution with an average composition typically corresponding to the chemical formulae $[(U,Pu)_{0.57}Zr_{0.40}(Fe,Cr)_{0.03}]O_{2\pm x}$.

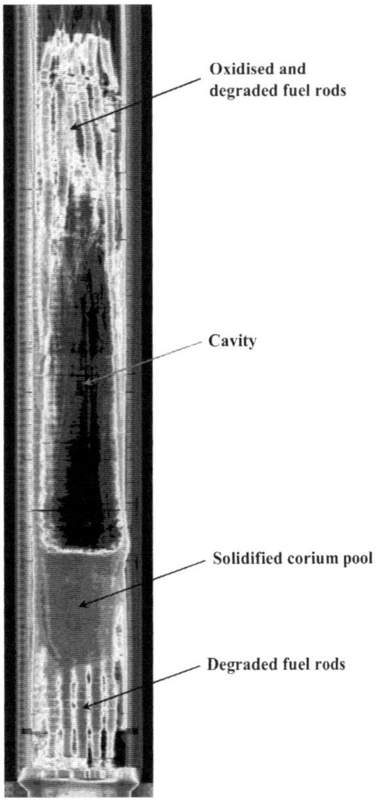

Figure 12. False colour X-ray radiograph showing the state of the melted FPT2 fuel rod assembly at the end of the test (courtesy of the Institut de Radioprotection et de Sûreté Nucléaire (IRSN), Cadarache, France). The material density increases in the order black-blue, green-yellow, red,

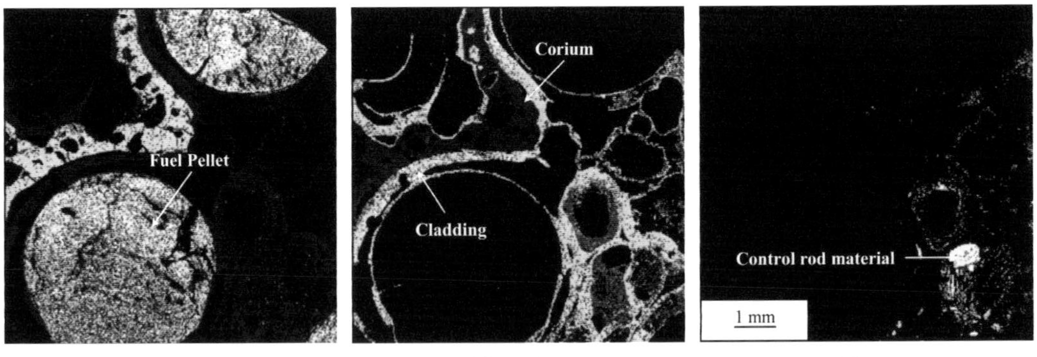

Figure 13. Fuel rods, Zircaloy cladding, corium and control rod material in the zone below the corium pool in the melted Phebus FPT2 fuel rod test assembly as revealed by large area, false colour, EPMA X-ray maps for U, Zr and Ag.

6. Epilogue

There has always been a synergetic relationship between EPMA and the nuclear power industry. Both were born in the 1950s, and in the 1960s EPMA and the civil use of nuclear power developed side by side. This was no coincidence. The small volume of material analyzed and the ease of quantification make EPMA the ideal analytical tool with which to study the chemical behaviour of fission products in nuclear fuel. EPMA supplied the first data on the local retention of Xe required to model fission gas release during a reactor power excursion, and has since provided a crucial insight into the mechanisms involved. Moreover, almost all that is presently known about the characteristics of the high burn-up structure of nuclear fuel; the mechanism of its formation and the processes contributing to its evolution, has been acquired from studies employing microbeam analysis techniques.

Today, EPMA and a range of other microbeam analysis techniques are providing fundamental information about the behaviour of nuclear fuel at high burn-up and about the performance of fuels and targets designed for the in-pile incineration or transmutation of plutonium and the minor actinides Np, Am and Cm. In recent years one of the important tasks of EPMA and SIMS at ITU has been the provision of experimental data for the validation of fuel performance codes used by fuel vendors and the operators of nuclear power plants to obtain licences from their national regulatory authority.

References

[1] Walker C T, Knappik P and Mogensen M 1988 *J. Nucl. Mater.* **166** 10
[2] Ray I L F, Thiele H and Matzke Hj 1992 *J. Nucl. Mater.* **188** 90
[3] Mogensen M, Walker C T, Ray I L F and Coquerelle M 1985 *J. Nucl. Mater.* **131** 162
[4] Spino J and Papaioannou D 2000 *J. Nucl Mater.* **281** 146
[5] Walker CT 1999 *J. Anal. At. Spectrom.* **14** 447
[6] Ronchi C and Walker C T 1980 *J. Phys. D: Appl. Phys.* **3** 2175
[7] Verwerft M 2000 *J. Nucl. Mater.* **282** 97
[8] Walker C T and Mogensen M 1987 *J. Nucl. Mater.* **149** 121
[9] Walker C T, Brémier S, Portier S, Hasnaoui R and Goll W 2009 *J. Nucl. Mater.* **393** 212
[10] Brémier S, Hasnaoui R, Portier S, Bildstein O and Walker C T 2006 *Microchim. Acta* **155** 113
[11] Equchi H, Tagata S, Yamada H 1974 *Proc. 8th Int. Congress Electron Microscopy (Canberra, Australia)* vol. 1 pp. 127
[12] Coquerelle M, Knappik P and Perrier J C 1984 *J. Physique* **45** C2 84
[13] Walker C T 1997 *J. Trace Microprobe Tech.* **218** 419
[14] Pöml P, Brémier S, Lahuerte F, Hasnaoui R and Walker C T 2010 *IOP Conf. Ser.: Mater. Sci. Engng.* **7** 01025
[15] Kleykamp H 1981 *Z. Naturforsch. A* **36** 1388
[16] Kleykamp H 1986 *Z. Naturforsch. A* **41** 681
[17] Lassman K, O'Carroll C, van de Laar J and Walker C T 1994 *J. Nucl. Mater.* **208** 223
[18] Lassmann K 1992 *J. Nucl. Mater.* **188** 295
[19] Kilgur W J, Turnbull J A, White R, Bull A J and Palmer I D 1991 *Proc. ENS Mtg. on LWR Fuel Performance (Avignon, France)* vol. 2 pp. 919
[20] Sontheimer F and Landkron H 2001 *IAEA Tech. Committee Mtg on Nuclear Fuel Behaviour at High Burn-up and its Experimental Support (Windermere, UK, June 2000)* Report IAEA-TECDOC-1233 pp. 105
[21] Lassmann K, Walker C T, van de Laar J and Lindström F 1995 J. *Nucl. Mater.* **226** 1
[22] Walker C T, Staicu D, Scheindlin M, Papaioannou D, Goll W and Sontheimer F 2006 *J. Nucl. Mater.* **350** 19
[23] Mogensen M, Bagger C and Walker C T 1993 *J. Nucl. Mater.* **199** 85
[24] Fischer S B, White R J, Cook P M A, Brémier S, Corcoran R C, Stratton R, Walker C T, Ivison P K and Palmer I D 2002 *J. Nucl. Mater.* **306** 153
[25] Walker C T, Kameyama T, Kitajima S and Kinoshita M 1992 *J. Nucl. Mater.* **188** 73
[26] Walker C T 1991 *Concerning the microstructure changes that occur at the surface of UO_2 pellets on irradiation to high burn-up.* (Karlsruhe: European Commission, Institute for Transuranium Elements) Report K0291141
[27] Nogita K and Une K 1995 *J. Nucl. Mater.* **226** 302
[28] Walker C T, Rondinella V V, Papaioannou D, Van Winckel S, Goll W and Manzel R 2005 *J. Nucl. Mater.* **345** 192
[29] Walker C T 1999 *J. Nucl. Mater.* **275** 56

[30] Schrire D and Pearce J H 1994 *Proc. Zirconium in the Nuclear Industry: 10th Int. Symposium,* (Philadelphia: ASTM) Garde A M and Bradley E R, eds. pp. 98
[31] Heikinheimo E, Goll W, Toscano E H and Walker C T 2001 *Proc. 34th Plenary Mtg. of the European Working Group on Hot Laboratories and Remote Handling (Madrid, Spain)* Diaz J L and Quinones J, eds. pp. 23
[32] Une K, Nogita K, Ishimoto S and Ogata K 2004 *J. Nucl. Sci. Technol.* **41** 731
[33] Gebhardt O 1999 *Fresenius J. Anal. Chem.* **365** 117
[34] Krischer W and Rubinstein M C 1990 *The Phebus fission product project: Presentation of the experimental programme and test facility.* (Routledge)
[35] Bottomley P W D, Schlutig S, Brémier S, Barrachin M et al. 2007 *Proc. ICAPP 2007 (Nice France, May 13-18)* Paper 7333
[36] Bottomley P D W, Stalios A D, Glatz J P, Sätmark B and Walker C T 2000 *J. Nucl. Mater.* **278** 136

Institute of Physics
Dirac House, Temple Back
Bristol BS1 6BE UK

proceedings.com